History, Philosophy and Theory of the Life Sciences

Volume 11

W0037309

More information about this series at http://www.springer.com/series/8916

Pierre-Alain Braillard • Christophe Malaterre

Explanation in Biology

An Enquiry into the Diversity of Explanatory
Patterns in the Life Sciences

 Springer

Pierre-Alain Braillard
Independant Scholar
Peyregrand, Drulhe
France

Christophe Malaterre
Département de philosophie & CIRST
Chaire de recherche UQAM en Philosophie
des sciences
Université du Québec à Montréal (UQAM)
Montréal, Québec
Canada

ISSN 2211-1948   ISSN 2211-1956   (electronic)
History, Philosophy and Theory of the Life Sciences
ISBN 978-94-017-7992-0   ISBN 978-94-017-9822-8   (eBook)
DOI 10.1007/978-94-017-9822-8

Springer Dordrecht Heidelberg New York London
© Springer Science+Business Media Dordrecht 2015
Softcover reprint of the hardcover 1st edition 2015

Printed on acid-free paper

Springer Science+Business Media B.V. Dordrecht is part of Springer Science+Business Media (www.
springer.com)

Contents

Chapter 1
Explanation in Biology: An Introduction

Pierre-Alain Braillard and Christophe Malaterre

Abstract Explanation in biology has long been characterized as being different from explanation in other scientific disciplines, in particular from explanation in physics. One of the reasons was the existence in biology of explanation types that were unheard of in the physical sciences: teleological and functional explanations, historical and evolutionary explanations. More recently, owing in part to the rise of molecular biology, biological explanations have been depicted as mechanisms. This profusion of explanatory patterns is typical of biology. The aim of the present volume *Explanation in Biology. An Enquiry into the Diversity of Explanatory Patterns in the Life Sciences* is to shed some new light on the diversity of explanation models in biology. In this introductory chapter, we recall the general philosophical context of scientific explanation as it has unfolded in the past seven decades, and highlight the specific issues that models of explanation have faced in biology. We then show how the different essays gathered in this collective volume tackle aspects of this important debate.

Keywords Scientific explanation • Biological explanation • Nomological explanation • Causal explanation • Mathematic explanation • Mechanistic explanation • Explanatory pluralism

1 Introduction

Among the different achievements of science – such as prediction, control or simply description – explanation is generally understood as occupying a very central and special location on the epistemic chessboard. Yet, explanation has also

P.-A. Braillard
Independent Scholar, Peyregrand, 12350 Drulhe, France
e-mail: brailla6@hotmail.com

C. Malaterre (✉)
Département de philosophie & CIRST, Chaire de recherche UQAM en Philosophie des sciences, Université du Québec à Montréal (UQAM), Case postale 8888, Succursale Centre-Ville, Montréal, QC, H3C 3P8, Canada
e-mail: malaterre.christophe@uqam.ca

© Springer Science+Business Media Dordrecht 2015
P.-A. Braillard, C. Malaterre, *Explanation in Biology*, History, Philosophy and Theory of the Life Sciences 11, DOI 10.1007/978-94-017-9822-8_1

raised numerous questions and controversies among scientists and philosophers, especially since the development of the deductive-nomological model (Hempel and Oppenheim 1948; Hempel 1965). Though not everybody agrees, it is largely assumed that it is part of the task of science to *explain* natural phenomena, and that its methods enable it to do so incomparably better than common sense. But then what *is* a scientific explanation? What distinguishes an *explanation* from something that is not an explanation? And what distinguishes an explanation that is *scientific* from one that is not? Understanding what scientific explanations consist of has been one of the most central issues in the philosophy of science as it has developed throughout the twentieth century, and more recently as analyzed in the realm of biology. Part of the agenda has been to identify necessary and sufficient criteria that would enable distinguishing genuine scientific explanations from pseudo-explanations. Hence the issue is not only a matter of descriptive adequacy (accounting for the practice of science), but also has normative ambitions. Another strongly debated topic is whether there exists a single model of scientific explanation or several different ones, possibly depending on specific explanatory practices across disciplines. If what is accepted as an explanation varies across disciplines, if what is sought after in elaborating an explanation also varies, as well as the reasons why one pursues such an explanatory quest, then wouldn't all of this be good reasons to be skeptical about the existence of a single model of explanation? This debated issue has proven to be of much relevance in biology, both in light of the specificities of this multifaceted discipline compared to physics – the latter being often taken as a discipline of reference in philosophy of science – and in light of the heterogeneity of explanatory practices within biology itself. In this seventh decade since Hempel and Oppenheim's "Studies in the Logic of Explanation", the present volume *Explanation in Biology. An Enquiry into the Diversity of Explanatory Patterns in the Life Sciences* proposes a collection of essays that discuss some of the most recent philosophical perspectives of scientific explanation in light of the specificities of modern biology, including such a broad range of sub-disciplines as molecular biology, systems biology, evolutionary biology or developmental biology. In this introductory chapter, our objective is to set the context within which these latest developments on philosophical models of explanation have taken place. The chapter is organized as follows. We first provide a brief overview of the major models of scientific explanation that have been developed in the past six decades, and their "universalist" aspirations (Sect. 2). We then review the key issues that these models generally face in biology (Sect. 3). These issues are often taken as good reasons for developing philosophical models of explanation that are more narrowly tailored to the claimed specificity of biology and of its sub-disciplines; incidentally, they are also taken as good reasons against any universalist model of explanation. They therefore shape the debate – which is notably salient in biology – between "universalists" who argue in favor of a unified account of explanation, and "pluralists" who argue in favor of a profusion of explanatory patterns. We then show (Sect. 4) how the different contributions of the volume address facets of this important debate, be it by documenting a profusion of explanatory patterns, analyzing the specific heuristics at work in biology, or critically assessing mechanistic-type explanations and exploring alternative approaches.

2 Six Decades of Scientific Explanation

Elaborating a model of scientific explanation is a long-lasting aim of philosophy of science. However, accounting for what a scientific explanation really is has turned out to be much more difficult than one might have initially expected. As reviewed by Salmon in his *Four Decades of Scientific Explanation* (Salmon 1989), the literature on the topic is vast, triggered in one way or another by Hempel and Oppenheim's (1948) essay. Two decades later, scientific explanation is still a matter of much debate. Since Hempel's *DN* model, the general question of characterizing scientific explanation has branched out to address three types of problems that have structured the debate in the past six decades, and still structure it today: (1) Are there unique characteristics to scientific explanations? Can we define a set of necessary and sufficient conditions for scientific explanations? (2) Is causation a primitive notion for scientific explanation? (or should causation be construed on the basis of explanation?) And if yes, which account of causation is suitable? (3) Which role does context play in explanation, if any? The debate also concerns the ways models can or cannot address canonical counter-examples that range from flagpole shadows (as an illustration of explanatory asymmetry) to hexed table salt (explanatory relevance), syphilitic mayors (low probability explanations) and storms and barometers (correlations). The current debate about explanation in biology has inherited from all of these questions. Our aim here is to map the general context within which this debate arose, and to survey some of the most salient models of scientific explanation.

Typically addressing problems of type (1) above, the model of scientific explanation that has been the most discussed and that has influenced nearly almost all subsequent work is the deductive-nomological (*DN*) model, mainly developed by Carl Hempel (Hempel and Oppenheim 1948; Hempel 1965). As is well known, the *DN* model construes explanation as consisting of a deductive argument, the premises of which must contain at least one law-like generalization. According to this model, scientific explanation involves two epistemic elements: an *explanandum* – the phenomenon to be explained – that appears as the conclusion of the argument, and an *explanans* that consists of the set of premises that do the explaining. Several criteria constrain the *explanandum* and the *explanans*. In particular, the *explanans* must deductively entail the *explanandum*, and the deduction must make essential use of general laws; the *explanans* must also have empirical content. The decision to define explanation as a logical argument involving laws (or law-like generalizations) originated in part from empiricist worries about other concepts such as causation, found to be metaphysically too loaded. But of course, the whole approach depends on the ability to properly define what a law is, and this has proven to be far from straightforward (e.g., Ayer 1956; Dretske 1977; Cartwright 1980). We will see how this question is especially problematic in biology. Because many cases involving statistical generalizations are not covered by the *DN* model (and most special sciences are not based on deterministic laws), Hempel added an account of what he called inductive-statistical (*IS*) explanations (Hempel 1965). Similarly to a *DN* explanation, an *IS* explanation takes the form of an argument. Yet contrary to a *DN* explanation where the argument is deductive, the argument of an *IS* explanation is inductive. Explanation of the *IS* type thereby works by showing that the *explanans* confers a high probability onto the *explanandum*. This, however, has somehow proved to be a weakness of the model, as there happen to be many cases where explanation is deemed satisfactory despite involving statistical generalizations that only confer a low probability to the *explanandum* (e.g., Salmon 1965). The *DN* and the *IS* models share one essential feature: the general idea that explaining is showing that the occurrence of a phenomenon is to be expected on the basis of lawful generalizations.

As is well-known, the *DN* and *IS* models of scientific explanation have faced many counterexamples, which have suggested that they do not properly capture what a scientific explanation really is (see Salmon 1989 for a review). Two of the more salient problems are the problem of explanatory asymmetry and the problem of explanatory relevance. The first problem refers to cases where it is possible to switch the *explanandum* and a premise of the *explanans* while still fulfilling the formal criteria of the *DN* model, thereby leading a counter-intuitive explanatory argument. This is illustrated, among others, by the famous example of the flagpole and its shadow: the height of the flagpole and the laws of optics and trigonometry explain the length of its shadow, but one would not say that the length of the shadow and the laws of optics and trigonometry explain the height of the flagpole, despite the fact that this argument also fulfills the *DN* criteria (e.g., Bromberger 1966; van Fraassen 1980). This suggests that the *DN* model lacks criteria to capture the directionality we see in many explanations. The second problem refers to arguments that are valid according to the *DN* criteria despite the fact that the *explanans* includes factors that are totally non-relevant to the *explanandum*, thereby leading to counter-intuitive explanations. For instance, the dissolving of a sample of table salt would be explained by the fact that it has been hexed and that all samples of hexed salt dissolve in water (Kyburg 1965). What these examples seem to suggest is that the *DN* model is at best incomplete. As we will see below, for several philosophers, the problems faced by the *DN* and *IS* models come from a very foundational issue: the fact that these models do not make causation a central feature of explanation (e.g., Scriven 1962; Salmon 1978). Much of subsequent work on explanation can be seen as addressing these type (2) questions.

A first answer to this problem has been the statistical-relevance (*SR*) model, chiefly developed by Wesley Salmon (1971). The idea behind the *SR* model of explanation is to include a condition that captures the causal relevance of a factor, causation being understood in probabilistic terms. According to the model, explanation is a matter of identifying the set of factors that are statistically-relevant to the explanandum. Generally stated, the criterion of statistical relevance offered by Salmon is the following: given some class or population characterized by A, an attribute C will be statistically relevant to another attribute B if and only if the probability of B conditional on A and C is different from the probability of B conditional on A alone. This is meant to capture the fact that C makes a difference with regards to elements of the population characterized by A having the property

B. In other words, *C* causally explains why elements characterized by *A* have *B*. The obvious advantage of this model of explanation is to solve the problem of explanatory relevance that affects the *DN* model. To cite another famous example, taking birth control pills is non-relevant when it comes to explaining a man's failure to get pregnant, because it is statistically non-relevant to a person becoming pregnant if this person is a man (but it is statistically relevant if this person is a woman). Similarly, the *SR* model solves the problem of explanatory asymmetry that affects the *DN* model (see Salmon 1971 for details). Another advantage of the *SR* model is that, unlike the *IS* model, it can account for explanations of low probability events.

The *SR* model however also faces a number of problems. One of the consequences of the SR model that Salmon endorses but that many take to be counter-intuitive is the fact that, in some cases, a set of statistically relevant factors can be used to explain both an event and its opposite (take the simple example of why someone catches a cold or why he/she does not, given that there is a certain probability of catching a cold in given circumstances). More fundamentally, there is the question whether causal relationships can indeed be captured by statistical relevance relationships (Cartwright 1979; Salmon 1984; Spirtes et al. 1983). Indeed, it appears that causal relationships are greatly underdetermined by statistical relevance relationships. This can happen, for instance, when the causal relata are characterized as too coarsely grained. And it has also been shown that the same statistical relevance relationships can account for different sets of causal relationships. These deficiencies have led Salmon to develop another type of causal model of explanation: the causal-mechanical model.

The central idea of the causal-mechanical (*CM*) model of explanation is to construe explanation in terms of causation, and causation in terms of interactions of causal processes (Salmon 1978, 1984). Within this framework, explaining an event is showing how it fits into a causal nexus, this causal nexus being constituted by causal processes that interact at certain points. According to Salmon, causal processes are physical processes that are capable of propagating marks or modifications imposed upon them (see Reichenbach's mark criterion in his 1958). This characteristic makes it possible to distinguish them from pseudo-processes, such as moving shadows, that cannot transmit a mark and are hence explanatorily irrelevant. A car traveling along a road with a scraped fender is a paradigmatic example of a causal process. Causal interactions are then construed as spatio-temporal intersections of causal processes that produce modifications in the very causal processes that intersect. And a car accident is just such a causal interaction. According to the *CM* model, explaining an event consists of tracing the causal processes and interactions leading to that event.

The *CM* model certainly has several merits, including the fact of fitting quite nicely with intuitive ideas about what explains an event. It also proposes a satisfying criterion for distinguishing causal processes from pseudo-processes. However, it has been markedly criticized for not providing a precise enough characterization of the very causal processes and interactions that do the explaining, as opposed to those that carry little or no explanatory force (the bug hitting the car fender a few seconds before the accident certainly constitutes a causal interaction, but one that has no explanatory import with regards to the accident itself). In short, the fact that a causal process can transmit a mark says nothing about its explanatory relevance. This general problem of sorting out the explanatorily relevant processes and interactions from the irrelevant ones has been identified as one of the major flaws of the *CM* model (e.g., Hitchcock 1995).

In more recent work, and in order to counter some of the problems identified above, Salmon developed a modified version of the *CM* model in which he replaced the mark-transmission characterization of causal processes by a conserved-quantity characterization (e.g., Salmon 1994; see also Dowe 2000). In this modified version, a causal process is a physical process that propagates a non-zero amount of a physical conserved quantity (such as momentum, charge, energy etc.). And a causal interaction is a spatio-temporal intersection of physical processes during which there is an exchange of a conserved quantity. This modified version of the *CM* model, however, has been shown to face the same problem of explanatory relevance as the previous one (Hitchcock 1995).

Another worry that is especially salient for explanation in biology is that the *CM* model locates explanation at the level of the physical causal nexus. A consequence is that explaining higher-level phenomena with higher-level processes, for instance biological, is not an option since these processes do not correspond to spatio-temporally continuous causal processes in Salmon's sense (e.g., Woodward 1989). This is, of course, particularly important for biology, because what seem to be explanatorily relevant factors in many cases do not fulfill the *CM* model.

On the other hand, this is no argument against any general theory of causal explanation. Though it might be the case that the theory of causation that Salmon's *CM* model builds upon is not adequate, this is not to say that no theory of causation can serve as a basis for developing other causal models of explanation. Indeed, the topic of causation has seen several key developments lately, including Woodward's influential interventionist account (2003), upon which novel models of explanation are being elaborated.

While both the *SR* and the *CM* models of explanation have been developed to counter the problems faced by Hempel's *DN-IS* model, and do so by making the notion causation central to that of explanation, an alternative route has been explored by Kitcher (1981, 1989 – see also Friedman 1974) who has proposed construing explanation in terms of unification. It is argued that, of the most central virtues of science, is its ability to unify phenomena that look completely unrelated at first, and thereby providing some understanding of the workings of nature. In biology for instance, Darwin's theory of natural selection unifies phenomena as diverse as fossil records, animal development and animal instinct, and it gives a general explanatory scheme able to account for all these various features of the living world. The unificationist model of explanation thereby defines explanation on the basis of the epistemic notion of unification.

More specifically, whereas, for Hempel, explaining is deriving the *explanandum* from an *explanans* (that includes laws of nature), for Kitcher explaining consists in showing that the derivation that leads to the *explanandum* is made according to an argument pattern that belongs to a very specific set of patterns: the "explanatory store". The key characteristic of this explanatory store – and the reason why, according to Kitcher, argument patterns that belong to it have explanatory force – is that it consists solely of argument patterns that maximally unify the set of beliefs that are accepted at a particular time in science. In other words, explaining is a matter of deriving as many *explananda* as possible from as few argument patterns as possible, these argument patterns belonging to the explanatory store.

As with its predecessors, this model has not won general agreement, because it raises various problems. One of these problems is that it is not clear how causal relationships can be derived from the concept of explanatory unification. This problem shows especially in cases that hinge on the problem of explanatory asymmetry. For instance, it has been argued that retrodictive derivations might be done using as few argument patterns as predictive derivations, hence making retrodictions as explanatory as predictions (e.g., Barnes 1992). Most importantly, it seems that unification is a broader notion than explanation, thereby implying that not all unifications are explanatory. This is illustrated by the use of a common mathematical formalism to describe different kinds of unrelated systems, or by the elaboration of broader classificatory schemes: such formal unification that is achieved by using the same argument patterns is often observed in science, and yet does not have any explanatory import (e.g., Morrison 2000; Sober 2003).

Parallel to the development of these models of explanation, some have sought to investigate the way explanations are provided so to speak in real life. This has led to developments that take type (3) problems – about the role context plays in explanation – as central. A classical objection to the *DN* model is that actual explanations are rarely set as a formal deductive argument (e.g., Scriven 1962). For instance, we tend to accept some historical narratives as explanatory despite the fact that these narratives do not make it possible to elaborate an argument that would show that the *explanandum* follows from a set of premises. As a response, some have looked into explanatory acts and the pragmatics of explanation as another means to characterize scientific explanation (e.g., van Fraassen 1980; Achinstein 1983). For instance, for van Fraassen, solving the puzzles of explanation can only be done if we have a clear understanding of the why-questions that are at the origin of the requests for explanation. Van Fraassen therefore construes explanations as answers to why-questions that he defines as ordered triples of the form $<P_k, X, R>$, P_k being the topic of the why-question, X its contrast class, and R the relevance relation that the answer A to the why-question must bear to $<P_k, X>$. Modeling requests for explanation in this way highlights the contrastive nature of why-questions (why P_k rather than P_m or P_n). It also highlights the contextual nature, made apparent both in the contrast class and in the relevance relation.

This model, however, has been criticized for not offering enough constraints as to what should really count as a proper explanation. Because the relevance relation, in particular, is not strictly specified, van Fraassen's account of explanation suffers from a risk of trivialization that would make any statement count as a proper explanation of a well-chosen why-question (Kitcher and Salmon 1987). On the other hand, it has been proposed that van Fraassen's non-formal criteria as to what should count as a *good* explanation could be rendered more formal, thereby alleviating parts of Kitcher and Salmon's criticisms (e.g., Richardson 1995). In any case, van Fraassen's account is often viewed as a key contribution that takes into account the pragmatics of explanation by specifying how we elaborate requests for explanation, yet one that fails to appropriately characterize what is offered as explanation.

So, what are we left with, six decades after Hempel and Oppenheim's essay? The harsh reality is that there does not seem be much consensus on how best to characterize scientific explanation. Several models have been proposed. And yet, each one has been shown to be plagued with imperfections. As a response, at least two different strategies can be pursued.

The first is to continue searching for an even better model of scientific explanation, be it one developed on the basis of yet another theory of causation, or one that would combine several features of existing models, or that would even explore radically novel avenues. In a way, this is still trying to solve the three major types of problems identified above. A presupposition of this strategy is that scientific explanation comes in one sort. In turn, this presupposition makes it possible to envision a set of conditions that would be jointly necessary and sufficient to single out scientific explanations from things that do not deserve such labeling (recall that there is also a normative aspect in this philosophical project). However, it is not at all clear that what counts and/or should count as a scientific explanation in domains as different as fundamental physics, genetics, ecology, sociology or economics, can be accounted for by a single model of explanation.

Arguing that this is indeed not the case leads to a second strategy for pursuing the debate: the strategy of defending some form of pluralistic view about scientific explanation. In short, the idea is to say that accounting for scientific explanation requires different models, possibly depending on the scientific disciplines or the types of *explananda* or both. Explanatory pluralism raises several questions: Are there good reasons to endorse pluralism, beyond the fact that so far no unique model of scientific explanation has been found? If one accepts explanatory pluralism, how are the different models related to each other? Do they stand in competition with each other, possibly offering different *explanantia* to the same *explanandum*? Do they complement each other by targeting specific types of *explananda*, or by being linked to scientific disciplines or fields, for instance via their methodological components? Should they be considered components of a (yet to come) more unified model of explanation that would be capable of capturing all these points while avoiding fatal counterexamples? Or should we rather acknowledge that no universal model might be able to capture all the dimensions of what a scientific explanation is?

Investigating how scientific explanation works in biology is one way to tackle the questions that arise when one adopts either one of these two strategies. In particular, there is a general worry as to whether the models of scientific explanation that have been mostly developed with physics as paradigmatic source of inspiration are indeed applicable to the biological sciences. As it has often been noted by commentators, until recently contemporary philosophy of science has been very much influenced by the physical sciences and many discussions have been somehow biased and possibly non-transposable to biology. A relative lack of interest in the special sciences in general has led to the ignoring of their peculiarities, including when it comes to investigating models of scientific explanation. Yet, there are obvious differences between explanations in physics and in biology. After all, contrary to physics and to most of chemistry, biology is held by many to have few very general theories based on universal laws and that make fundamental use of mathematical language. So the worry is that even if one (or several) of these models of explanation were able to capture adequately how explanation works in the physical sciences, it is not clear that the same could be said when it comes to biology or other special sciences.

Furthermore, because of a profusion of different schools of thoughts and research traditions *within* biology, different types of explanations seem to be pervasive across all of biology, ranging from historical narratives in evolutionary biology, to functional explanations in anatomy or to causal-mechanisms in molecular biology. This situation can be taken as an argument in favor of explanatory pluralism, not just by making explanation models specific to scientific disciplines, but by making explanation models even vary within a discipline, and possibly depending on finer elements such as problem types, heuristics, methodologies, cognitive and epistemic context and so forth (e.g., Sterelny 1996; Plutynski 2004; Brigandt 2013). Of course, it might just turn out that what superficially looks like rather peculiar forms of biological explanations are in fact only particular cases that would all fit a more general model in the end. Nevertheless, biological explanatory practice requires specific attention in this respect, and has indeed started to become the focus of a rich array of work.

3   Thinking About Explanation in Biology

Thinking about explanation in biology with the background mentioned above raises several kinds of questions and suggests at least that all of the proposed models of explanation need some revision if they are to work properly in biology. We will outline four of the most salient problems in the current debate. These problems are related to (1) whether natural laws exist in biology, (2) whether causation plays a specific explanatory role in biology, (3) whether other forms of explanation – e.g., functional or teleological – are also needed, and (4) whether the recent mechanistic-type model of explanation that brings together some form of law-like generalizations and of causation fulfill all expectations.

A major potential problem for the application of nomological models of explanation in biology is indeed the rarity (or perhaps even the absence) of natural laws, at least as they have been often conceived in philosophy from the study of the physical sciences. Of course, we find many generalizations in biology that appear to be involved in explanation, but they do not meet most criteria of lawfulness. These generalizations usually admit exceptions; they are spatio-temporally limited, or do not support counterfactuals (e.g., Smart 1962; Ruse 1970; Rosenberg 1994; Brandon 1997). Even the few generalizations that are called laws in biology, such as Mendel's laws, are problematic for *DN* types of models of explanation.

One reason for the absence of strict laws in biology stems from the nature of the biological world. Biological entities are the product of a long history, partly driven by natural selection and dependent on historical contingencies. Although this is a very difficult question in evolutionary biology, it is arguable that evolutionary history could have been different and that, as a consequence, the biological generalizations we now have could have been different too. This argument, made forcefully by John Beatty (1995), highlights the fact that evolutionary contingency undermines the very possibility for biological laws. All generalizations that are distinctively biological describe contingent outcomes of evolution. And yet, natural laws must be more than just contingently true. For instance, it is a very general biological fact that genetic heredity is encoded in nucleic acids (DNA or RNA). However, this might be only one possible solution that has been retained at the beginning of evolutionary history and that has then become universal (at least on our planet). If we think about other possible forms of life that might exist elsewhere in the universe, heredity might be handled in different ways and it is not clear at all whether our most general biological models (about say heredity or metabolism) would apply to them. Contrary to physics, biological generalizations seem to lack nomic necessity.

Another way to describe the problem is to recognize that many biological explanations are historical in nature. Explanations are obviously possible in the historical sciences, but they have peculiar features and raise several issues (e.g., Dray 1977; Clayton 1996). Most notably, *explananda* consist of unique events situated in the past. This does not preclude their explanation, but likely requires a different style of explaining, typically based on narratives. Authors like David Hull (1992) have argued that historical narratives in biology have a strong explanatory force, that does not depend much on general laws but more on particular circumstances: they tell stories that describe causal sequences of events. Historical explanations are legitimate, yet the impossibility to have direct empirical access to these causal chains give them a problematic status, which has led some scientists and philosophers to deny them the status of proper scientific explanation or at least to give them a lower explanatory status (e.g., Schaffner 1993). A particular and much-discussed case of historical explanation in biology is the explanation of traits as adaptation, i.e., as the product of natural selection. Following Gould and Lewontin (1979) famous criticisms, many have come to consider them as unfalsifiable "just-so-stories", which do not satisfy the criteria for genuine scientific explanation. It is indeed easier to imagine possible adaptive scenarios than to test them. This is of course not a fatal flaw for these explanations, but a sign that historical explanations must be offered with caution.

Coming back to the role of laws in explanation, their apparent absence in biology has led to various answers. One possibility is to relax the criteria for lawfulness and accept that biological generalizations are laws, but different from what we know from physics or chemistry (e.g., Sober 1993; Lange 1995; Mitchell 1997). In other words, the concept of scientific law can be redefined so as to accommodate the generalizations found in biological explanations. This move can save conceptions in line with the *DN* model, that make explanations depend on laws. Another response is to propose to drop the requirement for laws and accept that other kinds of generalizations can support genuine explanations (e.g., Woodward 2000). According to Woodward's account, the important feature for a generalization to be used in an explanation is not its lawfulness but rather its invariance. Invariance of a generalization means that it would continue to hold under a relevant class of changes. Conceiving explanation in terms of invariant generalizations has the advantage of avoiding several problems, including the restricted validity of generalizations in biology and the existence of exceptions. By acknowledging different degrees of contingency between laws and accidents, this strategy offers a more nuanced account of how generalizations work in scientific explanation. Yet another line of response might still be to argue that the arguments against laws in biology are mistaken, and that there exist indeed distinctively biological laws (Elgin 2006). In any case, the debate is still open and all the more so as the very notion of 'law of nature' is a delicate one to tackle (e.g., DesAutels 2009; Haufe 2013).

The problems raised by nomological accounts in biology give strong reasons to turn to alternative models of explanation, and in particular to those that center on causation. Indeed, causal accounts of explanation have been offered as a solution to many of the problems traditionally encountered by nomological models. Furthermore, many explanations in biology do involve citing causes and causal regularities (e.g., Schaffner 1993; Waters 1998). The question then becomes whether existing causal models of explanation fit the explanatory practice as found in biology and in all its sub-disciplines.

Concerns have been raised as to the applicability of Salmon's causal-mechanical (*CM*) model, be it under its mark-transmission form or under its more recent conserved-quantity version. As noted above, the mark-transmission model faces serious difficulties, leaving the conserved-quantity model as the only likely contender. Yet, the conserved-quantity model focuses very heavily on the physical level, in particular by situating causation at the level of conserved physical quantities. In this respect, explanations that fit this model must include physical details that are usually not considered relevant in biological explanations. Of course, causal relations at higher biological levels should in principle be analyzable in terms of physical processes. Yet, this is almost always unmanageable in practice. Furthermore, explanations that would be so framed at the level of conserved physical quantities would differ notably from what is usually taken as explanatorily relevant in biology (e.g., Woodward 1989; Glennan 1996).

An alternative solution would be to turn to the recently developed interventionist model of causation (Woodward 2003) as a basis for construing a causal model of explanation that would suit the practice of most – if not all – domains of biology. On this conception, a causal explanation consists in the exhibition of patterns of dependency between the factors cited in the *explanans* (causal factors) and those cited in the *explanandum*. These patterns of dependency are revealed by means of interventions onto the different causal factors – variables in a causal model – and the identification of subsequent changes in the effect factors. The main idea behind the interventionist account of causation is that causal relationships are revealed by the fact that when one intervenes on a given factor – while holding fixed a proper set of background conditions – one witnesses changes in another factor. One of the key motivations for such an account is to capture the practice of experimental science, which is characterized by specific interventions onto systems that are placed in well controlled set-ups. Another advantage is that interventionism does not require causal relationships to be exclusively located at the physical level. Rather, causal relationships are possible at any level of investigation, provided the variables that enter the relevant causal models fulfill the formal conditions of interventionism. Furthermore, as Woodward argues by looking at specific biological explanations, causal relationships that fulfill additional conditions of stability, proportionality and specificity are those that are usually called upon in proper causal explanations (Woodward 2010; see also Waters 2007). While the interventionist account of causation has generally been well adopted by proponents of mechanistic explanations in biology (see below – yet see also Weber 2008), the feasibility of developing a satisfactory model of explanation on this basis hinges on the viability of interventionism in general as a theory of causation. This is a matter of intense debate, with such questions concerning the circularity of the account (causation is defined by means of interventions, that are themselves causal), the possibility of interventions in cases of supervenient properties or foundational assumptions that relate to modularity and the Causal Markov Condition (see for instance Cartwright 2006; Glymour 2008; Mitchell 2008; Strevens 2008; Baumgartner 2009).

Whatever the merits of various theories of causal explanation, one ought not assume that all explanations are necessarily causal. One example that has received much attention in the philosophy of biology is the case of equilibrium explanation (e.g., Sober 1983; Potochnik 2007). Equilibrium is a stable state of a system that has a domain of attraction larger than the state itself, so that when the system is perturbed it returns to this equilibrium state. Using Fisher's explanation of sex-ratio equilibria, Sober (1983) has argued that equilibrium explanations are not causal. They are explanatory because they show that many initial conditions lead to this state, but they are not causal because they do not cite the actual causes that have produced that state. Tracing the actual causal history is not explanatorily relevant here. It seems that the same could be argued about a number of mathematical explanations in biology and in science (e.g., Baker 2009). Among them, models of patterns and allometric scaling laws, which describe how processes scale with body size and with each other, seem to capture essential properties of living systems without appealing to causes. However, it is not exactly clear what explanatory work they actually perform. Moreover, rather than genuine explanations they might be considered as observations in need of explanation, for example by physical principles (West et al. 1997).

Another type of explanation that immediately comes to mind in biology is functional explanations. Indeed a central part of the explanation of an organism structure and of its parts (traits) features and organization involves the concept of function. Organs are explained by their function and this explanatory pattern is omnipresent down to the molecular level (genes and proteins). Some have argued that this is even part of what gives biology its explanatory autonomy, since more fundamental sciences such as physics and chemistry simply do not possess such explanatory concepts (e.g., Hull 1974; Mayr 1988). Though they seem indispensable in biology, functional explanations are problematic because they imply the notion of goal, hence of teleology. And teleologically describing the world is not acceptable since the rise of modern science. Of course, by showing that the apparent teleological nature of organisms can be explained by the action of natural selection, Darwin has offered a solution to this problem. And philosophers such as Nagel (1961) and Hempel (1965) have accordingly tried to define "function" in terms that make functional explanations unproblematic. However, things have turned out to be more complicated and competing views about how to best define biological function have been proposed. In particular, two main accounts have been defended. The etiological account (e.g., Wright 1973; Millikan 1989) defines the function of specific traits by referring to what those traits were selected for doing in the organisms' ancestors. Such a construal of function is thus fundamentally historical. On the other hand, the systemic account of function (e.g., Cummins 1975) defines the function of some trait in terms of what this trait does in the organism that possesses it (for instance the role it plays in maintaining the overall organization of the organism in its present state). These two accounts are not necessarily incompatible, but a unified theory of function and of functional explanation still is the focus of much debate. The corresponding literature needs not be reviewed here (but see Wouters 2005), but it shows that the nature of this central type of biological explanation still raises questions, in particular regarding its relation to other models of explanation.

We have noted above that explanation in biology is often claimed to incorporate a strong causal component. In some domains of biology – such as molecular biology, cell biology, or physiology just to name a few – these causal explanations often take the form of mechanistic models. This has led a growing number of philosophers in the last 20 years to develop mechanistic models of explanation (e.g., Bechtel and Richardson 1993; Glennan 1996; Machamer et al. 2000; Craver 2007; Bechtel and Abrahamsen 2012; Craver and Darden 2013).

One of the motivations was to bring philosophical analysis closer to the reality of scientific research. Examining the explanatory practices of biologists reveals that phenomena are often explained by identifying the mechanisms that produce these phenomena. Although several alternative definitions of mechanism (and of mechanistic explanation) have been offered in the recent literature, they tend to converge on their most critical features. A mechanism can be thought of as being composed of parts that interact causally (usually through chemical and mechanical interactions) and that are organized in a specific way. This organization determines largely the behavior of the mechanism and hence the phenomena that it produces. Explaining a phenomenon in a mechanistic way involves decomposing the system that is at the origin of that phenomenon into interacting parts, and giving a description of how the organization and activities of these parts produce the phenomenon to be explained. A key difference from nomological models of explanation is that neither laws of nature nor logical derivations play any significant role in mechanistic explanation. Mechanisms can be formalized in different ways, including with the help of diagrams and schemas, and are usually supplemented by causal narratives that describe how the mechanisms produce the very phenomena to be accounted for. Explaining results from rehearsing how the different parts of a mechanism causally produce the *explanandum*.

The interest for mechanistic explanation also corresponds to a broadening of the discussions on scientific explanation. While analyses have traditionally focused on the context of justification, a lot of recent work has been devoted to elucidating how mechanistic explanations are actually developed (but see also Schaffner 1993). Close attention to heuristics and experimental methods (manipulation of biological systems) has thus enriched the understanding of explanatory practices, and especially in biology.

Since causation plays a significant role in mechanistic explanation, a theory of causation is an implicit assumption of this model of explanation. As noted above, several mechanistic philosophers tend to endorse an interventionist account of causation (e.g., Woodward 2002; Craver 2007 – but see also Bogen 2005). It is debated however whether mechanistic explanation requires such an account of causation. It is also debated whether mechanistic explanation so construed fits the increasingly complex and dynamic systems that are now uncovered in many domains of biology, and whether the mechanistic model of explanation needs to be extended in some way or another (e.g., Kaplan and Bechtel 2011). As many of the contributions to this present volume show (see below), the topic of mechanistic explanation currently receives a lot of attention.

4   The Seventh Decade of Explanation: Insights from Biology

When we look back at research on scientific explanation in philosophy of science, we cannot help but see a gap between two lines of investigation. On the one hand, there is a very central objective that was and still is pursued in general philosophy of science that consists in characterizing *any* scientific explanation, be it by addressing type-1, type-2 or type-3 problems as outlined above in Sect. 2 (Are there necessary and sufficient conditions for any scientific explanation? What role does causation play in explanation? What role does context play in explanation?). Underlying most approaches is the assumption that there *must be* a general model of explanation, and that one of the goals of philosophy of science precisely is to find that model. On the other hand, most of the work on explanation that has been done in the philosophy of biology has followed a less ambitious path, mainly attempting to characterize particular types of explanations as found in the practice of biology at large, from evolutionary biology to molecular biology, including developmental biology, systems biology or synthetic biology to name a few, and somehow structuring itself around the four major questions identified above in Sect. 3 (Are there natural laws in biology or other forms of explanatorily relevant generalizations? Which role does causation play in biological explanation? Are other forms of explanation needed? What are the strengths and weaknesses of the recent mechanistic-type model of explanation in biology?). For instance, all proponents of the new mechanism-based model of explanation are cautious to emphasize that they do not claim that all explanations in biology, let alone in other scientific domains, ought to take the form of mechanisms. Similarly, the same caution characterizes the debate about functional and teleological explanations and other types of explanations identified in biology. The objective that is generally pursued is to propose models of explanation that capture the specificities of each important sub-class of explanation we find here or there across all of biology. Recent debates have shown that this objective is already difficult to reach. However, it also gives the impression that the original question about explanation pursued by general philosophy of science has been given up by philosophers looking at biology. For this reason, rethinking the relationships between the different models of biological explanation and of general scientific explanation is crucial.

There is another reason for extending further the recent lines of research about explanation in biology. As mentioned earlier, mechanistic analyses have come to dominate the scene in the last decade. Based on the intuition that many explanations we find in the biological sciences fulfill mechanism-based models, several authors have offered various accounts of how these models are characterized, how they get their explanatory force, how they are built, and so forth. Though probably no philosopher would seriously argue that all biological explanations are mechanistic, the frequency of such mechanistic explanations in biology makes it tempting to adopt a rather expansionist attitude and develop an extremely broad conception of mechanistic explanation that would apply to an extended domain of science. For instance, some authors wonder whether historical and evolutionary explanations, despite their characteristic contingency and their populational character, might be considered particular cases of mechanism-based explanation (Skipper and Millstein 2005; Barros 2008; Glennan 2009). Because not everyone agrees on this view, it is worth clarifying once again the core of the concept of mechanism-based explanation, and, most importantly, determining its specific application domain and limits. In other words, it is needed to have a clearer view of what mechanistic explanations can really explain and of what falls outside of their explanatory range. It is also crucial to identify the reasons why certain biological phenomena might resist mechanistic explanations, and by the same token, to investigate whether the mechanism-based model of explanation might be extended to handle such difficult cases. Another question that arises is whether mechanism-based explanations might not simply be also subsumed under a broader model of explanation, such as one of the more general models investigated in general philosophy of science.

Such questions about the unity vs. diversity of explanation are particularly pressing in biology for several reasons. Although biology is certainly not completely unique from that point of view, it does offer a striking variety of explanatory practices and can be seen as the locus where very different explanatory patterns are indeed put to work, sometimes jointly, sometimes independently of each other. This specificity of biology with regards to the concept of scientific explanation originates from at least three sets of reasons.

First, the biological sciences cover a huge variety of phenomena studied at broad range of levels, from molecular entities all the way up to ecosystems. Models of explanation must therefore be able to adapt to the different scales at which these phenomena unfold. Note that these variations of scales are also temporal, as biology encompasses phenomena that span over billion of years, such as the evolution of unicellular organisms, as well as phenomena that unfold in much less than a second, as is the case, for instance, of a synaptic neuro-transmission.

Second, biology is a scientific discipline in which numerous diverse approaches to science coexist. Some may advocate reductionist methodologies, as is often said to be the case in molecular biology, while others favor more holistic views, as is the case in some approaches to developmental biology. Analysis is said to drive most of the research done in general biology, while at the same time synthesis is claimed as major methodological approach in some novel branches such as synthetic biology. The diversity of methodologies or heuristics in biology is also visible in the diversity of complementary disciplines that contribute to the development of biological knowledge, and that range from applied mathematics and computer science to engineering and complex systems science. Because such heuristics determine how biological phenomena are approached, it is important to understand also to which extent they might in turn determine shifts in what is taken to be explanatory.

Third, biology is also characterized by a lack of theoretical unifying principles or laws, except perhaps for the principle of evolution by natural selection. As a result, explanation takes the form of a patchwork of different explanatory practices that are related to each other in complex ways that require clarification. Moreover, as sub-disciplines within biology appear and disappear, reorganize, merge or split, these relationships between explanatory practices often change rapidly. Recently for instance, two interesting cases have attracted much attention: on the one hand, evo-devo considers as central the question of articulating evolutionary and developmental explanations; on the other hand, systems biology pursues the explanatory integration of mathematical and computational modeling with biologically relevant considerations.

The present volume *Explanation in Biology. An Enquiry into the Diversity of Explanatory Patterns in the Life Sciences* aims at addressing parts of these complex questions. Though some issues are tackled in several essays across the volume – for instance, the limitations of mechanistic explanation – we have organized the essays into five parts that echo the broad range of questions that the notion of explanation addresses in biology today, in this seventh decade of philosophical investigations about scientific explanation. The ordering could have been done otherwise, but we have tried to sort out the essays according to their main philosophical theses (e.g., explanatory pluralism, emendations of mechanistic explanation, role of mathematics and of heuristics in explanation, new theories of explanation) as opposed to the biological sub-disciplines from which they borrow examples (e.g., molecular biology, evolutionary biology, systems biology, environmental biology etc.).

Part I of the volume explores explanatory pluralism in biology, be it by looking at research traditions in biology or by investigating some of the explanatory specificities of particular branches of biology, such as systems biology, evolutionary biology, or developmental and molecular biology.

Michel Morange (Chap. 2) argues that explanations are plural in biology and often compete against each other. This has been amply documented, for instance, in the field of evo-devo, but the situation is common to all branches of biology. For Morange, this multiplicity of explanations has three sources. First, many of the explanation-seeking questions raised in biology are ambiguous and can be interpreted in different ways. This leads unavoidably to the search for different answers, hence different explanations. Second, the objects investigated by biology – e.g., biomolecular processes, cells, organisms, ecosystems etc. – are the product of a historical evolutionary process. As a result, and despite similarities, these different objects exhibit unique features that require unique explanations, hence resulting in a plurality of explanations of apparently similar phenomena. Third, biology is a discipline in which long lasting explanatory traditions compete against each other, such as the reductionist tradition versus the holistic one. This leads to a multiplicity of approaches when it comes to explaining biological phenomena.

By reflecting on case studies borrowed from systems biology, Constantinos Mekios (Chap. 3) argues that this biological sub-discipline is the locus of a rich plurality of types of explanations. These types of explanations include a profusion of mechanistic explanations that target specific minute domains of enquiry, but also of more systemic explanations that aim at providing a broader integrative view, as well as of explanation patterns borrowed from other sciences, physics in particular as well as mathematics. While mechanisms do play an important explanatory role, they are not sufficient to capture the richness of the explanatory endeavor in systems biology, be they bottom-up or top-down. As Mekios argues, integrating or patching together many explanatory schemes of different nature, some of them of ambiguous status, is the only method of reaching intelligibility in systems biology. Mekios also shows that explanatory pluralism is not just a matter of abstract theoretical considerations, but is driven by real practical problems that arise in the actual practice of systems biology.

As Derek Turner reminds us (Chap. 4), many explanations in evolutionary biology are of historical nature. These have received some attention from philosophers in the past, but Turner takes a fresh look at paleontology and macroevolutionary theory in order to clarify the explanatory relation between two central issues, the contingency of evolutionary history and the passive vs. driven nature of large-scale evolutionary trends. These issues are important when we think about explanation in biology because the role of contingency in evolution has direct consequences for our ability to predict evolutionary outcomes and for the possibility to make generalizations, and also because the existence of trends has been accounted for very differently, both in adaptive and non-adaptive terms. Turner explores the hypothesis that the contingency of evolutionary history explains why some evolutionary trends are passive. He also discusses what sense of contingency is adequate in this context. According to Turner, the relation between these two issues reveals a deeper unity in macroevolutionary theory than it is usually recognized. This is important for the potential explanatory contribution of macroevolutionary models and concepts in evolutionary biology, a question that has been much debated since the 1970s. Turner's argument thus supports the view that macroevolutionary theory has a genuine and irreducible explanatory role to play.

In her contribution to this volume, Francesca Merlin (Chap. 5) examines how explanation is a matter of linking an *explanandum* to an *explanans*. Interestingly, specific phenomena sometimes act as *explanandum* and sometimes as *explanans*. Merlin points at the particular case of developmental noise as studied by developmental biology. Sometimes, developmental noise is considered an *explanans*, for instance when it comes to explaining the particular physical characteristics of individual organisms. And at other times, Merlin argues, developmental noise is considered an *explanandum* that is addressed typically from a selective-evolutionary history, but also from a physico-chemical perspective in terms of noise-reducing constraints. Such a case study highlights in particular the plurality of possible explanations linked to a same phenomenon.

Part II of the volume gathers contributions that investigate the applications of the mechanism-based model of explanation in biology, while also looking at some possible emendations. These limits are apparent, for instance, in molecular biology when faced with the tantamount complexity of the systems that are analyzed, but also in other areas of biology such as developmental biology where classic mechanisms are supplemented with other models when used to formulate explanations. Such situations can in turn be used as stepping-stones to elaborate extensions of the mechanism-based model of explanation.

Molecular biology, one of the most successful biological domains of the last century, has provided philosophers with numerous examples of how explanations work in biology and has been interpreted as supporting the mechanistic framework. However, by looking at recent research on genetic regulation by microRNA, Frédérique Théry (Chap. 6) argues that explanation in molecular biology (and functional biology in general) cannot be reduced to mechanistic explanation, as it has been characterized in recent philosophical discussions. Biologists are increasingly interested in properties of living systems that the concept of mechanism fails to capture. Théry describes two alternative types of explanation involved in microRNA research that go beyond the limits of mechanistic explanations. First, quantitative explanations fully take into account the fact that molecules are present in cells as populations. Mechanistic explanations primarily deal with the qualitative component of causal relations and largely ignore that the function of many molecular components depends on the relative cellular concentrations of the different molecules involved in the process. Second, systemic explanations go beyond the idealization of mechanisms producing a function autonomously and show how different causal processes are interconnected at the cellular level. These alternative explanations do not contradict classical mechanistic models, but are rather meant to complement them. By recognizing the specificities of the various explanatory schemes in functional biology Théry thus invites us to embrace a more pluralistic perspective.

Ingo Brigandt (Chap. 7) explores in turn the limits of philosophical accounts of mechanistic explanation in evolutionary developmental biology. Contrary to the classical account that distinguishes explanatory from phenomenological models, and that classifies mathematical models among the latter, Brigandt argues that mathematical models are often indispensable components of mechanistic explanations, and therefore have specific explanatory value. In particular, Brigandt argues that quantitative models are required as a complement to mechanisms to provide proper explanations not only of quantitative features but also of qualitative features of mechanisms. To support his case, Brigandt draws on the scientific literature of the evo-devo field. For instance, he shows that mechanistic explanations – when conceived in the classical fashion – cannot account for the evolutionary origin of morphological features like the cusp number and shape of the teeth in mammals, or the development of segments in vertebrates: indeed, explanations of such phenomena cannot be achieved by rehearsing the causal roles of the entities of the underlying mechanisms, but do require mathematical models that make it possible to understand the emergence of the spatio-temporal behaviors that constitute the explananda. Brigandt extends his case to other evo-devo phenomena, like the explanation of robustness, of plasticity and of modularity.

As Fridolin Gross recalls (Chap. 8), the mechanistic explanatory framework is often taken to consist of entities and activities characterized by change-relating relationships construed along a manipulationist account of causation. These change-relating relationships occupy a central role in explaining why the mechanisms behave the way they do, and this picture of mechanistic explanation assumes that only relationships of dependence have explanatory value. But, as Gross points out, relationships of *non*-dependence may sometimes also play a crucial explanatory role in systems biology. Gross argues that this is typically the case when it comes to explaining such behaviors as stability at an equilibrium point, transition from one stable equilibrium to another, or robustness in the face of perturbations. In this contribution, Gross argues that non-dependence relationships cannot be deemed explanatorily irrelevant simply on the basis of not being change-relating relationships (as the classic account of mechanism would require). Rather, depending on contexts, non-dependence relationships also play an explanatory role alongside change-relating relationships.

A peculiarity of mechanistic explanation is its apparently restricted scope. Biologists study particular mechanisms in particular model organisms, which makes generalization difficult and always uncertain. However, generality seems to be an important explanatory virtue. William Bechtel (Chap. 9) shows how the use of new tools, in particular graph-theoretic representations of mechanisms, provides a basis for developing more general accounts of mechanisms organization and behaviour. Bechtel focuses on two strands of research, the first concerns the broad topological organization of large networks, while the second is interested in the structure of small network motifs, which are specialized for specific types of processing. By enabling the formalization of organization patterns, these tools facilitate the identification of a common abstract graph structure in different mechanisms and hence allow generalizing knowledge gained on particular mechanisms.

Because of the complexity of many biological phenomena, explanations in biology often make ample use of mathematics and mathematical tools, and, hence, the question whether mathematics play a specific role in explanation arises. Part III consists of a set of essays that investigate this question through different angles. Biology appears to be an interesting discipline to carry out such investigations: the types of mathematical problems are highly diverse; these problems are also complex but not too complex, and they trigger tools and solutions from several branches of mathematics. Biology therefore is a most relevant place to ask the question whether there exist or not distinctly mathematical explanations in science, but also to investigate the relationships between "abstract" mathematical theorems and derivations on the one hand, and "empirical" biological explanations on the other, be they of the causal-mechanical type or otherwise.

Alan Baker (Chap. 10) focuses on the explanatory role of mathematical objects, especially when mobilized in the context of biological phenomena. This is for instance the case when mathematics play a critical role in explaining the periodical life-cycle of certain North American species of cicada, or the hexagonal shape of bee honeycombs. More generally, Baker proposes to distinguish three types of genuine mathematical explanations in science (*MES*): (1) *Constraint MES* that explain why some physical outcomes are impossible by showing that they are mathematically impossible, (2) *Equilibrium MES* that explain why some physical outcomes occur by showing that they are mathematically inevitable across a range of starting conditions, and (3) *Optimization MES* that explain why some physical outcomes occur by showing that they are mathematically optimal. Baker argues that the crucial role that mathematics play in some recent and controversial explanations of puzzling biological phenomena calls for an even finer characterization of these different types of genuine mathematical explanations that permeate science.

By looking at the case of systems biology, Tobias Breidenmoser and Olaf Wolkenhauer (Chap. 11) argue that *organizing principles* – which typically take the form of specific mathematical theorems – are an indispensable complement to mechanistic explanation when it comes to explaining key behavioral features of biological systems. Breidenmoser and Wolkenhauer distinguish phenomenological models –that only "save the phenomena" by providing curve fitting of some sort – from mechanistic models – that explain by describing underlying molecular and cellular processes. Yet they argue that mechanistic models are limited in that they fail to account for specific *explananda* that are of major interest in systems biology, such as the origin of robustness. Breidenmoser and Wolkenhauer argue that explanation of such features must appeal to analytic theorems, such as the "robustness theorem". More broadly, they propose that certain types of *explananda* require the use of organizing principles as a complement to mechanistic models.

For Tarik Issad and Christophe Malaterre (Chap. 12), the plurality of explanatory patterns should reflect in the ways these explanatory patterns get their explanatory force. Issad and Malaterre investigate the explanatory force of (classically construed) mechanistic explanations (e.g., Machamer et al. 2000; Glennan 2002) and of dynamic mechanistic explanations (e.g., Bechtel and Abrahamsen 2012). They argue that, whereas mechanistic explanations get their explanatory force from rehearsing a causal story, dynamic mechanistic explanations are explanatory in virtue of displaying mathematical warrants that show how the explanandum follows from a mathematical model. Dynamic mechanistic explanations therefore are not causal explanations, even if elements of the models they rely on may receive causal interpretations. Despite this apparent difference in explanatory force, Issad and Malaterre argue that mechanistic explanations and dynamic mechanistic explanations can be construed as limit cases of a more general pattern of explanation that they name "Causally Interpreted Model Explanations" and in which the explanatory force of causation and of mathematical derivation are redistributed. Such pattern of explanation draws its explanatory force from a model, a causal interpretation that links the model to biological reality (but does not necessarily extend into a causal story), and a mathematical derivation that links the model to the explanandum phenomenon (and that may, in simple cases but not in complex ones, be reformulated as a causal story).

Several essays stress the role that different heuristic strategies in biology play when it comes to formulating explanations. We have gathered them in Part IV of the volume. The debate concerns the question whether current models of explanations are sufficient or not to account for how explanations are developed in science, and particularly in biology. In also concerns the question whether heuristic strategies impact or not the ways explanations are formulated. These questions arise in the context of the discovery of more and more complex systems in biology that require novel conceptual tools for their investigation, be they from mathematics, computer science or engineering.

Carlos Zednik (Chap. 13) recalls that much of the philosophical debate about mechanisms has focused on heuristic strategies used to build mechanistic explanations. Bechtel and Richardson's analysis in terms of decomposition and localization has been particularly influential. However, Zednik argues that the classical examples on which most discussions have been based offer only a limited view on the strategies actually used by scientists. In particular, the application of mathematical and computational modelling has deeply changed heuristic methods and explanatory practices. Using examples from contemporary research in neurobiology and evolutionary robotics, Zednik shows that the discovery and description of mechanisms involve more than the principles of decomposition, localization and diagrammatic representation. He focuses in particular on the use of tools from graph theory and dynamical systems theory, which have been recently, increasingly put to work in different domains of biology. An important consequence of these new heuristic strategies is that they increase the scope and power of mechanistic explanation, by enabling the discovery and description of very complex mechanisms that were beyond the reach of classical models.

Pierre-Alain Braillard (Chap. 14) explores how transfer of methods, models and concepts from engineering in the context of the emergence of systems biology, has influenced biologists' explanatory practices. In particular, mathematical modelling methods developed by engineers have been increasingly put to use in order to capture biological systems complex dynamic properties that are difficult or even impossible to analyze and represent with traditional mechanistic approaches. But more than simple mathematical tools, Braillard shows that it is a whole view of how to best decompose and explain complex systems that engineering-oriented scientists have brought to biology. Based on the assumption that biological systems are modular in the same way that many complex engineered systems are, this heuristic and explanatory framework offers promising solutions to deal with the complexity biologists are facing. However, part of the validity of these approaches depends on important assumptions made about biological systems structure, which remain hypothetic.

Tudor Baetu (Chap. 15) argues that models of mechanisms and mathematical models, rather than being antagonistic, play a complementary role in explaining biological phenomena such as those stemming from molecular networks in systems biology. Baetu's thesis can be understood as mediating the positions of Craver who argues that mathematical models are explanatory only to the extent that they identify the physico-chemical entities that causally produce the phenomena (e.g., Craver 2007), and of Weber who argues that explanation requires the derivation of the phenomena from mathematically formulated law-like regularities (e.g., Weber 2005). As Baetu explains, (quantitative) mathematical models cannot replace (qualitative) mechanistic models, but rather complement them in so far as they are required for formulating a broad range of predictions that mechanistic models alone cannot formulate, and as they offer insights into the temporal dynamics of certain systemic parameters. This complementarity, Baetu explains, also shows when considering the development of biological explanations over time and the associated heuristics: because molecular networks are abstract representations of molecular mechanisms, revisions of the latter entail revisions of the former. But the converse is also true in that mathematical models may reveal unsuspected anomalies or holes in previously accepted mechanisms, thus prompting also their revision. For Baetu, the molecular networks studied in systems biology show that a mixed approach to explanation is needed, one that associates models of mechanisms with mathematical derivation from law-like regularities.

An alternative path to probing the limits of mechanistic explanation and elaborating emendations to this model is to explore models of explanation within which mechanisms are not central. Part V of the volume gathers several essays that propose to think anew some of the classical models of explanation – such as Hempel's covering law model – in light of biological specificities, or that investigate, for instance, how causation is mobilized in explanation, in particular when several causal factors act together.

Despite the apparent heterogeneity of explanations in biology, especially when compared to physics, Joel Press (Chap. 16) argues that an underlying unity in all scientific explanations can be found in the form of a "cursory covering law" model of explanation. As the name suggests, the cursory covering law model builds upon Hempel's covering law – or deductive-nomological – model of explanation, and the criticisms this latter model received when it was tentatively applied to biology. Because biology lacks generalizations of the sort required by laws of nature, it has often been argued that the covering law model cannot account for the types of explanations that one finds in biology, thereby prompting either a revision of what laws are in biology (e.g., Brandon 1997; Mitchell 1997), or a dismissal of laws as explanatorily relevant in biology (e.g., Machamer et al. 2000). Press proposes to reconsider the covering law model and relax some of the constraints that bear on the *explanans* so as to include approximating statements about laws, somehow extending Hempel's idea of partial explanations. The proposed "cursory covering law" model subsumes, Press argues, alternative accounts of explanation in biology, be they based on modified accounts of what laws are or articulated without appeal to the notion of laws of nature.

For Melinda Bonnie Fagan (Chap. 17), the analysis of research on stem cells provides illuminating examples of explanations-in-progress linked to developmental processes. Stem cell research broadly construed aims at explaining the branching process of cell development, from a single initiating "stem", through intermediate stages, to one or more termini. By looking more specifically at the case of stem cell reprogramming, Fagan investigates how explanations are constructed in this domain of biology. In light of this analysis, she reviews three major accounts of explanation: a broad interventionist account (based on Woodward 2003), a more focused gene-centered account (based on Waters 2007), and a mechanistic account (based on Craver 2007). She argues that all three accounts encounter limitations when it comes to explaining stem cell reprogramming in so far as the *explananda* typically require the *joint* action of several causal factors, and not just their separate listing as causal contributors. She proposes that explanation – at least in the case of stem cell reprogramming – requires appeal to a notion of *jointness* which appears to be key in explaining how experimental interventions onto already differentiated cells result in pluripotent cells, through the joint action of several factors.

Laura R. Franklin-Hall (Chap. 18) also takes causation to be a central feature of explanation, and in particular of explanation in biology, but she tackles a different issue: she addresses the question of how we select *some* causal factors as being explanatory relevant rather than others. This causal selection problem has two facets: a horizontal one that concerns the omission of background conditions, and a vertical one that concerns the omission of low-level details. Franklin-Hall proposes a Causal Economy account according to which explaining an event is citing precisely the causal factors that "cost less" in virtue of being more abstract, and "deliver more" in virtue of making the event to be explained more stable under variations of other causal influences. While applying her account to the explanation of biological phenomena such as signalling systems and biological development, she also suggests that the Causal Economy account could offer a single principle that would guide explanation construction across the sciences, and in particular across the sub-disciplines of biology, thereby subsuming their apparent heterogeneity and plurality.

5 Beyond the Seventh Decade

As a collection of essays, *Explanation in Biology. An Enquiry into the Diversity of Explanatory Patterns in the Life Sciences* is bound to be exploratory. It would be unrealistic to ask that all problems raised by such a rich question as that of "explanation in biology" be solved. Nevertheless it is our hope that the volume addresses at least some of these problems and contributes interesting and challenging ideas about how to tackle them. There is, no doubt, much more to be said about explanation in biology, and each one of the volume's essays branches out further to novel research directions. The plurality of explanatory patterns and practices in biology remains puzzling. We see it as a springboard for future investigations about whether such pluralism reveals something deeper about the notion of scientific explanation. If this is so, then the quest for a unified account of scientific explanation is likely to be ill grounded. If not, such a unified account might still be possible, yet the role of pragmatics in shaping the explanatory context would certainly deserve further studying. One possible way to go is to continue investigating the limitations of the mechanistic model of explanation and, by so doing, to possibly identify means of extending this model beyond its current limitations. Some may prefer going the route of exploring alternative models, either in the direction of nomological models (hence focusing on the role law-like generalizations may play in explanation), or in the direction of causal models (thereby addressing questions about the necessity of causation in any model of explanation as well as about the modalities of its explanatory force). Others still may opt for investigating the heuristics of explanation as a means to a deeper analysis of explanation itself, and in this respect, looking at the actual practice of science and how this practice currently evolves under the pressure of more and more complex objects of investigation might very well also generate radically novel insights about how science now explains nature. The complexity of biological objects may also point to the need for better articulation of explanation and understanding, and all the more so as some phenomena – say the behavior of complex molecular networks – may somehow challenge our capacity for understanding nature despite being amenable to some form of explanation or another. In any case, scientific explanation remains a key target for philosophical analysis, and its study from the biological angle continues to uncover issues that were previously not considered.

Acknowledgements We thank two anonymous reviewers for their comments and their suggestions of additional references. CM acknowledges support from UQAM research chair in Philosophy of science.

References

Achinstein, P. (1983). *The nature of explanation.* New York: Oxford University Press.

Ayer, A. J. (1956). What is a law of nature? *Revue Internationale de Philosophie, 36,* 144–165.

Baetu, T. (2015). From mechanisms to mathematical models and back to mechanisms: Quantitative mechanistic explanations. In P.-A. Braillard & C. Malaterre (Eds.), *Explanation in biology. An enquiry into the diversity of explanatory patterns in the life sciences* (pp. 345–363). Dordrecht: Springer.

Baker, A. (2009). Mathematical explanation in science. *The British Journal for the Philosophy of Science, 60*(3), 611–633. doi:10.1093/bjps/axp025.

Baker, A. (2015). Mathematical explanation in biology. In P. A. Braillard & C. Malaterre (Eds.), *Explanation in biology. An enquiry into the diversity of explanatory patterns in the life sciences* (pp. 229–247). Dordrecht: Springer.

Barnes, E. (1992). Explanatory unification and the problem of asymmetry. *Philosophy of Science, 59*, 558–571.

Barros, D. B. (2008). Natural selection as a mechanism. *Philosophy of Science, 75*, 306–322.

Baumgartner, M. (2009). Interventionist causal exclusion and non-reductive physicalism. *International Studies in the Philosophy of Science, 23*(2), 161–178.

Beatty, J. (1995). The evolutionary contingency thesis. In G. Wolters & J. G. Lennox (Eds.), *Concepts, theories, and rationality in the biological sciences, the second Pittsburgh-Konstanz colloquium in the philosophy of science* (p. 1995). Pittsburgh: University of Pittsburgh Press.

Bechtel, W. (2015). Generalizing mechanistic explanations using graph-theoretic representations. In P.-A. Braillard & C. Malaterre (Eds.), *Explanation in biology. An enquiry into the diversity of explanatory patterns in the life sciences* (pp. 199–225). Dordrecht: Springer.

Bechtel, W., & Abrahamsen, A. (2012). Thinking dynamically about biological mechanisms: Networks of coupled oscillators. *Foundations of Science, 18*(4), 707–723. doi:10.1007/s10699-012-9301-z.

Bechtel, W., & Richardson, R. C. (1993). *Discovering complexity: Decomposition and localization as strategies in scientific research*. Princeton: Princeton University Press.

Bogen, J. (2005). Regularities and causality; generalizations and causal explanations. *Studies in History and Philosophy of Science Part C, 36*(2), 397–420.

Braillard, P.-A. (2015). Prospect and limits of explaining biological systems in engineering terms. In P.-A. Braillard & C. Malaterre (Eds.), *Explanation in biology. An enquiry into the diversity of explanatory patterns in the life sciences* (pp. 319–344). Dordrecht: Springer.

Brandon, R. (1997). Does biology have laws? The experimental evidence. *Philosophy of Science, 64*, S444–S457.

Breidenmoser, T., & Wolkenhauer, O. (2015). Explanation and organizing principles in systems biology. In P. A. Braillard & C. Malaterre (Eds.), *Explanation in biology. An enquiry into the diversity of explanatory patterns in the life sciences* (pp. 249–264). Dordrecht: Springer.

Brigandt, I. (2013). Explanation in biology: Reduction, pluralism, and explanatory aims. *Science & Education, 22*, 69–91.

Brigandt, I. (2015). Evolutionary developmental biology and the limits of philosophical accounts of mechanistic explanation. In P.-A. Braillard & C. Malaterre (Eds.), *Explanation in biology. An enquiry into the diversity of explanatory patterns in the life sciences* (pp. 135–173). Dordrecht: Springer.

Bromberger, S. (1966). Why questions. In R. Colodny (Ed.), *Mind and cosmos: Essays in contemporary science and philosophy*. Pittsburgh: University of Pittsburgh Press.

Cartwright, N. (1979). Causal laws and effective strategies. *Noûs, 13*, 419–437.

Cartwright, N. (1980). Do the laws of physics state the facts? *Pacific Philosophical Quarterly, 61*, 75–84.

Cartwright, N. (2006). From metaphysics to method: Comments on manipulability and the causal Markov condition. *The British Journal for the Philosophy of Science, 57*(1), 197–218. doi:10.1093/bjps/axi156.

Clayton, R. (1996). *The logic of historical explanation*. University Park: Pennsylvania State University Press.

Craver, C. F. (2007). *Explaining the brain: Mechanisms and the mosaic unity of neuroscience*. Oxford: Oxford University Press.

Craver, C. F., & Darden, L. (2013). *In search of mechanisms: Discoveries across the life sciences*. Chicago: University of Chicago Press.

Cummins, R. (1975). Functional analysis. *Journal of Philosophy, 72*, 741–765.

DesAutels, L. (2009). Sober and Elgin on laws of biology: A critique. *Biology & Philosophy, 25*(2), 249–256. doi:10.1007/s10539-009-9182-x.

Dowe, P. (2000). *Physical causation*. Cambridge: Cambridge University Press.

Dray, W. (1977). Les explications causales en histoire. *Philosophiques, 4*(1), 3–34.

Dretske, F. I. (1977). Laws of nature. *Philosophy of Science, 44*(2), 248–268.

Elgin, M. (2006). There may be strict empirical laws in biology, after all. *Biology & Philosophy, 21*(1), 119–134. doi:10.1007/s10539-005-3177-z.

Fagan, M. B. (2015). Explanatory interdependence: The case of stem cell reprogramming. In P.-A. Braillard & C. Malaterre (Eds.), *Explanation in biology. An enquiry into the diversity of explanatory patterns in the life sciences* (pp. 387–412). Dordrecht: Springer.

Franklin-Hall, L. R. (2015). Explaining causal selection with explanatory causal economy: Biology and beyond. In P.-A. Braillard & C. Malaterre (Eds.), *Explanation in biology. An enquiry into the diversity of explanatory patterns in the life sciences* (pp. 413–438). Dordrecht: Springer.

Friedman, M. (1974). Explanation and scientific understanding. *Journal of Philosophy, 71*, 5–19.

Glennan, S. S. (1996). Mechanisms and the nature of causation. *Erkenntnis, 44*(1), 49–71.

Glennan, S. S. (2002). Rethinking mechanistic explanation. *Philosophy of Science, 69*(3), S342–S353.

Glennan, S. (2009). Ephemeral mechanisms and historical explanation. *Erkenntnis, 72*(2), 251–266. doi:10.1007/s10670-009-9203-9.

Glymour, B. (2008). Stable models and causal explanation in evolutionary biology. *Philosophy of Science, 75*(5), 571–583.

Gould, S. J., & Lewontin, R. C. (1979). The spandrels of San Marco and the Panglossian paradigm: A critique of the adaptationist programme. *Proceedings of the Royal Society of London B, 205*, 581–598.

Gross, F. (2015). The relevance of irrelevance: Explanation in systems biology. In P.-A. Braillard & C. Malaterre (Eds.), *Explanation in biology. An enquiry into the diversity of explanatory patterns in the life sciences* (pp. 175–198). Dordrecht: Springer.

Haufe, C. (2013). From necessary chances to biological laws. *British Journal for the Philosophy of Science, 64*(2), 279–295.

Hempel, C. (1965). *Aspects of scientific explanation and other essays in the philosophy of science.* New York: Free Press.

Hempel, C., & Oppenheim, P. (1948). Studies in the logic of explanation. *Philosophy of Science, 15*, 135–175.

Hitchcock, C. (1995). Discussion: Salmon on explanatory relevance. *Philosophy of Science, 62*, 304–320.

Hull, D. (1974). *The philosophy of biological science.* Englewood Cliffs: Prentice-Hall.

Hull, D. (1992). The particular circumstance model of scientific explanation. In M. H. Nitecki & D. V. Nitecki (Eds.), *History and evolution.* Albany: State University of New York Press.

Issad, T., & Malaterre, C. (2015). Are dynamic mechanistic explanations still mechanistic? In P.-A. Braillard & C. Malaterre (Eds.), *Explanation in biology. An enquiry into the diversity of explanatory patterns in the life sciences* (pp. 265–292). Dordrecht: Springer.

Kaplan, D. M., & Bechtel, W. (2011). Dynamical models: An alternative or complement to mechanistic explanations? *Topics in Cognitive Science, 3*(2), 438–444.

Kitcher, P. (1981). Explanatory unification. *Philosophy of Science, 48*, 507–531.

Kitcher, P. (1989). Explanatory unification and the causal structure of the world. In P. Kitcher & W. Salmon (Eds.), *Scientific explanation* (pp. 410–505). Minneapolis: University of Minnesota Press.

Kitcher, P., & Salmon, W. (1987). Van Fraassen on explanation. *The Journal of Philosophy, 84*(6), 315–330.

Kyburg, H. (1965). Comment. *Philosophy of Science, 32*, 147–151.

Lange, M. (1995). Are there natural laws concerning particular species. *Journal of Philosophy, 112*, 430–451.

Machamer, P., Darden, L., & Craver, C. (2000). Thinking about mechanisms. *Philosophy of Science, 67*, 1–25.

Mayr, E. (1988). *Towards a new philosophy of biology: Observations of an evolutionist.* Cambridge: Harvard University Press.

Mekios, C. (2015). Explanation in systems biology: Is it all about mechanisms? In P.-A. Braillard & C. Malaterre (Eds.), *Explanation in biology. An enquiry into the diversity of explanatory patterns in the life sciences* (pp. 47–72). Dordrecht: Springer.

Merlin, F. (2015). Developmental noise: Explaining the specific heterogeneity of individual organisms. In *Explanation in biology. An enquiry into the diversity of explanatory patterns in the life sciences* (pp. 91–110). Dordrecht: Springer.

Millikan, R. G. (1989). In defense of proper functions. *Philosophy of Science, 56*, 288–302.

Mitchell, S. (1997). Pragmatic laws. *Philosophy of Science, 64*(4), S468–S479.

Mitchell, S. (2008). Exporting causal knowledge in evolutionary and developmental biology. *Philosophy of Science, 75*(5), 697–706.

Morange, M. (2015). Is there an explanation for... the diversity of explanations in biological sciences? In P.-A. Braillard & C. Malaterre (Eds.), *Explanation in biology. An enquiry into the diversity of explanatory patterns in the life sciences* (pp. 31–46). Dordrecht: Springer.

Morrison, M. (2000). *Unifying scientific theories*. Cambridge: Cambridge University Press.

Nagel, E. (1961). *The structure of science. Problems in the logic of explanation*. New York: Harcourt, Brace & World, Inc.

Plutynski, A. (2004). Explanation in classical population genetics. *Philosophy of Science Proceedings, 71*, 1201–1215.

Potochnik, A. (2007). Optimality modeling and explanatory generality. *Philosophy of Science, 74*(5), 680–691. doi:10.1086/525613.

Press, J. (2015). Biological explanations as cursory covering law explanations. In P.-A. Braillard & C. Malaterre (Eds.), *Explanation in biology. An enquiry into the diversity of explanatory patterns in the life sciences* (pp. 367–385). Dordrecht: Springer.

Reichenbach, H. (1958). *The philosophy of space and time*. New York: Dover Publications.

Richardson, A. (1995). Explanation: Pragmatics and asymmetry. *Philosophical Studies, 80*(2), 109–129.

Rosenberg, A. (1994). *Instrumental biology or the disunity of science*. Chicago: The University of Chicago Press.

Ruse, M. (1970). Are there laws in biology? *Australasian Journal of Philosophy, 48*(2), 234–246.

Salmon, W. (1965). The status of prior probabilities in statistical explanations. *Philosophy of Science, 33*, 137–146.

Salmon, W. (1971). Statistical explanation. In W. Salmon (Ed.), *Statistical explanation and statistical relevance* (pp. 29–87). Pittsburgh: University of Pittsburgh Press.

Salmon, W. (1978). Why ask "why?"? – an inquiry concerning scientific explanation. *Proceedings and Addresses of the American Philosophical Association, 51*(6), 683–705.

Salmon, W. (1984). *Scientific explanation and the causal structure of the world*. Princeton: Princeton University Press.

Salmon, W. (1989). *Four decades of scientific explanation*. Minneapolis: University of Minnesota Press.

Salmon, W. (1994). Causality without counterfactuals. *Philosophy of Science, 61*, 297–312.

Schaffner, K. (1993). *Discovery and explanation in biology and medicine*. Chicago: University of Chicago Press.

Scriven, M. (1962). Explanations, predictions, and laws. In H. Feigl & G. Maxwell (Eds.), *Minnesota studies in the philosophy of science, scientific explanation, space, and time* (Vol. 3, pp. 170–230). Minneapolis: University of Minnesota Press.

Skipper, R. A., & Millstein, R. L. (2005). Thinking about evolutionary mechanisms: Natural selection. *Studies in History and Philosophy of Biological and Biomedical Sciences, 36*, 327–347.

Smart, J. C. C. (1962). *Philosophy and scientific realism*. London: Routledge and Kegan Paul.

Sober, E. (1983). Equilibrium explanation. *Philosophical Studies, 43*(2), 201–210.

Sober, E. (1993). *Philosophy of biology*. Boulder: Westview.

Sober, E. (2003). Two uses of unification. In F. Stadler (Ed.), *Institute Vienna circle yearbook 2002*. Dordrecht: Kluwer.

Spirtes, P., Glymour, C., & Scheines, R. (1983). *Causation, prediction and search*. New York: Springer.

Sterelny, K. (1996). Explanatory pluralism in evolutionary biology. *Biology and Philosophy, 11*(2), 193–214.

Strevens, M. (2008). Comments on Woodward, making things happen. *Philosophy and Phenomenological Research, 77*(1), 171–192.

Théry, F. (2015). Explaining in contemporary molecular biology: Beyond mechanisms. In P.-A. Braillard & C. Malaterre (Eds.), *Explanation in biology. An enquiry into the diversity of explanatory patterns in the life sciences* (pp. 113–133). Dordrecht: Springer.

Turner, D. (2015). Historical contingency and the explanation of evolutionary trends. In P.-A. Braillard & C. Malaterre (Eds.), *Explanation in biology. An enquiry into the diversity of explanatory patterns in the life sciences* (pp. 73–90). Dordrecht: Springer.

van Fraassen, B. (1980). *The scientific image*. Oxford: Clarendon.

Waters, C. K. (1998). Causal regularities in the biological world of contingent distributions. *Biology and Philosophy, 13*, 5–36.

Waters, C. K. (2007). Causes that make a difference. *The Journal of Philosophy, 104*(11), 551–579.

Weber, M. (2005). *Philosophy of experimental biology*. Cambridge: Cambridge University Press.

Weber, M. (2008). Causes without mechanisms: Experimental regularities, physical laws, and neuroscientific explanation. *Philosophy of Science, 75*(5), 995–1007. doi:10.1086/594541.

West, G. B., Brown, J. H., & Enquist, B. J. (1997). A general model for the origin of allometric scaling laws in biology. *Science, 276*(5309), 122–126.

Woodward, J. (1989). The causal/mechanical model of explanation. In P. Kitcher & W. Salmon (Eds.), *Scientific explanation. Minnesota studies in the philosophy of science* (Vol. 13, pp. 357–383). Minneapolis: University of Minnesota Press.

Woodward, J. (2000). Explanation and invariance in the special sciences. *British Journal for the Philosophy of Science, 51*, 197–254.

Woodward, J. (2002). What is a mechanism? A counterfactual account. *Philosophy of Science, 69*, S366–S377.

Woodward, J. (2003). *Making things happen: A theory of causal explanation*. Oxford: Oxford University Press.

Woodward, J. (2010). Causation in biology: Stability specificity, and the choice of levels of explanation. *Biology and Philosophy, 25*, 287–318.

Wouters, A. (2005). The function debate in philosophy. *Acta Biotheoretica, 53*, 123–151.

Wright, L. (1973). Functions. *The Philosophical Review, 82*, 139–168.

Zednik, C. (2015). Heuristics, descriptions, and the scope of mechanistic explanation. In P.-A. Braillard & C. Malaterre (Eds.), *Explanation in biology. An enquiry into the diversity of explanatory patterns in the life sciences* (pp. 295–317). Dordrecht: Springer.

Part I
Exploring Explanatory Pluralism in Biology

Chapter 2
Is There an Explanation for ... the Diversity of Explanations in Biological Studies?

Michel Morange

Abstract The multiplicity of explanations in the biological sciences has already been amply discussed by philosophers of science. The field of Evo-Devo has been a focus of much attention, with the obvious coexistence and competition of evolutionary and developmental explanations. In this contribution I borrow examples from hugely different areas of biological research to show that this multiplicity of explanations is common to all branches of biology. I will emphasize three explanations for this diversity. The first is the ambiguity of the questions raised, which can be understood in different ways and require different answers. One recurring ambiguity concerns the local or general nature of the questions (and answers). The second explanation is in the historicity of life, which makes every situation unique, and may require different models for the explanation of apparently similar situations. Another cause of this plurality is the existence of long-lasting competing traditions of explanations. These traditions result from the existence of distinct approaches to reality in scientific thinking, such as the opposition between reductionism and holism, and from a complex history of scientific ideas, models, and theories proper to each biological field. The multiplicity of explanations in the biological sciences therefore has a heterogeneous origin, both epistemic and ontological.

Keywords Historicity • Holism • Plurality of explanations • Research traditions • Reductionism • Themata

1 Introduction

This contribution has a dual objective. It aims to understand the origin of the multiplicity of explanations of a singular phenomenon in science, but also to explain why this diversity is particularly high in the biological sciences. This question is not new. In 1961 Ernst Mayr clearly distinguished two types of explanations in

M. Morange (✉)
Centre Cavaillès, République des savoirs: Lettres, sciences, philosophie USR 3608, Ecole normale supérieure, 29 rue d'Ulm, 75230 Paris Cedex 05, France
e-mail: morange@biologie.ens.fr

© Springer Science+Business Media Dordrecht 2015
P.-A. Braillard, C. Malaterre, *Explanation in Biology*, History, Philosophy and Theory of the Life Sciences 11, DOI 10.1007/978-94-017-9822-8_2

biology, those answering to "how" questions by describing mechanisms, and those responding to "why" questions by proposing evolutionary scenarios (Mayr 1961). The distinction was not new, but it was clearly outlined by Ernst Mayr. Later, Nikolaas Tinbergen distinguished two kinds of questions within each category, one concerning the appearance of the trait, and the other its present state (Tinbergen 1963). These first contributions underlined the relation between the diversity of explanations and the multiplicity of the questions that can be raised concerning a unique phenomenon. In 1983, John Dupré launched a wide philosophical debate about the ontological origin of the "disunity of science" (Dupré 1983) that persisted for many years (Dupré 1993). This debate had a wider scope than the one I want to address in this contribution, and a different one since it raised ontological issues that will not be discussed here. Nevertheless, interesting clues emerged during this debate about the origin and nature of the plurality of explanations in the biological studies that were nicely summarized and extended by Sandra Mitchell in 2003 (Mitchell 2003). One of the most demonstrative examples of the plurality of explanations in biology presented in her book concerned the origin of the division of labour in insect colonies.

A very important perspective on the issue underlined by Sandra Mitchell and the one that I will describe first is that the diversity of explanations originates in part in the diversity, explicit or more often implicit, of questions that are raised. The second point that I will discuss is the relation between the diversity of explanations in biology and the historical nature of the biological objects.[1] This question was present in Mitchell's book through the notion of biological complexity, but I prefer to focus the discussion on the origin of this complexity, that is, the historical process that generated it, in order to show that interesting lessons can be borrowed from the work of historians. But the diversity of explanations in the biological sciences also emerges from the existence of different types of explanations. They are not Kantian categories, but instead belong to long-lasting scientific traditions. The time when these traditions reveal themselves in the most obvious way is when the question to be answered remains vague and strategies to address it are progressively elaborated. Conceiving them in different ways illuminates more about the plurality of explanatory schemes than the simple observation of competing explanations.

To discuss these three points, I will use examples from different branches of biology, not from Evo-Devo, where this plurality of explanations has been the most extensively scrutinized,[2] but from other fields, from research on ageing and cancer, and even from well-established and apparently non-problematic disciplines as biochemistry. I will also use many historical examples.

In contrast, three alleged reasons for the plurality of explanations in biological sciences will receive only limited attention in this contribution. I consider not only that they have already been amply studied, but also that they are peripheral to the

---

[1]The historical and contingent nature of biological objects and processes is also discussed by Turner (2015, this volume).

[2]See Brigandt (2015, this volume) for a discussion of the explanatory diversity in Evo-Devo.

main issue: the involvement of different disciplinary approaches within biology; the existence of "levels" in biological objects; and the importance of mechanistic explanations in biology. While different disciplines within biology may favour different types of explanations, one discipline, and even one unique type of research within this discipline, such as the study of social insects discussed by Sandra Mitchell, may harbour many different coexisting and/or competing explanations. So appealing to multi-disciplinarity to explain the plurality of explanations is not sufficient. Certain explanations are limited to one level of organization of biological objects, and different explanations correspond to the numerous levels at which the phenomenon can be studied, but different explanations can also compete at the same level. In addition, the distinction between levels is nothing less than problematic. Finally, the importance of mechanistic explanations in biological sciences is obvious, but mechanistic explanations are a heterogeneous category: a reference to machines can be a metaphor or a precise comparison, and very different types of machines can be used.[3]

2   In Many Cases, the Diversity of Explanations Originates from the Diversity of Questions That Are Raised

This point was raised by Ernst Mayr (1961) and discussed at length by Sandra Mitchell (2003). I will illustrate it by the case of the explanation of cancer. Different explanations are currently in competition. In the somatic mutation theory, cancer results from an accumulation of mutations in tumour cells, whereas in the stem cell theory cancer results from the proliferation of a small fraction of the cells present in the tumour called stem cells, and these stem cells are responsible for the tumours' resistance to the treatments tested so far. To explain the origin of cancer by the occurrence of somatic mutations or by the existence of a small subpopulation of stem cells does not exactly answer the same set of questions, although these two sets are overlapping. To be simple, let us say that the somatic mutation theory explains the progression of tumours, whereas the stem cell model mainly explains why treatments have so far been unsuccessful. In fact, the two explanations are fully compatible, and can be combined: these mutations occurred in a population of stem cells.

The existence of two different explanations can be the result of the nature – general or particular – of the answer that is expected. There is a heated debate about the function of what some have called "junk" DNA: the large fraction of the genome that does not give rise to functional RNAs or proteins and has no known regulatory functions. Two explanations coexist. The first is that this DNA has no functions and

---

[3]For critical discussions of mechanistic explanations see (Mekios 2015, this volume; Théry 2015, this volume; Zednik 2015, this volume; Baetu 2015, this volume; Issad and Malaterre 2015, this volume).

persists for generations because its presence is neutral, and is therefore not sieved by natural selection. The second type of explanation is that this DNA has a role in organizing the genome or controlling its expression in one way or another that has not yet been discovered. Since it has been recently shown that a large part of this DNA is copied into RNAs, a new debate has emerged concerning the role of this transcriptional process.

The answer may be different for different fragments of this "junk" DNA. One fragment may have an essential role in the regulation of the expression of a gene localized elsewhere in the genome; and another fragment may have no obvious role, its presence being the result of contingent DNA rearrangements, and its persistence the consequence of its invisibility to the control exerted by natural selection.

If the answers can be different, depending upon the fragment that is examined, an overall quantitative answer nevertheless has its place. It would, at least theoretically, be possible to estimate quantitatively the percentage of "junk" DNA that has a function. For instance, 10 % of the "junk" DNA could be shown to have a function, and 90 % to have no function. In this case, it would be possible to state that, in general, "junk" DNA has no function. It is interesting to notice that, in the case of "junk" DNA, the two explanations are incompatible, whereas they were compatible in the case of cancer.

An additional layer of complexity comes from the fact that this "junk" DNA was introduced in the genome at a specific time in the evolutionary history of organisms. To the question of the present function of such a fragment of "junk" DNA can be added another: What explains the progressive invasion of the genomes of eukaryotic organisms by "junk" DNA? According to Michael Lynch, the increase in the size of the genomes of eukaryotes, concomitant with the increasing percentage of "junk" DNA, was the result of contingent events, and of the neutrality of this increase towards natural selection (Lynch 2007). This event is independent of the later acquisition by some of these sequences of regulatory functions, by a process of exaptation whose importance was emphasized by Stephen Jay Gould (Gould and Vrba 1982). Therefore, as I emphasize in the next section, the diversity of explanations is also the result of the historical trajectories of organisms: the extant presence of a sequence of "junk" DNA can be explained both by a neutral process that allowed its acquisition, and the selection of its present function.

Another example – in which the historical dimension has not yet been introduced – is the origin of the catalytic power of enzymes. Traditionally, the catalytic efficiency of enzymes has been explained by their capacity to stabilize, through the formation of specific weak bonds, the transition state of the reaction, and by their direct participation in the catalytic mechanism, for instance, by providing protons at some specific steps in the catalytic process. More recently, it has been proposed that it was necessary to appeal to quantum physics, to strange phenomena such as the tunnel effect, to explain the catalytic efficiency of enzymes. Other quantum effects would explain the high efficiency of photon capture by photosynthetic systems (Scholes et al. 2011), or the measurement by migrating animals of the weak magnetic field of Earth in order to position themselves (Arndt et al. 2009). Do these recent results require a "revisionist's view" of enzymology? Is enzymology at a turning point? And is it correct to say that the "origin of enzyme catalysis has remained unresolved"? (Nagel and Klinman 2009, 543). In this case, as in the previous one, it seems possible to reconcile the two opposite visions of enzymatic catalysis: the catalytic power of most enzymes can be accounted for by "traditional" chemistry, but in some cases, quantum effects have a dominant role that gives the previous explanatory schemes a limited contribution to the catalytic power. Some of the parties to these controversies present the debate as the necessary choice between two opposed and incompatible explanations; this can only be understood by the fact that the two types of explanations belong to long-lasting, different, and opposing scientific traditions (see below).

The coexistence of different explanations is also the consequence of the often long causal chains responsible for the production of the phenomena under study. In the case of cancer, the explanation of the formation of a particular lung tumour is the accumulation of mutations in the cells of the lung tissues. But the explanation of cancer is also the numerous cigarettes that the patient has smoked. The two explanations are in fact a single one, but the attention is focused on different steps in the long causal chain that led from the cigarettes to the development of the tumour. The explanation by the smoking habit will be probably favoured, not only because smoking precedes the occurrence of mutations in the causal chain, but also because this explanation offers concrete ways to reduce the incidence of lung cancer.

In the previous example, different explanations concern different steps in a linear causal chain. The situation can be more complex. One explanation may target one step in the causal chain, whereas the other describes the context that permitted the deployment of the causal chain. In a search for an explanation of the emergence of life on Earth, some researchers will put forward the role played by self-replicating macromolecules (RNAs or other types of nucleic acids). Others will argue that the explanation of the emergence of life on Earth is to be found in the abundance of liquid water on its surface. The two explanations are of a different nature. The first explanation concerns what is considered as the most important step in the formation of the first organisms; the second is simply a description of the conditions without which life would not have been possible – and that explain what specialists call the "habitability" of Earth. Both explanations are partial and necessary, and they are not in competition. The first may appear more central, but the second is also essential. If the causal chain leading to the formation of the first organisms is a deterministic one, the second type of explanation will receive more attention because, once the conditions for life were present, life would automatically arise. It is the present *credo* that guides the work of astrobiologists. But if the causal chain leading to the formation of organisms is a complex one, a mixture of deterministic and historical, contingent, events, the formation (or not) of life will be explained only through a precise description of the succession of events that led to the first organism, and not by a description of the conditions that obviously were not sufficient to explain its later formation.

Thermodynamic explanations belong to the second type of explanation. The laws of thermodynamics explain the whole of metabolism. The role of ATP, as well as the mechanisms of its production, is explained by thermodynamics. But the characteristics of the metabolic map cannot be deduced from knowledge of thermodynamics: a molecule different from ATP might have been selected by evolution, and ATP can be produced by different pathways (and may well have been produced by an even greater number). Without water or the existence of thermodynamic laws, life or the central role of ATP would not have been possible. This type of explanation corresponds to the *sine qua non* rule that lawyers use to ascribe responsibilities in trials. Often, in human affairs, the *sine qua non* explanations are more important than the direct causal explanations: when an airplane crashes, the errors of the pilot will be considered as less important, at least less likely to be sanctioned, than the indirect unfavourable conditions that made the occurrence of the catastrophe possible.

In this case, as in the previous one, it is obvious that the way the question is formulated will lead to one or another type of explanation. If the question is: "What explains the formation of life on Earth?" the answer will be automatically directed towards a description of the physicochemical conditions at the surface of Earth that favoured the emergence of organisms. If the question is: "How can we explain the formation of the first organism?" the answer will seek to describe the causal chain leading to this organism.

The situation is exactly the same in the field of cancer research, with the supposed opposition between the somatic mutation theory and the tissue organization field theory of cancer (Soto and Sonnenschein 2011). In the latter, disorganization of the tissues is considered as *the* cause of cancer. Its authors refuse to accommodate the observations in favour of their theory within the somatic mutation theory, convinced that the two theories are incompatible. In contrast, I have the feeling that the two explanations are fully compatible but of a different nature. Somatic mutations are the cause of cancer, but the disorganization of the tissue creates the conditions favourable for the occurrence of the mutations, their selection and their expression.

I will discuss a final example of an ambiguous question that can elicit different answers, and different explanations, because of the importance it had in the history of biological thought. The (simple) question is: "What is the explanation for the presence in organisms of a particular mutation?" It is a question that is often raised in Evo-Devo. This question has two different meanings: what is the explanation for the occurrence of such a mutation? And what is the explanation for its continued presence? The first question could (theoretically) receive a deterministic physicochemical answer, such as the specific impact of radiation on a precise part of the genome, the wrong incorporation of a nucleotide at a specific place in the genome, the absence of a repair process during replication, etc. The answer to the second question will be of a very different nature: this mutation gave the organisms in which it was present a specific advantage in the conditions in which they were living at that time, or, possibly, this mutation had no obvious effect, but by contingent sorting during reproduction has invaded the population. In both cases, the second explanation leaves much room for contingency: the occurrence of mutations and the effects they have on organisms belong to different causal chains, the meeting of which is unpredictable. The ambiguous nature of the question generated a lot of confusion in the answers. The wish of many biologists to give a non-deterministic explanation for the presence of mutations in organisms led them, as did Jacques Monod in *Chance and Necessity*, to propose that the contingency of evolution had its origin in the quantum indeterminacy of nucleotide electronic structure and the errors it generates during DNA replication (Monod 1971). While quantum indeterminacy can explain some mutations, it is clearly not the one and only cause of mutation, and obviously does not explain the contingency of the evolutionary process.

3   The Diversity of Explanations as the Natural Consequence of the Historicity of Life

Sandra Mitchell nicely described the competition between evolutionary and developmental explanations in Evo-Devo. Whereas evolutionary explanations were dominant until the 1970s, Stephen Jay Gould, Pere Alberch and many others argued in the following years in favour of the importance of developmental explanations: many traits cannot be explained by the action of natural selection, but are a consequence of the programme of development.

Sandra Mitchell suggests a piecemeal approach: a choice between the two explanations may be possible – at least in some cases – but has to be made independently in each case. Ideally, every situation might be located somewhere on a line running from "fully explainable by evolutionary explanations" to "fully explainable by developmental ones". This impossibility of establishing general rules originates in the complexity of the developmental process as well as in the diversity of situations faced by different organisms. The diversity of explanations is the consequence of the diversification process generated by the complex history of life.

The situation is even more complex since each of the characteristics of organisms that have to be explained is the result of a causal chain formed of different steps, each of which may receive a different explanation. For instance, the origin of a trait can be due to developmental constraints, the side effect of another modification in the developmental programme; but the precise characteristics of the trait observed today can be the result of a selective process that has altered later steps in development, a situation similar to that we described in part one.

This situation is familiar to historians. When they try to explain a historical event, such as the Battle of Waterloo, they know that what happened that day was the result of long heterogeneous causal chains, containing events of a very different nature and apparent importance. The error against which historians had to fight was the introduction of an a priori hierarchy between the different steps and their explanations (Veyne 1984). For instance, for Marxists, economic and social transformations were used to explain events such as the French Revolution. But the errors made by the Governor of the Bastille had a major role: no one knows what would have happened if the Bastille had not been taken!

The absence of a predetermined hierarchy in the explanation of historical events is a lesson that not all biologists have yet learned. The development of an epidemic or of a pandemic is the result of long causal chains, including events that differ greatly in nature and importance. However, these tiny events can have dramatic consequences. Many epidemiologists overlook this when they are questioned about whether or not an epidemic or pandemic will develop, and their answers are awkward and incautious.

Historians have also learned to be very cautious in the way they ask (or do not ask) questions. Most would be reluctant to ask general questions such as: "What is the explanation of the French Revolution?" They would either provide multiple explanations in parallel, or limit their scope to one episode, and give a precise explanation of it. This back-and-forth movement between general and local explanations is also valid, as we have seen, in the biological sciences.

Another reason for the multiplicity of explanations in the biological sciences is the place occupied by etiological explanations. Initially, this type of explanation was put forward in medicine to explain the development of a disease. More generally, they consist in the construction of likely scenarios to account for the transformations that occurred in the past, and are no longer observable. The constraints on the elaboration of these scenarios are weak; this, in conjunction with the highly developed spirit of contradiction of scientists, leads to the proliferation of conflicting hypotheses.

The resulting multiplicity of explanatory hypotheses will not endure indefinitely. New observations, or new possibilities to test some of these scenarios by synthetic experimental evolution (Erwin and Davidson 2009), will limit the number of competing scenarios to one or a small set.

4   The Diversity of Explanations as a Consequence of the Existence of Competing Types of Explanations

The diversity of explanations in the biological sciences also stems from the coexistence of different types of explanations. What appears at a given time as transiently competing explanations may on a longer timescale be seen as different ways of looking at reality and of searching for explanations. One example will illustrate what I mean by "types of explanations". Accounting for the functions of organisms by the existence within them of mechanisms has a long tradition initiated in antiquity by Aristotle and Galen, pursued in the Renaissance by William Harvey, Galileo Galilei, and René Descartes, and continued today in the life sciences through the mechanistic explanations of molecular biology.

But another tradition has coexisted with the previous one, which might retrospectively be called a "chemical" tradition. As early as antiquity, a comparison was made between the progressive formation of organisms and the action of ferments responsible for the production of wine and bread. Present during the Renaissance within the alchemical tradition, it gained a dominant position in the first part of the twentieth century with the so-called "enzyme theory of life" (Olby 1974).

Epigenetics is probably an extant avatar of this tradition. Such an opposition is reminiscent of the themata studied by Gerald Holton (1978). However, I will diverge from the approach of Gerald Holton on two points. The first is that I will emphasize their continuity rather than their alternance. The second is that I will show that they may coexist at the same time, and even in the same person. For instance, the mechanical and chemical explanations of the properties of organisms coexist in Aristotle's writings.

In most cases, it is difficult to precisely designate these types of explanations. They do not fit perfectly with the opposition of categories favoured by philosophers such as the one between holism and reductionism. The reason is that they are a combination of historical traditions and categories of thinking. They certainly correspond to different "styles" and "epistemic cultures", but they transcend these categories. They are the result of this complex history of disciplines, and for this reason their characteristics are partially contingent. The best way to characterize them and to show their importance is through examples, and by looking at an early step in the formation of scientific knowledge, at the time when scientists, faced with a phenomenon they wanted to explain, looked for the best strategies to discover these explanations.

A good example is the search for the nature of the gene in the 1930s. After the "rediscovery" of Mendel's laws at the beginning of the twentieth century, genes played an increasingly important role in the explanations of biological phenomena over the following three decades. They were considered the "atoms of biology" localized on the chromosomes. The nature (and structure) of genes had become a central issue in biology.

The first approach to this question, developed by Max Delbrück, one of the founders of the American Phage Group and later a leader in the young science of molecular biology, was indirect. It consisted in targeting the genes with X-rays and deducing the gross characteristics of the gene from the effects of this irradiation. Results of this experimental approach were published in what was called the "Three-Man Paper" (Sloane and Fogel 2011). The second strategy, developed the same years by the Russian biologist Nikolai Koltzoff (Kol'tsov), was simply to use the chemical knowledge accumulated on the constituents of the cells and, in particular, of the nucleus, to elaborate a reasonable model of the gene (Koltzoff 1928, 1939; Morange 2011a). Both attempts were unsuccessful, but for different reasons. However, the discovery of the double helix structure of DNA by Jim Watson and Francis Crick in 1953, based on a precise chemical knowledge of the constituents of this macromolecule, obviously belonged to the same chemical tradition as the one favoured by Koltzoff.

This dual approach was not limited to the gene. In the same period, two different methodologies were also developed to gain information about the structure of proteins. One, initiated by Emil Fischer at the beginning of the twentieth century, consisted in progressively improving the description of the building blocks of the proteins, the amino acids, and the way they are linked within proteins. The second physicochemical approach was more global and external, and focused on the study of the shape of proteins and of their electric properties.

This second tradition did not become immediately obsolete with the development of X-ray diffraction studies, and the elucidation of the first precise three-dimensional structures of proteins. In the 1960s, a new category of proteins, the allosteric proteins, was identified. The activity of these proteins is regulated by interaction with molecules distinct from their substrates. This interaction triggers a conformational change that alters the activity of these proteins. What was the best experimental approach to explain the remarkable properties of allosteric proteins? One strategy was to describe their three-dimensional structure, i.e., the precise position of the different atoms forming them. The second strategy, favoured by Jacques Monod, the father of the allosteric theory, was indirect. It consisted in altering the equilibrium between the different conformations of these proteins to extract from the results of these perturbation experiments information on the global organization of the proteins.

These two related examples illustrate the persistence of similar approaches despite transformations in the nature of the objects under study, and dramatic changes in the state of knowledge. Interestingly, as seen previously, the same scientist could successively choose the two different approaches. In 1941, Max Delbrück proposed a model of gene duplication in which he identified the gene with a long protein chain (Delbrück 1941).

Despite the fact that these experimental approaches differ greatly, they cannot easily be put into simple categories. The first, that of Delbrück and Monod, was more "physical", and the second, Koltzoff's, more "chemical". The first was global, whereas the second paid more attention to the details of the chemical structure. The proponents of the first were more interested by the search for principles of organization, the existence of symmetries and rules that would have simplified the description of the structure, while the proponents of the second were not intimidated by the terrible chemical complexity of the objects under study. But to contrast the two attitudes as holist versus reductionist is clearly not the appropriate way to investigate their differences.

This does not mean that these general categories cannot be useful in some cases to contrast different types of explanations. Let us consider again the question of the origin of the catalytic power of enzymes. Those who support the hypothesis that quantum effects play a major role clearly belong to a long-lasting reductionist tradition that seeks to discover the secrets of life at the most fundamental level. In the 1960s, some researchers unsuccessfully advocated the conversion of biochemistry to quantum biochemistry (Pullman and Pullman 1963). Recent observations on the "tunnel effect" in catalysis can be viewed as the accomplishment of this reductionist programme, simply postponed by the absence of appropriate technologies.

Symmetrically, the emphasis placed by Emmanuel Farge and other researchers on the role of mechanical deformations in development (Brouzés and Farge 2004; Farge 2011) aims to show that the molecular descriptions are not sufficient, and that a more global view of development is required to understand it fully.

As we have already seen, these contrasts are hardened by the participants. Both sides demand that a choice be made between the different explanations. In most cases, the need for a choice vanishes when knowledge of the system under study increases. Since their first description, mechanical deformations and tensions now have a recognized place in cell biology, and the discontinuity between mechanical and molecular explanations is progressively disappearing: mechanical effects can alter the production or the activity of molecular signals, or change protein conformations, two categories of phenomena familiar to molecular biologists. Another interesting contrast that provides some clues to understanding many debates in biology is between a static and a dynamic vision of biological phenomena. At the beginning of the twentieth century, the gene was considered by many geneticists not as an object, but as a dynamic process. Another illustration of the contrast between these two different views is to be found in the explanation of the phenomenon of memory. The first mechanisms of memory that were favoured in the 1950s were dynamic: memories corresponded to the stabilized circulation of the nerve impulse in neuron circuits. This dynamic view was challenged in the 1960s by the identification of memories and behaviours with certain types of macromolecules, RNAs, or proteins (Morange 2006). After some years of intense debate, these static models of memory disappeared, and a dynamic model of memory re-emerged – although the dynamics of the circulation of the nerve impulse may be dependent on the structural modifications occurring at the synaptic junctions between neurons.

The way I have presented the debate about the existence of memory macromolecules is not the usual one. However, it casts some new light on the controversy, and in particular explains the major role played by George Ungar at the beginning of the 1970s. Many observers conclude that Ungar's results prolonged a debate that, without him, would have ceased earlier, with the abandonment of the idea that macromolecules can be bearers of memories. Is this prolongation explained by the quality of Ungar's results? The harsh criticisms of his experiments obviously show that the answer is no. The reason is different: Ungar proposed a model in which the peptides (small proteins) that he characterized as the bearers of memories were "signposts" present in the membranes of neurons. With this model he apparently reconciled the static and dynamic views, and made the hypothesis of "bearers of memories" acceptable to all.

This contrast between the static and dynamic visions of biological phenomena is still present in current biological debates. One of the reasons for the present success of epigenetics – the study of the modifications of DNA and surrounding proteins that do not alter the sequence of the DNA but have an effect on its expression – is that epigenetic marks are unstable, dynamically added and erased in response to signals from the organism and from the environment, whereas genetic information appears, in part erroneously, as static. To harden this opposition, the supporters of epigenetics include in it dynamical genetic regulatory models such as the operon model, although this model was considered to fully belong to genetics when it was initially proposed!

Similarly, the static vision of protein structure is progressively being replaced by a dynamic one in which proteins oscillate between different conformations (Morange 2012a). The present interest in this dynamic view of protein (and macromolecular, in general) structure is the consequence of the development of new technologies, such as NMR, which permit direct access to molecular dynamics.

But these technologies were developed because some researchers were convinced that the static, rigid view of proteins was unable to account for their properties. Philosophers have recently wondered whether dynamic models are alternatives or complements to mechanistic explanations (Kaplan and Bechtel 2011). Both types of models are related: what distinguishes them is simply the emphasis put on structures, or on their variations.

To conclude this survey of the long-lasting competing types of explanations, I want to discuss the recurrent opposition between models and explanations in which the final state reached by a system is the result of a "positive" process, and those in which the system evolves "by default", simply through the loss of its previous characteristics.

One historical example of this "evolution by loss" was the model proposed by Theodor Boveri at the beginning of the twentieth century to explain the role of chromosomes in development and differentiation. For Boveri, development was linked with the progressive loss of chromosomes during cell division. The specialization of the cells was the direct consequence of this loss of genetic material.

A similar model "by loss" explains ageing (Morange 2011b). In the simplest evolutionary models, ageing results from the progressive loss of functions with age, or increase in dysfunctions, defects that have not been eliminated by natural selection because they occur too late in the life of organisms to alter reproduction.

Interestingly, various positive models of ageing have been proposed regularly since the Renaissance, such as ageing being the result of poisoning of the body by toxins that accumulate during life. At the beginning of the twentieth century, Elie Mechnikov thought that he had discovered the source of those toxins in the microbes present in the gut, and he designed a special diet to eliminate them. Today, many researchers believe that the formation of protein aggregates during ageing, which is well demonstrated in Alzheimer's and Parkinson's diseases in the brain but that in fact occurs in all the tissues of ageing organisms, has toxic effects. Negative and positive explanations of ageing are not incompatible: the toxic effects of these protein aggregates might explain the dysfunctions observed in ageing. As previously seen, what is significant is negative (the effects) or positive (the existence of a toxic substance) emphasis of the explanations.

In the case of evolution, the situation is the same, but the dominant place of the positive explanations – evolution by the emergence of novelty – is obvious. This has not prevented recurrent support among biologists for an alternative model, evolution by loss. In the 1940s, the French biologist André Lwoff proposed a model of evolution by loss of biochemical functions (Lwoff 1944). More recently, human evolution was explained by the loss, in our ancestors, of genes involved in specialized functions (Olson 1999) or of their regulatory sequences (McLean et al. 2011). Gene loss is seen as the way for the organism to free itself from the constraints accumulated during evolution.

Explanations by loss have also recurrently been used to explain the origin of tumours. Cancer has been considered as a loss of the differentiated characteristics of the cells, their return to a "primitive" stage and/or a loss of regulation – cancer cells escaping the body's control (Morange 2012b). The vision of cancer as a deregulation was dominant in the 1970s, before the adoption of the somatic mutation theory (Morange 1997).

An explanation by loss does not require further explanations. It can be limited to a description of the obstacles that had to be overcome to reach the basal level. This is obvious in the case of cancer. When John Cairns looked for explanations of cancer in his book *Cancer: Science and Society* published in 1978 (Cairns 1978), he found them in the failure of the mechanisms that prevent cells from returning to an uncontrolled state of proliferation.

The fact that explanations by loss are self-sufficient has been contested by Daniel McShea and Robert Brandon who propose that the loss of homogeneity between different cells – a process characteristic of multicellular organisms called cell differentiation – is in fact the result of what they called "a zero force" or "biology's first law" (McShea and Brandon 2010). A similar debate concerns innovation in evolution. Must innovation be specifically explained, as Marc Kirschner and John Gerhart among others propose (Kirschner and Gerhart 2005), or is it the unavoidable consequence of any process of evolution by variation (and selection)?

I have, in previous publications, tried to list the different types of explanations (Morange 2009, 2012c). Today, I am more sceptical about the success of such a project. Some types of explanation are relatively well defined, such as the mechanistic and Darwinian types, but others, such as what I have called "chemical" explanations, are much more difficult to delineate. The reason is simple: types of explanations are also the result of a complex historical process. Some have acquired a well-defined shape, whereas others have successively existed under different, more or less well defined, avatars.

5 Conclusion

The plurality of explanations in the biological sciences has different causes. One is ontological, linked with the historical dimension of biological objects. Two others are epistemic. The first is mundane, the consequence of the formulation of ambiguous questions. In particular, it is not always obvious whether a question concerns a particular phenomenon, or an ensemble of similar phenomena. The second epistemic origin of the plurality of explanations is to be found in the existence of long-lasting traditions of explanations that persist and recurrently conflict with one another, despite the transformations of scientific knowledge. These two epistemic causes of plurality are not totally independent. The description of the conditions that allowed the occurrence of the phenomena under study, what I have called the *sine qua non* explanations, answers a particular type of question. But it can also be considered as a specific type of explanation, competing in some cases with other explanations such as mechanistic explanations. This complex conjunction gives to the landscape of explanations in biology its richness and diversity. My description is clearly at odds with the too simple vision that many scientists have that knowledge is acquired through the simple fitting of scientific facts to alternative theories and models (Soto and Sonnenschein 2012). Data have a major role in the evolution of scientific knowledge, not simply as a means of choosing between competing explanations, but through the far more indirect process of reshaping the complex landscape of biological explanations.

Acknowledgements I am indebted to Dr. David Marsh for critical reading of the manuscript, to Pierre-Alain Braillard and Christophe Malaterre for inviting me to participate in this collective enterprise, and for the numerous remarks they did on the first version of the manuscript, and to the two anonymous reviewers.

References

Arndt, M., Juffmann, T., & Vedral, V. (2009). Quantum physics meets biology. *HFSP Journal, 3*, 386–400.

Baetu, T. (2015). From mechanisms to mathematical models and back to mechanisms: Quantitative mechanistic explanations. In P.-A. Braillard & C. Malaterre (Eds.), *Explanation in biology. An enquiry into the diversity of explanatory patterns in the life sciences* (pp. 345–363). Dordrecht: Springer.

Brigandt, I. (2015). Evolutionary developmental biology and the limits of philosophical accounts of mechanistic explanation. In P.-A. Braillard & C. Malaterre (Eds.), *Explanation in biology. An enquiry into the diversity of explanatory patterns in the life sciences* (pp. 135–173). Dordrecht: Springer.

Brouzés, E., & Farge, E. (2004). Interplay of mechanical deformation and patterned gene expression in developing embryos. *Current Opinion in Genetics and Development, 14*, 367–374.

Cairns, J. (1978). *Cancer: Science and society*. San Francisco: WH Freeman & Co.

Delbrück, M. (1941). A theory of autocatalytic synthesis of polypeptides and its application to the problem of chromosome reproduction. *Cold Spring Harbor Symposia on Quantitative Biology, 9*, 122–124.

Dupré, J. (1983). The disunity of science. *Mind, 92*, 321–346.

Dupré, J. (1993). *The disorder of things: Metaphysical foundations of the disunity of science*. Cambridge, MA: Harvard University Press.

Erwin, D. H., & Davidson, E. H. (2009). The evolution of hierarchical gene regulatory networks. *Nature Reviews Genetics, 10*, 141–148.

Farge, E. (2011). Mechanotransduction in development. *Current Topics in Developmental Biology, 95*, 243–265.

Gould, S. J., & Vrba, S. (1982). Exaptation – A missing term in the science of form. *Paleobiology, 8*, 4–15.

Holton, G. (1978). *The scientific imagination: Case studies*. Cambridge: Cambridge University Press.

Issad, T., & Malaterre, C. (2015). Are dynamic mechanistic explanations still mechanistic? In P.-A. Braillard & C. Malaterre (Eds.), *Explanation in biology. An enquiry into the diversity of explanatory patterns in the life sciences* (pp. 265–292). Dordrecht: Springer.

Kaplan, D. M., & Bechtel, W. (2011). Dynamical models: An alternative or complement to mechanistic explanations? *Topics in Cognitive Science, 3*, 438–444.

Kirschner, M. W., & Gerhart, J. C. (2005). *The plausibility of life: Resolving Darwin's dilemma*. New Haven: Yale University Press.

Koltzoff, N. K. (1928). Physikalisch-chemische grundlage der morphologie. *Biologisches Zentralblatt, 48*, 345–369.

Koltzoff, N. K. (1939). *Les molécules héréditaires*. Paris: Hermann.

Lwoff, A. (1944). *L'évolution physiologique: Etude des pertes de fonction chez les micro-organismes*. Paris: Hermann.

Lynch, M. (2007). *The origins of genome architecture*. Sunderland: Sinauer.

Mayr, E. (1961). Cause and effect in biology. *Science, 134*, 1501–1506.

McLean, C. Y., Reno, P. L., Pollen, A. A., Bassan, A. I., Capellini, T. D., et al. (2011). Human-specific loss of regulatory DNA and the evolution of human-specific traits. *Nature, 471*, 216–219.

McShea, D. W., & Brandon, R. (2010). *Biology's first law: The tendency for diversity and complexity to increase in evolutionary systems*. Chicago: The University of Chicago Press.

Mekios, C. (2015). Explanation in systems biology: Is it all about mechanisms? In P.-A. Braillard & C. Malaterre (Eds.), *Explanation in biology. An enquiry into the diversity of explanatory patterns in the life sciences* (pp. 47–72). Dordrecht: Springer.

Mitchell, S. D. (2003). *Biological complexity and integrative pluralism*. Cambridge: Cambridge University Press.

Monod, J. (1971). *Chance and necessity: An essay on the natural philosophy of modern biology*. New York: Knopf.

Morange, M. (1997). From the regulatory vision of cancer to the oncogene paradigm, 1975–1985. *Journal of the History of Biology, 30*, 1–29.

Morange, M. (2006). The transfer of behaviours by macromolecules. *Journal of Biosciences, 31*, 323–327.

Morange, M. (2009). Articulating different modes of explanation: The present boundary in biological research. In A. Barberousse, M. Morange, & T. Pradeu (Eds.), *Mapping the future of biology: Evolving concepts and theories*. Heidelberg: Springer.

Morange, M. (2011a). The attempt of Nikolai Koltzoff (Kol'tsov) to link genetics, embryology and physical chemistry. *Journal of Biosciences, 36*, 211–214.

Morange, M. (2011b). From Mechnikov to proteotoxicity: Ageing as the result of an intoxication. *Journal of Biosciences, 36*, 769–772.

Morange, M. (2012a). A new life for allostery. *Journal of Biosciences, 37*, 13–17.

Morange, M. (2012b). What is really new in the current evolutionary theory of cancer? *Journal of Biosciences, 37*, 609–612.

Morange, M. (2012c). *Les secrets du vivant: Contre la pensée unique en biologie*. Paris: La Découverte.

Nagel, Z. D., & Klinman, J. P. (2009). A 21st century revisionist's view at a turning point in enzymology. *Nature Chemical Biology, 5*, 543–550.

Olby, R. (1974). *The path to the double helix*. London: Macmillan.

Olson, M. V. (1999). When less is more: Gene loss as an engine of evolutionary change. *American Journal of Human Genetics, 64*, 18–23.

Pullman, B., & Pullman, A. (1963). *Quantum biochemistry*. New York: Interscience Publ., Wiley.

Scholes, G. D., Fleming, G. R., Olaya-Castro, A., & van Grondelle, R. (2011). Lessons from nature about solar light harvesting. *Nature Chemistry, 23*, 763–774.

Sloane, P. R., & Fogel, B. (2011). *Creating a physical biology: The three-man paper and early molecular biology*. Chicago: The University of Chicago Press.

Soto, A., & Sonnenschein, C. (2011). The tissue organization field theory of cancer: A testable replacement for the somatic mutation theory. *Bioessays, 33*, 332–340.

Soto, A. M., & Sonnenschein, C. (2012). Is systems biology a promising approach to resolve controversies in cancer research? *Cancer Cell International, 12*(1), 12.

Théry, F. (2015). Explaining in contemporary molecular biology: Beyond mechanisms. In P.-A. Braillard & C. Malaterre (Eds.), *Explanation in biology. An enquiry into the diversity of explanatory patterns in the life sciences* (pp. 113–133). Dordrecht: Springer.

Tinbergen, N. (1963). On the aims and methods of ethology. *Zeit Tierpsychologie, 20*, 410–433.

Turner, D. (2015). Historical contingency and the explanation of evolutionary trends. In P.-A. Braillard & C. Malaterre (Eds.), *Explanation in biology. An enquiry into the diversity of explanatory patterns in the life sciences* (pp. 73–90). Dordrecht: Springer.

Veyne, P. (1984). *Writing history: Essay on epistemology*. Middletown: Wesleyan University Press.

Zednik, C. (2015). Heuristics, descriptions, and the scope of mechanistic explanation. In P.-A. Braillard & C. Malaterre (Eds.), *Explanation in biology. An enquiry into the diversity of explanatory patterns in the life sciences* (pp. 295–317). Dordrecht: Springer.

Chapter 3
Explanation in Systems Biology: Is It All About Mechanisms?

Constantinos Mekios

Abstract As the mechanistic understanding of how biological systems function continues to become more refined, the power of scientists to manipulate and control them for the sake of achieving concrete practical goals increases. Proponents of the recent movement for a new mechanistic philosophy of biology have taken up the task of defining the theoretical conditions that would promote such understanding. In this paper, the case of systems biology serves to highlight some of the limitations of the explanatory models that are grounded on their philosophical accounts of mechanism. In light of the prominent role played by mechanisms in top-down and bottom-up systems biology, I argue that a purely mechanism-based framework that complies with the constraints stipulated by the new mechanists is adequate for producing useful explanations of biological phenomena in the context of the systems approach. Nevertheless, this framework remains limited in its capacity to carry out the comprehensive explanatory integration demanded for the holistic understanding of complex biological systems whose attainment constitutes systems biology's most ambitious objective. In addition, in their current formulation the models endorsed by the new mechanists are not sufficient for capturing the rich explanatory pluralism that characterizes the practice of systems biology. The fact that a pragmatic view of the practice of systems biology reveals its pluralistic character suggests that moving beyond theoretical considerations provides useful insights into the nature of explanation. I propose, however, that philosophers do not need to make a choice between adopting either a theoretical or a pragmatic attitude: both perspectives contribute in a mutually complementary way to constructive thinking about explanation in biology.

Keywords Systems biology • New mechanistic philosophy • Mathematical models • Limitations of mechanism • Explanatory pluralism • Intelligibility • Integration

C. Mekios (✉)
Department of Philosophy, Stonehill College, Easton, MA, USA
e-mail: cmekios@stonehill.edu

© Springer Science+Business Media Dordrecht 2015
P.-A. Braillard, C. Malaterre, *Explanation in Biology*, History, Philosophy and Theory of the Life Sciences 11, DOI 10.1007/978-94-017-9822-8_3

1 Introduction

A general presupposition made in the philosophical discourse concerning explanation is that to explain a phenomenon means to render it intelligible. Regardless
of whether the explanation is correct or not, it establishes a relation between the
explanans and the explanandum which gives rise to some understanding of the
phenomenon under consideration (Machamer, Darden, and Craver 2000; MDC
hereafter). For scientists, to understand the phenomena that they study constitutes
the ultimate end of their work and the main motivation behind it. Hence, in the
introductory paragraph of his recent book on scientific explanation Michael Strevens
(2008) writes:

> If science provides anything of intrinsic value, it is explanation. Prediction and control are
> useful, and success in any endeavor is gratifying, but when science is pursued as an end
> rather than as a means, it is for the sake of understanding – the moment when a small,
> temporary being reaches out to touch the universe and makes contact. (Strevens 2008, 3)

The verb "to understand" dominates biological talk, since biologists use it
regularly to define their experimental objectives. Statements to the effect that a
proposed set of experiments could help us "understand the regulation of glucose
metabolism in yeast," or that the experimental data may afford us "a novel
understanding of wing development in Drosophila," are as commonplace in the
literature as they are in informal exchanges among biologists. In the context in which
such statements are used, the claim that something is understood usually means that
an answer to the question about how that thing might possibly work can now be
provided. Contemporary biologists generally treat the postulation of a mechanism
as the most suitable method for fulfilling this explanatory requirement. Mechanism-
based explanations have, therefore, acquired a central role in the biological sciences:
they have come to be regarded as the primary sources of intelligibility.

In this essay I set out to show how the case of systems biology (SB) draws
attention to the explanatory limitations of the types of mechanism-based models
offered by proponents of the growing movement for a new mechanistic philosophy
of biology.[1] The widespread appeal to mechanisms in SB has produced practically
useful and experimentally confirmable insights into the behavior of complex biological systems. A strictly mechanism-based explanatory strategy, however, defined
by theoretical constraints like those found in the various articulations of the new
mechanistic philosophy, relies on models that can only capture a fraction of the rich
pluralism that characterizes the practice of SB. The paper begins (Sect. 2) with a
brief overview of the origins of mechanistic explanation in biology, as well as of the
main philosophical features of its new mechanistic agenda. In Sect. 3, the focus turns
on SB: an analysis of top-down and bottom-up versions of the approach shows that

---

[1]For related discussions of possible limitations of mechanistic models in systems biology, see
Baetu (2015), Breidenmoser and Wolkenhauer (2015), Issad and Malaterre (2015), and Théry
(2015).

in both of these cases mechanisms play an important role for providing explanations at different levels of a system's organization. The discussion that follows (Sect. 4) consists in a critical assessment of the implications of these examples from the practice of SB for the adequacy and completeness of purely mechanism-based explanations. Finally, in Sect. 5, after exploring the limited capacity of such theoretical schemes for making sense of biological complexity in the context of SB's pluralistic implementation, I examine the contributions of some recent pragmatic philosophical approaches to thinking about what SB's methodological pluralism means for the fulfillment of its explanatory objectives. By means of this analysis, I attempt to draw attention to the value of taking into account both theoretical and pragmatic constraints for addressing questions about explanation in biology.

2 Mechanism-Based Thinking in Biology: Conceptual Origins, the New Mechanistic Philosophy, and a First Taste of Explanatory Pluralism

The influence of mechanism-based thinking on the development of science has a long and thoroughly studied history. Scholars generally agree that it goes as far back as the seventeenth century, when the mechanical philosophy of eminent thinkers such as Rene Descartes, Galileo Galilei, Pierre Gassendi, Robert Boyle, and Isaac Newton, among others, propelled the scientific revolution which gave rise to a mechanistic natural science. The late seventeenth century also marks the beginning of mechanistic biology, which has its origins in the same philosophical tradition. But the relationship between mechanism and biology is not simple, nor is its history as straightforward as one might have anticipated on the basis of what seems to be the case with other natural sciences. This is primarily because in biology the term "mechanism" has not been used consistently. Instead, it has taken different meanings depending on the context of its use (Allen 2005; Nicholson 2012). Accordingly, mechanistic biology has had various incarnations in the 350 years since its emergence (Nicholson 2012, 153). The different conceptions of mechanism that remain operative in today's biology and in the philosophical discourse pertaining to it correspond to distinct approaches to explanation, all of which can legitimately be claimed to be mechanism-based. This constitutes a first manifestation of explanatory pluralism within the generally construed realm of mechanistic biology. The potential for a more fundamental kind of explanatory pluralism, however, emerges from the features of recently articulated philosophical theories of biological mechanism that will be examined later in this section.

For Nicholson (2012), a tripartite distinction between "machine mechanism," "mechanicism," and "causal mechanism," accurately captures the variability in the historical use of the term "mechanism" in biology. To directly relate the notion of mechanism to that of an actual machine is to most closely adhere to the literal meaning of the term. A relation of precisely this kind, Nicholson suggests, is at the heart of what he calls the machine mechanism understanding of mechanism-based biology. Implicit in the conception that living beings admit of descriptions and obey principles analogous to those that apply to clocks, steam engines, or computers, is a strong ontological claim: organisms, that is, are not only completely explainable in the same manner that machines are – i.e., by means of an account of their parts and their interactions – as the proponents of mechanicism would have it, but they constitute themselves mechanical contrivances. Placing the emphasis on epistemological considerations rather than on ontological determinations is what distinguishes mechanicism from machine mechanism. Hence, mechanicism stands as a philosophical program preoccupied with the task of providing explanations by appealing to the ontological picture painted by machine mechanism. According to Nicholson's analysis, throughout its history mechanistic biology has identified with this program and its influence remains strong, as indicated by the methodological features of biological fields such as molecular and cell biology, which have flourished during the twentieth century, or of the currently emerging field of synthetic biology. Lastly, causal mechanisms represent the most recently articulated alternative sense of "mechanism" in biology. They capitalize on the explanatory power of elucidating the causal relations among the activities and structures that give rise to biological phenomena. Because of their attention to the explanatory role of causal sequences, Allen (2005, 263) chooses to label these mechanisms "operative or explanatory": they consist in a "step-by-step description/explanation of how the components in a system interact to yield a process or outcome."

The prospect of providing adequate accounts of biological phenomena through the investigation of causal mechanisms has recently sparked new interest in the subject among philosophers. Despite the fact that it includes a variety of perspectives, the overall discourse has come to be known as the "new mechanistic philosophy" after Skipper and Millstein (2005) baptized it using this term. But contrary to the earlier description of the mechanistic approach as involving the study of machine mechanisms, this new movement is concerned primarily with the examination of causal mechanisms. For this reason, Nicholson takes issue with the use of the designation "mechanistic" in this context and suggests instead that the new program for the study of causal mechanisms in biology be called "mechanismic" (Nicholson 2012, 154–155). In order to avoid the conflation of meaning that could result from failing to notice this distinction, I adopt Nicholson's recommendation and refrain from using the term "mechanistic" to refer to explanatory approaches that rely on causal mechanisms for the rest of this essay, with the exception of instances in which I am either quoting philosophers who take a more flexible approach to its use, or refer directly to the implications of the popular designation "new mechanistic philosophy."

The new philosophical discourse concerning the role of mechanisms in biology has developed around two distinct conceptions of causal mechanism: one articulated by Glennan (1996, 2002a, b, 2005) and the other by MDC (2000). Glennan takes mechanisms to be actual entities: they are "complex-systems" and should therefore be treated as "things (or objects)" (Glennan 2002a, S345). In addition, mechanisms as complex-systems consist of parts whose interactions produce the behavior(s) of these organized entities (Glennan 2002b, 126). These observations are incorporated in the following revised version of the definition of mechanism that Glennan had first offered in 1996:

> A mechanism for a behavior is a complex system that produces that behavior by the interaction of a number of parts, where the interactions between parts can be characterized by direct, invariant, change-relating generalizations. (Glennan 2002a, S344)

The behavior of a mechanism, "Glennan (2005, 445) clarifies, is simply "what the mechanism does." Thus, he continues, using language that brings to mind functionalist accounts,[2] "[a] heart is a mechanism for pumping blood, a Coke machine is a mechanism for dispensing Cokes in return for coins, and so on (Glennan 2005, 445).

Among the aspects that stand out in this brief overview of Glennan's conception of mechanism, the most relevant for the purposes of our discussion is that his account does not break with the ontological assumptions of machine mechanism and mechanicism. While the examination of the causal interactions among the parts of mechanisms is indispensable for characterizing the brand of mechanism-based behaviors described by Glennan, his analysis depends fundamentally on an onto-logical commitment to the claim that mechanisms are things: the complex-systems whose behaviors are to be understood by means of mechanisms are themselves these mechanisms. The emphasis, then, remains on highlighting the ontological status of mechanisms rather than their explanatory capacities. Also deserving attention in Glennan's definition of mechanism is the central role reserved for interactions of parts. Since systems are "collections of parts" (Glennan 2002b, 128) – i.e., they amount ontologically to the sum of their constitutive elements – their behavior, as well, arises from the interactions between their parts, whose relations are governed by invariant generalizations. This description implies, moreover, that a bottom-up view is required for explaining how system behavior is produced. Taken together these features of Glennan's vision of causal mechanism indicate that the spirit of reductionism remains deeply engrained within it, both with respect to its ontological claims and in so far as explanation is concerned.

The second and most frequently cited conception of causal mechanism in biology was provided by MDC (2000), who criticized Glennan for his interactionist view of mechanism. As their definition shows, for MDC the key to thinking about mechanisms is to concentrate on activities rather than on interactions of parts:

> Mechanisms are entities and activities organized such that they are productive of regular changes from start or set-up to finish or termination conditions. (MDC 2000, 3)

The mechanism of chemical neurotransmission, for example, consists in the description of the way by which the termination condition – the depolarization of the

---

[2]Glennan admits that his analysis resembles in many respects functional analysis but notes that, while it is tempting to substitute the term "behavior" by the term "function," he avoids it because "(1) the function of an entity is often construed as its role within a larger system, and (2) it is also sometimes assumed that entities can have functions only if they have this function as a result of design" (Glennan 2005, 127).

postsynaptic neuron – is produced by the set-up conditions and intermediate stages – the release of neurotransmitters into the synapse by the presynaptic neuron and their binding to receptors of the postsynaptic cell, respectively (MDC 2000, 3). MDC recognize that a description of this mechanism could not have been provided without reference to entities – e.g., neurons, neurotransmitter molecules, and receptors – and to their properties. Ultimately, however, activities – in this case the release of the neurotransmitters and their binding to receptors – are the effectors of the changes leading up to the transmission of the signal. Their function within the mechanism renders them essential to its description. In addition, MDC stipulate that "to give a description of a mechanism for a phenomenon is to explain that phenomenon" (2000, 3). Both entities and activities, then, play an important explanatory role in MDC's account of causal mechanism, which has for this reason been characterized as dualistic.

In contrast to Glennan's interactionist paradigm, the MDC model establishes the conditions for a mechanism-based biology with a non-reductive orientation.[3] By capitalizing on the study of activities in multilevel mechanisms it provides an explanatory alternative to classical reductive strategies that rely exclusively on the study of the interactions between a mechanism's elementary components. Furthermore, Craver (2005) and Darden (2005) have both argued that methods involving mechanisms at different levels of biological organization can allow for interfield integration without formal reduction of theories. At the same time, the MDC model represents a turn away from the traditional deductive-nomological explanatory paradigm of mechanistic science. In particular, MDC relax the relatively stringent condition for "laws," or other "invariant generalizations," that persists throughout Glennan's articulations of his definition of mechanism. Mechanisms are regular, they write, "in that they work always or for the most part in the same way under the same conditions" (MDC 2000, 3). Lacking references to laws or generalizations,

---

[3]Mechanistic biology is often placed firmly within the tradition of reductionism. Indeed, the conceptual roots of reductionism overlap with those of mechanistic biology: they can be traced back to the corpuscularian doctrine of the pioneers of mechanical philosophy, who sought to account for physical phenomena by referring exclusively to the "interactions between impenetrable particles of matter" (Sarkar 1998, 16). In addition, reductionism and mechanism share a similarly tortuous conceptual history, primarily because of the various meanings that both concepts have acquired through their participation in long-standing philosophical debates about the nature of biology. In these debates, "mechanism" has often played the role of the polar opposite of "vitalism," while, similarly, "reductionism" has been construed as the antagonist of "holism" and "organicism." But these juxtapositions alone cannot fully illuminate the nature of the relationship between the concepts involved: they capture the roles that these concepts have assumed within a specific historical and philosophical context, but at the same time they conceal the context-dependence of these roles. Thus, with respect to the causal mechanisms under consideration, to the extent that Glennan's model retains many of the assumptions of mechanicism it remains more closely linked to a traditional conception of reductionism, which regards biological systems as explainable through a description of the interactions among their components, than MDC's account, which distinguishes itself more clearly from the mechanicist conception of mechanism by emphasizing the role of activities. In that sense, the MDC model may be taken to provide the conditions for non-reductive mechanism-based explanations, if the conception of reductionism that is used to make this pronouncement is properly qualified.

this is a weaker appeal to regularity in comparison to Glennan's. Following along the trajectory set by MDC, other proponents of the new mechanistic philosophy have taken further steps towards a fully irregularist conception of mechanism: Bogen (2005) proceeded to explicitly reject the need for regularities in mechanists' accounts, Bechtel and Abrahamsen (2006) proposed a definition of mechanism free from appeals to regularities of any type, and, interestingly, Machamer (2004) himself has argued in favor of dropping the condition that genuine mechanisms must be regular (DesAutels 2011).

Taken together, the elements of the conceptual history of mechanism and the recent theoretical developments concerning causal mechanisms that were considered in this section offer us a first illustration of pluralism in biological explanation and in the philosophical discourse surrounding it. The evidence showing that in contemporary biology mechanism-based explanatory strategies are as dominant and ubiquitous as is talk about mechanisms would appear to undermine this assessment; it would seem to indicate that a univocal sense of mechanism prevails. Nevertheless, the preceding examination of the different senses of mechanism that continue to be operative in biological practice and in the accounts of the new mechanists, suggests instead that mechanistic explanation itself is pluralistic. At a first level, the reference to pluralism concerns the simple observation that several distinct conceptions of mechanism are integral to different mechanism-based explanations and to theories about mechanism-based explanation in biology. In addition, however, the possibility of a second level of explanatory pluralism presents itself in the details of the MDC account and of the accounts that have followed in its footsteps. Notably, this possibility appears to be borne out by MDC's break with regularist, interactionist, or strictly mechanicist conceptions of mechanism, as well as by their generally non-reductive orientation. Specifically, the flexibility of MDC's mechanism, owed to its somewhat abstract emphasis on activities, the relaxation of the requirement for regularity, the often incomplete "mechanism schemata" or "mechanism sketches" (MDC 2000, 15–16) that it employs, and its capacity to accommodate multiple-levels of explanation, renders it compatible with a pluralistic explanatory framework for biology. While frequently the characterization "pluralistic" is reserved for a scenario in which different types of explanation are taken to provide equally legitimate but separate insights into a given phenomenon, in the context of MDC pluralism consists in the accommodation of multiple explanatory accounts of the same type within a larger mechanism-based scheme. In this case, all the explanations are mechanisms. Each one of them, however, accounts for a different level of a given biological system's organization. One might refer to this state of affairs as deep-level explanatory pluralism, not as a way of suggesting that it is more authentic than other instances of pluralism but simply to convey the sense in which mechanism-based explanation itself encompasses a plurality of distinct and irreducible partial mechanisms. We proceed with a study of the case of SB, in order to evaluate the suitability of the multilevel mechanisms of the new mechanists for adequate explanation of complex biological systems and to determine the implications of their use for explanatory pluralism.

3   Mechanism-Based Explanation in Top-Down and Bottom-Up SB

The recently coined term "systems biology" has been used somewhat loosely to characterize a wide range of biological research projects, as well as an emerging multidisciplinary approach for the study of biological systems. In light of this observation, Fischbach and Krogan (2010, 1) have pointed out that "systems biology means different things to different people." Nevertheless, a number of shared methodological features, along with an underlying theoretical commitment to the idea that a global view of biological systems could provide new insights into their complex organization, stand out upon closer examination of SB's diverse practice. More specifically, in pursuit of its primary objective to produce an understanding of "biological systems as systems" (Kitano 2001, 2), SB in its various incarnations seeks the formalization, quantification, and integration of the properties of the systems that it studies. In a review regarding the practice and the challenges faced by SB, one of its pioneers, Alan Aderem, strongly echoes the assessment that integration is central to the systems approach and provides a description of the kind of integration that it must bring to fruition:

> [I]f one had to summarize the essence of the ISB [Institute for Systems Biology], it would be integration: integration of the various technologies, integration of the various hierarchical levels of biological information, integration of technology and biology, integration of crossdisciplinary scientists, integration of industry and academia, and, finally, integration of discovery and hypothesis driven science. (Aderem 2005, 513)

In *Systems Biology in Practice* (2005), Edda Klipp and her coauthors concur and amplify:

> Systems biology relies on the integration of experimentation, data processing, and modeling. Ideally, this is an iterative process. (Klipp et al. 2005, 4)

Despite the general consensus on the indispensability of integration, iteration, and model construction for SB among its proponents, however, other conceptual differences have produced divergent views regarding the optimal implementation of the approach. Those who perceive and practice SB as a "top-down" approach begin the examination of the system studied by adopting a bird's eye view (Bruggeman and Westerhoff 2006). They first consider the whole system – the top – and use the high-throughput, genome-wide experimentation that characterizes the "-omics" analyses (genomic, transcriptomic, proteomic, etc.) that initiated the systems approach in biology to elucidate and characterize lower level biological mechanisms. A different strategy is adopted by those who maintain that the role of SB is to characterize biological systems using the principles of physical chemistry and starting from the molecular components at the bottom of these systems (Hyman 2011). Instead of using induction to gain insights into the higher levels of biological organization, "bottom-up" SB aims to deduce the functional properties of organisms from a detailed mechanistic analysis of the behavior of their smallest manageable parts. The subsequent integration of formal expressions of component properties by means of mathematical and computational methods, leads to the construction of a mechanism-based model that predicts behavior at higher levels of the system's organizational hierarchy.

The methodological shift towards the global perspective that characterizes SB coincides with the emergence of the new mechanistic philosophy of biology. In addition, the two movements are linked by their common preoccupation with integration. Granted, the kind of integration the new mechanists had originally set out to account for differs from the nature of integration currently pursued by SB, both qualitatively and in terms of scope. The new mechanists were motivated by philosophical interest in defining the theoretical features of a mechanism for explanatory integration within the relatively narrow framework of molecular and cell biology, whereas systems biologists have been seeking larger scale integration as a means to the fulfillment of their principal research objective: the attainment of a comprehensive view of complex biological systems. To the extent, however, that practice can help put theory in perspective and vice versa, the development of a closer relationship between these two distinct integrative projects might prove mutually beneficial.[4] Thus, SB has the potential to serve as an informative test case for philosophers inquiring about the limits of the new mechanists' explanatory models with the intention of producing more accurate theories of mechanism. Correspondingly, SB itself stands to benefit from theoretical developments in mechanism-based explanation, to the degree that normative statements about the best explanatory models could facilitate the pursuit of its practical objectives. In what follows, the nature of the relationship between mechanism and SB is examined in more detail with the help of two case studies. The first illustrates the role of mechanism in a "top-down" approach to SB, while the second highlights the strong influence of mechanistic thinking to the practice of "bottom-up" SB. The anticipated philosophical gain from this exploration is that it will put us in a better position to address questions about the sufficiency of the new mechanists' models for supporting the broad explanatory integration sought by the systems approach in biology.

3.1 Top-Down SB Relies on Mechanisms and on Abstract Phenomenological Models

A relatively early example of the implementation of a top-down strategy for SB is the Baliga et al. (2004) study of the response to UV radiation in the Halobacterium NRC-1. Preliminarily, the authors gathered any previously available information

---

[4]Should this prove to be the case, this example would illustrate paradigmatically what some regard as the ideal relationship between philosophy and applied science: the agenda of the former is tested and defined by the pragmatic demands of the latter and, conversely, the development of the latter is shaped and facilitated by theoretical contributions of the former.

pertaining to the stress response to UV radiation both in eukaryotic organisms and in bacteria. Although the precise mechanism by which the *Halobacterium* NRC-1 strives to counteract the effects of UV radiation was practically unknown at the beginning of the project, the few facts about the process that were at the researchers' disposal allowed them to identify target genes and other macromolecules that could be perturbed in the course of a systems analysis.

The study's next step involved registering and integrating the effects of UV irradiation at various levels of organization, both in selected mutant strains of *Halobacterium* NRC-1 and in wild type cells. The researchers relied primarily on data from the mRNA level, obtained through micro-array experiments. Algorithmic computational methods were used to integrate system-wide experimental results at the mRNA level with data about inferred interactions at the protein level and to visualize them together on a network of interconnected nodes and edges (Fig. 3.1). Specifically, Baliga et al. used Cytoscape, a software tool that permits the construction of "integrated models of biomolecular interaction networks" (Shannon et al. 2003, 2498), for the visualization of the protein data. The program displayed the interactions between proteins as edges between nodes that represent the genes encoding for these proteins. The mRNA data were then incorporated into this model. Changes in mRNA levels were visualized by using different node colors (displayed in Fig. 3.1 as different shades of grey), whose intensity depended on the magnitude and the direction (up-regulation or down-regulation) of each change. Moreover, protein function annotation allowed the grouping of the nodes of the network using functional proximity as a criterion. The model generated, combined with other predictions about the function of various halobacterial proteins and previous observations about interactions between them, led to the identification of "several putative repair mechanisms in *Halobacterium NRC-1*" (Baliga et al. 2004, 1026). Moreover, the integration of the obtained data permitted not only the identification of 33 proteins participating in the stress response to UV radiation, but also the assignment of specific putative functions to 26 among them, which were previously unannotated.

The protocol described in this case is typical of the most widely applied version of SB. It faithfully follows basic prescriptions for the implementation of the systems approach in biology that were offered by some of its pioneers (e.g., Kitano 2001; Hood 2002), capitalizing on integration, iteration, data processing, and construction of models. Especially relevant to our discussion, however, is the observation that mechanisms retain an important role in the application of top-down SB as carried out by Baliga et al. The identification of putative mechanisms at the level of macromolecular interaction, ultimately permitting the assignment of functions to the macromolecules that participate in them, is explicitly mentioned by the researchers as a measure of the success of their experimental work. The implicit conclusion is that an understanding of the stress response of Halobacteria to UV radiation – of the way that this process works – could be reached by appealing to such mechanisms, generated by integration of the data that reflect behavior at the system level, with data concerning the properties of the system's components. It is worth noticing that, on the one hand, a tentative representation of the whole system as a network,

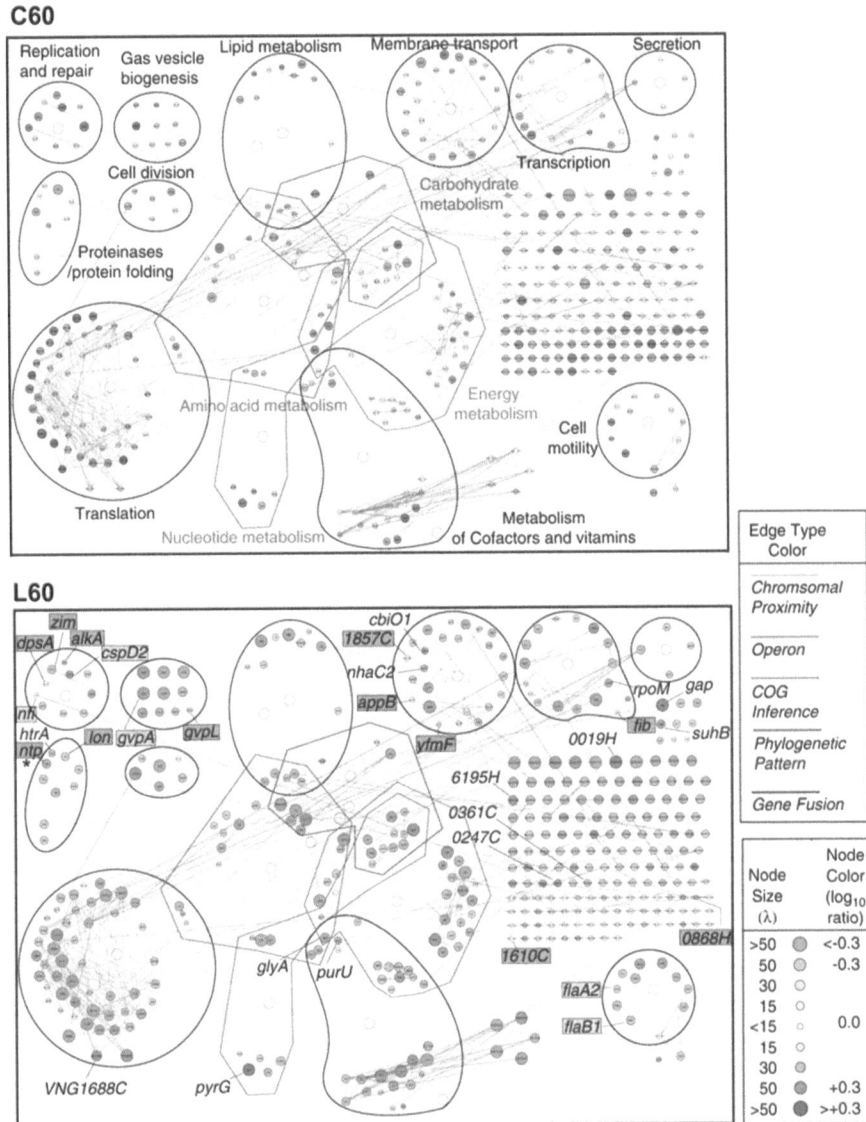

Fig. 3.1 System-level view of the *Halobacterium* NRC-1 response to UV irradiation (Baliga et al. 2004)

generated on the basis of computational processing of system-wide measurements of behavior following irradiation, precedes the elucidation of mechanisms that could explain how it works. On the other hand, the availability of preliminary putative mechanisms that are based on prior knowledge and account for causal interactions between the system's components is an essential prerequisite for the construction of a high-level model of the system. Since the approach is iterative, more sophisticated high-level models give rise to increasingly precise mechanisms at lower levels of the system's hierarchy, which are, in turn, used in further global analyses. In addition, each high-level model is phenomenological (Bruggeman and Westerhoff 2006). It constitutes, that is, a relatively abstract schematic representation of the whole system in its complexity, which highlights the frequently non-linear and non-sequential relationships between its components without characterizing their precise causal interactions. This suggests that the models simulating the system's high-level function and organization in top-down SB lack the basic features of traditional mechanisms. Only if used in a broad sense could the term "mechanism" be applied to such models, which are primarily of mathematical origin. The conclusion that characterization of system function in the special case of top-down SB hinges on representations whose mechanistic status is ambiguous, raises questions about the suitability of pure mechanisms for the explanatory purposes of SB, in general: Could an explanatory strategy that, like many of the explanatory schemes proposed by the new mechanists in other biological contexts,[5] would depend exclusively on successful interlevel integration of mechanisms produce adequate explanations in the context of SB? Furthermore, could such explanations suffice for rendering complex biological systems intelligible? Since these related questions are not only relevant to the case of top-down SB but they concern the explanatory role of mechanisms in all varieties of SB, we should be in a better position to address them after considering the following example from bottom-up SB.

3.2 Bottom-Up SB: From Lower-Level Mechanisms to Stochastic Models of Complex System Behavior

To illustrate the main methodological features of bottom-up SB we appeal to a study of gene regulation in the bacterium *E. coli* by Guido et al. (2006). The authors, who openly proclaim that they will be taking a bottom-up approach, describe the general objective of their work by stating that they seek to demonstrate how "the properties of regulatory subsystems can be used to predict the behavior of larger, more complex regulatory networks" (Guido et al. 2006, 856).

The regulatory subsystem in question is a synthetic gene network, under the control of a promoter that was designed so that it could be activated and repressed in a complex pattern within the cellular environment of *E. coli*. This subsystem's behavior was tested under various conditions: in the absence of regulatory molecules, in the presence of repressors only, in the presence of activators only and, finally, in the presence of both repressors and activators. Quantitative data of gene expression under these conditions were collected. In addition, a deterministic mathematical model predicting the in vivo behavior of the modular system was created. The model simulated the expected mean transcriptional response of the three

---

[5]See Craver (2005) for a relevant example from neuroscience.

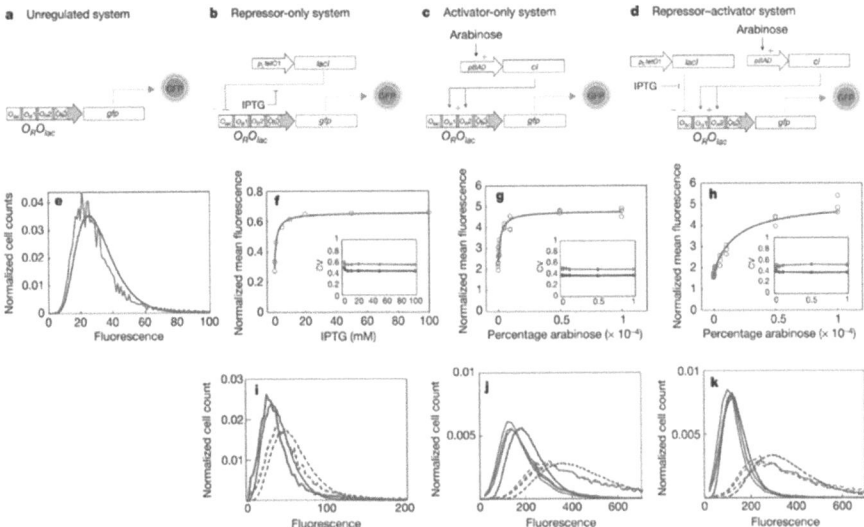

Fig. 3.2 (i) (**a–d**) Are causal mechanisms of transcriptional regulation of the synthetic gene network. (ii) (**e–k**) Represent mathematical models of the system's behavior under different regulatory conditions (unregulated, repressor-only, regulator-only, and repressor-activator). In these graphs, model predictions are represented by *darker grey lines* whereas *lighter grey lines* are used to indicate the actual experimental data (Guido et al. 2006, Reprinted by permission from Macmillan Publishers Ltd: Nature Reviews Molecular Cell Biology, copyright (2006))

regulated systems (repressor-only, activator-only, repressor-activator) and it was based on formal expressions of the kinetics of their components under conditions of equilibrium. When the transcriptional data actually gathered by the researchers in the laboratory were fitted to these expressions, which were derived theoretically, an estimate of the mean transcriptional response of the system under each of the conditions tested was generated (Fig. 3.2). The fitting involved computer software permitting the solution of non-linear least squares problems using a gradient-based optimization algorithm (Guido et al. 2006, Supplementary Information 5). With the help of algorithmic computational methods, once more, the initial deterministic model of this modular subsystem was ultimately extended to include stochastic effects in order to become capable of predicting quantitatively the behavior of a more complex system which involves regulatory feedback (Fig. 3.3).[6] The authors conclude by presenting experimental evidence that verifies the accuracy of the predictions made on the basis of the extended model.

---

[6]As stated by Adalsteinsson et al. (2004), the developers of the software package used for simulating the extended regulatory network studied in this example, stochastic models are necessary to predict the activity of engineered gene networks that include promoters regulated through various types of feedback due to the stochastic nature of the biochemical reactions involved. The stochastic effects encountered in complex systems such as these synthetic gene networks cannot be ignored when attempting to model their behavior because small changes in the number of the regulatory molecules have a major impact on transcriptional activity.

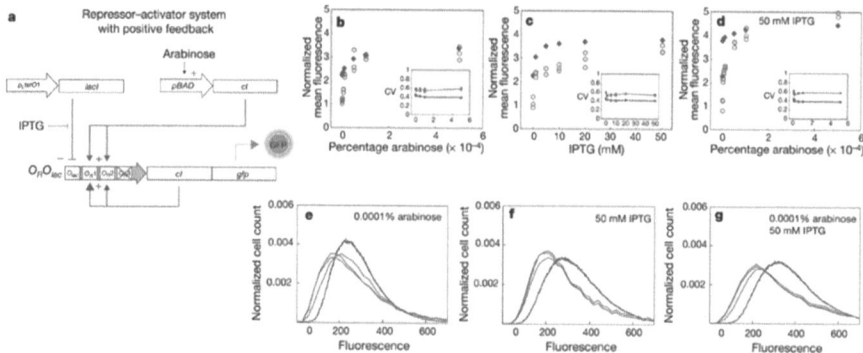

Fig. 3.3 Model predictions (*darker grey lines*) and experimental results (*lighter grey lines*) for the extended repressor-activator regulatory system, which includes positive feedback, under different regulatory conditions. Image (**a**) represents the system schematically. The next three plots show the variation in normalized mean fluorescence (which is proportional to the expression of the gfp reporter gene) in the presence of different levels of molecules that either induce or repress transcription in the context of this system: arabinose (**b**), IPTG (**c**), and arabinose with 50 mM of IPTG (**d**). *Light grey circles* indicate experimental data whereas *dark grey diamonds* represent levels predicted by the mathematical model. Finally, (**e**), (**f**), and (**g**) are graphs of normalized cell count versus fluorescence level for 0.0001 % arabinose, 50 mM IPTG, and 0,0001 % arabinose plus 50 mM IPTG, respectively. Once again, both experimental results (*light grey lines*) and model predictions (*dark grey lines*) are shown in these graphs (Guido et al. 2006, Reprinted by permission from Macmillan Publishers Ltd: Nature Reviews Molecular Cell Biology, copyright (2006))

As this example demonstrates, bottom-up applications of SB are strongly committed to the mechanistic methodology of molecular biology. Initially, at least, the concern is with the mechanistic details of the behavior of lower-level components of the system studied. Guido et al. begin their project with an experimental practice that follows the principles of mechanistic molecular biology: they test hypotheses about the regulatory properties of the promoter that controls the transcription of their synthetic gene network by correlating gene expression levels to different regulatory conditions (i.e., to interactions of the promoter with different repressors or activators). Reference to a preliminary basic mechanism, representing a causal relationship between the interaction of the promoter with specific regulatory molecules and its ensuing repression or activation, is a prerequisite for proceeding with the design of the experiment's next steps. The appeal to a causal story, that is, is very much what directs the experimental procedure early on.

The transition from characterizing the properties of this relatively simple subsystem on the basis of straightforwardly causal mechanisms to thinking about the larger, more complex regulatory network to which it belongs, however, involves the development of models by means of mathematical and computational techniques. First, a deterministic model that predicts the subsystem's behavior on the basis of theoretical expressions of the kinetics of its parts is required. Subsequently, an extended stochastic model that, as the authors claim, can provide insights into the behavior of a complex system with regulatory feedback must be developed. Neither of these models is causal in nature. They both might be better characterized as mathematical. Elsewhere in this volume, Issad and Malaterre (2015) argue that models of this type confer explanatory power to dynamic mechanistic explanations, which must be regarded as distinct from conventional mechanisms, whose explanatory power is derived from their capacity to provide a causal story. In the way that Guido et al. utilize them, mathematical models seem to be well-suited for prediction of complex system behavior. If they are not ultimately reducible to mechanisms, then the sufficiency of mechanism-based models for explanation in bottom-up SB is called into question.

Philosophical accounts that aim to accommodate the mathematical explanations of complex systems exhibiting non-sequential, cyclic organization within a mechanism-based framework have recently been offered (see Bechtel and Abrahamsen 2010; Bechtel 2011). Bechtel, more specifically, claims that "the additional tools needed to understand such systems are computational modeling and dynamic systems analysis" (Bechtel 2011, 552). Both of these methods, in Bechtel's view, still have their foundations in a mechanistic account. Nevertheless, just as with top-down SB, finding purely mechanism-based ways to integrate lower-level mechanisms with higher-level models, even when ensuring that the latter could also meet the criteria for qualifying as mechanisms, remains a significant challenge for bottom-up SB.

4 Adequacy and Completeness of Mechanism-Based Explanation in Systems Approaches

The challenges identified in the previous section point directly to the issues of adequacy and completeness of mechanisms for explanation in SB. More precisely, since integration of multiple explanatory approaches is a condition for successful explanation in SB, we must consider whether the requirements for such integration to occur could be fully satisfied within a purely mechanism-based theoretical framework. In what follows, I argue that although the multi-level causal models of the new mechanists can serve as heuristically adequate tools for applied SB, in their current formulation they remain limited in their capacity to carry out the broad integration demanded for understanding not only how complex biological systems work, but also why they work in a given way.

To recapitulate, the evidence considered shows that appeals to mechanisms for explanation at each of the various levels of biological organization are commonplace in SB. But models of ambiguous status, that may not qualify as mechanisms even when evaluated on the basis of relatively flexible standards such as those stipulated by the MDC account, are also integral to the approach. If the interlevel experiments – bottom-up or top-down – by which SB seeks to produce a comprehensive understanding of biological systems as systems generate results that cannot be fully accommodated by mechanisms, then could the requisite explanatory integration take place? Furthermore, could this task be accomplished without stipulating an overarching mechanism for bringing together mechanism-based insights from the system's different levels?

According to the MDC account, the specification of an all-inclusive mechanism is the first condition for any adequate explanation. For a consistent mechanist, SB cannot be treated as an exception to the rule. Hence, in this context, too, a continuously productive mechanism must do the explanatory work by relating the system's levels. For Richardson and Stephan (2007), MDC-type mechanisms that serve this purpose are not only required, but also sufficient for successful explanatory integration in SB. Their suitability for SB, they argue, is owed to their multilevel constitution which enables them to explain without referring exclusively to proximate causes. Their constitutive features, in turn, allow them to account for context-dependent system behaviors that are often described as emergent. But Richardson and Stephan's position in favor of the explanatory sufficiency of mechanisms for SB also depends on the presumption that explanation at each of a biological system's distinct levels is nothing but mechanism-based. I suspect, therefore, that their response to the question that was posed first would be negative: explanatory integration in systems approaches to biology cannot be carried out successfully unless the system-wide networks of top-down SB, as well as the deterministic and the stochastic models of bottom-up SB, qualify as legitimate mechanisms.

It is unclear whether the aforementioned models that are essential to the methodology of SB could indeed be regarded as mechanisms on the basis of mainstream interpretations of the conditions set by the MDC account. In particular, the extent to which they could be described as representations consisting of entities and activities that bring about regular changes in a sequential order is what is at stake here. Because they constitute idealizations of dynamic complex systems, whose function is not characterized by productive regularity or linearity, their capacity to strictly comply with the MDC criteria is questionable at best. Besides, the mathematical origins of these models[7] indicate that they might be better understood as mathematical explanations rather than as mechanisms. Mathematics is frequently thought to amount to nothing more than a language for describing phenomena or objects. According to this nominalist view, the category "mathematical explanations" is redundant: since mathematical expressions are merely descriptions, they could be fully integrated within the explanatory framework that is appropriate for the phenomena that they describe. Thus, when a mathematical formalism describes a mechanism, it ought to be possible to accommodate it within a mechanism-based framework. This conception of the role of mathematics in explanation has been challenged by philosophers who, like Colyvan (2001) and Baker (2005, 2015), have defended the claim that in some cases mathematical formalisms may be explanatory in their own right. In such cases, it would be justified to speak of genuine mathematical explanations. If the networks and stochastic models of SB depend indeed on mathematical components that play a distinct and essential explanatory role, then their description as mechanisms can be legitimately disputed.

---

[7]Two different mathematical modeling approaches have been used extensively in SB: one involves ordinary differential equations (ODE) and the other relies on stochastic process algebras (SPA).

The sufficiency of an exclusively mechanism-based framework for explanation in SB becomes problematic, as well, in this case, unless it could somehow integrate genuinely mathematical explanations of biological phenomena.

For some philosophers of the new mechanistic movement who are convinced that biological explanation should depend on a purely mechanism-based framework, however, the move that would guarantee that such an outlook could be maintained in contemporary biology is the expansion of the notion of mechanism. As mentioned in Sect. 3, Bogen (2005), as well as Bechtel and Abrahamsen (2006), took significant steps in this direction by offering accounts of mechanism that could accommodate models of systems that do not display regularities or sequential execution of operations. More recently, Bechtel (2011) has proposed that the mathematical explanations provided by dynamic systems analysis ought to be included within a sufficiently expanded mechanistic framework. This broader conception of mechanism is, therefore, intended to lift theoretical obstacles to an exclusively mechanism-based understanding of biological systems.

Mechanisms can be regarded as adequate explanations, if adequacy is measured by the fruitfulness of the explanation. Capitalizing on the causal relations among entities and activities in biological phenomena, rather than on ontological claims about the real constitution of organisms, the multilevel mechanisms to which contemporary biology appeals serve as powerful explanatory tools with concrete practical applications. By means of such models, we gain insights into how complex biological systems work, enabling us to manipulate, control, and predict their behavior with increasing reliability and efficiency. On the other hand, much like the classical mechanistic models of molecular biology, the various mechanism-based explanatory schemes offered by proponents of the new mechanistic philosophy are limited in their capacity to produce answers to questions that go beyond the issue regarding how things work but also deal with the problem concerning why things work in that particular way. Consequently, MDC (2000) are correct in contending that to describe a mechanism for a phenomenon amounts to explaining it, only if what they mean is that mechanism-based descriptions are adequate explanations. Such explanations can be legitimately characterized adequate in the narrow sense that they can be used reliably in a practically beneficial way, to the extent that they inform us about the manner in which the phenomenon in question was produced. The stronger claim that mechanism-based explanations of phenomena are not only adequate but also exhaustive or complete, however, would have required additional support: it would have required evidence showing that mechanism-based accounts can also be used to address the question regarding why the phenomenon was produced in that particular way. In biology such questions typically entail evolutionary and developmental explanations.

Skipper and Millstein (2005) have argued convincingly that Glennan's and MDC's theories of mechanism are insufficient for supplying mechanism-based explanations of evolutionary processes, given that they cannot characterize natural selection. Nevertheless, they have also explicitly stated their assessment that "the basic resources for characterizing the mechanism of natural selection may be found in the new mechanistic philosophy" (Skipper and Millstein 2005, 345) and urged its proponents to explore the possibilities for developing a mechanism that could accomplish this goal. Similarly, a recent argument by Braillard (2010) defends the position that mechanism as conceived by the new mechanistic philosophers cannot fully capture a type of non-causal explanations – design explanations[8] – that is important for explanation in the context of SB. As understood by Braillard, design explanations are not concerned with explaining "how a dynamical process occurs but why a certain structure is present" (Braillard 2010, 50). In both of these cases, the evidence presented suggests that the main philosophical views of mechanism fail to account for types of explanation required for responding to "why questions" about biological systems. The comprehensive understanding of living beings that is sought by strategies like SB requires such explanations. Without them, accounts of the organization and the function of the systems examined that are heuristically valuable for science can still be provided via the currently available mechanism-based models, but explanations that fully capture their behavior remain out of reach. The current doctrine of mechanism appears, therefore, to be both instrumental to the progress of SB and limited in its capacity to produce the explanatory integration that would be required for the actualization of its more ambitious objectives.

Braillard, as well as Skipper and Millstein, are careful to point out that such conclusions are not meant to undermine the explanatory power of mechanism,[9] nor are they meant to suggest that fully mechanism-based explanations for evolution and SB are in principle impossible. I concur with these authors that quite the opposite is the case: expanded mechanisms that would succeed in this task might be developed. Nevertheless, it is important to be aware of the pitfalls of invoking expanded mechanisms as the antidote to the limitations of versions such as MDC's or Glennan's without specifying their detailed characteristics and the precise way by which they could overcome the current explanatory insufficiencies. Insufficiently defined mechanisms whose scope is expanded at the expense of explanatory clarity and reliability, would not help in disclosing the phenomena by making them more intelligible but could further obscure them, hindering the explanatory task. Consequently, manifesto statements such as Bechtel's (2011), proclaiming that expanded mechanisms need to include dynamic systems analysis and computational modeling, accurately identify the challenges faced by the new mechanists but must be followed by additional work in the direction of developing a full account of the specifics of mechanisms that could address them. Their proponents must also

---

[8]For a detailed discussion of "design explanations" and of their role, see Arno Wouters' (2007) "Design Explanations: Determining the Constraints on What Can Be Alive."

[9]In this anthology, Braillard (2015) elaborates on ways in which traditional mechanisms can be simplified and thus rendered more powerful in dealing with the task of explaining complex biological systems, thanks to the contributions of mathematical dynamical models. The construction of the latter is made possible by the application of methods originating from engineering.

consider whether such mechanisms would advance understanding of the systems represented or they would simply amount to models that are as complex as the systems under scrutiny themselves.

In short, the lack of conclusive evidence that mechanisms substituting for types of explanation that have traditionally been considered non-mechanistic are in principle unattainable does not diminish the force of criticisms concerning the fitness of the current models of the new mechanists for achieving complete explanatory integration in SB. Systems approaches in biology aim to attain a global perspective from which complex organisms could be understood as whole systems. To construe this explanatory task as dependent exclusively on the integration of mechanisms at different levels of a biological system's organizational hierarchy would be to ignore the limits of the current mechanismic framework. In addition, it would amount to a disregard for the significance of the richer pluralism that is revealed by the practice of SB.

5 Integration of Many Explanatory Perspectives in the Practical Implementation of SB

The constraints imposed on mechanism-based explanation by the theories of the new mechanistic philosophers are compatible with a non-reductive explanatory approach that allows for pluralism. As argued in Sect. 2, this is a pluralism that consists of different explanatory mechanisms, each focused' on a different system level and together all contributing to the same all-encompassing mechanism.[10] It is also not a trivial or fleeting pluralism since the distinct contributions at various stages of the system are not eventually eliminated, or substituted by a lower-level description, but their role is preserved within the larger explanatory mechanism. To the extent that multi-level causal mechanisms allow for non-reductive, pluralistic explanation, they fit well within the holistic agenda and the multidisciplinary strategy of SB. At the same time, however, their potential as non-reductive alternatives to classical reductionist mechanisms is limited due to their mechanistic roots and the scope of the pluralism that they can support is relatively narrow. More specifically, multi-level mechanisms are only one step removed from reductionism because of their continued reliance on basic principles of mechanistic explanations, which "are inherently reductionistic insofar as they require specifying the parts of a mechanism and the operations the parts perform" (Bechtel 2011, 538). Furthermore, the degree of pluralism of which they can admit is restricted by their current inability to include evolutionary and other historical explanations, whereas it is unclear whether they could efficiently incorporate mathematical explanations. In addition, the adoption of multi-level mechanisms in biology might ultimately promote a type of pluralism

---

[10]What I have in mind here is a mechanism-based explanatory scheme for non-reductive interlevel integration like the one described by Craver (2005).

similar to that designated by Sandra Mitchell (2004) as "isolationist pluralism,"[11] if it simply accepts different mechanism-based explanations of phenomena at various levels of analysis as independent and equally legitimate but fails to effectively integrate them by showing how they are related. Precisely this kind of effective explanatory integration is, of course, the goal of the new mechanistic philosophers – as well as of systems biologists – but the mechanisms that they have at their disposal for achieving it, as well as the theoretical framework upon which these mechanisms are based, appear insufficient in light of the limitations mentioned above. In Mitchell's view, the desired integrative pluralism is an alternative to both reduction and isolation and its attainment requires a diversity of approaches: "[n]o single theoretical framework, no simple algorithm, will suffice" (2004, 87). Accordingly, the attention of philosophers must shift from theories of single, overarching, mechanism-based models, to theoretical developments in the direction of diversification of explanation.[12]

At this point, it is worth noting that effective explanation in biology is not entirely a matter of compliance with certain well-articulated theoretical requirements. Most working biologists, at least, do not treat it as such. Instead, they regard the degree to which biological explanations reflect and promote the practical aims of the discipline to be the most relevant measure of their effectiveness. This view may be criticized, but its implication for thinking about the question of explanation in SB is clear: the explanatory method that it employs should also reflect the pluralism that characterizes its application. Since SB is practiced as a multi-disciplinary science and explanatory integration in its context is pursued by means of different strategies that are followed by different researchers, one could argue that the systems approach in biology demands a richer sense of pluralism than the mechanism-based framework of the new mechanistic philosophers could support.

The popularity of the view that a more pragmatic approach, concentrating on the implications of scientific practice for explanation, is more productive for the discussion of epistemological problems than inquiries that develop around purely theoretical concerns, has been increasing among philosophers of biology. Alan Love (2008), for example, has proposed an epistemological framework within which multidisciplinary contributions are coordinated around problem agendas. Criteria of explanatory adequacy are essential for the integration and prioritization of these contributions, or for the exclusion of certain explanations, but these standards, Love argues, are not universal but rather associated with particular problem domains. According to this account, pluralism in explanation is not defined by abstract theoretical requirements but it is driven by real practical problems demanding multidisciplinary solutions. Applied to SB, a problem-centered epistemological scheme would treat explanation as a matter that depends on the integration of independent explanatory contributions from disciplines such as genetics, chemistry,

---

[11]For Mitchell (2004, 86) the strategy of isolationist pluralism is to encourage "the pluralism of questions and the consequent independence of answers."

[12]Morange (2015) offers a discussion of explanatory diversity in biology.

mathematics, computer science, and evolutionary biology. In this framework, the relative weight of the contributions is determined by the set of problems that is being addressed in each experimental project rather than by the theoretical constraints of a mechanism for interfield integration.

From a pragmatic point of view, the exploration of the wide range of practices that can be encountered in the various projects that are carried out under the banner of SB is crucial for understanding how explanatory integration might take place in the process of conducting scientific research within this growing domain of biological inquiry. O'Malley and Soyer (2012), who embrace this perspective, base their account of integration in SB on a thorough investigation that capitalizes on the exposition of the multidimensionality of the integrative activity and the diversity of practices that characterize the developing systems approach in biology. Examples that illustrate the diversity of exploratory questions and the plurality of data, methods, and models encountered in various fields of molecular systems biology, can be used to paint a more accurate picture of integration than the one offered by studies of the theoretical aspects of this process, the authors argue. The picture of integration that emerges from these examples is a compelling representation of the explanatory pluralism that permeates every case of SB's application. Consider, for instance, the study of vesicle transport by Schuster et al. (2011) that O'Malley and Soyer present in order to illustrate the nature of integration that occurs in molecular systems biology. The problem explored in the selected study belongs traditionally to the domain of cell biology. But a conventional cell biology project is transformed into a case of molecular systems biology – of "mathematical cell biology," as O'Malley and Soyer (2012, 62) prefer to describe it – by importing a mathematical model that was previously used in the context of the physical sciences to simulate the process of "particle hopping" and using it in order to conceptualize data from biochemical assays. As a result the transport of vesicles along microtubules in fungal hyphae, which was originally regarded as a deterministic process, is now understood more accurately as a stochastic process. The description "multidisciplinary" would not have accurately conveyed what is distinctive about the systems approach that produced this new understanding. Instead of integration of independent disciplines, in this case the strategy involves transferring an explanatory model from one discipline to another to achieve a more fundamental kind of integration, transforming the discipline that adopts it and its traditional methods of explanation. The more fundamental integration identified in this example suggests that in its practice SB is not merely a multidisciplinary approach but it is characterized by a deeper pluralism that penetrates the individual disciplines that participate in its frame.

Several contributors to this collection of papers discuss either in passing or in detail the issue of explanatory pluralism in biology. Regardless of their conclusions regarding the current status of explanation in various biological disciplines – regardless, that is, of whether they read the evidence as indicative of pluralism – many of them include considerations about the role of mathematical models in their arguments. Baetu (2015), Brigandt (2015), and Théry (2015), for example, maintain that mathematical models[13] play an important role in complementing mechanistic explanations. Similarly, Braillard (2015) and Zednik (2015) demonstrate the notable heuristic function of dynamic mathematical models for improving the efficiency of traditional biological mechanisms, while Issad and Malattere (2015) observe that in biology explanatory power can be drawn both from causal interpretation and from mathematical derivation. In all of these accounts, the examination of different aspects of the relationship of mathematical representations to mechanisms provides a relevant basis for addressing the question regarding the state of affairs of explanation in biology. In this paper, the evidence considered supports the claim that mathematical models are indispensable for explanation in both top-down and bottom-up varieties of SB. The possibility of full integration of these models into a mechanism-based framework of explanation in SB remains open, but at this stage of development of the systems approach such integration remains incomplete. In addition, the analysis I have presented does not disqualify the view that in certain occasions mathematical explanations are not merely descriptive of biological phenomena but they constitute stand-alone explanations in their own right (Baker 2015). Should the latter be conclusively established, what must also be acknowledged is that explanation in SB involves at least one more type of explanation besides mechanisms. But even the prospect of mathematical models becoming successfully subsumed under expanded mechanisms does not necessarily negate explanatory pluralism. Integration of mathematical models in a mechanism-based framework that preserves the independence of their explanatory contributions alongside with those of causal mechanisms – and potentially of other irreducible explanatory sources – is still compatible with an integrative pluralism of the kind that was proposed by Mitchell (2004) and was discussed earlier in this section. Therefore, both of these alternatives regarding the role that mathematical models could play in the context of contemporary SB are compatible with the conclusion that explanation in biology, more generally, is a pluralistic affair.

Overall, our examination of the case of SB suggests that one possible way of dealing with the difficulties involved in explaining complex systems is by further investing on theoretical developments. For instance, an expanded theoretical account, capable of specifying how historical and mathematical explanations may be incorporated into mechanisms, could open up new possibilities for overcoming practical obstacles to understanding non-linear biological phenomena. The example of SB, however, also alerts us to the useful insights that are gained when we move beyond strictly theoretical considerations and think about the problem of explanation from a pragmatic viewpoint. As discussed, by shifting the focus on the practice of SB, the pluralism that characterizes the approach – a pluralism that may be too broad to be captured by any purely mechanism-based idealization –

---

[13]Théry (2015) distinguishes between quantitative and qualitative explanations in the context of the special case of molecular biology. I take mathematical models to be integral to the quantitative explanations, which, as Théry argues, complement qualitative explanations provided by mechanisms.

is revealed and can be properly appreciated. Construing the choices available to those who think about explanation in biology as a dilemma between pragmatic and theoretical approaches, however, constitutes a misrepresentation of their options. Besides, this dilemma appears to be artificial: in the context of an inquiry that is motivated by a genuine philosophical attitude, both perspectives contribute in a mutually complementary way to a richer understanding of the phenomena studied. Therefore, with respect to the issue of explanation in SB, on the one hand, a pragmatic orientation produces fruitful insights into the practical influence of pluralism and its significance for explanatory integration. Theoretical reflection and abstraction, on the other hand, are the means by which we reach awareness of the epistemological limitations of explanatory mechanisms and of the conditions for the possibility of explanatory integration. In addition, a theoretical attitude allows us to identify and evaluate the criteria for distinguishing between good and bad explanations. And yet, it could be legitimately argued that practice is a better guide for adjudicating these matters and that, in any case, theory has little real impact on the actual methodological choices made by scientists. Both of these claims are defensible and they may often be demonstrated to be factually true. But the most important contribution of theories of explanation in science is perhaps less related to their impact on the sciences themselves than it is linked to their significance for thinking about the ways in which humans explain and about the limits of intelligibility. The epistemological implications of theories of explanation, then, should not be neglected, regardless of whether the understanding gained from studying them is capable of directly promoting successful scientific practice.

6   Conclusions

In summary, the preceding discussion suggests that the new mechanisms proposed by philosophers of biology hold promise for advancing the purposes of SB, especially because of their potential to accommodate the many levels of analysis that are characteristic of its methodology. In their current constitution, however, mechanisms are limited in their capacity to explain important aspects of the behavior of complex biological systems because, despite recent progress, they have not yet successfully incorporated evolutionary, more generally historical, or even mathematical explanations into their framework. In addition, the practical applications of SB are pluralistic in a sense that is incommensurable with the pluralism that could be captured by such mechanism-based explanations.

The identification of some of the limitations of current accounts of mechanism does not preclude the possibility that expanded mechanisms that could serve as fully adequate explanations of complex system behavior may eventually be developed. The evidence does not show that such mechanism-based models are in principle impossible; it simply suggests that, as they now stand, the models of the new mechanists alone are insufficient for carrying out the explanatory integration demanded by SB. But the question that remains open has to do with the normative implications of this assessment for the future direction of SB with respect to explanatory strategy. The results of this study do not commit us to a strong normative conclusion in favor of either pluralism or unification of explanatory perspectives. They do, however, indicate that at this stage of its development SB can offer practically adequate explanations of biological phenomena by relying heavily on mechanisms, but also depending on a variety of other types of explanation. If taken as evidence of the viability of a tolerance for explanatory pluralism, then this observation could also be regarded as an argument against the usefulness of a dogmatic commitment to explanatory unification through theoretical accounts of mechanism. On the other hand, mere reference to the currently diverse status of explanation in SB does not constitute sufficient proof in favor of the opposite dogmatic view, which urges the pursuit of pluralism at any cost for attaining the holistic understanding sought by systems approaches. The picture that emerges, therefore, concerns more the present rather the future of SB. It reveals a deeply pluralistic explanatory landscape, which is tolerated by scientists insofar as it can support productive research. The latter presupposes that this framework both allows for successful integration of diverse explanations and averts the threat of uncritically endorsing all explanations, described by Sandra Mitchell (2004) as "anything goes" pluralism. Theoretical constraints and pragmatic guidelines play an equal part in securing that these conditions for a productive pluralism in the biological sciences can be satisfied. Similarly, theoretical reflection and a pragmatic view of scientific practice can contribute in a mutually complementary manner to more constructive philosophical thinking about explanation in biology.

Acknowledgments I am grateful to the editors of this volume, Pierre-Alain Braillard and Christophe Malaterre, for inviting me to contribute, as well as for their many insightful comments on earlier versions of this paper. In addition, I would like to thank two anonymous reviewers for their input on the manuscript and for their recommendations for improving it.

References

Adalsteinsson, D., McMillen, D., & Elston, T. C. (2004). Biochemical network stochastic simulator (BioNetS): Software for stochastic modeling of biochemical networks. *BMC Bioinformatics, 5*, 24.

Aderem, A. (2005). Systems biology: Its practice and challenges. *Cell, 121*, 511–513.

Allen, G. E. (2005). Mechanism, vitalism and organicism in late nineteenth and twentieth- century biology: The importance of historical context. *Studies in History and Philosophy of Biological and Biomedical Sciences, 36*, 261–283.

Baetu, T. (2015). From mechanisms to mathematical models and back to mechanisms: Quantitative mechanistic explanations. In P.-A. Braillard & C. Malaterre (Eds.), *Explanation in biology. An enquiry into the diversity of explanatory patterns in the life sciences* (pp. 345–363). Dordrecht: Springer.

Baker, A. (2005). Are there genuine mathematical explanations of physical phenomena? *Mind, 114*, 223–238.

Baker, A. (2015). Mathematical explanation in biology. In P.-A. Braillard & C. Malaterre (Eds.), *Explanation in biology. An enquiry into the diversity of explanatory patterns in the life sciences* (pp. 229–247). Dordrecht: Springer.

Baliga, N. S., Bjork, S. J., Bonneau, R., Pan, M., Iloanusi, C., Kottemann, M. C. H., Hood, L., & Di Ruggiero, J. (2004). System level insights into the stress response to UV radiation in the Hallophilic Archaeon Halobacterium NRC-1. *Genome Research, 14*, 1025–1035.

Bechtel, W. (2011). Mechanism and biological explanation. *Philosophy of Science, 78*, 533–557.

Bechtel, W., & Abrahamsen, A. (2006). Phenomena and mechanisms: Putting the symbolic, connectionist, and dynamic systems debate in broader perspective. In R. Stainton (Ed.), *Contemporary debates in cognitive science*. Oxford: Basil Blackwell.

Bechtel, W., & Abrahamsen, A. (2010). Dynamic mechanistic explanation: Computational modeling of circadian rhythms as an exemplar for cognitive science. *Studies in History and Philosophy of Science A, 41*, 321–333.

Bogen, J. (2005). Regularities and causality; generalizations and causal explanations. *Studies in History and Philosophy of Biological and Biomedical Sciences, 36*, 397–420.

Braillard, P.-A. (2010). Systems biology and the mechanistic framework. *History and Philosophy of the Life Sciences, 32*, 43–62.

Braillard, P.-A. (2015). Prospect and limits of explaining biological systems in engineering terms. In P.-A. Braillard & C. Malaterre (Eds.), *Explanation in biology. An enquiry into the diversity of explanatory patterns in the life sciences* (pp. 319–344). Dordrecht: Springer.

Breidenmoser, T., & Wolkenhauer, O. (2015). Explanation and organizing principles in systems biology. In P.-A. Braillard & C. Malaterre (Eds.), *Explanation in biology. An enquiry into the diversity of explanatory patterns in the life sciences* (pp. 249–264). Dordrecht: Springer.

Brigandt, I. (2015). Evolutionary developmental biology and the limits of philosophical accounts of mechanistic explanation. In P.-A. Braillard & C. Malaterre (Eds.), *Explanation in biology. An enquiry into the diversity of explanatory patterns in the life sciences* (pp. 135–173). Dordrecht: Springer.

Bruggeman, F. J., & Westerhoff, H. V. (2006). The nature of systems biology. *Trends in Microbiology, 15*, 45–50.

Colyvan, M. (2001). *The indispensability of mathematics*. Oxford: Oxford University Press.

Craver, C. F. (2005). Beyond reduction: Mechanisms, multifield integration and the unity of neuroscience. *Studies in History and Philosophy of Biological and Biomedical Sciences, 36*, 373–395.

Darden, L. (2005). Relations among fields: Mendelian, cytological and molecular mechanisms. *Studies in History and Philosophy of Biological and Biomedical Sciences, 36*, 349–371.

DesAutels, L. (2011). Against regular and irregular characterizations of mechanisms. *Philosophy of Science, 78*, 914–925.

Fischbach, M. A., & Krogan, N. J. (2010). The next frontier of systems biology: Higher-order and interspecies interactions. *Genome Biology, 11*, 208.

Glennan, S. (1996). Mechanisms and the nature of causation. *Erkenntnis, 44*, 49–71.

Glennan, S. (2002a). Rethinking mechanistic explanation. *Philosophy of Science, 69*, S342–S353.

Glennan, S. (2002b). Contextual unanimity and the units of selection problem. *Philosophy of Science, 69*, 118–137.

Glennan, S. (2005). Modeling mechanisms. *Studies in History and Philosophy of Biological and Biomedical Sciences, 36*, 443–464.

Guido, N. J., Wang, X., Adalsteinsson, D., McMillen, D., Hasty, J., Cantor, C. R., Elston, T. C., & Collins, J. J. (2006). A bottom-up approach to gene regulation. *Nature, 439*, 856–860.

Hood, L. (2002). *My life and adventures integrating biology and technology*. Commemorative Lecture for the 2002 Kyoto Prize in Advanced Technologies. http://www.systemsbiology.org/extra/2002kyoto.pdf. Accessed 20 Jan 2012.

Hyman, A. A. (2011). Whither systems biology. *Philosophical Transactions of the Royal Society B, 366*, 3635–3637.

Issad, T., & Malaterre, C. (2015). Are dynamic mechanistic explanations still mechanistic? In P.-A. Braillard & C. Malaterre (Eds.), *Explanation in biology. An enquiry into the diversity of explanatory patterns in the life sciences* (pp. 265–292). Dordrecht: Springer.

Kitano, H. (2001). Systems biology: Toward system-level understanding of biological systems. In H. Kitano (Ed.), *Foundations of systems biology* (pp. 1–36). Cambridge: MIT Press.

Klipp, E., Herwig, R., Kowald, A., Wierling, C., & Lehrach, H. (2005). *Systems biology in practice.* Weinheim: Wiley-VCH.

Love, A. C. (2008). Explaining evolutionary innovations and novelties: Criteria of explanatory adequacy and epistemological prerequisites. *Philosophy of Science, 75,* 874–886.

Machamer, P. (2004). Activities and causation: The metaphysics and epistemology of mechanisms. *International Studies in the Philosophy of Science, 18,* 27–39.

Machamer, P., Darden, L., & Craver, C. F. (2000). Thinking about mechanisms. *Philosophy of Science, 67,* 1–25.

Mitchell, S. D. (2004). Why integrative pluralism? *Emergence: Complexity and Organization Special Double Issue, 6*(1–2), 81–91.

Morange, M. (2015). Is there an explanation for… the diversity of explanations in biological sciences? In P.-A. Braillard & C. Malaterre (Eds.), *Explanation in biology. An enquiry into the diversity of explanatory patterns in the life sciences* (pp. 31–46). Dordrecht: Springer.

Nicholson, D. J. (2012). The concept of mechanism in biology. *Studies in History and Philosophy of Biological and Biomedical Sciences, 43,* 152–163.

O'Malley, M. A., & Soyer, O. S. (2012). The roles of integration in molecular systems biology. *Studies in History and Philosophy of Biological and Biomedical Sciences, 43,* 58–68.

Richardson, R. C., & Stephan, A. (2007). Mechanism and mechanical explanation in systems biology. In F. Boogerd & F. Bruggeman (Eds.), *Theory and philosophy of systems biology* (pp. 123–144). Amsterdam: Elsevier.

Sarkar, S. (1998). *Genetics and reductionism.* Cambridge: Cambridge University Press.

Schuster, M., Kilaru, S., Ashwin, P., Lin, C., Severs, N. J., & Steinberg, G. (2011). Controlled and stochastic retention concentrates dynein at microtubule ends to keep endosomes on track. *EMBO Journal, 30,* 652–664.

Shannon, P., Markiel, A., Ozier, O., Baliga, N. S., Wang, J. T., Ramage, D., Amin, N., Schwikowski, B., & Ideker, T. (2003). Cytoscape: a software environment for integrated models of biomolecular interaction networks. *Genome Research, 13,* 2498–2504.

Skipper, R. A., Jr., & Millstein, R. L. (2005). Thinking about evolutionary mechanisms: Natural selection. *Studies in History and Philosophy of Biological and Biomedical Sciences, 36,* 327–347.

Strevens, M. (2008). *Depth: An account of scientific explanation.* Cambridge: Harvard University Press.

Théry, F. (2015). Explaining in contemporary molecular biology: Beyond mechanisms. In P.-A. Braillard & C. Malaterre (Eds.), *Explanation in biology. An enquiry into the diversity of explanatory patterns in the life sciences* (pp. 113–133). Dordrecht: Springer.

Wouters, A. G. (2007). Design explanations: Determining the constraints on what can be alive. *Erkenntnis, 67,* 65–80.

Zednik, C. (2015). Heuristics, descriptions, and the scope of mechanistic explanation. In P.-A. Braillard & C. Malaterre (Eds.), *Explanation in biology. An enquiry into the diversity of explanatory patterns in the life sciences* (pp. 295–317). Dordrecht: Springer.

Chapter 4
Historical Contingency and the Explanation of Evolutionary Trends

Derek Turner

Abstract One "big question" of macroevolutionary theory is the degree to which evolutionary history is contingent. A second "big question" is whether particular large-scale evolutionary trends, such as size increase or complexity increase, are passive or driven. Showing that a trend is passive or driven is a way of explaining it. These two "big questions" are related in both a superficial and a deep way. Superficially, defending historical contingency and showing that major trends are passive are two complementary ways of downplaying the importance of natural selection in evolutionary history. A passive trend is one that's not explained by selection. In order to appreciate the deeper connection between the two issues, it is necessary to distinguish different senses of contingency (especially sensitivity to initial conditions vs. unbiased sorting). It's plausible that passive trends are generally due to unbiased sorting processes. Serendipitously, thinking of contingency as unbiased sorting also helps to clarify its relationship to species selection, which some think of as biased sorting. Macroevolutionary theory thus turns out to have considerable unity.

Keywords Contingency • Convergence • Evolutionary trends • Historical explanation • Macroevolution • Passive trends • Species sorting

1 Introduction

In this contribution, I explore the relationship between two major issues in macroevolutionary theory. The first issue, which Stephen Jay Gould highlighted in his 1989 book, *Wonderful Life*, is the degree to which evolutionary history is contingent. This issue has been the subject of some debate among paleontologists. Simon Conway Morris—who, ironically, was one of the heroes of Gould's book on the Burgess Shale—has argued that convergence, rather than contingency, is the hallmark of evolutionary history (Conway Morris 2003a, b). The issue has inspired

D. Turner (✉)
Connecticut College, 270 Mohegan Ave., New London, CT 06320, USA
e-mail: Derek.turner@conncoll.edu

© Springer Science+Business Media Dordrecht 2015
P.-A. Braillard, C. Malaterre, *Explanation in Biology*, History, Philosophy and Theory of the Life Sciences 11, DOI 10.1007/978-94-017-9822-8_4

some experimental work by Richard Lenski and colleagues (Lenski and Travisano 1994; Travisano et al. 1995). Gould's argument about contingency has also had a significant, albeit indirect impact on the philosophy of biology, by way of John Beatty's classic paper arguing that the contingency of evolutionary history means that there are no distinctively biological laws (Beatty 1995). The debate about the degree to which evolutionary history is contingent is also a good example of what Beatty (1995, 1997) has called a "relative significance debate" in evolutionary biology. No one would ever deny that there are some fascinating examples of evolutionary convergence, and even the most loyal partisans of convergence would admit that some aspects of evolutionary history are contingent. They disagree over which is the dominant theme in the history of life.

The second major issue has to do with explanation in macroevolution and with the distinction between passive and driven large-scale evolutionary trends (McShea 1994; for a good introduction, see Rosenberg and McShea 2007, Ch. 5). When paleontologists identify a large-scale trend in the fossil record—size and complexity increase are good examples, but see McShea (1998) for some others—one of the first questions they ask about it is whether the trend is passive or driven. Where a trend is driven, the assumption is that some force is "pushing" evolutionary change. A passive trend, by contrast, involves a random walk away from some fixed starting point. The passive/driven distinction marks two different ways of explaining trends (Grantham 1999). Here, too, one can find a Beatty-style relative significance debate. Most scientists who work on these issues try to focus on tractable questions, such as whether a morphological trend in a given clade, over a given time interval, is driven. Stephen Jay Gould (1996, 1997) has suggested that larger-scale evolutionary trends are mostly passive. Everyone acknowledges that some trends in evolutionary history are driven. And there are also some well-documented cases of passive trends (see, e.g., Jablonski 1997). But the existing body of empirical work on the subject leaves plenty of room for disagreements of emphasis.

How are these two issues—contingency *vs.* convergence, and passive *vs.* driven trends—related? One possibility is that the two issues are largely independent of one another. For example, suppose we knew that evolutionary history is highly contingent. That might make no difference at all to the question whether passive *vs.* driven trends are more prevalent. Or suppose we determine that passive trends are the rule. That might be compatible with either the convergentist or the contingentist pictures of macroevolution. In these cases, we'd have a theory of macroevolution that is compartmentalized. We'd have no story to tell about how our answers to different questions about macroevolution are related. It might turn out that this outcome is just unavoidable. But my aim here is to show that there really is an interesting and heretofore underappreciated connection between these two issues. An initial clue is that Stephen Jay Gould was consistent in his defense of both contingency and passive trends. Though he nowhere (to my knowledge) explicitly discusses the possible connections between these ideas, Gould is nothing if not a systematic thinker.

Is there a special connection between the concept of historical contingency and the concept of a passive trend? I'll begin, in Sect. 2, by showing how this rather narrow issue is related to broader questions about the nature of explanation in historical science. Section 3 focuses on a superficial connection between the notions of contingency and passive trends. Both are, broadly speaking, non-selectionist ideas, and so defending them might be one way of downplaying the importance of natural selection. Section 4 introduces, as a working hypothesis, the suggestion that the contingency of evolutionary history explains why some evolutionary trends are passive. One immediate problem with this working proposal is that 'contingency' has different meanings. In Sect. 5, I argue that the plausibility of the working proposal depends on which sense of 'contingency' we have in mind. If we think of contingency as sensitivity to initial conditions, then the working proposal does not fare too well. However, things look better if we think of contingency as unbiased sorting. Section 6 explores the relationship between unbiased sorting and passive trends in greater detail and argues that unbiased sorting can (partly) explain passive trends.

2   Explaining Evolutionary Patterns

When we think about historical science, it's natural to suppose that the explanatory targets must be particular events. For example, one might think that the explanation of the end-Cretaceous mass extinction in terms of an asteroid impact is paradigmatic. In many cases, though, paleontologists are more interested in explaining patterns and trends that show up in the fossil record, rather than particular events. One example of such a pattern is the so-called "Lilliput effect" (Harries and Knorr 2009). The animals that show up in the fossil record in the aftermath of mass extinction events tend to have smaller body sizes than the animals that preceded the mass extinction. One potential explanation of this pattern invokes species sorting. It could be that large body size increases extinction risk during calamitous times. Another possible explanation invokes within-population sorting processes: Perhaps, in the lineages that survive mass extinction events, there is strong selection pressure in favor of smaller body size (Harries and Knorr 2009, p. 7). Of course, these two hypotheses about process are not necessarily incompatible. Species sorting is compatible with natural selection working to cause dwarfism within populations. It's possible that both of these processes contribute in some way to explaining the Lilliput effect. I offer this merely as an example of the sort of explanation that is quite common in paleobiology, where the goal is to explain pattern in terms of process.

An evolutionary trend is just one kind of pattern in the fossil record. A trend is usually defined as a persisting, directional change in some variable of interest. In his classic (1994) paper, McShea treats the passive/driven distinction as a distinction between two different "mechanisms" or processes that can generate trends, so in claiming that a trend is, say, driven, one is trying to explain a pattern in terms of an underlying process. McShea's use of the term "mechanism" differs somewhat from the more familiar usage (e.g. in the contributions to Part II of this volume).

The rough idea is that if you wanted to design a computer simulation that produced a directional trend, there are two fundamentally different ways of doing so (i.e. two kinds of trend mechanisms). The more general point is just that the investigation of passive vs. driven trends fits neatly into the larger paleontological practice of explaining patterns in terms of evolutionary processes.

One would think that if evolutionary processes were contingent, as Gould, Beatty, and others have claimed, that would matter somehow to the overall project of explaining patterns in terms of evolutionary processes. My goal here is to explore some of the ways in which contingency might matter to that explanatory project. Would the contingency of evolutionary processes make any difference at all to the way in which those processes help explain patterns and trends?

It's easier to see how the contingency of evolutionary processes might make a difference to the explanation of particular events. There are at least two ways in which contingency might constrain the explanatory options when it comes to particular events. First, suppose that contingency does imply (as Beatty 1995 argued) that there are no distinctively biological laws. That, presumably, would mean that explaining particulars by subsuming them under general biological laws is not in the cards. And some philosophers have indeed tried to show there are distinctively historical or narrative styles of explanation that do not invoke laws (Gallie 1959; Hull 1975). Second, Sterelny (1996) draws a distinction between actual sequence explanations and robust process explanations. A robust process explanation claims, counterfactually, that a given event would have happened even if initial conditions had been different. If history is highly contingent, where 'contingency' is taken to mean that later outcomes are sensitive to variations in initial conditions, then that would seem to rule out robust process explanations by definition. Although historical contingency has the potential to constrain explanation in at least these two ways, these issues both concern the explanation of particular events.

So the goal in what follows is to see if contingency matters at all to the larger project of explaining evolutionary patterns in terms of evolutionary processes. I'll pursue this by zeroing in on the somewhat narrower issue of how contingency is related to passive trends. We'll see in the next section that there is a superficial connection between the two. In order to get to the bottom of things, however, it will be necessary to examine some of the different possible meanings of 'contingency'.

3 The Non-selectionist Spirit

It's well known that Gould was skeptical about the prospects for using natural selection to explain many of the patterns of evolutionary history (see especially Gould and Lewontin 1979, but also Gould 2002). Much of Gould's thinking about evolution is anti-selectionist—or perhaps just non-selectionist—in spirit. Gould was also a consistent critic of the idea of evolutionary progress. One can hold that natural selection is the major cause of evolutionary change without necessarily believing in evolutionary progress. For example, one could argue that selection usually only adapts populations to local environmental conditions, without driving evolutionary change in any particular direction over longer timescales. But it is much harder to see how one could believe in evolutionary progress without emphasizing the power of natural selection. So one way to attack the notion of progress is to attack selectionism. Because progress is (partly) a normative notion, it raises further issues that might render the discussion unmanageable. So I won't say much more about progress here (but see Turner 2011, Ch. 6, for an overview). One can see contingency and passive trends as two components of a larger Gouldian critique of the power of natural selection.

To begin with, convergent evolution is usually thought to occur when distantly related lineages evolve similar adaptations under similar selection regimes, or come to occupy similar ecological niches. For example, ichthyosaurs and dolphins have remarkably similar body designs. Both have features that look like adaptations for sustained, fast swimming. Both evolved from terrestrial ancestors that returned to the water where they confronted the same biomechanical constraints. Defenders of convergence, such as Simon Conway Morris, are explicitly pro-selectionist. Clearly, one way of downplaying the importance of natural selection in evolutionary history is to downplay evolutionary convergence, and to argue that contingency is relatively more significant.

In addition, most paleontologists take it as obvious that there is a connection between natural selection and driven trends. For example, if we were to find that size increase in a given group is driven (see, e.g. Alroy's 1998 study of mammals), it would seem reasonable to conclude that natural selection is the driver. That is, the trend toward larger size is driven because natural selection is, for whatever reason, favoring larger body size. Hone and Benton (2005) discuss some of the ways in which larger body size might confer survival or reproductive advantage. Thus, for many paleontologists, investigating whether a trend is passive or driven is just an indirect way of investigating natural selection's role in evolution. There are some exceptions. McShea (2005) has argued that driven trends need not be driven by selection, and this idea also shows up in McShea and Brandon's (2010) argument that complexity increase is a driven trend, but one that's not driven by selection. Nevertheless, in spite of these recent developments, it's fair to say that most paleontologists just assume that selection is what "drives" large-scale trends. In this context, arguing that many trends are passive is another way of downplaying the importance of natural selection.

Here, then, is one interesting connection between passive trends and contingency: both ideas pose challenges to strong selectionist or adaptationist views of evolutionary history. Or coming at things from the other direction, we might say that both convergent evolution and driven trends are typical signatures of natural selection. Having made this observation, it might be tempting to stop here. But I want to press on in hopes of being able to say something deeper about the relationship between contingency and passive trends. We can say that the greater the significance of contingency and passive trends, the less important natural selection is in shaping evolutionary history. But that leaves open whether there is any tighter conceptual connection between contingency and passive trends.

4 Is There a Deeper Connection Between Contingency and Passive Trends?

Consider, as a first pass, the following working proposal, which is suggested by Gould's defense of both passive trends and contingency:

WP Passive evolutionary trends generally result from contingent evolutionary processes. Driven evolutionary trends generally result from non-contingent evolutionary processes

My plan of attack will be to investigate WP in a systematic way. It will turn out to require some clarification and revision. But I'll argue that with some adjusting, tweaking, and disambiguation, WP gets things basically right. Moreover, if WP is indeed correct, then contingency does matter to the explanation of larger scale evolutionary patterns and trends. Where trends are passive, the contingency of historical processes would help explain why.

Although my main concern here is with the unity of macroevolutionary theory, WP might prove to be interesting for another reason. Scientists have developed several good empirical methods for determining whether evolutionary trends are passive vs. driven (McShea 1994). These include (i) the ancestor/descendant test, (ii) the subclade test, and (iii) the stable minimum test. Suppose, for example, that we know that mean body size increases in a certain clade over a specific time interval. Is that because there is an underlying bias toward larger size? One way to check this is by comparing ancestors with their descendants (assuming we know the phylogenetic relationships). Are the descendant species typically larger than their ancestors? Another way to check this is by looking at the size of the smallest members of the clade. Does the size of the smallest-bodied lineage of the clade increase over time? If not—if the minimum is stable—then that might suggest that the trend is passive, and that the clade is "bumping against" a fixed minimum size boundary. A third approach is to look at the size distributions in a subclade (or a sample of subclades) that is far removed from the hypothesized minimum size boundary. A pattern of size increases in those subclades suggests that the trend is driven. These methods give paleobiologists a great deal of traction when it comes to the question whether trends are passive vs. driven. There may be other questions about the causes of trends that are more difficult to answer (Turner 2009), but it is often possible to determine whether the basic dynamic is passive or driven.

Notice that WP makes specific reference to evolutionary processes. That reflects a decision to restrict the focus here to evolutionary theory. However, evolution is by no means the only context in which it makes sense to ask whether history is contingent or convergent. One can ask this about any historical processes. One especially interesting example is the history of science. Many realist philosophers of science share C.S. Peirce's faith that if scientists apply their methods consistently over the indefinitely long run, they will eventually converge on the same results no matter where their inquiries started. Anti-realists see more contingency in the history of science, and they are more open to the thought that our current scientific picture might have been very different, and yet equally well supported by the evidence (Radick 2005). Furthermore, there are many different fields of study in which researchers seek to document and explain the occurrence of historical trends, from economics to climate science. The passive/driven distinction is entirely general: any time we identify a trend, we might ask whether it is passive or driven. Thus, the relationship between historical contingency and trend dynamics is one that could be of interest to researchers in virtually any field that's concerned with historical processes. Although I will continue to focus rather narrowly on evolutionary theory, because that's where these issues have been discussed, the issues themselves could be said to belong to the philosophy of historical science more generally, an area that has received a lot of philosophical attention in recent years (Cleland 2001, 2002, 2011; Jeffares 2008; Tucker 2004, Turner 2007; among others).

If WP is correct, then these methods of determining whether a trend is passive or driven might give scientists an indirect way of addressing questions about contingency. Contingency is not easy to get a grip on empirically. Conway Morris's (2003a) approach is to provide a lengthy catalogue of examples of evolutionary convergence. But Sterelny (2005) has observed that one problem with that approach is that it's not always clear whether two traits in different lineages ought to count as the same: Is the evolution of agriculture in humans and ants an instance of convergent evolution? Experimental work has yielded fascinating results, but it's not clear what conclusions we can draw about macroevolutionary history on the basis of observations of *E. coli* populations in the lab. Even more pressingly, it's not entirely clear how to assess counterfactual claims about evolutionary history (e.g., "If this or that had been different in the past, humans would never have evolved.") If, as WP says, passive trends result from contingent processes, then our methods for determining whether trends are passive could also be put to use on the question of contingency. Thus, WP could conceivably have some empirical payoff.

5 Varieties of Historical Contingency

As it stands, however, WP contains a crucial ambiguity because there are several different senses of 'contingency' (Beatty 2006; Turner 2010; DesJardins 2011). The ambiguity even shows up in Gould's original presentation of the thought experiment of replaying the tape of history. At times, Gould has us imagine what it would be like to rewind the tape of life to some point in the distant past—say, the Cambrian period, some 540 mya—and then play it back again *after tweaking the initial conditions.* In his 1989 book, Gould describes in rich detail some of the bizarre creatures of the Burgess Shale, including the unimpressive *Pikaia*, a possible ancestor of the chordates. Suppose that *Pikaia* had gone extinct, but that some of the other strange organisms of the Cambrian had persisted—say, *Anomalocaris* or *Wiwaxia*? Things on Earth would look very, very different today. This idea—namely, that seemingly minor changes in upstream conditions can have major downstream consequences— is what many writers mean by contingency (see, e.g., Ben Menahem 1997).

It is also very close to Beatty's (2006) notion of contingency as causal dependence. It's what we have in mind when we say that later outcomes are contingent upon earlier historical conditions. The claim that evolution is contingent, in this sense, is also what Conway Morris (2003a) takes himself to be arguing against. His view is precisely that evolution by natural selection is usually *in*sensitive to variations in initial conditions. Let natural selection start out with insects, reptiles, birds, or mammals, and it will find a way to design wings every time.

There are also moments when Gould imagines replaying the tape of history multiple times from *the same starting point*, without altering the initial conditions. He suggests that if history were contingent, one would still see different downstream outcomes. This version of the thought experiment comes very close to a defense of causal indeterminism. One concern here is that if the contingency *vs.* convergence issue is supposed to be a matter of empirical scientific investigation, we might want to avoid equating contingency with causal indeterminism. The latter at least seems like a metaphysical, rather than a scientific issue, although some naturalists might think that the latest physics is our best guide to the question whether determinism is true. At any rate, it's not entirely clear why issues in evolutionary biology should hinge on the truth or falsity of causal determinism. Beatty (2006) suggests that it may be helpful here to think of contingency as *unpredictability*. On this interpretation, to say that history is contingent is merely to say that downstream outcomes are not predictable given knowledge of the upstream conditions. However, whether outcomes are predictable depends on various facts about who is doing the predicting: how much background knowledge does the predictor have, for example? How good is the predictor at processing large amounts of data? Interestingly, the degree to which the outcomes of historical processes are predictable could also depend on the degree of sensitivity to initial conditions, and perhaps also the degree of sensitivity to external disturbances. So it's possible that both versions of Gould's famous thought experiment help illustrate the view that the outcomes of evolutionary processes are unpredictable. The two versions just give different reasons for the unpredictability. For present purposes, though, I will follow Beatty's usage, and say that the unpredictability sense of contingency has to do with the second sort of case, where different outcomes result from the same initial conditions.

Gould was likely influenced by the MBL model, an early attempt by a group of paleontologists meeting at the Marine Biological Laboratory at Woods Hole, Massachusetts, to simulate macroevolution using computers (Raup et al. 1973; Raup and Gould 1974; Sepkoski 2012, Ch. 7). The MBL model treats species sorting as a stochastic process, so that it's a matter of chance whether a species goes extinct, speciates, or persists with no change during each time step of the model. This is a bit speculative, but it seems plausible that when Gould talks about "rewinding the tape of history" he may well have in mind the early magnetic tape that was used in computers. This suggests another way of thinking about contingency as unbiased species sorting (Turner 2010, 2011, Ch. 8). On this reading, when Gould says that evolutionary history is contingent, he's making a very specific claim about macroevolutionary mechanisms—namely, that macroevolution works in somewhat the same way as the MBL model. This interpretation has the advantage of drawing a connection between

Gould's views about contingency and his defense of what is sometimes known as the hierarchical expansion of evolutionary theory, or his view that in order to understand macroevolutionary patterns, we need to consider distinctively macro-level processes such as species selection. Although different theorists think of species selection in different ways (see Turner 2011, Chs. 4 and 5), one view is that species selection is merely biased species sorting. Biased *vs.* random species sorting would then be analogous to the processes of natural selection and drift at the population level. The precise nature of the relationship between natural selection and random drift has been the subject of much controversy in recent philosophy of biology, and so one must tread very carefully here. However, some do think of the difference between selection and drift as a matter of biased *vs.* unbiased sorting (or sampling). Putting all of this together, Gouldian contingency might be a kind of macro-evolutionary, species-level drift. I'll call this *contingency as unbiased sorting*.

We have, then, three different senses of historical contingency: sensitivity to initial conditions, unpredictability, and unbiased sorting. What happens when we plug these different senses of contingency into WP?

5.1 Contingency as Unpredictability

WP says that passive trends usually result from contingent processes. If we plug in the unpredictability sense of contingency, WP does not fare very well. The problem is that passive trends are eminently predictable. If you take a computer simulation designed to generate a passive trend—that is, one where increases and decreases in the value of the variable being tracked are equally probable, and where the state space is structured by a fixed boundary—and let it run over and over, you will see much the same pattern on each run, even though it's impossible to predict whether the value being tracked will increase or decrease over one "turn" or time interval. The simulation will do a random walk away from the fixed boundary. Passive trends are predictable in precisely the same sense in which the distribution of heads and tails in a series of coin tosses is predictable: We may not be able to predict the outcome of a single toss, but we can predict that as the number of trials increases, the ratio of heads to tails will approximate 50:50. If we think of contingency as unpredictability, then WP comes out false.

5.2 Contingency as Sensitivity to Initial Conditions

WP fares much better when we think about contingency as sensitivity to initial conditions. The issues here are a little more complicated, however, because the notion of sensitivity to initial conditions has more moving parts. There are, as it were, a variety of different initial conditions that one could hypothetically manipulate:

(i) The starting value of the trait.

(ii) The strength of the directional bias in the state space. (The directional bias is akin to the probability that an unfair coin will come up heads. The probability of heads might be 0.6, or it might be 0.8, or it could have any number of other values.)

(iii) The location of the fixed boundary in the state space.

With respect to (i), sensitivity to variation in the starting value of the trait, there is an interesting difference between passive and driven trends. A passive trend with respect to some variable, such as mean body size, is indeed sensitive to the starting value of the variable. A passive trend requires that the initial setting for the variable be at or near the fixed boundary in the state space, so that there is nowhere to go but up. A driven trend, on the other hand, is less sensitive to variations in the starting value of the variable in question. If natural selection always drives increases in body size, then it doesn't really matter what the starting body size of the clade might be; whatever the starting size, the mean body size will subsequently increase. Thus, there does seem to be a connection between passive trends and contingency, in the sense that passive trends are sensitive to variations in the starting value of the trait.

What about (ii) sensitivity to variations in the strength of the directional bias in the state space? Note that in principle, the strength of the bias could remain constant, but that it could also change over time, creating what I have elsewhere called a "shifting bias" trend (Turner 2009). It might help to think of a passive trend as the limiting case in which the bias is set to zero. If you introduce a bias at all, the trend is technically no longer passive. Changing the strength of the bias can certainly also change the shape of the resulting driven trend. Differences in the strength of the directional bias could reflect differences in the strength of natural selection. For example, it's possible that while organisms are small, there is strong selection in favor of larger body size, but that as size increases, the strength of selection flags. The shape of the resulting trends will be sensitive to changes in the strength of the bias. By analogy, the expected distribution of heads and tails in a series of coin tosses will differ depending on whether one is using a mildly weighted coin (Prob heads = 0.6) *vs.* a strongly weighted one (Prob heads = 0.8). Pulling all of this together, driven trends are sensitive to changes in the directional bias in a way that passive trends are not.

A third parameter that can be set in different ways is (iii) the fixed boundary in the state space. There is, plausibly, a minimum size for mammals, below which the mammalian *Bauplan* just isn't biomechanically or physiologically feasible. The smallest mammals living today—the pygmy shrew or the bumblebee bat—are probably at or near that size minimum. What if the size minimum were different? For example, what if it were a great deal smaller? Imagine a hypothetical scenario in which mammals start out the size of house cats, but where much smaller, insect sized mammals are biologically possible. In that case, the starting body size would be far away from the fixed boundary, and so the conditions would not be right for a passive trend. For passive trends, what matters is the distance (in the state space) between the fixed boundary and the starting value of the trait, so passive trends are sensitive to changes in both of those variables. Driven trends could, however, also be sensitive to the presence or absence of a fixed boundary in the state space. Instead of a size minimum, imagine that there is a maximum size for a given clade that's similarly imposed by biomechanical or physiological constraints. We can imagine that natural selection would drive the mean body size of the clade upward until it hits the ceiling of biological impossibility, so to speak. Change the location of the ceiling, or the floor, and you can change the resulting trend. Elsewhere I call such trends shifting boundary trends (Turner 2009).

The relationship between passive trends and sensitivity to initial conditions is thus rather messy and complicated. There are three different sorts of initial conditions to which trends might be sensitive. It's only in case (i) that we get a result that looks good for WP—namely the result that passive trends result from contingent evolutionary processes while driven trends do not. But (ii) and (iii) also represent potentially interesting kinds of sensitivity to initial conditions.

There is another obstacle that confronts any attempt to understand WP in terms of sensitivity to initial conditions. Sensitivity to initial conditions comes in degrees (Ben-Menahem 1997). In the limiting case, virtually any initial condition would result in the same outcome. This limiting case is well illustrated by a simple equilibrium model provided by Sober (1988), who imagines a ball released from some point along the rim of a bowl. It makes no difference what the release point is; the ball will always come to rest at the same place at the bottom of the bowl. Think of this as a simplified landscape (Inkpen and Turner 2012 develop the landscape metaphor as a way of thinking about historical contingency). In a more complex landscape—say, a basin with an uneven floor—the point of release would make more of a difference to the outcome. The sensitivity of the outcome is a matter of degree, and it depends on the nature of the topography. The trouble is that the distinction between passive and driven trends does not seem to be a matter of degree. Because the strength of the "driver" can vary, we can distinguish between stronger and weaker driven trends (Rosenberg and McShea 2007, p. 148). And of course, trends themselves can be more or less pronounced. But the passive/driven distinction seems like an either/or distinction.

One scientist has argued that passive and driven trends "represent extremes of a continuum" (Wang 2001, p. 851). However, Wang's point has to do with scaling effects (see Turner 2011, pp. 134–5 for discussion). For example, if mean body size increase in a clade is driven, it might be driven in 75 % of the subclades and passive in the other 25 % of subclades. So we might want to say that the trend is "mostly" driven. Contingency is also scale-dependent in much the same way (Inkpen and Turner 2012). If we resolve to focus on a given clade at a particular temporal scale, the question whether the trend is passive *at that scale* will often have a yes/no answer. For example, the stable minimum test might show that the trend in a particular clade at a particular level of resolution is passive. By contrast, contingency, understood as sensitivity to initial conditions, always comes in degrees.

Because contingency as sensitivity to initial conditions is a matter of degree while the passive/driven distinction is not, the contingency/inevitability distinction does not map onto the passive/driven distinction in the way that WP suggests. In many contexts, it is illuminating to think of contingency as sensitivity to initial conditions, but if we want to understand how contingency is related to passive trends, we should keep looking.

5.3 Contingency as Unbiased Sorting

Some proponents of species selection take the view that all that's required for species selection is biased sorting of lineages. That just means that different lineages have different probabilities of branching, persisting, or going extinct. If it helps to do so, we can think of those probabilities as species-level fitnesses. Some theorists think that more than this is required for species selection (see Turner 2011, Ch. 5, for an overview). For present purposes, we don't need to worry about what does and does not count as species selection. It's enough to work from some simple and hopefully uncontroversial observations: First, lineage sorting does in fact occur in nature. Lineage sorting is the process that generates the Darwinian tree of life. Second, lineage sorting can (indeed, must) be either biased or unbiased. The MBL model of the early 1970s shows what unbiased lineage sorting would look like. One thing we might mean when we say that history is contingent is that it involves unbiased sorting processes. Indeed, I suspect that this is one thing that Gould (1989) did mean by contingency. One major theme of *Wonderful Life* is that the sorting of *Baupläne* in the early history of the metazoans was unbiased. Beatty (2006, p. 345) has also observed that Gould seems to be talking about something like "sampling error at the level of lineages." Gould (1993, p. 307) also writes sympathetically about the "random model" of lineage sorting during mass extinction events, though he doesn't use the term 'contingency' in that context.

Some of the puzzles that arose when we tried to understand the second version of Gould's thought experiment disappear if we think of contingency as unbiased sorting. If one were to "rewind" the MBL model and play it back again in a series of trials—and that, in fact, is just what the MBL group did—then the unbiased sorting process would lead to different outcomes on every run, even when the model runs from the same starting conditions. And although it might look like Gould was flirting with causal indeterminism in the second version of his thought experiment (and maybe he was), Millstein (2000) points out that unbiased sorting processes can occur in a metaphysically deterministic universe. A series of fair coin tosses is an unbiased sorting process. But no one thinks it possible to refute determinism by tossing a coin.

At the very least, there is a strong analogy between (a) the distinction between biased and unbiased sorting processes, and (b) the distinction between driven and passive trends. The claim that some historical trend is passive amounts to the claim that history is chancy or unbiased, or more precisely, that changes in one direction are no more or less probable than changes in another. The claim that history is contingent, on the current proposal, also amounts to the claim that history is chancy or unbiased. This suggests that there is a more general issue about whether historical processes are biased or unbiased. Unbiased sorting and unbiased directional change are just two ways in which history can be chancy.

My view is that *both* unbiased sorting and sensitivity to initial conditions are extremely interesting and important concepts in evolutionary biology. And both of these concepts are in play in Gould's classic (1989) discussion. Both are interesting and important because of the way in which they generate empirical questions. Once we appreciate that historical processes can vary with respect to their degree of sensitivity to initial conditions, then it's an empirical question how much sensitivity to initial conditions is exhibited by any particular historical process. Likewise, once we appreciate that sorting processes can be biased or unbiased, then it's a live empirical question whether any particular sorting process is unbiased.

The main claims of this section can be stated without using the term 'contingency' at all: The relationship between the notion of a passive trend and the notion of a process that exhibits sensitivity to initial conditions is not at all straightforward. We cannot explain why a trend is passive by pointing out that the historical processes that generated it are sensitive to initial conditions. By contrast, I'll argue in the next section that the relationship between passive trends and unbiased sorting processes is much more straightforward. I tend to think of both of these notions—unbiased sorting and sensitivity to initial conditions—as different senses of 'contingency', and part of the reason for that is historical. Gould used the term 'contingency' in both ways. Since Gould, however, philosophers and scientists alike have tended more to use 'contingency' to refer to sensitivity to initial conditions. If someone wanted to insist on that usage and to deny that unbiased sorting is really one kind of historical contingency, that terminological disagreement would just require me to reformulate the take-home message of this paper. For those who insist that contingency is sensitivity to initial conditions, the claim would be that it's not quite right to say (without a lot of qualifying and clarifying) that passive trends result from contingent processes. The claim that passive trends result from unbiased sorting processes has much more going for it.

6 Passive Trends and Unbiased Sorting

Substituting "unbiased sorting processes" for "contingent evolutionary processes" in WP yields the following:

WP* Passive evolutionary trends generally result from unbiased evolutionary sorting processes with proper fixed boundary conditions. Driven evolutionary trends generally result from biased evolutionary sorting processes

WP* seems plausible. I won't try to offer a full defense of it here; I just want to suggest that it has some advantages, and that it represents an improvement over WP. To begin with, the second half of WP* coheres extremely well with the standard understanding of driven trends. As noted earlier, most scientists assume that the "driver" of driven trends is cumulative natural selection. And natural selection is readily understood as a biased sorting process. Body size increase will be a driven trend if body size makes a difference to (or biases) the sorting processes going on at the population level.

WP* also preserves the analogy between micro- and macroevolutionary theory that animated some of the efforts of the scientists who participated in the "paleobiological revolution" of the 1970s and 1980s (see Stanley 1975 for an especially clear example). Trends in gene frequencies in a population can be passive or driven. Passive trends in gene frequencies are usually associated with random genetic drift, whereas driven trends are associated with selection. Some philosophers of biology have at times seemed to want to identify selection with driven trends in gene (or trait) frequencies, arguing that is not a cause of evolutionary change but rather a certain sort of statistical trend (Matthen and Ariew 2002; Walsh et al. 2002). One could similarly say that drift just is a passive trend. Or one could say that driven trends in gene frequencies are caused by selection, whereas passive trends are caused by drift. The precise nature of the relationship between selection and drift is much contested. For present purposes, all that matters is that at the micro-level, it's plausible to say that biased sorting generates driven trends in trait frequencies, whereas unbiased sorting generates passive trends.

I argued above that the distinction between passive and driven trends does not map neatly onto the distinction between historical contingency (understood as relative sensitivity to initial conditions) and historical inevitability or convergence. That's because sensitivity to initial conditions comes in degrees. We avoid this problem completely when we think of contingency as unbiased sorting. The distinction between biased and unbiased trends perfectly parallels the distinction between passive and driven trends. Biased sorting does come in degrees, since the strength of the bias can vary, but this, as we saw earlier, is also true of driven trends.

In Sect. 2, I observed that no one had really shown how the contingency of evolutionary processes might be relevant to the explanation of evolutionary patterns and trends. WP* makes this much clearer. Suppose we find that a trend (in a given clade, over a given time interval) is passive. This tells us something about the processes that generated the trend, but we might wish to know more. Why is the trend passive? One part of the answer to that question will be a story about the fixed boundary in the state space. For example, why might there be a minimum size for mammals? Is that lower bound policed by natural selection? Is it due rather to biomechanical or physiological constraints? But even if we knew the cause of the fixed boundary, we would not have the whole story about why the trend is passive. We might also want to know why there is no directional bias in the state space. For example, why are size decreases and increases equally probable? One answer to that question is that the relevant sorting processes are unbiased with respect to size. That might be true of the sorting processes occurring within the relevant populations (at the "micro" level), and it might also be true of the sorting among lineages. If we think of contingency as unbiased sorting, then it turns out that the contingency of history really is relevant to the larger project of explaining patterns in terms of processes.

As noted at the end of Sect. 5, some philosophers might prefer to reserve the term "contingency" for sensitivity to initial conditions. That's just a terminological issue. What I hope to have shown here is that from the perspective of explanation in paleobiology, the concept of unbiased sorting is more helpful than the concept of sensitivity to initial conditions.

7 Conclusion

There are four or maybe five big ideas in macroevolutionary theory, depending on how you count: punctuated equilibria, species selection, historical contingency (a term that may cover two rather different concepts), the distinction between passive and driven trends, and perhaps also idea that evolutionary history is characterized by major transitions (Maynard Smith and Szathmary 1995). To the extent that scientists wish to develop a unified account of how evolution works at large scales, it is important to explore the relationships among these ideas. This is a highly theoretical undertaking to which philosophers of biology may have something to contribute. Here I have tried to shed some light on the connections between two of these big ideas: historical contingency and the distinction between passive and driven trends. The issues are complicated owing to the different senses of contingency. Although it's common—and in many contexts, very useful—to think of contingency as sensitivity to initial conditions, I've argued that we can make more headway by thinking of contingency as unbiased macro-level sorting. Not only does that help illuminate the connection between contingency and passive trends, but it serendipitously highlights a tight conceptual connection with yet another of the big ideas of macroevolutionary theory: species selection. Recall that on some of the looser conceptions of species selection, it is nothing other than biased species sorting. Roughly: Contingency is to species selection as drift is to selection. Eldredge and Gould's theory of punctuated equilibria is also related to these ideas, though as I've argued elsewhere (Turner 2010), that connection is a weaker one that's best understood in terms of theoretical "suggestiveness." The connections between these ideas and the major transitions have yet to be explored (though Calcott and Sterelny 2011 provide a good starting point). The theory of macroevolution that paleontologists have forged and refined over the last couple of decades has its share of unity and elegance.

Beginning in the 1970s and 1980s, in an episode in the history of science that David Sepkoski (2012) has documented, a number of paleontologists self-consciously undertook to show that their discipline had something significant to contribute to evolutionary theory. In order to accomplish that goal, these paleontologists—including Gould, Niles Eldredge, David Raup, Jack Sepkoski, Thomas Schopf, Elisabeth Vrba, Steven Stanley, and others—had to do two things. First, they had to make some theoretical innovations. Because the fossil record is the special purview of paleontology, and the fossil record also happens to be our main source of evidence concerning macroevolution, most of the new theory had to do with macroevolution. Second, they had to show that these new ideas could do some explanatory work that could not be done in any other way, and that you really need punctuated equilibria, species selection, etc., in order to make sense of patterns in the fossil record. One of the things at issue here is the overall explanatory power of macroevolutionary theory. If I am right, and (some of) the theoretical innovations that paleobiology has seen over the last few decades turn out to hang together very well, then it might make sense to talk of a single body of macroevolutionary theory that can be deployed to explain a wide variety of patterns in the historical record.

Acknowledgments I shared earlier versions of this paper at the PSA meeting in San Diego in November, 2012, and at the AAAS meeting in Boston in February, 2013. Thanks to those audiences and especially to John Beatty, Eric DesJardins, Marc Ereshefsky, and David Sepkoski, for helpful feedback. I am grateful to the editors of this volume, Christophe Malaterre and Pierre-Alain Braillard, for their detailed comments on an earlier draft and their help improving the paper. The paper also benefitted from feedback from two anonymous reviewers.

References

Alroy, J. (1998). Cope's rule and the dynamics of body mass evolution in North American fossil mammals. *Science, 280*, 731–734.

Beatty, J. (1995). The evolutionary contingency thesis. In G. Wolters & J. Lennox (Eds.), *Concepts, theories, and rationality in the biological sciences* (pp. 45–81). Pittsburgh: University of Pittsburgh Press.

Beatty, J. (1997). Why do biologists argue like they do? *Philosophy of Science, 64*(4 supp), S432–S443.

Beatty, J. (2006). Replaying life's tape. *Journal of Philosophy, 103*(7), 336–362.

Ben-Menahem, Y. (1997). Historical contingency. *Ratio, 10*, 99–107.

Calcott, B., & Sterelny, K. (Eds.). (2011). *The major transitions in evolution revisited*. Cambridge, MA: MIT Press.

Cleland, C. (2001). Historical science, experimental science, and the scientific method. *Geology, 29*, 987–990.

Cleland, C. (2002). Methodological and epistemic differences between historical and experimental science. *Philosophy of Science, 69*, 474–496.

Cleland, C. (2011). Prediction and explanation in historical natural science. *British Journal for the Philosophy of Science, 62*(3), 551–582.

Conway Morris, S. (2003a). *Life's solution: Inevitable humans in a lonely universe*. Cambridge/New York: Cambridge University Press.

Conway Morris, S. (Ed.). (2003b). *The deep structure of biology: Is convergence sufficiently ubiquitous to give a directional signal*. West Conshohocken: Templeton Foundation Press.

DesJardins, E. (2011). Historicity and experimental evolution. *Biology and Philosophy, 26*, 339–364.

Gallie, W. B. (1959). Explanations in history and the genetic sciences. In P. Gardiner (Ed.), *Theories of history* (pp. 386–402). Glencoe: The Free Press.

Gould, S. J. (1989). *Wonderful life: The Burgess shale and the nature of history*. New York: W.W. Norton.

Gould, S. J. (1993). *Eight little piggies*. New York: W.W. Norton.

Gould, S. J. (1996). *Full house: The spread of excellence from Plato to Darwin*. New York: W.W. Norton.

Gould, S. J. (1997). Cope's rule as psychological artefact. *Nature, 385*(6613), 199–200.

Gould, S. J. (2002). *The structure of evolutionary theory*. Cambridge, MA: Harvard University Press.

Gould, S. J., & Lewontin, R. (1979). The spandrels of San Marco and the Panglossian paradigm: A critique of the adaptationist programme. *Proceedings of the Royal Society B, 205*, 581–598.

Grantham, T. (1999). Explanatory pluralism in paleobiology. *Philosophy of Science, 66*(supp), S223–S 236.

Harries, P. J., & Knorr, P. O. (2009). What does the "Lilliput effect" mean? *Paleogeography, Paleoclimatology, Paleoecology, 284*, 4–10.

Hone, D. W. E., & Benton, M. J. (2005). The evolution of large size: How does Cope's rule work? *Trends in Ecology and Evolution, 20*(1), 4–6.

Hull, D. L. (1975). Central subjects and historical narratives. *History and Theory, 14*(3), 253–274.

Inkpen, R., & Turner, D. (2012). The topography of historical contingency. *Journal of the Philosophy of History, 6*, 1–20.

Jablonski, D. (1997). Body size evolution in cretaceous mollusks and the status of Cope's rule. *Nature, 385*, 250–252.

Jeffares, B. (2008). Testing times: Regularities in the historical sciences. *Studies in History and Philosophy of Biological and Biomedical Sciences, 39*, 469–475.

Lenski, R. E., & Travisano, M. (1994). Dynamics of adaptation and diversification: A 10,000 generation experiment with bacterial populations. *Proceedings of the National Academy of Sciences, 91*, 6808–6814.

Matthen, M., & Ariew, A. (2002). Two ways of thinking about fitness and natural selection. *The Journal of Philosophy, 99*(2), 55–83.

Maynard Smith, J., & Szathmary, E. (1995). *The major transitions in evolution*. Oxford: W.H. Freeman.

McShea, D. W. (1994). Mechanisms of large-scale evolutionary trends. *Evolution, 48*, 1747–1763.

McShea, D. W. (1998). Possible largest-scale trends in organismal evolution: Eight 'live hypotheses'. *Annual Review of Ecology and Systematics, 29*, 293–318.

McShea, D. W. (2005). The evolution of complexity without natural selection: A possible large-scale trend of the fourth kind. *Paleobiology, 31*(supp), 146–156.

McShea, D. W., & Brandon, R. N. (2010). *Biology's first law*. Chicago: University of Chicago Press.

Millstein, R. (2000). Chance and macroevolution. *Philosophy of Science, 67*(4), 603–624.

Radick, G. (2005). Other histories, other biologies. *Royal Institute of Philosophy Supplements, 80*(56), 21–47.

Raup, D. M., & Gould, S. J. (1974). Stochastic simulation and the evolution of morphology—towards a nomothetic paleontology. *Systematic Zoology, 23*, 305–322.

Raup, D. M., Gould, S. J., Schopf, T. J. M., & Simberloff, D. (1973). Stochastic models of phylogeny and the evolution of diversity. *Journal of Geology, 81*, 525–542.

Rosenberg, A., & McShea, D. W. (2007). *Philosophy of biology: A contemporary introduction*. New York: Routledge.

Sepkoski, D. (2012). *Rereading the fossil record: The growth of paleontology as an evolutionary discipline*. Chicago: University of Chicago Press.

Sober, E. (1988). *Reconstructing the past: Parsimony, evolution, and inference*. Cambridge: MIT Press.

Stanley, S. (1975). A theory of evolution above the species level. *Proceedings of the National Academy of Sciences, 72*(2), 646–650.

Sterelny, K. (1996). Explanatory pluralism in evolutionary biology. *Biology and Philosophy, 11*, 193–214.

Sterelny, K. (2005). Another view of life. *Studies in History and Philosophy of Biology and Biomedical Sciences, 36*, 585–593.

Travisano, M., Mangold, J. A., Bennett, A. F., & Lenski, R. E. (1995). Experimental tests of the roles of adaptation, chance, and history in evolution. *Science, 27*(5194), 87–90.

Tucker, A. (2004). *Our knowledge of the past: A philosophy of historiography*. Cambridge: Cambridge University Press.

Turner, D. (2007). *Making prehistory: Historical science and the scientific realism debate.* Cambridge: Cambridge University Press.

Turner, D. (2009). How much can we know about the causes of evolutionary trends? *Biology and Philosophy, 24*, 341–357.

Turner, D. (2010). Gould's replay revisited. *Biology and Philosophy, 26*, 65–79.

Turner, D. (2011). *Paleontology: A philosophical introduction.* Cambridge: Cambridge University Press.

Walsh, D. M., Ariew, A., & Lewens, T. (2002). The trials of life: Natural selection and random drift. *Philosophy of Science, 69*(3), 452–473.

Wang, S. C. (2001). Quantifying passive and driven large-scale evolutionary trends. *Evolution, 55*(5), 849–858.

Chapter 5
Developmental Noise: Explaining the Specific Heterogeneity of Individual Organisms

Francesca Merlin

Abstract Recent research in molecular developmental biology has shown that the stochastic character of development (i.e., developmental noise) can produce phenotypic heterogeneity even in the absence of any other source of change (genetic and environmental). More precisely, developmental noise triggers phenotypic heterogeneity amongst the members of a clonal population (synchronic heterogeneity) and even within an individual organism over time (diachronic heterogeneity), in a stable and homogeneous environment. This paper deals with such stochasticity in order to explore its epistemological relevance and role, both as *explanans* and as *explanandum*. First, I investigate whether developmental noise is part of the explanation of the physical characteristics of individual organisms (i.e., the phenotypic outcome of development). Then, I try to assess whether or not heterogeneity due to stochastic events in development can be explained by a selective-evolutionary history. My final aim is to argue for the two following theses. First, from the developmental point of view, I argue that developmental biologists need to take into account developmental noise in order to explain the uniqueness of each individual organism and its own heterogeneity over time, at the phenotypic level at least, that genetic and environmental changes cannot explain alone. Second, from the evolutionary point of view, I critically evaluate explanations of developmental stochasticity in term of adaptation, in particular the idea that noise is a trait that has been selected to increase the capacity of natural populations to evolve ("evolvability"). Then, I identify other ways in which biologists should try to explain developmental noise. I conclude by highlighting the limits of any univocal explanatory approach in biology.

Keywords Developmental noise • Explanation • Stochasticity • Development • Individual heterogeneity • Evolvability • Adaptationist program

F. Merlin (✉)
CNRS UMR 8590, IHPST – Institut d'histoire et de philosophie des sciences et des techniques, Paris, France

Université Paris 1 Panthéon-Sorbonne, Paris, France
e-mail: francesca.merlin@gmail.com

© Springer Science+Business Media Dordrecht 2015
P.-A. Braillard, C. Malaterre, *Explanation in Biology*, History, Philosophy and Theory of the Life Sciences 11, DOI 10.1007/978-94-017-9822-8_5

1   Introduction

Recent research in molecular developmental biology has started to acknowledge the stochastic character of development, also called "developmental noise", showing that it can produce phenotypic heterogeneity even in the absence of any other source of change (genetic and environmental). More precisely, developmental noise triggers phenotypic heterogeneity amongst the members of a clonal population (synchronic heterogeneity) and even within an individual organism over time (diachronic heterogeneity), in a stable and homogeneous environment (Johnston and Desplan 2010). The present paper deals with such stochasticity in order to explore its epistemological relevance and role, both as *explanans* and as *explanandum*, respectively from the developmental and the evolutionary point of view.

I will address the two following questions. The first is about whether developmental noise participates in the explanation of the characteristics of individual organisms. In other words, is developmental noise part of the *explanans* of why each individual organism is as it is? In light of recent advances in molecular developmental biology, the answer seems to be yes. But, more specifically, what does developmental noise account for? I will address this more specific question as well.

The second question is about developmental noise as *explanandum* (i.e., the thing to be explained); more specifically, it is about the hypothesis of an explanatory link between developmental noise and the capacity of living systems to evolve (evolvability). In this context, I will try to assess whether or not heterogeneity due to stochastic events in development could provide an advantage at some level of biological organization (genes, organisms, groups, etc.), and so be explained by a selective-evolutionary history. This will lead me to critically evaluate current predominant explanations of developmental noise in terms of adaptation, and to point to other possible kinds of explanation biologists could and should appeal to in order to account for it.

The paper is organized as follows. I start by briefly introducing recent results of research on developmental noise in multicellular and unicellular organisms. In fact, biologists have started to talk of a developmental process, broadly speaking, for colonies of unicellulars as well. I then examine whether or not noise is part of the *explanans* of the developmental outcome. Finally, I critically evaluate noise as an *explanandum* of the theory of evolution by natural selection. I conclude the paper by underlining the limits of any univocal explanatory approach in the life sciences.

2   Past and Recent Research on Developmental Noise

What do biologists refer to when they talk about developmental noise (developmental stochasticity or instability)? A first, loose definition can be formulated as follows: little random perturbations occurring during the developmental process, affecting its normal progression and its outcome at the organismic level, i.e., the physical characteristics of the individual organism (its phenotype). These perturbations are hard to predict and are not due to genetic nor environmental changes, but are internal to the individual organism concerned. Note that this definition does not say anything about the origin of these internal random perturbations of development.

I do not intend to provide here a revisited account of development, even though it would be interesting to evaluate traditional definitions of this process.[1] Rather, in this context, I uncritically assume the classical view of what development is: the set of molecular and cellular mechanisms initiating and constructing the individual organism, from a single cell (the zygote) to the adult. This definition of development applies to multicellular organisms. Biologists have started talking about the development of unicellular organisms, too. More precisely, after Shapiro's proposal to talk about bacterial multicellularity in 1998, which was based largely on the study of colony morphology (Shapiro 1988), the discovery of cell-cell communication mechanisms (e.g., "quorum sensing") has led more and more microbiologists to think about unicellular colonies as multicellular organisms characterized by elaborate morphological architectures (Shapiro and Dworkin 1997; Aguilar et al. 2007).

Biologists started investigating the phenomenon of developmental noise at the end of the first half of the twentieth century. In particular, in the 1950s and 1960s, the observation of morphological asymmetries between the two sides (left and right) of bilateral organisms – for instance, differences between the number of bristles on the left and on the right side of the body in drosophila (Wright 1952; Reeve and Robertson 1953; Thoday 1956; Latter 1964) and between the geometrical patterns on the left and on the right wing of butterflies (Mason et al. 1967; Soulé 1982) – led them to formulate the following hypothesis: as these differences, which they called "fluctuating asymmetries", were not due to genetic nor environmental modifications, they must be produced by some internal perturbation or instability affecting the developmental process. Indeed, observations and experimental results were based on isolated singular individual organisms or groups of clonal organisms in homogeneous and stable environments.[2]

From the 1970s, and on a larger scale starting from the end of the 1980s, results of (stochastic) mathematical modeling and computer simulation of the developmental processes came to support the hypothesis that "fluctuating asymmetries" in morphological characteristics of clonal organisms, and even between specular parts of the same organism, are due to molecular random events taking place inside the cells. More precisely, microscopic random events affecting the process of gene expression are considered one of the most important components of developmental noise and of the variation it produces at the phenotypic level (Rigney and Schieve 1977;

---

[1] For a critical analysis of traditional definitions of development, see Pradeu et al. (2011), Minelli and Pradeu (2014).

[2] For a complete bibliography of the study of developmental noise, see Markow 1993; Palmer 1996; Polak 2003; Hallgrimsson and Hall 2005.

Rigney 1979; Berg 1978). Gene expression is the mechanism whose final result is the production of different types of proteins that are the building blocks of living organisms: it is involved in cell fate decision and, more broadly, in the developmental process.

From the 1990s until now, the hypothesis that the origin of developmental noise lies in microscopic random events affecting gene expression has been massively corroborated by theoretical and experimental research and is now well warranted (McAdams and Arkin 1997; Samoilov et al. 2006; Maheshri and O'Shea 2007; Johnston and Desplan 2010; see also Klingenberg 2005), but is still to be tested in certain specific cases. Nowadays, a slightly more precise definition of developmental noise is the following: non-genetic, non-environmental random fluctuations taking place at different levels inside the individual organism (at the level of molecules, cells, tissues, etc.), in particular at the intracellular level of the process of gene expression that is involved in cell fate decision (differentiation). These random fluctuations affect the normal progress of development and its phenotypic outcome.

What does happen inside the cell during gene expression (protein synthesis) and how do biologists deal with fluctuations affecting this process?

The two main steps of gene expression are the following: (1) transcription, i.e., the synthesis of a variable quantity of RNA molecules from a single DNA strand and (2) translation, i.e., the synthesis of a variable quantity of proteins from single ARN strands. Each of these two steps includes many intermediary stages consisting in specific physico-chemical interactions between two (or more) molecular species.

A variety of microscopic events can take place during transcription and translation: they produce stochastic and unpredictable fluctuations, and so have an impact on gene expression, in particular on the number of proteins produced. For instance, errors can occur during the transcription of DNA to RNA and during the translation of RNA to proteins. Moreover, during transcription, genes are activated and repressed by the association-dissociation of transcription factors, which occurs randomly, depending on the concentration, the localization, and the state of different molecular species inside the cell. Indeed, all these variables change randomly and in an unpredictable way over time because of the following reasons. First, the fact that a low number of molecules are involved in gene expression and the non-homogeneous character of the intracellular environment produce time delays in some steps of the process. Second, the fact that the molecules in the intracellular milieu are always in thermal agitation, i.e., continuously moving around and colliding randomly, is another source of such random and unpredictable changes. Third, quantum fluctuations could also have some effects at the molecular-biochemical level and so produce stochasticity in gene expression.

As I have noted above, biologists started to take noise in gene expression into account in the 1970s, even though the phenomenon of noise was well known since the beginning of the twentieth century, at least in physics. This has led biologists to acknowledge the limits of deterministic models of intracellular processes (e.g., gene expression) taken from deterministic chemical kinetics and to complement them with probabilistic models taken from stochastic chemical kinetics (Gillespie 2007) and with stochastic simulations too (see Fig. 5.1).

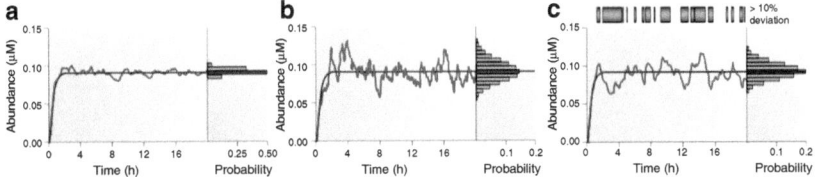

Fig. 5.1 The *broken line* in this figure (From Kaern et al. 2005, Reprinted by permission from Macmillan Publishers Ltd: Nature Reviews Genetics, copyright (2005)) represents time series of proteins concentrations generated from stochastic simulations taking into account noise in gene expression and the resulting variation around the average protein abundance produced in an individual cell over time. The *straight line* represents the result of deterministic simulations. (**a**) Low amplitude fluctuations in protein abundance; (**b**) & (**c**) Large fluctuations in protein abundance due to a lower level of ARNs expression

Nowadays, they use the expression "noise in gene expression" (or "intracellular noise") to refer to microscopic random events taking place inside the cell and producing fluctuations in one or more steps of this molecular intracellular process. More precisely, biologists talk about noise in gene expression to designate the random variation around the mean value of this process (i.e., around the average amount of proteins produced inside the cell, in the absence of any genetic and environmental change). They statistically measure it, calculating the coefficient of variation of protein abundance (i.e., the standard deviation by the mean) at the populational level at some point in time (synchronically) or at the individual level during a certain lapse of time (diachronically).

It is also interesting to say a few words about the consequences of noise in gene expression and so, at a higher level, of developmental noise (Raser and O'Shea 2005). Broadly speaking, noise produces phenotypic differences in isolated individual organisms over time or between members of clonal populations, even in the absence of genetic and environmental modifications. For instance, look at the two cats in the figure below (Fig. 5.2).

They are clones (they have the same genome), they developed in the same environment, but are phenotypically different. Why is it so? The answer given by researchers working on developmental noise is that the phenotypic differences between these two cats are due to stochastic fluctuations in the expression of some of their genes (i.e., noise).

Developmental noise is currently the object of a large amount of studies performed by biologists and by physicists as well.[3] All investigate, from bacteria to humans, the fact that some cell fate decisions are made at random (or stochastically) due to microscopic stochastic events taking place inside the cell and affecting gene expression. Let us briefly look at two representative studies as examples.

---

[3]For a review, see Johnston and Desplan (2010).

Fig. 5.2 Example of possible stochastic influences on the phenotype (From Raser and O'Shea 2005, Reprinted by permission from the author): the first cloned cat and its genetic mother display different coat patterns and personality

The first is about what biologists call the "bet-hedging" competence in bacteria *Bacillus subtilis*. Competence is the ability of taking up DNA from the environment and incorporating it into the chromosome by recombination. Theoretical and experimental research (for instance, see Maamar et al. 2007; for a review, see Johnston and Desplan 2010) has shown that, in *B. subtilis* colonies, each bacterium stochastically "choose" the competence or non-competence state. What does the expression "stochastic" mean in this context? It refers to the fact that the cell state depends on the level of the expression of some gene that randomly fluctuates from cell to cell, and even in the same cell over time. In the specific case of *B. subtilis* competence state, it depends on the expression of the *comK* gene, involved in the production of the binding protein *ComK*. More precisely, it is the number of *ComK* proteins produced that randomly varies. So, as far as the cell state (competent or not) depends on the fluctuating abundance of this protein (indeed, above a certain threshold, the cell becomes competent), the choice is made stochastically: it is not univocally determined but the competent or non-competent state can happen with some specific probability. That is why, in a colony of isogenic *B. subtilis*, in a homogeneous and stable environment, we can observe competent and non-competent cells (see Fig. 5.3).

The second study is about the development of *Drosophila melanogaster*'s color vision system. Drosophila is characterized by compound eyes, which are composed of multiple units (around 800) called "ommatidia". In each ommatidium, a stochastic "choice" is made in one of the eight photoreceptor cells to become one of two possible cell subtypes. For instance, the two ommatidial subtypes Rh3 ("pale") and Rh4 ("yellow") (Fig. 5.4a, on the left), are stochastically distributed in R7 neurons; the subtypes Rh5 ("blue") and Rh6 ("green") (Fig. 5.4a, on the right), are stochastically and exclusively distributed in R8 neurons and their expression is coupled to the subtype specification in R7.

Fig. 5.3 Stochastic distribution of cell states in *B. subtilis* bacteria (From Losick and Desplan 2008, Reprinted with permission from AAAS). The fluorescence reveals the subpopulation of bacteria that are competent (they are active for the expression of *comK* gene)

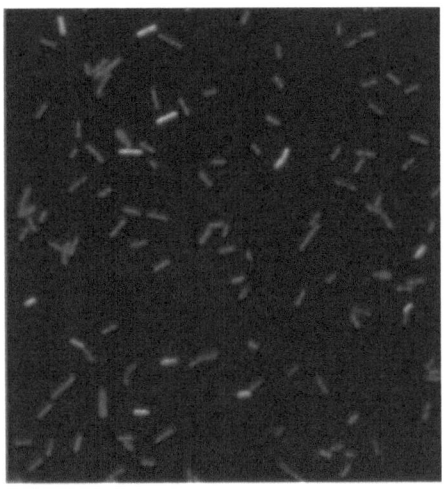

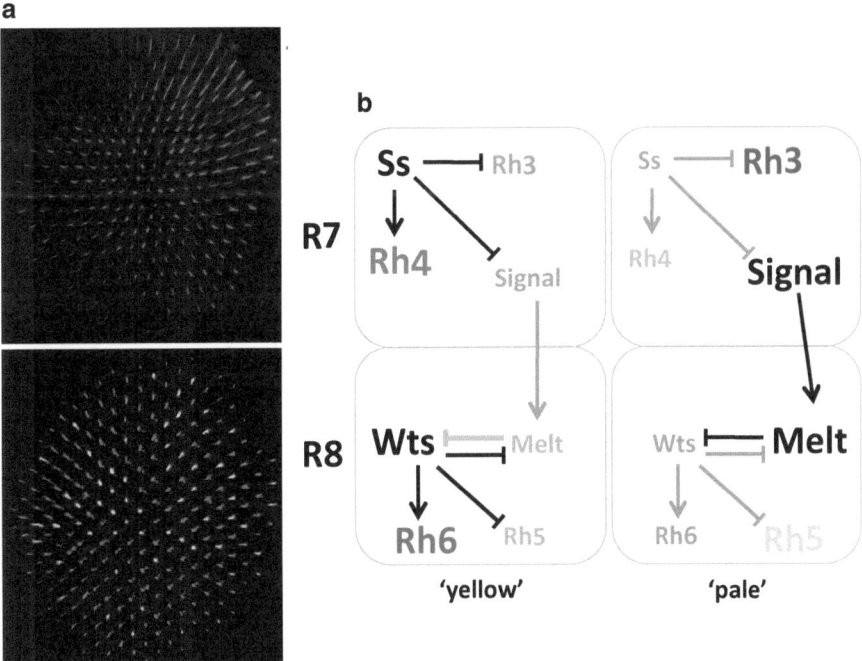

Fig. 5.4 Stochastic distribution of cell states in drosophila photoreceptors (© Dr. Brent Wells and Dr. Jens Rister). On the *left*, stochastic distribution of Rh3 and Rh4 in R7 neurons and of Rh5 and Rh6 in R8 neurons. On the *right*, stochastic differentiation of ommatidia in R7 neurons and its influence on the differentiation of ommatidia in R8 neutrons

It has been shown (e.g., see Wernet et al. 2006; Bell et al. 2007; for a review, see Johnston and Desplan 2010) that variation in the expression of some specific genes involved in the production of sensory visual receptors (Rh3 and Rh4 in R7, Rh5 and Rh6 in R8) has a causal role in the differentiation of the cells composing drosophila's compound eyes. For instance, in the case of R7 subtype specification, the differentiation of each ommatidium into the yellow or the pale subtypes depends on stochastic fluctuations in the expression of Rh3 and Rh4 genes. Then, the stochastic ommatidial subtype decision made in R7 is conferred upon R8 and comes to specify the distribution of the blue and green subtypes in R8 neurons (Fig. 5.4b, at the bottom). More precisely, in the pale subtype, Rh3 is expressed in R7 coupled with Rh5 expression in R8. In the yellow subtype, Rh4 is expressed in R7 coupled with Rh6 expression in R8. Thus, as in the case of the "bet-hedging" competence in bacteria, the cell state "decision" depends on random fluctuations in gene expression (i.e., noise).

So, why is it important to investigate the explanatory role of noise and its explanation, both from the developmental and the evolutionary point of view? Because, as shown by these two studies, developmental noise can produce "new" variation randomly ("new" in the sense of "ignored before" the introduction of stochastic modeling in molecular biology at the end of the 1970s), variation which is more easily reversible (and so less risky for the organisms concerned) and less stable than variation due to genetic changes, but still potentially important both from the point of view of the fate of the individual organism and/or of the population it belongs to. In other words, developmental noise has been shown to play an important role in the origin of differences in the phenotypic characteristics of each individual organism over time and in the origin of phenotypic variation in populations, even when they are isogenic and in homogeneous and stable environments. This is a big novelty with respect to the traditional way to account for the origin of phenotypic variation (by genetic changes or by the environment).

These considerations naturally lead us to the first question I want to address in this paper, i.e., to consider noise as part of the *explanans* of the developmental outcome.

3   The Explanatory Role of Developmental Noise

Does developmental noise have any explanatory role? Is it part of the explanation of the physical characteristics of the individual organism, that is, the explanation of why each individual organism is as it is? One way to answer this question is to assess whether developmental noise plays any particular role in the way a developmental system develops, in other words, whether noise is part of it. Oyama (2000) in particular defines a developmental system (DS) as containing all the factors ("developmental resources") explaining the construction of the characteristics of the individual organism, i.e., one isolable organism. This includes both individual traits that are common to other organisms (e.g., species-specific traits) and unique traits (note that "individuality" is different from "uniqueness"). It is clear that Oyama epistemologically defines the DS in terms of explanation.[4] The question at stake here is the following: is developmental noise included in such a system of explanatory factors of development?

Let us give a simple but meaningful example in order to understand which kind of developmental resources belong to the DS of an individual organism. For instance, let us look at my DS, that is, the set of factors explaining why I am as I am. My genes belong to it, as they are involved in the production of proteins inside the cells of my body. My mother's immune system is part of it too, because it was transmitted to me when I was a fetus in her womb and triggered the development of my own immune system. With Oyama, we can also say that the "surgeon's knife" also belongs to my DS because it explains the scar I have on my knee. And even Chopin is part of my DS because it is one of the resources explaining why I am as I am, in particular, why I'm sensitive to a certain kind of music (i.e., romantic music). The question at stake here is the following: is developmental noise part of my DS (according to Oyama's account) and, more generally, of the DS of an individual organism? And, if that is the case, what does it explain?

In the light of recent research advances on developmental noise, the answer to the first part of the question seems to be: definitely yes! In fact, noise has been shown to be at the origin of phenotypic random variation we can observe in individual organisms over time and in populations of organisms, even in the absence of genetic and environmental changes. But what does developmental noise account for more precisely? I argue that it contributes to explanations of the phenotypic uniqueness of each individual organism, i.e., its unique characteristics or traits (remember that, as I said before, an individual organism is characterized both by common traits, e.g., species-specific, and by unique traits). In other words, with genetic and environmental differences, noise explains why two individual organisms are not phenotypically identical to one another. I would say that it even provides the ultimate explanation of that fact as far as it is the only factor that can account for the phenotypic uniqueness of individual organisms in the absence of any genetic and environmental difference amongst them (e.g., as in the case of two or more clonal organisms developing in a homogeneous and stable environment).

The explanatory role of developmental noise is clearer if we look at the way biologists investigate it. Remember that they use single individual organisms that do not genetically change over time and are in a stable environment or isogenic populations in a stable and homogeneous environment. In this context, they observe phenotypic variation (over time in individual organisms or at some point in time in populations). How can they account for it? Not by invoking gene mutation

---

[4]For a critical evaluation of Developmental Systems Theory, in particular its formulation by Oyama, on the one hand, and Griffiths and Gray, on the other hand, see Barberousse et al. (2009).

nor environmental changes, but developmental noise (i.e., stochastic fluctuations perturbing the normal progression of the development of each individual organism). In other words, we need to include developmental noise in the DS if we want to explain such a variation and, more precisely (from the point of view of development), both the unique characteristics of each individual organism and its inherent heterogeneity over time.

One could ask why it is important to account for the uniqueness of each isolated individual organism. Does it have to be one of the tasks of developmental biologists? Traditionally, variation between individual organisms is not of interest, and is even an annoyance, in developmental biology. As Lewontin (2000, p. 9–10) claims, the focus is rather "the set of mechanisms that are common to all individuals and preferably to all species" and are at the origin of their common traits. In other words, as he critically continues: "Developmental biology is not concerned with explaining the extraordinary variation in anatomy and behavior even between offspring of the same mother and father, which enables us to recognize individuals as different. Even the large differences between species are not within the concerns of the science. (. . .) The concentration on developmental processes that appear to be common to all organisms results in a concentration on those causal elements which are also common" (p. 10). Nevertheless, as far as unique (phenotypic) characteristics can be relevant for the fate of each individual organism – from the developmental perspective and, as we will see later, in the evolutionary fate of the natural population the organism belongs too – I maintain, with Lewontin and Oyama, that biologists involved in developmental issues have to account for the uniqueness of individual organisms in order to fulfill their task in a satisfactory way, i.e., in order to account for the construction of the characteristics of an individual organism throughout its life. I argue that biologists should integrate developmental noise into the DS of the individual organism to account for its characteristics (both common and unique).

Last but not least, taking into account developmental noise and the "new" ("ignored before") phenotypic variation it can produce revives the traditional task of developmental biology, i.e., explaining the common characteristics of individual organisms and the common set of mechanisms at their origin. More explicitly, research on developmental noise could represent a challenge for the question of the regularity (e.g., morphological) of developmental outcomes across organisms belonging to the same species despite stochastic and unpredictable events perturbing the development of each individual. In other terms, the fact that microscopic random events continuously take place inside individual organisms and affect the course of their development renews the question of how it comes about that different organisms share a certain number of common (species-specific) characteristics. Moreover, research on developmental noise calls us to reconsider the explanation of developmental canalization, defined by Waddington (1942) as genetically controlled, and more broadly, of developmental robustness. More precisely, it represents a further challenge for the explanation of how the process of development is able to pass through all its stages, maintaining the function of each of them, regardless of variability due to internal (i.e., noise) and external perturbations.[5]

Before turning the focus on developmental noise as an *explanandum*, let us look at some consequence, apparently problematic, of considering it as part of the *explanans* of the unique characteristics of each individual organism. Indeed, one may object that, as far as developmental noise (i.e., stochasticity or, in other words, chance) plays an explanatory role, chance is considered an *explanans* of the developmental outcome. Can it really be the case? Can chance play an explanatory role? The answer to this question depends on what chance is; in other words, on the way chance is defined.

Do biologists characterize developmental noise as stochastic (chancy, random) fluctuations because they do not know the causal origin of such fluctuations (their causes), and so they cannot precisely predict them but only make predictions in probabilistic terms? Or, on the contrary, does it mean that such fluctuations are fundamentally random, i.e., non deterministic, and so the probability biologists attribute to them are objective? In other words, does stochasticity refer to their ignorance about the variety of factors perturbing the developmental process (i.e., some subjective chance), or does it refer to some specific property of the mechanisms involved in development (i.e., some objective chance)?

I suggest that if the stochastic character of noise refers to some kind of subjective chance (e.g., chance as the ignorance of the causes of developmental perturbations and their phenotypic effects), noise does not seem to be a good candidate to explain such perturbations and their consequences at the level of the developmental outcome (i.e., the characteristics of individual organisms). The underlying causes, once discovered, would play such an explanatory role.

This issue becomes more delicate if the stochastic character of noise refers to some objective chance. Indeterminism at the quantum level, often called "pure chance", is a typical example of it. At first glance, it seems possible to attribute an explanatory role to developmental noise because chance is conceived here as a property of fluctuations affecting development (e.g., quantum fluctuations at the molecular-biochemical level), that is something happening in the natural world preceding and producing some further event.

However, with Paul Humphreys (1989), some could object that chance, and the probabilities used to mathematically describe it, leaves a gap in the explanation, a gap that cannot be fulfilled because it is by definition, in the case of objective chance,

---

[5]Suggesting that noise could also play the role of "capturing some typical features, stages, or mechanisms of development", an anonymous reviewer pointed at the fact that developmental noise is common to all organisms: rather than a challenge, it could represent an additional, relevant, element for the explanation of developmental processes that different organisms have in common. Moreover, it could provide an explanation of why development is so robust to external perturbations. For instance, it could be argued that, as source of internal variability, developmental noise could provide living systems with a certain amount of flexibility, allowing them to cope with external disturbances, and so reliably develop regardless of environmental contingencies.

a gap of indetermination in the sense of absence of causation. And if explanation is conceived in causal terms, as Humphreys does, this implies that chance cannot explain anything.

In response to such an objection, one could deny that biological explanations are causal. However, it seems difficult to find convincing arguments in favor of such thesis, especially in the light of the large amount of publications in the last 10 years that argue for (causal) mechanistic explanation as the distinctive kind of explanation in biology and, more specifically, in molecular biology (e.g., see Machamer et al. 2000; Tabery 2009).

A more promising strategy to face Humphrey's objection consists in showing that objective chance does not necessarily correspond to the idea of absence of causation. Actually, the history of chance reveals that this idea is just one way to conceive it. Furthermore, one could argue that defining chance as the absence of causation is misleading. Amongst others, Millstein (2011) makes this point by using the example of radioactive decay, which is usually considered a fundamentally indeterministic phenomenon of the quantum level (in other terms, an event without any cause at its origin). She argues that the half-life of a given element (for instance, uranium-238), and so the fact that it decays at a particular moment in time, are *caused* by the structure of the atom in question. Thus, she argues that objective chance does not correspond to the absence of causation, even when it refers to something happening at the quantum level.[6]

In the same article, Millstein shows that chance can be characterized in causal terms by considering seven different notions of it. This represents a further, powerful argument against Humphreys' idea that chance is not a cause and, that therefore cannot provide causal explanations. Actually, according to Millstein, the definition of each notion of chance consists in identifying different types of causes she calls "considered" causes, "ignored" causes, and "prohibited" causes. Let us take as an example the specific (biological) notion of "evolutionary chance", which is at the core evolutionary theory since Darwin (Merlin 2010), in order to understand Millstein's thesis. To say that a genetic mutation occurs by "evolutionary chance" means that, starting from a specified sub-set of causes, the mutation can be advantageous, disadvantageous, or neutral for the organism concerned in a given environment. The considered subset of causes includes any cause that does not proceed primarily in an adaptive direction ("considered" causes) and excludes causes that, on the contrary, act in an adaptive way and so would produce directed mutations ("prohibited" causes). Moreover, some causes can be ignored, for instance, for pragmatic reasons, but still belong to the subset of causes considered. Thus, the notion of "evolutionary chance" is defined in causal terms – primarily by the causes it forbids – and so can play an explanatory role, i.e., accounting in causal terms for the fact that genetic mutations do not occur in an directed adaptive way.

---

[6]For a direct critical argument against Humphreys' position in the specific case of the explanatory role of random genetic drift, see Millstein (1997).

I maintain that Millstein's argument for the causal characterization of every notion of chance represents a good argument against Humphrey's denial of any role of chance in causal explanations and so a good point for attributing explanatory relevance to it. Therefore, developmental noise can play a role in the explanation of why each individual organism is as it is, in particular accounting for its unique traits and its intrinsic heterogeneity over time.

4 The Explanation(s) of Developmental Noise

Let us now look at the explanation of developmental noise, in particular at noise as *explanandum* of the theory of evolution by natural selection. In recent studies, biologists tend to treat and explain developmental noises as an adaptive trait whose presence is due to evolution by natural selection. More precisely, we can find the two following accounts of it. First, sometimes developmental noise is interpreted as a by-product of selection for something else, another trait which plays a causal role in differential survival and reproduction (the selective process). This is the case of noise at the origin of the "bet-hedging" competence in *B. subtilis* bacteria (see Sect. 1). In other words, the idea is that there is a correlation, necessary or contingent, between the selected trait ("selection for") and noise ("selection of") (Sober 1984). The second account holds that, other times, developmental noise is conceived as an adaptation for what biologists call "evolvability", literally speaking, the capacity of living organisms to evolve. Actually, by providing a certain amount of random variability, developmental noise could enhance the chances of natural populations to survive and reproduce over time, in particular when the environmental conditions fluctuate in a certain way. Adopting the "insurance hypothesis", which comes from the diversity-stability debate in community ecology (see Diamond and Case 1986), it could be considered as a sort of "insurance policy" or "pre-adaptation" to environmental contingencies (Thattai and Oudenaarden 2004).

Are such explanations of developmental noise in terms of a selective-evolutionary history, in particular, the second one in terms of evolvability, legitimate and well warranted or, on the contrary, do they represent the kind of questionable explanations characterizing the adaptationist program? With the expression "adaptationist program", I refer to Gould and Lewontin's critique (1979) of the following evolutionary trend: the tendency biologists have to consider each trait separately from the others, attributing to it a function, and to invent a selective (adaptive) history – what Gould and Lewontin critically call "just-so-stories" – in order to explain its presence. I argue that's exactly what actually happens in the study of developmental noise.

Here are two examples of this tendency. The first is a study of intracellular noise in yeast *Saccaromyces cerevisiae*. Fraser and his colleagues (2004) have evaluated the noise level in the expression of each gene of this unicellular organism. On this basis, they have observed that noise in the expression of essential (housekeeping) genes and of genes involved in the production of multi-proteins complexes is lower than noise affecting genes with no vital role. They conclude that noise in gene expression is subject to natural selection, is detrimental to organismal fitness, and so its minimization in the expression of important genes is a selected adaptation.

The second study is about noise in *E. coli* bacteria. Kussel and Leibler (2005) have examined the switch mechanism from antibiotic-sensible to antibiotic-resistant phenotype: in this way, clonal individual bacteria can become resistant to antibiotics even in the absence of any mutational event and so in a more easily reversible way. Kussel and Leibler have theoretically and experimentally shown that such a switch mechanism is due to noise in the expression of some particular genes and mimics the distribution of environmental changes over time. More precisely, they maintain that, when the environment changes infrequently, the stochastic switching due to noise is likely to be favored (i.e., selected) over a different strategy consisting in sensing the environmental changes and reacting to them.

These two studies provide two clear examples of the presence of adaptationist explanations in research on developmental noise: actually, their common conclusion is that noise in gene expression is an isolated trait and is likely to evolve by natural selection. In other words, both of these two groups of biologists suggest an explanation of noise in terms of a selective-evolutionary history by considering noise from the point of view of the advantage/disadvantage it could provide to the living system concerned, i.e., from the point of view of its adaptive value.

If this kind of explanation is right, it follows that the conditions for the principle of evolution by natural selection to occur (Lewontin 1970) apply in the case of "developmental noise" conceived as a trait. More explicitly this means that, first, in a population of organisms there's variation in terms of developmental noise, i.e., differences between organisms in the level of fluctuations perturbing their developmental process (phenotypic variation). Second, organisms characterized by different levels of developmental noise have different survival and reproductive propensities (fitness differences) or, in causal terms, differences in developmental noise cause differences in survival and reproduction among organisms. Third, there is a correlation between differences in the level of developmental noise of parents and of offspring (heritability).

Do these three conditions apply in the case of developmental noise considered as a trait? And, more generally, are selective-evolutionary explanations of developmental noise well grounded (and even possible), both from the theoretical and the experimental point of view? I maintain that the answer to this question depends on a certain number of issues, fundamental in the evolutionary context, that have rarely been addressed by biologists.

The first issue is about the level of selection: at which level does the selective process at the origin of noise take place? This is the question of the units of selection, which can be, for instance, a gene, an individual organism, or a group. This first question has to come with the identification of the entity that is the beneficiary of the adaptation. In fact, the beneficiary does not necessarily correspond to the biological unit of selection. As an example, selection for some level of developmental noise can take place at the level of individual organisms (some of them survive and reproduce more than others because of the selective advantage provided by the characteristic noise of the expression of their genes); nevertheless, at the same time, the real beneficiary of the adaptation can be the group because the presence of variation among the individual organisms composing it (the phenotypic variants due to developmental noise) enhances its chances to persist and to evolve more rapidly in changing environmental conditions.

Another fundamental question in order to legitimately suggest a selective-evolutionary explanation of developmental noise is about whether developmental noise is a heritable trait, that can be transmitted from one generation to the next and, if so, how (genetically, epigenetically, etc.). This sends us back to the third condition for evolution by natural selection to take place. In fact, if a certain level of noise is the result of evolution by natural selection, it means that it has to be heritable over generations and likely to be cumulatively selected over time. The question of the heritability of developmental noise is rarely addressed; nevertheless, note that some evidence in favor of this hypothesis has been produced recently (e.g., see Kaufmann et al. 2007).

More specifically, if we consider developmental noise as a result of selection at the level of groups and not at the individual level, other fundamental issues should be addressed. First, we need to show that the selective process at the level of groups can really take place and be effective. This is an empirical question, of course, and it needs to be addressed each time one argues that some particular trait (e.g., developmental noise) is the result of evolution by natural selection at the group level. In fact, group selection is effectively possible if a certain number of conditions are fulfilled. For instance, the advantage that the trait provides to the group has to be higher than any conflicting selective pressure at the individual level; in other words, the selective pressure between the individual organisms of the group cannot be stronger that the selective pressure between groups (Sober and Wilson 1998). Moreover, remember that group selection is still a controversial notion in biology, even though in the 1960s–1970s it was unanimously rejected (Williams 1966). This is another reason to address this question.

Second, and more particularly, if we want to explain developmental noise as an adaptation for evolvability, we also need to answer the question of what evolvability is and, above all, the question of how there can be selection for a capacity that does not provide any immediate advantage but may potentially in the long term and at the group level (as far as populations, and not individual organisms, evolve). Many different and ambiguous definitions of what evolvability is have been suggested during the last 20 years (Pigliucci 2008): it would require another paper to introduce and analyze them in detail. With regard to the aim of the present paper, just note that there is no real consensus on this notion: this makes the issue of evolvability a real urgent one, in particular if biologists want to account for developmental noise as an adaptation for evolvability.

In sum, I argue that biologists cannot legitimately account for developmental noise by formulating a selective-evolutionary history if they have not addressed and answered the questions I have just mentioned (there are surely others I've not pointed out). If it is not the case, their explanations, in particular explanations appealing to evolvability, cannot escape the qualification of "just-so-stories", exposing themselves to the vigorous critique addressed to the adaptationist program more than 30 years ago (Gould and Lewontin 1979). This is, I maintain, the weakness of predominant (adaptive) accounts of developmental noise: they are not legitimate explanations in terms of adaptation because they are not warranted enough.

Are there alternative explanations to developmental noise and, if so, do they have to be considered first with respect to selective-evolutionary explanations? Yes, there are. I want to point at two of them.

The first possible alternative is represented by non selective-evolutionary explanations. They still consist in telling an evolutionary history to account for the trait considered. However, in this context, noise is not considered as the result of a selective process, but as due to chance: it is the result of evolution by random genetic drift. As far as I know, this kind of explanation is not present in the literature on developmental noise.

The second alternative explanations can be called explanations in terms of constraints, where the trait in question is considered as due to different kinds of restrictions on the possible structural and dynamical organization inside the cell and at the intra-organismal level. More precisely, constraints can be due to physical laws but can also be developmental (i.e., principles of construction that have been historically retained) and architectural (i.e., necessary consequences of materials and designs selected to fulfill some function); they act on intracellular networks of genes and proteins and on intra-organismal networks of cells. Contrary to evolutionary explanations by chance, explanations in terms of constraints are rare but present in the scientific literature on developmental noise – in particular, in papers describing theoretical models and simulations of this biological phenomenon (Hasty et al. 2001; Rao et al. 2002). Nevertheless, in these studies, physical factors constraining noise (e.g., the structural and dynamical organization of gene networks) as noise itself are always treated, in the end, as an *explanandum* of evolutionary biology, and not as the result of some other physical process. Why is this so? It would be worth investigating the reason of such a tendency and assessing whether or not this is inevitable.

I argue that biologists should consider these two alternative kinds of explanation of developmental noise prior to selective-evolutionary ones. Why? It is, first of all, a matter of parsimony. An explanation that is more parsimonious is simpler, more economical (i.e., it makes less assumptions) than other less parsimonious explanations, which are consistent with the given empirical data. This is the case for non selective-historical explanations with respect to selective ones: they do not postulate any possible function for developmental noise, but look at what happens by default, in a null hypothesis where no causes but only chance has a role. The same is true for explanations in terms of constraints: biologists should consider them first because, again, they are more parsimonious than selective-evolutionary explanations (they do not require to postulate any direct selective process). Moreover, explanations in terms of constraints allow delineating the space of physico-chemical possibilities where evolution, and more specifically, evolution by natural selection, can take place. This is the main reason why biologists should explore and test them before appealing to other possible evolutionary explanations.

Finally, these alternative explanations need to be taken into account in the present situation where selective-evolutionary explanations of developmental noise are not yet well warranted, the reason being to avoid the charge of "just-so-stories".

To conclude, another limit of actual research on developmental noise that I want to mention is that biologists rarely distinguish between the origin, the diffusion, and the maintaining of noise in a population. That's why they rarely consider more than one possible explanation of this trait and so they fail to explore the possible articulation of different ways to account for it. In contrast to such current explanatory practices, I want to strongly underline the importance of articulating different types of explanation, whenever it is possible, of a given biological phenomenon (e.g., developmental noise). For instance, the origin of noise in gene expression could be a simple by-product of physico-chemical laws, and so the way networks of genes and proteins are structurally and dynamically organized inside the cell. The diffusion of a certain level of noise in the expression of some gene could be due just be a matter of chance (i.e., random genetic drift), especially in small populations; and its maintaining over time could be due to a selective process because the variability it provides at the populational level has turned out to be adaptive in fluctuating environmental conditions. These three explanations are not incompatible: they look at different aspects of developmental noise and respectively answer a specific question about its presence in a natural population.

The limit of any univocal explanatory approach is primarily due to the fact that biologists having different disciplinary backgrounds and practicing in distinct fields tend to limit their analysis of a given phenomenon to a specific biological perspective: the explanation changes depending on the level at which biologists describe the phenomenon, on the question they raise about it, on the explanatory aim motivating their research, and on the methodology they adopt. The limited scope of each disciplinary perspective in biology is just one of the reasons to maintain that different types of explanation, and their articulation, are needed in biology in order to provide a comprehensive and satisfactory explanation of the living world.[7]

5 Conclusion

This paper examines two questions. The first is about whether or not noise is an *explanans* of the outcome of development at the level of individual organisms (for instance, as part of Oyama's DS). The answer is yes! More precisely, we need developmental noise in order to explain the uniqueness of each individual organism, that genes and the environment cannot account for, and its intrinsic heterogeneity (phenotypic, at least), that genetic and environmental changes cannot explain alone. One possible objection to the attribution of an explanatory role to developmental noise is to argue, with Humphreys, that noise – i.e., some form of chance – cannot

---

[7]For a discussion of such explanatory diversity see Morange (2015, this volume).

play such a role because it has no causal power. I maintain that it is possible to resist such an objection showing that chance can always be characterized in causal terms.

The second question is about current prevalent explanations of developmental noise in terms of a selective-evolutionary (i.e., adaptive) history. Are they legitimate or not? I argue that they are not warranted enough (at least, not yet) because the biologists suggesting them have not addressed a certain number of fundamental issues, in particular the issue of the units of selection, such explanations should be based on. Moreover, other possible explanations of developmental noise have to be taken into account before selective-evolutionary explanations in order to fully account for this developmental phenomenon.

Acknowledgments Many thanks to Claude Desplan and Mathias Wernet for providing Fig. 5.4.

References

Aguilar, C., Vlamakis, H., Losick, R., & Kolter, R. (2007). Thinking about *Bacillus subtilis* as a multicellular organism. *Current Opinion in Microbiology, 10*, 638–643.

Barberousse, A., Pradeu, T., & Merlin, F. (Eds). (2009). Developmental systems theory, thematic session. *Biological Theory, 5*(3), 199–222.

Bell, M. L., Earl, J. B., & Britt, S. G. (2007). Two types of *Drosophila* R7 photoreceptor cells are arranged randomly: A model for stochastic cell-fate determination. *Journal of Comparative Neurology, 502*, 75–85.

Berg, O. G. (1978). A model for statistical fluctuations of protein numbers in a microbial population. *Journal of Theoretical Biology, 71*, 587–603.

Diamond, J., & Case, T. J. (Eds.). (1986). *Community ecology*. London: Harper & Row.

Fraser, H. B., Hirsh, A. E., Glaever, G., Kumm, J., & Elsen, M. B. (2004). Noise minimization in eukaryotic gene expression. *PLoS Biology, 2*, 0834–0838.

Gillespie, D. T. (2007). Stochastic simulation of chemical kinetics. *Annual Review of Physical Chemistry, 58*, 35–55.

Gould, S. J., & Lewontin, R. C. (1979). The spandrels of San Marco and the Panglossian paradigm: A critique of the adaptationist programme. *Proceedings of the Royal Society of London Series B, 205*(1161), 581–598.

Hallgrimsson, B., & Hall, B. K. (Eds.). (2005). *Variation. A central concept in biology*. San Diego: Elsevier Academic Press.

Hasty, J., McMillen, D., Isaacs, F., & Collins, J. J. (2001). Computational studies of gene regulatory networks: *In numero* molecular biology. *Nature Reviews Genetics, 2*, 268–279.

Humphreys, P. (1989). *The chances of explanation: Causal explanation in the social, medical, and physical sciences*. Princeton: Princeton University Press.

Johnston, R. J., Jr., & Desplan, C. (2010). Stochastic mechanisms of cell fate specification that yield random or robust outcomes. *Annual Review of Cell and Developmental Biology, 26*, 16.1–16.31.

Kaern, M., Elston, T. C., Blake, W. J., & Collins, J. J. (2005). Stochasticity in gene expression: From theories to phenotypes. *Nature Reviews Genetics, 6*, 451–464.

Kaufmann, B. B., Qiong, Y., Jerome, T., Mettetal, J. T., & van Alexandre, O. (2007). Heritable stochastic switching revealed by single-cell genealogy. *PLoS Biology, 5*(9), e239. 1973–1980.

Klingenberg, C. P. (2005). Developmental constraints, modules, and evolvability. In B. Hallgrimsson & B. K. Hall (Eds.), *Variation. A central concept in biology* (pp. 219–247). San Diego: Elsevier Academic Press.

Kussel, E., & Leibler, S. (2005). Phenotypic diversity, population growth, and information in fluctuating environments. *Science, 309*, 2075–2078.

Latter, B. D. H. (1964). Selection for a threshold character in *Drosophila*. I. An analysis of the phenotypic variance of the underlying scale. *Genetic Research, 5*, 198–210.

Lewontin, R. (1970). The units of selection. *Annual Reviews of Ecology and Systematics, 1*, 1–18.

Lewontin, R. (2000). *The triple helix: Gene, organism, and environment*. Cambridge: Harvard University Press.

Losick, R., & Desplan, C. (2008). Stochasticity and cell fate. *Science, 320*, 65–68.

Maamar, H., Raj, A., & Dubnau, D. (2007). Noise in gene expression determines cell fate in *Bacillus subtilis*. *Science, 317*, 526–529.

Machamer, P., Darden, L., & Craver, C. (2000). Thinking about mechanisms. *Philosophy of Science, 67*, 1–25.

Maheshri, N., & O'Shea, E. K. (2007). Living with noisy genes: How cells function reliably with inherent variability in gene expression. *Annual Review of Biophysics and Biomolecular Structures, 36*, 413–434.

Markow, T. A. (Ed.). (1993, June 14–15). *International conference on developmental instability: its origin and evolutionary implications*, Tempe, Arizona. Dordrecht: Kluwer Academic Publications.

Mason, L. G., Ehrlich, P. R., & Emmel, T. C. (1967). The population biology of the butterfly, *Euphydryas editha*. V. Character clusters and asymmetry. *Evolution, 21*, 85–91.

Mcadams, H. H., & Arkin, A. (1997). Stochastic mechanisms in gene expression. *Proceedings of the National Academy of Sciences USA, 94*, 814–819.

Merlin, F. (2010). Evolutionary chance mutation: A defense of the modern synthesis' consensus view. *Philosophy & Theory in Biology, 2*(e103). http://hdl.handle.net/2027/spo.6959004.0002.003

Millstein, R. L. (2011). Chances and causes in evolutionary biology: How many chances become one chance. In P. McKay Illari, F. Russo, & J. Williamson (Eds.), *Causality in the sciences* (pp. 425–444). Oxford: Oxford University Press.

Minelli, A., & Pradeu, T. (Eds.). (2014). *Towards a theory of development*. Oxford: Oxford University Press.

Morange, M. (2015). Is there an explanation for... the diversity of explanations in biological sciences? In P.-A. Braillard & C. Malaterre (Eds.), *Explanation in biology. An enquiry into the diversity of explanatory patterns in the life sciences* (pp. 31–46). Dordrecht: Springer.

Oyama, S. (2000). *The ontogeny of information. Developmental systems and evolution*. Durham: Duke University Press.

Palmer, A. R. (1996). Waltzing with asymmetry. *Bioscience, 46*, 518–532.

Pigliucci, M. (2008). Is evolvability evolvable? *Nature Reviews Genetics, 9*, 75–82.

Polak, M. (Ed.). (2003). *Developmental instability. Causes and consequences*. New York: Oxford University Press.

Pradeu, T., Laplane, L., Morange, M., Nicoclou, A., Vervort, M. (Eds.) (2011). The boundaries of development. Thematic session. *Biological Theory, 6*(1), 1–88.

Rao, C. V., Wolf, D. M., & Arkin, A. P. (2002). Control, exploitation and tolerance of intracellular noise. *Nature, 420*, 231–237.

Raser, J. M., & O'Shea, E. K. (2005). Noise in gene expression: Origins, consequences, and control. *Science, 309*, 2010–2013.

Reeve, E. C. R., & Robertson, F. W. (1953). Analysis of environmental variability in quantitative inheritance. *Nature, 171*, 874–875.

Rigney, D. R. (1979). Stochastic model of constitutive protein levels in growing and dividing bacterial cells. *Journal of Theoretical Biology, 76*, 453–480.

Rigney, D. R., & Schieve, W. C. (1977). Stochastic model of linear, continuous protein synthesis in bacterial populations. *Journal of Theoretical Biology, 69*, 761–766.

Samoilov, M. S., Gavin, P., & Arkin, A. P. (2006). From fluctuations to phenotypes: The physiology of noise. *Science STKE, 366*(re 17), 1–9.

Shapiro, J. (1988). Bacteria as multicellular organisms. *Scientific American, 258*, 82.

Shapiro, J., & Dworkin, M. (Eds.). (1997). *Bacteria as multicellular organisms*. Oxford: Oxford University Press.

Sober, E. (1984). *The nature of selection. Evolutionary theory in philosophical focus*. Chicago: The University of Chicago Press.

Sober, E., & Wilson, D. S. (1998). *Unto others: The evolution and psychology of unselfish behavior*. Cambridge: Harvard University Press.

Soulé, M. E. (1982). Allomeric variation. 1. The theory and some consequences. *American Naturalist, 120*, 751–764.

Tabery, J. (2009). Difference mechanisms: Explaining variation with mechanisms. *Biology and Philosophy, 24*(5), 645–664.

Thattai, M., & van Oudenaarden, A. (2004). Stochastic gene expression in fluctuating environments. *Genetics, 167*, 523–530.

Thoday, J. M. (1956). Balance, heterozygosity, and developmental stability. *Cold Spring Harbor Symposia on Quantitative Biology, 21*, 318–326.

Waddington, C. H. (1942). Canalization of development and the inheritance of acquired characters. *Nature, 150*, 563–565.

Wernet, M. F., Mazzoni, E. O., Celik, A., Duncan, D. M., Duncan, I., & Desplan, C. (2006). Stochastic spineless expression creates the retinal mosaic for color vision. *Nature, 440*, 174–180.

Williams, G. C. (1966). *Adaptation and natural selection*. Princeton: Princeton University Press.

Wright, S. (1952). The genetics of quantitative variability. In E. C. R. Reeve & C. H. Waddington (Eds.), *Quantitative inheritance* (pp. 5–41). London: Stationery Office.

Part II
Mechanistic Explanation: Applications and Emendations

Chapter 6
Explaining in Contemporary Molecular Biology: Beyond Mechanisms

Frédérique Théry

Abstract Over the past decade, the concept of mechanism has drawn considerable attention in the philosophy of biology. This interest stemmed from the recognition that mechanistic explanations are central to the practice of biologists. So far, most discussions have aimed at defining the mechanism and characterizing mechanistic explanations, rather than assessing the genuine significance of these explanations in the overall explanatory activity of biologists. This reinforced the view that in functional biology, and in particular in molecular biology, explaining a phenomenon mostly consists in showing how this phenomenon is produced by its causes, by describing the mechanisms that maintain and underlie it. From this perspective, mechanistic explanations appear to be the most relevant causal explanatory scheme in functional biology. However, in this chapter, I argue that causal explanations have been mistakenly reduced to mechanistic explanations. I focus on current research on the regulation of genetic expression by microRNAs to suggest that in contemporary molecular biology, explanations increasingly describe some features of causal processes that mechanistic explanations are not meant to grasp. Given this, I characterize two types of explanations – namely, quantitative explanations and systemic explanations – that do not rely on the concept of mechanism. Altogether, these considerations prompt the reconsideration of the status of the concept of mechanism in biological practice, as well as the development of a pluralistic approach of explanations in molecular biology.

Keywords Mechanism • Mechanistic explanation • Quantitative explanation • Systemic explanation • Pluralism • Network • MicroRNAs • Genetic regulation

F. Théry (✉)
IHPST (Institut d'Histoire et de Philosophie des Sciences et des Techniques), 13 rue du Four, Paris, France
e-mail: thery.frederique@gmail.com

© Springer Science+Business Media Dordrecht 2015
P.-A. Braillard, C. Malaterre, *Explanation in Biology*, History, Philosophy and Theory of the Life Sciences 11, DOI 10.1007/978-94-017-9822-8_6

1   Introduction

Since the twentieth century, the mechanistic perspective has been used successfully to explain how a wide range of biological phenomena is causally brought about. This perspective has distinctly dominated research in the field of molecular biology, following its rise in the 1950s. Recently, the realization by philosophers of science that the concept of mechanism plays a crucial role in biological practice has motivated discussions about the characteristics and the explanatory value of mechanistic explanations. These discussions hold the concept of mechanism as central to explanations in neuroscience, molecular biology, cell biology, and genetics. Remarkably, the debate intersects with major philosophical issues, including causation, reduction, and function, bringing interesting new perspectives to these issues.[1]

Here, I wish to examine how the mechanistic perspective fits into the explanatory activity in contemporary molecular biology,[2] following the recent development of postgenomic biology and of systems biology. To address this issue, I focus on recent work on microRNAs, a class of small non-coding RNAs that appear to be key players in the regulation of gene expression. Over the past decade, research on microRNAs has spawned an abundant literature that overlaps with many biological issues. This research has already been proven to be an excellent case study to show the need for exploratory experimentation in post-genomic molecular biology (Burian 2007), and to develop a pluralistic model of scientific inquiry (O'Malley et al. 2010). I shall argue that microRNA research also provides solid ground for defending a pluralistic account of explanations in contemporary molecular biology.[3] Indeed, central explanations of the regulatory role of microRNAs do not rely on the concept of mechanism. They have their own explanatory specificities, which lead me to define what I call quantitative explanations and systemic explanations. These types of explanations are becoming increasingly important in molecular biology, because they compensate for some limits of the concept of mechanism to grasp particular features of molecular processes.[4] In the end, my analysis leads to a reconsideration of the status that has so far been ascribed to the concept of mechanism by philosophers, as well as to the development of a pluralistic account of explanations that more accurately fits biological practice.

---

[1] See for example: Machamer et al. (2000), Machamer (2004), Craver (2001), and Craver (2005).

[2] I define molecular biology broadly as the study of biological processes at the molecular level.

[3] The issue of explanatory pluralism in contemporary biology is also discussed in this volume: Mekios (2015), Morange (2015), Brigandt (2015).

[4] For related discussions of possible limitations of mechanistic models in systems biology, see in this volume: Baetu (2015), Breidenmoser and Wolkenhauer (2015), Issad and Malaterre (2015).

2 Explaining with Mechanisms in Contemporary Molecular Biology

2.1 The Concept of Mechanism in the New Mechanistic Philosophy

In the philosophy of science, issues about the nature of scientific explanation have long been dominated by the deductive-nomological model, according to which explanation takes the form of a deductive argument with at least one natural law among its premises. The debate took a decisive turn when Machamer et al. (2000) claimed that much of the explanatory activity in the biological sciences implies the discovery and the description of mechanisms, which they defined as "entities and activities organized such that they are productive of regular changes from start or set-up to finish or termination conditions" (p. 3). Their analysis triggered important discussions about the characteristics of mechanisms and mechanistic explanations. More specifically, philosophers have debated the possibility of providing a unified characterization of the concept of mechanism across biology, and several accounts of this concept have been proposed (Bechtel and Abrahamsen 2005; Glennan 2002; Torres 2009). Despite their differences, these accounts all share the following core elements (reviewed in Craver and Bechtel 2006):

(i) A mechanism explains how a *phenomenon* is produced, how some *task* is carried out, or how some *function* is performed.

(ii) A mechanism is composed of *parts*, *entities*, or *component parts*, with their properties.

(iii) In virtue of their properties, the parts of the mechanism *interact*, engage in *activities*, or perform *component operations*. Activities or operations are the causal components of mechanisms.

(iv) The parts of the mechanism and their causal relations are spatially and temporally organized, so that they produce the phenomenon.

This new mechanistic philosophy arose from the realization by philosophers of science that the concept of mechanism plays a central role in biological practice. Therefore, the development of the mechanistic framework is meant to grasp scientific practice more adequately, especially in the fields of neuroscience, genetics, and molecular and cell biology, where mechanism-talk is pervasive. However, this recent interest in the concept of mechanism does not stem from new developments within biological sciences, since the mechanistic perspective has been used to explain biological phenomena throughout the twentieth century. Rather, the introduction of the concept of mechanism in philosophical debates was prompted by the need to offer an alternative to the traditional law-based approaches of scientific explanation that fail to account for most of biological practice (Bechtel and Abrahamsen 2005).

A widely shared assumption within the new mechanistic philosophy is that describing mechanisms is the bulk of explanatory activities in biology. Although the possibility that researchers may draw on other types of explanations is not denied, it seems to be taken for granted that, at least in functional biology, explaining a phenomenon usually requires the identification of the mechanism responsible for this phenomenon. For instance, Bechtel (2006) asserts that in many domains, "the aim of inquiry leads to meticulous accounts of complex mechanisms. This is particularly true in the functional domains of biology" (p. 2). Craver, who is interested in explanations in neuroscience, makes a similar claim: "in the final analysis, even if it is false to state that all explanations must describe mechanisms, many of them do" (2007, p. xi). In what follows, I offer a critical examination of this widely shared, mechanism-centered account of biological explanations by examining current research on the regulation of gene expression by microRNAs. As I shall argue, this case study provides an illuminating insight into explanatory activities in contemporary molecular biology.

2.2 Explaining miRNA Regulation with Mechanisms

MicroRNAs (miRNAs) are a class of small non-coding RNAs, approximately 21–23 nucleotides long, found both in plants and animals. They have emerged over the last decade as key regulators of gene expression that act post-transcriptionally to inhibit gene expression. The total number of human miRNAs is estimated to be close to 800 (Bentwich et al. 2005), and about one third of all genes are predicted to be under miRNA regulation (Lewis et al. 2005). MiRNAs are involved in the regulation of virtually all biological processes, including development, cell proliferation, immune responses, and metabolism, to give but a few examples.[5]

After miRNAs were reported to represent a wide class of small non-coding RNAs in animals (Lau et al. 2001; Lee and Ambros 2001; Lagos-Quintana et al. 2001), research efforts mainly concentrated on the mechanisms responsible for their biogenesis and their regulatory role.[6] It appears that in animals, miRNA genes are transcribed into primary miRNA transcripts that undergo processing in the nucleus by the Microprocessor complex to generate precursor miRNAs. Precursor miRNAs are then exported into the cytoplasm, where they are further processed by the Dicer enzyme to form mature miRNAs. MiRNAs associate with the RISC complex (RNA-Induced Silencing Complex) and bind to mRNA targets in a sequence-specific manner, thereby guiding the RISC to these mRNAs. This results in either translational repression or mRNA degradation by the Argonaute protein of the RISC complex.

This description is a simplistic outline that does not reflect the complex accounts of miRNA biogenesis and regulatory action found in scientific literature. However, it provides sufficient grounds to claim that at least part of the research on miRNAs fits

---

[5]See for example: Stefani and Slack (2008), Lodish et al. (2008), Rottiers and Näär (2012).

[6]For a review on miRNA biogenesis and function, see: Ghildiyal and Zamore (2009), Pasquinelli (2012).

the mechanistic framework. Indeed, the explanation above involves parts, such as miRNAs, Dicer, Argonaute, that engage in activities, such as binding or processing. These parts and activities are organized so that their productive order is responsible for miRNA biogenesis or miRNA regulation. However, although the mechanistic perspective is undoubtedly successful in explaining miRNA-mediated regulation, I shall later demonstrate that explanations in the field of miRNA research encompass far more than describing mechanisms. Before doing so, I shall briefly review how philosophers conceptualize the diversity of explanatory schemes in biological sciences.

2.3 Mechanisms, Nothing But Mechanisms?

The new mechanistic philosophy has undeniably improved our understanding of scientific inquiry and explanatory practice in the biological sciences. Nevertheless, the claim that explaining biological phenomena consists in identifying and describing mechanisms seems to rule out, or at least downplay, the extent to which biologists draw on a variety of types of explanations. This perspective seems to be at odds with the one advocated by Morange (2005), who distinguishes between three explanatory schemes in biology: the mechanistic scheme, the Darwinian scheme and the physical non-causal scheme. Morange argues that these different types of explanations should be articulated in order to provide complete explanations of biological phenomena.

More recently, Braillard (2010) pointed out that part of the explanatory practice in systems biology cannot be accounted for by the mechanistic framework. He characterizes a non-causal type of explanation that consists in showing how a system's function determines its structure. Braillard's work highlights how the development of systems biology has influenced and transformed the explanatory activity of biologists. Following this analysis, it seems legitimate to wonder how the transition from classical molecular biology to postgenomic biology has impacted the way biologists explain phenomena at the molecular level. This issue is particularly relevant to our concern, since miRNA research has greatly benefited from genomic and transcriptomic analyses. Indeed, these analyses allow researchers to systematically identify all miRNAs and their targets in organisms, as well as study the impact of miRNA-mediated regulation on gene expression at a large scale, in different cell types and under various conditions. However, the consequences of the transition to post-genomic biology regarding the explanatory practice in molecular biology have been poorly investigated. More precisely, philosophers have insufficiently explored the possibility that the status of the concept of mechanism in scientific investigation may have significantly evolved during the last decade.[7] In this regard, the strong emphasis placed on mechanistic explanations by the new mechanistic philosophy

---

[7] However, see: Moss (2012).

seems problematic. The concept of mechanism has undoubtedly been at the core of research in classical molecular biology; however, the centrality of this concept in contemporary molecular biology has still to be assessed. I shall indeed suggest that the current role played by the concept of mechanism in biological practice has been overrated, and therefore, that the explanatory weight of this concept has been wrongly assessed, at least in molecular biology.

One reason for this failure, I contend, results from the misleading view that at the molecular level, causal explanations boil down to mechanistic explanations. This view is most salient in Craver's work. Following Salmon (1984), Craver defends a causal-mechanical account of explanation in neuroscience, according to which good explanations in neuroscience show how phenomena are situated within the causal structure of the world, by describing mechanisms. In line with this perspective, Craver alludes to causes and mechanisms in an interchangeable way: "Mechanistic models describe the relevant causes and mechanisms in the system under study" (Kaplan and Craver 2011, p. 608). Similarly, when mentioning the "virtue of the causal or mechanistic view of explanation" (Kaplan and Craver 2011, p. 606), he seems to conflate causal and mechanistic explanations. This opinion is not peculiar to Craver. Nicholson (2012) recently noticed that "it is increasingly the case that philosophers employ the term 'mechanistic' simply as a synonym for 'causal' when characterizing scientific explanations" (p. 154). Even Morange, who is most concerned with the diversity of explanatory schemes in biological sciences, seems to take for granted that causal explanations accounting for biological functions at the molecular level are basically mechanistic explanations. However, in what follows, I show that explanations of the regulatory roles of miRNAs do not solely consist in describing mechanisms. I sketch out a richer account of explanations in the field of miRNA research, by characterizing two explanatory contexts in which biologists avoid using the concept of mechanism. This leads me to define two kinds of non-mechanistic explanations, namely, quantitative and systemic explanations, which overcome some limits encountered by the concept of mechanism when describing causal processes at the molecular level.

3 Explaining miRNA Regulation with Quantitative Explanations

3.1 Beyond Mechanisms: Molecular Populations and Their Quantitative Properties

In molecular biology, the basic idea underlying the use of the concept of mechanism is that of a description of the causal patterns that bring about the phenomenon of interest. In that respect, the process of discovering a molecular mechanism always requires identifying the parts involved in the production of the phenomenon. To put it differently, it always involves a qualitative description of causal patterns.

Only some mechanistic explanations also describe quantitative features of molecular systems. By quantitative, I mean here exclusively either the number of copies or the concentration of a molecule or molecular complex.[8] My point is that in contemporary molecular biology, the concept of mechanism is basically used to place emphasis on the qualitative component of the causal structure of molecular systems, rather than on its quantitative component.[9] For instance, mechanistic explanations of miRNA biogenesis or miRNA regulatory action describe the molecular components responsible for these phenomena, but they do not specify the concentrations of these parts, such as miRNA or mRNA concentrations. Thus, in order to provide an account of the concept of mechanism that properly grasps contemporary biological practice, one should stress the qualitative component of causal relations, rather than their quantitative component.

Yet, at the molecular level, quantitative properties of living systems are of crucial importance, since most molecules inside cells are present in multiple copies. Therefore, it is appropriate to talk about 'populations' or 'pools' of molecules. For instance, many miRNAs are present at levels more than 1,000 molecules per cell, some of them exceeding 50,000 molecules per cell. As I shall now argue, quantitative features of miRNA regulation have become a major focus of attention in miRNA research. I will suggest that there has been a concomitant shift in the explanatory discourse toward explanations that depart from the mechanistic framework.

3.2 Quantitative Explanations of miRNA Regulation

Research on miRNAs encompasses far more than a mere qualitative description of the causal processes in which these RNAs are embedded. Many quantitative studies based on accurate assessments of miRNAs and mRNAs concentrations are currently performed. This interest in the quantitative features of miRNA-mediated regulation straightforwardly stems from the phenomenon to be explained, namely the regulation of gene expression. Indeed, explaining gene regulation can be addressed from two complementary perspectives: a qualitative one, which consists in describing how particular molecular components are responsible for the activation or inhibition of gene expression; and a quantitative one, which consists in explaining how a gene product is expressed at a given level. This quantitative aspect should not be deemed less important than its qualitative counterpart. Indeed, alterations in the precise level of most proteins can impair cell functions and cause diseases, such

---

[8]However, it is clear that many other features of molecular systems can be quantitatively characterized.

[9]This does not mean that mechanistic explanations never assume some quantitative characterizations of the activities, but rather that the qualitative description of causal patterns is the common core of all mechanistic explanations in molecular biology.

as cancers. Up to now, aberrant miRNA expression has been reported in various diseases, ranging from cancers to metabolic disorders (Lujambio and Lowe 2012; Rottiers and Näär 2012), and therapeutic strategies that aim at restoring normal miRNA levels are currently developed (Esteller 2011). Thus, quantitative analyses of miRNA-mediated regulation are fully relevant in order to gain insight into the regulatory roles of miRNAs, and to account for their involvement in diseases. These analyses have become easier to perform in the post-genomic era, due to the development and improvement of techniques allowing precise measurements of the expression levels of individual miRNAs and of their mRNA targets. Besides, the advent and refinement of microarray and sequencing technologies have made it possible to carry out high-throughput transcriptome analyses that provide an overview of miRNA expression on a large scale and under various conditions.

Quantitative studies have brought very interesting results regarding the quantitative features of miRNA regulation. First, the effect of miRNAs on the expression of their targets is usually modest, with most targets being down-regulated by less than a half (Baek et al. 2008; Selbach et al. 2008). This result is consistent with a role of miRNAs as fine-tuners of gene expression. Nevertheless, the effects of miRNAs on their targets expression actually exhibit some kind of diversity, and, based on these effects, miRNA-mRNA interactions can basically be categorized into three classes (Bartel 2009; Wu et al. 2009; Ebert and Sharp 2012). First, a miRNA can repress its target expression to inconsequential levels, thereby acting as a 'switch'. Switch interactions have been shown especially to help sharpen developmental transitions and maintain cell fates. Second, a miRNA can adjust the mean expression level of its target to a lower level, thereby acting as a 'fine-tuner' of gene expression. Third, a miRNA can reduce the variance of its target expression, thus contributing to buffer stochastic fluctuations in gene expression. A recent study has further clarified how miRNAs can act either as switches or as fine-tuners of gene expression, by demonstrating that miRNA regulation establishes a threshold level of target mRNA (Mukherji et al. 2011). Below this threshold, translational repression is high, and the miRNA acts as a switch; near the threshold, the protein output responds sensitively to target mRNA transcription, and the miRNA acts as a fine-tuner; above the threshold, translational repression is weak. Therefore, given a miRNA-mRNA interaction, the miRNA can either act as a switch or as a fine-tuner, depending on the mRNA concentration. This can be accounted for by titration effects: as target abundance increases, target mRNAs interact with available miRNAs and titrate them away, so that fewer free miRNAs are able to repress additional targets. This titration effect highlights that the relative concentrations of a miRNA and of all its targets are important to explain miRNA regulation. Indeed, since miRNAs partition among all their target mRNAs, miRNAs that have a higher number of available targets in a cell will repress each individual target to a lesser extent: this effect is known as the dilution effect (Arvey et al. 2010).

Recently, the discovery of competitive endogenous RNAs (ceRNAs) has further complicated the overall picture of the quantitative characteristics of miRNA regulation (Salmena et al. 2011). ceRNAs are targets of common miRNAs, and they can therefore compete with each other for the same pool of miRNAs, which results in a reduction of the inhibitory action of these miRNAs on other targets. Competition between different targets for binding to the same miRNA may widely influence the regulation efficiency for each individual target.

Altogether, these data show that a qualitative description of the interactions between miRNAs and their targets is not sufficient to provide a complete explanation of miRNA regulation. Indeed, the regulatory function of miRNAs crucially depends on the relative cellular concentrations of these small RNAs and of their targets. The basic reason underlying this property is that molecular components are present in cells as populations. However, the concept of mechanism is not primarily meant to grasp this property. In line with this interpretation, biologists tend to avoid using the concept of mechanism in this research context. They have to think of miRNAs and of their targets as populations of molecules, and this conceptual shift is mirrored in the literature by the idea that these molecules form 'pools' inside cells. I refer to the resulting explanations as quantitative explanations. Such explanations describe the relations between quantitative properties of populations of molecules that are causally related: changes in miRNA concentrations cause changes in the expression level of their targets.

Quantitative explanations have their own explanatory specificities. Whereas mechanistic explanations focus on the qualitative interactions between miRNAs and mRNAs, as well as the properties in virtue of which these interactions occur, quantitative explanations focus on the quantitative features of these interactions. To put it differently, mechanistic and quantitative explanations study biological phenomena from two different perspectives: the former are mostly interested in the qualitative relations between molecules, whereas the latter deal with the quantitative relations between populations of molecules. I contend that when the phenomenon to be explained includes a quantitative component, as it is the case for the regulation of gene expression, an adequate explanation of this phenomenon will require both a mechanistic explanation and a quantitative explanation.

3.3 Mechanistic Explanations and Quantitative Explanations

Quantitative explanations highlight features of miRNA-mediated regulation that are not accounted for by the descriptions of parts and activities that lie at the core of mechanistic explanations. As such, they bear genuine explanatory relevance, and they successfully overcome a major limit of the mechanistic framework. Since mechanistic and quantitative explanations provide insights into different aspects of biological phenomena at the molecular level, they complement each other, and are to be articulated. Thus, quantitative explanations should not be considered as an alternative to mechanistic explanations. They are interested in properties of molecular systems that are not the primary focus of mechanistic explanations. However, one could argue that instead of distinguishing between two different types of explanations, the mechanistic framework could be extended so as to include quantitative explanations. To support this view, it could be stressed that quantitative explanations describe relations between populations of molecules that are related in a mechanism. For instance, mRNA concentrations depend on miRNA concentrations because there is a mechanism of post-transcriptional gene silencing in which miRNAs causally interact with mRNAs to inhibit translation.

This solution has recently been advocated by both Craver and Bechtel (Kaplan and Craver 2011; Bechtel and Abrahamsen 2010; Kaplan and Bechtel 2011), who consider mathematical descriptions of dynamical systems as part of mechanistic explanations in cognitive neuroscience. Craver elaborates a model-to-mechanism-mapping constraint, according to which a dynamical or mathematical model explains a phenomenon if elements in the model plausibly map onto elements in the mechanism. In so doing, he states that "mechanisms are frequently described using equations that represent how the values of component variables change with one another" (Kaplan and Craver 2011, p. 606). He later supports the view that, in successful mechanistic explanations, "the (perhaps mathematical) dependencies posited among these variables in the model correspond to the (perhaps quantifi-able) causal relations among the components of the target mechanism" (Kaplan and Craver 2011, p. 611). Bechtel and Abrahamsen have developed a similar perspective. The growing use of the tools of quantitative computational modeling to investigate the dynamic behavior of mechanisms led them to extend the mechanistic framework to accounts of dynamics. Dynamic mechanistic explanations, as they refer to them, do not merely describe the parts, operations, and organization of the mechanisms responsible for biological phenomena; they also explain how the parts and operations of complex mechanisms are orchestrated in real time to produce dynamic phenomena. Bechtel clearly endorses the view that dynamic descriptions should be embedded in mechanistic explanations: "the mechanistic perspective on dynamical models is uniform and remarkably clear: dynamical explanations do not provide a separate kind of explanation; when they explain phenomena, it is because they describe the dynamic behavior of mechanisms" (Kaplan and Bechtel 2011, p. 440). The mechanistic account defended by Craver and Bechtel conflicts with my claim that quantitative explanations should be articulated rather than conflated with mechanistic explanations. I shall provide grounds for holding mechanistic and quantitative explanations distinct in molecular biology. Before doing that, though, it is worth pointing out that Craver and Bechtel have developed their 'quantitative mechanistic' framework in response to claims that dynamical models explain phenomena independently of the mechanisms underlying these phenomena, and consequently that dynamical models are an alternative to mechanistic explanations. There is no such need to defend the explanatory relevance of the mechanistic framework in molecular biology, where the concept of mechanism remains strongly entrenched in the explanatory practice. Instead, my analysis stems from a careful examination of the specific research contexts where biologists favor or avoid the use of the concept of mechanism. This examination is, I believe, necessary in order to develop a mechanistic perspective that is consistent with biological practice. However, the fact that biologists distinguish between mechanistic and quantitative descriptions in their explanatory discourse does not legitimate my claim that these descriptions should be held distinct in a philosophical account of biological explanations. Thus, I now provide some grounds for this claim.[10]

First, from a pragmatic standpoint, mechanistic and quantitative analyses are often carried out separately. Indeed, because they investigate different aspects of the same phenomena, they usually rely on different sets of techniques. Quantitative analyses of miRNA regulation involve techniques that allow accurate measurements of concentrations or copy numbers of miRNAs and mRNAs. They are not designed to investigate the precise mechanisms responsible for the regulatory role of miRNAs. Rather, they complement mechanistic studies by providing quantitative features associated with these mechanisms. Besides, quantitative studies have therapeutic implications on their own, since they lead to develop treatments that could restore adequate levels of miRNAs. The basic principle underlying such treatments is not to change the processes occurring in cells qualitatively, but rather to change their quantitative features.

Second, a hallmark of miRNA regulation lies in the diversity it exhibits in different cell types, at different developmental stages, and in various physiological and pathological conditions. This diversity is both qualitative, since different miRNAs may be expressed in different environmental conditions, and quantitative, since the same miRNA may be expressed at different levels in different conditions. In this latter case, the same qualitative pattern of entailment operates, but with various quantitative characteristics. However, if one chooses to extend the mechanistic framework so as to include quantitative relations between parts of mechanisms, then it is legitimate to wonder whether quantitatively different occurrences of the same interaction between parts should be treated as different mechanisms. Unfortunately, this issue has not been addressed by the new mechanistic philosophy. Nevertheless, it should be kept in mind that in molecular biology, the concept of mechanism is used to emphasize some kind of regularity in the occurrence of a causal sequence of events. This regularity requirement is central to the role played by mechanisms in biological explanations, since it provides the basis for generalizations from one instance to another (Andersen 2011). Because of the quantitative variability exhibited by biological processes, conflating mechanistic and quantitative explanations might weaken the ability of the concept of mechanism to stress such regularity, which in turn might lead to give up on a major epistemic interest of this concept.

Third, according to my analysis, mechanistic explanations at the molecular level can be regarded as idealized representations of causal processes, in which the idea of 'pools' of molecules is generally irrelevant. They describe how molecular components engage in activities, with little regard for the populations of molecules in which these components are embedded. It follows that mechanistic explanations are mostly interested in properties of molecules, whereas quantitative explanations

---

[10]In this volume, Baetu (2015) discusses how mathematical modeling relates to mechanistic explanations. Issad and Malaterre (2015) also offer an interesting critical examination of the concept of dynamic mechanistic explanation.

are interested in properties of populations of molecules. Philosophical accounts of the concept of mechanism rightly insist on the properties in virtue of which parts engage in activities. However, it is important to distinguish between properties of parts and properties of populations of parts: the quantity or concentration of a molecular component is a property of the population of this component, not of the component itself. As such, quantitative properties of molecular systems do not fall straightforwardly within the scope of the mechanistic framework. In the recent refinement of their account of mechanistic models, Bechtel and Craver inadvertently conflate these two kinds of properties. This is most apparent in Bechtel and Abrahamsen's revised definition of the mechanism, according to which "the orchestrated functioning of the mechanism, manifested in patterns of change over time in properties of its parts and operations, is responsible for one or more phenomena" (p. 323). In this definition, the properties of the parts actually refer to the properties of the populations of parts. This is probably why Bechtel fails to identify the different contributions of mechanistic and quantitative explanations to the explanatory practice of biologists.

Finally, the parts and operations involved in quantitative explanations may not be confined within the boundaries of a particular mechanism. For instance, explanations involving dilution effects, which rely on the relative concentrations of miRNAs and of all their targets, span multiple mechanisms, and could be seen as relating to the system level. In this regard, a major limit of the concept of dynamic mechanistic explanation developed by Bechtel is that it only fits cases where dynamical models are used to understand the behavior of a particular mechanism. Yet, if one takes into account the diversity of quantitative explanations, it becomes clear that it is not relevant to merge the quantitative and the mechanistic perspectives. When quantitative explanations relate to the system level, it could, however, still be argued that the mechanistic perspective adequately accounts for explanations at this level. I now discuss whether explanations at the system level should be considered mechanistic explanations.

4   Explaining miRNA Regulation with Systemic Explanations

4.1   Explaining miRNA Regulation at the System Level

After the discovery of miRNAs and of their regulatory action, specifying how cellular functions are regulated by miRNAs has become a pressing issue. As the connections between particular miRNAs and these functions have been uncovered, a new terminology has emerged to account for the roles of miRNAs. Explanatory texts describe how miRNAs "orchestrate" cellular functions, how they are "integrated" in cellular "networks" or "circuitry", how they "mediate" biological phenomena, how the miRNA pathway "crosstalks" and is "coordinated" with other cellular pathways. Concomitantly, the concept of mechanism tends to be less frequently used. This terminological shift has already been noticed by Moss (2012), who rightly relates it to a scientific interest in interrelatedness of function: "the hallmark of leading edge research in biomedical sciences has become that of seeking to understand how the very many low-level basic pieces of chemistry are responsively, flexibly, and contingently weaved together (*orchestrated*, *mediated*, *regulated*, etc.) into coherently global responses to developmental and environmental cues, internal and external perturbations." (p. 168). In miRNA research, the current interest in the interplay between miRNA-mediated regulation and cellular processes results in a shift from mechanistic explanations to a type of explanation that emphasizes how causal processes interact with one another. I shall outline the main features of this explanatory scheme, and highlight its importance with regard to miRNA biology.

Current research on miRNAs investigates how previously characterized mechanisms, those responsible for miRNA regulatory action and those involved in various cellular functions, are interconnected. Interactions between miRNA-mediated gene silencing mechanisms and other cellular mechanisms directly follow from the regulatory function of miRNAs. Indeed, miRNAs have an impact on the cellular functions in which their target genes are involved. Far from being incidental, this impact is widespread, due to the high number of genes that are targeted by miRNAs. Descriptions of the interactions between miRNAs and cellular functions are not detailed mechanistic descriptions. Rather, they span multiple mechanisms, and they characterize causal relations in cells at the systemic level. That is why I refer to them as systemic explanations. Surprisingly, Moss did not insist on the importance of a systemic thinking in this explanatory framework.

It has been suggested, quite provocatively, that the impact of miRNAs is always at the system level (Jost et al. 2011). The reason underlying this claim is that most miRNAs target multiple genes, which results in the coupling between the expression of genes involved in the same cellular function, or in a crosstalk between seemingly unrelated biological processes. Moreover, since many protein-coding genes are predicted to be targets of multiple miRNAs, distinct miRNAs can act cooperatively to regulate gene expression. This systemic perspective of miRNA regulation has implications for the explanatory discourse. Braillard (2010) has stated that both mechanistic and systemic approaches aim at explaining biological phenomena "by studying the molecular components, their properties and their interactions", and that "systems biology is clearly concerned with large and complex mechanisms, but they are still mechanisms" (p. 45). I wish to point out some limits of this assertion, by stressing important differences between mechanistic and systemic explanations in the context of miRNA research. Consider the following explanation of the role of miRNAs in the resolution of inflammation:

> Later, the induction of miR-146a[11] and miR-9 resolves the pro-inflammatory response by targeting TNFR-associated factor 6 (TRAF6) and IL-1R-associated kinase 1 (IRAK1), which are key components of TLR signaling pathways, and the nuclear factor-κB (NF-κB) subunit p50, respectively. miR-132 expression has also been shown to be increased by TLR signaling and would limit the antiviral response by targeting p300. (O'Neill et al. 2011, p. 172)

---

[11]Under the standard nomenclature system, the prefix «miR» refers to mature miRNAs.

This explanation differs from detailed mechanistic explanations in several important respects. First, it only reports two kinds of molecular components, miRNAs and their targets, without much interest in their binding, or in the properties by virtue of which this binding occurs. Moreover, most of the causal chain responsible for the inhibition of gene expression, such as the sorting of miRNAs into the RISC complex or the interaction between the RISC and mRNA targets, is not described. Rather, this explanation shows how previously characterized mechanisms (the mechanisms responsible for the expression of specific miRNAs and those involved in immune responses) interact with one another to resolve inflammation. In this respect, there is no need to describe the complete causal processes, since mechanistic descriptions already did so. Instead, in systemic explanations, biologists only focus on portions of the causal processes they deem relevant to explain how these processes interact to fulfill the biological function of interest. Here, in order to explain how cellular functions are regulated by miRNAs, it is only relevant to describe the set of miRNAs that regulate the expression of the genes involved in these functions. The resulting explanation is not at a higher hierarchical level than mechanistic explanations, but at a systemic level. More importantly, this explanation possesses its own explanatory specificities, especially regarding the features of the patterns of entailment that are picked out as relevant.

The concept of network is particularly well equipped to describe the interactions between mechanisms that lie at the core of systemic explanations. Accordingly, this concept has become ubiquitous in miRNA research. Sometimes, it is used in a loose way to convey the idea that biological processes are interconnected in an intricate way:

> Clearly, miRNAs cannot independently perform a single task in cells. Instead, miRNAs regulate cellular networks as network components in many cellular functions. (Li et al. 2011, p. 1)

More frequently, the concept of network is used to describe specific gene regulatory networks involving miRNAs. These networks usually depict the interactions either between miRNA regulation and cellular functions, or between the regulation by miRNAs and by transcription factors. They consist of nodes that can be genes, proteins, mRNAs, or miRNAs, and of links connecting the nodes that indicate the regulatory relations (activation or inhibition) between nodes. These networks do not exhaustively describe the causal processes of interest, but they show how several causal processes interact with one another.

4.2 Mechanistic Explanations and Systemic Explanations

In molecular biology, mechanistic explanations are used to describe continuous causal chains that produce or maintain the phenomenon to be explained. Consequently, they satisfy two requirements. First, the description of the component parts, their relevant properties, and the activities they engage in must be exhaustive.

Second, mechanistic explanations only describe the relevant causal relations that are responsible for the production of the phenomenon under investigation. Craver has formulated these requirements: "Complete explanatory texts are complete because they represent all and only the relevant portions of the causal structure of the world" (Craver 2007, p. 27). Systemic explanations describe how previously characterized processes that together perform a function interact with one another. In miRNA research, they tackle the issue of the biological functions of miRNAs from a different perspective than mechanistic explanations do, and accordingly, they often draw on a different set of investigative tools. As such, they fruitfully complement the mechanistic approach, but they should not be conflated with it.

In contemporary molecular biology, the concept of mechanism is primarily used to individualize some portions of the causal structure of the cell. This fragmentation of living systems into autonomous mechanisms has a heuristic interest that has already been put forward by Nicholson (2012): "The idea of autonomous causal mechanisms operating within the organism is, I suggest, nothing more than a pragmatic idealization that biologists appeal to in order to narrow their focus on the particular parts of the organism they happen to be investigating" (p. 159). However, in the case of miRNA-mediated regulation, the description of a regulatory mechanism does not provide, on its own, a complete explanation of how a cellular function is regulated by miRNAs at the molecular level. This is why researchers are currently increasingly interested in the interrelatedness of causal processes, which is better investigated and described with the concept of network than with the mechanistic framework.[12]

So far, I have characterized two explanatory schemes that do not rely on the concept of mechanism. This analysis has important implications regarding the status of the concept of mechanism in contemporary molecular biology.

5 Reassessing the Status of the Concept of Mechanism in Contemporary Molecular Biology

My account of the explanatory practice in miRNA research ensues from a careful examination of explanatory texts and diagrams found in scientific literature. This perspective has been motivated by the assumption that in order to provide a characterization of the mechanistic framework that is grounded in biological practice, one has to bear heavily on these texts and diagrams. This focus on explanatory texts, rather than on the causal structure of the world, departs from the perspective adopted by Craver. Indeed, Craver places much more emphasis on objective explanations, which are the mechanisms in the world, than on explanatory texts, which describe these mechanisms:

---

[12]For a different perspective regarding networks and the mechanistic framework, see Bechtel (2015) in this volume.

> There are perhaps many interesting things to be said about explanatory texts, but one crucial aspect of their adequacy has to do with whether explanatory texts accurately characterize the causal structure of the world. (Craver 2007, p. 27)

Giving more prominence to explanatory texts results in a mechanistic framework that differs from the one offered by the new mechanistic philosophy in two important respects. First, it shows that in biological practice, causal explanations do not boil down to mechanistic explanations. Second, it highlights the epistemic component of the concept of mechanism. In what follows, I provide more details about these claims.

5.1 Causal Explanations in Contemporary Molecular Biology

I previously argued that in the context of miRNA research, biologists tend to not use the concept of mechanism when quantitative or systemic properties of molecular systems are being investigated. Accordingly, I distinguished between three explanatory schemes in contemporary molecular biology: mechanistic, quantitative, and systemic explanations. New mechanistic philosophers have failed to notice the explanatory shift that resulted in this diversity of causal explanations in postgenomic biology. This may be because they embrace a causal-mechanical view of scientific explanation, according to which the causal structure of the world is characterized through the concept of mechanism. Besides, little attention has been paid to the specific research contexts in which researchers actually use the concept of mechanism and those in which they prefer employing other concepts. This could result in accounts of scientific explanation that somehow move away from actual explanatory practice.

I do not claim that explaining a phenomenon does not require showing how it is produced by its causes, although it should be kept in mind that not all biological explanations are causal (Morange 2005; Braillard 2010). Rather, my critical examination of some claims made by new mechanistic philosophers aims at developing an account of explanations that more adequately characterizes explanatory practice in contemporary molecular biology. Indeed, in this field, the concept of mechanism is less prevalent than it used to be in classical molecular biology. This is because biologists increasingly study molecular causal processes from different perspectives, each focusing on a specific feature of these processes: their qualitative component, their quantitative component, or their interrelatedness. Therefore, I advocate a pluralistic approach of biological explanations that accurately fits explanatory practice in contemporary molecular biology in two important respects. First, it emphasizes the fact that mechanistic, quantitative and systemic studies of molecular systems are often performed separately, and rely on different methodological and conceptual backgrounds. Second, it is consistent with explanatory texts, in which biologists prefer articulating the results coming from these studies, rather than merging them into a unique complete explanation.

The categorization of explanatory schemes I sketch out offers a basis for developing a richer and finer conceptual framework of explanations in molecular biology. However, it should be stressed that it is also simplistic, insofar as the different types of explanations overlap to some extent. Indeed, some explanations belong to more than one of these types. For instance, explanations involving the dilution effect can be regarded either as quantitative explanations or as systemic explanations. Indeed, they focus on the relative concentrations of miRNAs and of all their targets, and as such, they deal with the systemic level. This overlap suggests that the different types of explanations should not be conceived of as mutually exclusive, but rather as a continuum of explanatory practices.

5.2   The Epistemic Component of the Concept of Mechanism

I related the use of the concept of mechanism to specific research contexts where qualitative aspects of causal patterns are the primary (but not necessarily the sole) focus of attention. If this analysis is correct, it implies that the concept of mechanism is used to make salient some properties of the living system under investigation, and consequently that this concept has an epistemic component that should not be ignored. When molecular biologists study causal processes, they often decompose their properties into several categories for heuristic purposes. In this respect, the concept of mechanism makes it possible to put some properties of biological processes in the foreground, while other properties are overlooked. Concomitantly, it helps mobilize methodological and conceptual tools that are well equipped to study these properties; it also provides a conceptual background to analyze experiences and to draw up explanations. Therefore, the wrong appreciation of the status of mechanistic explanations by the new mechanistic philosophy may be related to its defense of an ontic view of the concept of mechanism. Craver and Bechtel, who consider mechanisms as real things in the world, explicitly advocate this view. This may be why they fail to realize that the concept of mechanism is mostly used to place emphasis on specific features of the patterns of entailment occurring inside cells. This may also account for their choice to include quantitative explanations into the mechanistic framework. As a matter of fact, their conceptualization of mechanisms includes the three features of causal patterns I distinguished. The problem is that in so doing, their definition of the mechanism does not grasp what most biologists mean when they use this concept.

Other philosophers have already stressed the epistemic component of the concept of mechanism. According to Nicholson, mechanisms are epistemic models used to explain how biological phenomena are produced. In this view, mechanisms are idealized representations of causal processes, which capture some features of living systems at the expense of others. More precisely, mechanisms are individuated both temporally and spatially in the specific context of the explanation, and consequently, mechanistic explanations abstract away the complexity of the organismic context. Powell (2012) also rejects an ontic conception of mechanisms. Instead, he develops a characterization of mechanisms that heavily relies on cognitive and psychological dispositions. Nicholson and Powell's views both have attractive features. Yet, the characterization of mechanisms I am most sympathetic to is the one embraced by Moss (2012). According to Moss, mechanism-talk in biology draws metaphorically on the knowledge of the workings of macroscopic machines. This helps biologists to grasp and make intelligible phenomena occurring in the microscopic world. Moss' assumption is supported by the fact that biologists never define (nor try to define) the concept of mechanism. Rather, this concept seems to be used quite intuitively, more as an explanatory metaphor than as a scientific concept. Accordingly, Moss denies the possibility (as well the philosophical interest) of offering a normative account of mechanistic explanations that would specify standards to evaluate explanations. Moss' thesis is helpful to make sense of the idea that mechanistic explanations are primarily used to foreground the qualitative component of causal patterns, rather than the quantitative features and interrelatedness. Indeed, if he is right, then the concept of mechanism cannot be satisfactorily accounted for without reference to the workings of machines. This view strongly contrasts with the new mechanistic discourse, which insists on differences between mechanisms and machines (Craver 2007; Craver and Darden 2005). Keeping in mind that biologists do not provide any definition of the concept of mechanism, it follows that the way biologists conceptualize mechanisms must be at least partly influenced by their individual experience of machines, and most certainly by their knowledge of emblematic machines such as the clock. In this respect, it is noteworthy that describing the working of a clock consists in a qualitative description of how its component parts interact to perform a task. A quantitative characterization would not be relevant, because these parts are not present as 'populations'. Besides, one can straightforwardly describe all the parts and operations of the mechanism of the clock, with no need to examine how different mechanisms are interconnected in an intricate way. Therefore, it is tempting to come to the conclusion that the working of emblematic mechanical devices has been heuristically transposed onto the descriptions of molecular mechanisms. In this respect, the current conceptual shift towards explanations that depart from the mechanistic discourse highlights some limits faced by the metaphor of the machine to accurately describe causal patterns occurring in cells.

6   Conclusion

Throughout the twentieth century, the mechanistic framework has been very successful in providing explanations of a wide range of biological functions. The new mechanistic philosophy has rightly shed light on this success. However, recent developments in molecular biology reveal that biologists are, with increasing frequency, investigating properties of living systems that the concept of mechanism captures only poorly, and that provide different insights into biological phenomena. This gives rise to explanatory schemes that stress the quantitative and systemic properties of molecular systems and that have their own explanatory specificities. Rather than considering these schemes as part of the mechanistic framework, a more fruitful standpoint consists in developing a pluralistic approach of explanations that better fits biological practice. Such a pluralistic approach requires us to give more emphasis to the epistemic properties of the concept of mechanism.

References

Andersen, H. K. (2011). Mechanisms, laws, and regularities. *Philosophy in Science, 78*, 325–331.

Arvey, A., Larsson, E., Sander, C., Leslie, C. S., & Marks, D. S. (2010). Target mRNA abundance dilutes microRNA and siRNA activity. *Molecular Systems Biology, 6*, 363.

Baek, D., Villén, J., Shin, C., Camargo, F. D., Gygi, S. P., & Bartel, D. P. (2008). The impact of microRNAs on protein output. *Nature, 455*, 64–71.

Baetu, T. (2015). From mechanisms to mathematical models and back to mechanisms: Quantitative mechanistic explanations. In P.-A. Braillard & C. Malaterre (Eds.), *Explanation in biology. An enquiry into the diversity of explanatory patterns in the life sciences* (pp. 345–363). Dordrecht: Springer.

Bartel, D. P. (2009). MicroRNAs: Target recognition and regulatory functions. *Cell, 136*, 215–233.

Bechtel, W. (2006). *Discovering cell mechanisms*. Cambridge: Cambridge University Press.

Bechtel, W. (2015). Generalizing mechanistic explanations using graph-theoretic representations. In P.-A. Braillard & C. Malaterre (Eds.), *Explanation in biology. An enquiry into the diversity of explanatory patterns in the life sciences* (pp. 199–225). Dordrecht: Springer.

Bechtel, W., & Abrahamsen, A. (2005). Explanation: A mechanist alternative. *Studies in History and Philosophy of Biological and Biomedical Sciences, 36*, 421–441.

Bechtel, W., & Abrahamsen, A. (2010). Dynamic mechanistic explanation: Computational modeling of circadian rhythms as an exemplar for cognitive science. *Studies in History and Philosophy of Science, 41*, 321–333.

Bentwich, I., Avniel, A., Karov, Y., Aharonov, R., Gilad, S., Barad, O., Barzilai, A., Einat, P., Einav, U., Meiri, E., Sharon, E., Spector, Y., & Bentwich, Z. (2005). Identification of hundreds of conserved and nonconserved human microRNAs. *Nature Genetics, 37*, 766–770.

Braillard, P.-A. (2010). Systems biology and the mechanistic framework. *History and Philosophy of the Life Sciences, 32*, 43–62.

Breidenmoser, T., & Wolkenhauer, O. (2015). Explanation and organizing principles in systems biology. In P.-A. Braillard & C. Malaterre (Eds.), *Explanation in biology. An enquiry into the diversity of explanatory patterns in the life sciences* (pp. 249–264). Dordrecht: Springer.

Brigandt, I. (2015). Evolutionary developmental biology and the limits of philosophical accounts of mechanistic explanation. In P.-A. Braillard & C. Malaterre (Eds.), *Explanation in biology. An enquiry into the diversity of explanatory patterns in the life sciences* (pp. 135–173). Dordrecht: Springer.

Burian, R. M. (2007). On microRNA and the need for exploratory experimentation in post-genomic molecular biology. *History and Philosophy of the Life Sciences, 29*, 285–311.

Craver, C. F. (2001). Role functions, mechanisms and hierarchy. *Philosophy in Science, 68*, 53–74.

Craver, C. F. (2005). Beyond reduction: Mechanisms, multifield integration and the unity of neuroscience. *Studies in History and Philosophy of Biological and Biomedical Sciences, 36*, 373–395.

Craver, C. F. (2007). *Explaining the brain: Mechanisms and the mosaic unity of neuroscience*. Oxford: Oxford University Press.

Craver, C. F., & Bechtel, W. (2006). Mechanism. In S. Sarkar & J. Pfeiffer (Eds.), *Philosophy of science: An encyclopedia*. New York: Routledge.

Craver, C. F., & Darden, L. (2005). Introduction. *Studies in History and Philosophy of Biological and Biomedical Sciences, 36*, 233–244.

Ebert, M. S., & Sharp, P. A. (2012). Roles for microRNAs in conferring robustness to biological processes. *Cell, 149*, 515–524.

Esteller, M. (2011). Non-coding RNAs in human disease. *Nature Reviews Genetics, 12*, 861–874.

Ghildiyal, M., & Zamore, P. D. (2009). Small silencing RNAs: An expanding universe. *Nature Reviews Genetics, 10*, 94–108.

Glennan, S. (2002). Rethinking mechanistic explanation. *Philosophy in Science, 69*, S342–S353.

Issad, T., & Malaterre, C. (2015). Are dynamic mechanistic explanations still mechanistic? In P.-A. Braillard & C. Malaterre (Eds.), *Explanation in biology. An enquiry into the diversity of explanatory patterns in the life sciences* (pp. 265–292). Dordrecht: Springer.

Jost, D., Nowojewski, A., & Levine, E. (2011). Small RNA biology is systems biology. *BMB Reports, 44*, 11–21.

Kaplan, D. M., & Bechtel, W. (2011). Dynamical models: An alternative or complement to mechanistic explanations? *Topics in Cognitive Science, 3*, 438–444.

Kaplan, D. M., & Craver, C. F. (2011). The explanatory force of dynamical and mathematical models in neuroscience: A mechanistic perspective. *Philosophy in Science, 78*, 601–627.

Lagos-Quintana, M., Rauhut, R., Lendeckel, W., & Tuschl, T. (2001). Identification of novel genes coding for small expressed RNAs. *Science, 294*, 853–858.

Lau, N. C., Lim, L. P., Weinstein, E. G., & Bartel, D. P. (2001). An abundant class of tiny RNAs with probable regulatory roles in *Caenorhabditis elegans. Science, 294*, 858–862.

Lee, R. C., & Ambros, V. (2001). An extensive class of small RNAs in *Caenorhabditis elegans. Science, 294*, 862–864.

Lewis, B. P., Burge, C. B., & Bartel, D. P. (2005). Conserved seed pairing, often flanked by adenosines, indicates that thousands of human genes are microRNA targets. *Cell, 120*, 15–20.

Li, Y., Li, Y., Zhang, H., & Chen, Y. (2011). MicroRNA-mediated positive feedback loop and optimized bistable switch in a cancer network involving miR-17-92. *PLoS One, 6*, e26302.

Lodish, H. F., Zhou, B., Liu, G., & Chen, C. Z. (2008). Micromanagement of the immune system by microRNAs. *Nature Reviews Immunology, 8*, 120–130.

Lujambio, A., & Lowe, S. W. (2012). The microcosmos of cancer. *Nature, 482*, 347–355.

Machamer, P. (2004). Activities and causation: The metaphysics and epistemology of mechanisms. *International Studies in the Philosophy of Science, 18*, 27–39.

Machamer, P., Darden, L., & Craver, C. F. (2000). Thinking about mechanisms. *Philosophy in Science, 67*, 1–25.

Mekios, C. (2015). Explanation in systems biology: Is it all about mechanisms? In P.-A. Braillard & C. Malaterre (Eds.), *Explanation in biology. An enquiry into the diversity of explanatory patterns in the life sciences* (pp. 47–72). Dordrecht: Springer.

Morange, M. (2005). *Les secrets du vivant: Contre la pensée unique en biologie*. Paris: La découverte.

Morange, M. (2015). Is there an explanation for… the diversity of explanations in biological sciences ? In P.-A. Braillard & C. Malaterre (Eds.), *Explanation in biology. An enquiry into the diversity of explanatory patterns in the life sciences* (pp. 31–46). Dordrecht: Springer.

Moss, L. (2012). Is the philosophy of mechanism philosophy enough? *Studies in History and Philosophy of Biological and Biomedical Sciences, 43*, 164–172.

Mukherji, S., Ebert, M. S., Zheng, G. X. Y., Tsang, J. S., Sharp, P. A., & van Oudenaarden, A. (2011). MicroRNAs can generate thresholds in target gene expression. *Nature Genetics, 43*, 854–859.

Nicholson, D. J. (2012). The concept of mechanism in biology. *Studies in History and Philosophy of Biological and Biomedical Sciences, 43*, 152–163.

O'Malley, M. A., Elliott, K. C., & Burian, R. M. (2010). From genetic to genomic regulation: Iterativity in microRNA research. *Studies in History and Philosophy of Biological and Biomedical Sciences, 41*, 407–417.

O'Neill, L. A., Sheedy, F. J., & McCoy, C. E. (2011). MicroRNAs: The fine-tuners of Toll-like receptor signaling. *Nature Reviews Immunology, 11*, 163–175.

Pasquinelli, A. E. (2012). MicroRNAs and their targets: Recognition, regulation and an emerging reciprocal relationship. *Nature Reviews Genetics, 13*, 271–282.

Powell, A. (2012). Biological mechanisms: A case study in conceptual plasticity. In J. F. Davies & F. Michelini (Eds.), *Frontiere della Biologia: Prospettive Filosofiche Sulle Scienze Della Vita.* Mimesis: Milano.

Rottiers, V., & Näär, A. M. (2012). MicroRNAs in metabolism and metabolic disorders. *Nature Reviews Molecular Cell Biology, 13*, 239–250.

Salmena, L., Poliseno, L., Tay, Y., Kats, L., & Pandolfi, P. P. (2011). A ceRNA hypothesis: The Rosetta stone of a hidden RNA language? *Cell, 146*, 353–358.

Salmon, W. C. (1984). *Scientific explanation and the causal structure of the world.* Princeton: Princeton University Press.

Selbach, M., Schwanhäusser, B., Thierfelder, N., Fang, Z., Khanin, R., & Rajewsky, N. (2008). Widespread changes in protein synthesis induced by microRNAs. *Nature, 455*, 58–63.

Stefani, G., & Slack, F. J. (2008). Small non-coding RNAs in animal development. *Nature Reviews Molecular Cell Biology, 9*, 219–230.

Torres, P. J. (2009). A modified conception of mechanisms. *Erkenntnis, 71*, 233–251.

Wu, C. I., Shen, Y., & Tang, T. (2009). Evolution under canalization and the dual roles of microRNAs: A hypothesis. *Genome Research, 19*, 734–743.

Chapter 7
Evolutionary Developmental Biology and the Limits of Philosophical Accounts of Mechanistic Explanation

Ingo Brigandt

Abstract Evolutionary developmental biology (evo-devo) is considered a 'mechanistic science,' in that it causally explains morphological evolution in terms of changes in developmental mechanisms. Evo-devo is also an interdisciplinary and integrative approach, as its explanations use contributions from many fields and pertain to different levels of organismal organization. Philosophical accounts of mechanistic explanation are currently highly prominent, and have been particularly able to capture the integrative nature of multifield and multilevel explanations. However, I argue that evo-devo demonstrates the need for a broadened philosophical conception of mechanisms and mechanistic explanation.

Mechanistic explanation (in terms of the qualitative interactions of the structural parts of a whole) has been developed as an alternative to the traditional idea of explanation as derivation from laws or quantitative principles. Against the picture promoted by Carl Craver, that mathematical models describe but usually do not explain, my discussion of cases from the strand of evo-devo which is concerned with developmental processes points to qualitative phenomena where quantitative mathematical models are an indispensable part of the explanation. While philosophical accounts have focused on the actual organization and operation of mechanisms, properties of developmental mechanisms that are about how a mechanism reacts to modifications are of major evolutionary significance, including robustness, phenotypic plasticity, and modularity. A philosophical conception of mechanisms is needed that takes into account quantitative changes, transient entities and the generation of novel types of entities, feedback loops and complex interaction networks, emergent properties, and, in particular, functional-dynamical aspects of mechanisms, including functional (as opposed to structural) organization and distributed, system-wide phenomena. I conclude with general remarks on philosophical accounts of explanation.

I. Brigandt (✉)
Department of Philosophy, University of Alberta, 2-40 Assiniboia Hall,
Edmonton, AB T6G2E7, Canada
e-mail: brigandt@ualberta.ca

© Springer Science+Business Media Dordrecht 2015
P.-A. Braillard, C. Malaterre, *Explanation in Biology*, History, Philosophy
and Theory of the Life Sciences 11, DOI 10.1007/978-94-017-9822-8_7

Keywords Evolutionary developmental biology • Mechanistic explanation • Mathematical models • Mechanisms • Scientific explanation • Integration

1 Introduction

Evolutionary developmental biology (evo-devo) is sometimes hailed as a "mechanistic science" (Cañestro et al. 2007, p. 940; Wagner et al. 2000, p. 819). The notion of the "mechanistic framework of evolutionary developmental biology" (Laubichler 2010, p. 208) stems from the fact that evo-devo does not just lay out phylogenetic transformation sequences of morphological characters, but offers a causal explanation of how those character transformations occurred by means of changes in developmental mechanisms. Advances in developmental genetics endow evo-devo with an enormous degree of scientific promise. Moreover, evo-devo is clearly an integrative approach, in that its explanations make reference to entities and processes on several levels of organismal organization and use contributions from several fields in an interdisciplinary fashion (Love 2013).

Multilevel and multifield explanation can be captured by current philosophical accounts of *mechanistic explanation* (Bechtel and Abrahamsen 2005), as mechanisms contain entities on several levels (where the entities stand in relations permitting systematic accounts), and different fields contribute to elucidating different components of a mechanism (Craver 2007). Thereby philosophical accounts of mechanistic explanation offer a model of epistemic integration as opposed to the traditional idea of reduction (Brigandt 2013a; Brigandt and Love 2012b; Craver 2005; Craver and Darden 2013; Darden 2005).[1] And rather than just analyzing the result of science (reductive or integrative explanations), such philosophical approaches also take into account the process of scientific research, such as the change between reductive episodes and integrative strategies (Bechtel 2010; Craver 2005).

While philosophical accounts of mechanistic explanation have been developed in the context of molecular biology, cell biology, and neuroscience (Bechtel 2006; Craver 2007; Darden 2006), evo-devo is another scientific domain exhibiting interdisciplinary research and multilevel, mechanistic explanations. However, in this chapter I argue that some aspects of evo-devo mandate a revised, broadened philosophical conception of mechanisms and mechanistic explanation. First, philosophical accounts tend to give a stereotypical portrayal of mechanisms (Machamer et al. 2000). The image conveyed is that a mechanism consists of a fixed stock of entities, it has structural parts in a spatial organization, the activities among the parts are qualitative, there is a linear causal sequence from start to termination state, and what has to be studied is the actual organization and regular operation of the

---

[1]A complementary epistemological way of articulating integration is in terms of problem agendas that structure how contributions from different fields are to be coordinated (Brigandt 2010; Brigandt and Love 2010, 2012a; Love 2008a, b).

mechanism. Based on my evo-devo case studies, in the concluding section I will lay out how this stereotypical construal is erroneous and what important aspects of mechanisms it omits.

Second, accounts of mechanistic explanation have been developed as an alternative to the covering-law model, according to which an explanation is the derivation from laws or quantitative principles. As one of the main proponents of the mechanistic approach, Carl Craver (2006, 2007, 2008) has argued that quantitative models can describe and predict, but they usually do not explain. The main part of my discussion will take issue with this, as I will point to cases where mathematical models are explanatorily indispensable. Some strands of evo-devo show it is possible to integrate mechanistic explanation (in terms of the concrete structural components of a developing organism) and mathematical modeling. This thesis is in line with William Bechtel and Adele Abrahamsen's notion of 'dynamic mechanistic explanation' (Bechtel 2011, 2013; Bechtel and Abrahamsen 2010, 2011). I improve upon previous philosophical accounts which claim that mathematical modeling plays some epistemic roles by specifically arguing that mathematical models can be indispensable in biological *explanations* (see also Baker 2015 on mathematical explanations in biology). Moreover, besides the cases from chronobiology that Bechtel and Abrahamsen address, I look at evo-devo as a distinct biological domain. I have recently made analogous points in the context of systems biology (Brigandt 2013c; on systems biology and its use of molecular data and mathematical models see also Baetu 2015; Gross 2015; Isaad and Malaterre 2015; Mekios 2015; Théry 2015).

The following section lays out a description of evo-devo, emphasizing its interdisciplinary nature and the fact that its explanatory frameworks go beyond the study of gene regulation. Section 3 discusses how equations can be explanatory components of mechanistic accounts. Since mathematical models play an obvious explanatory role in evolutionary genetics, my case studies on evo-devo focus on its *developmental* aspects. It is unsurprising that an account of all quantitative aspects and the full temporal dynamics of a developmental process mandates the use of quantitative models. However, my case studies point to *qualitative* explananda where equations are still required. Section 4 analyzes mathematical models of the development and evolutionary origin of morphological structures. Developmental properties of major significance for morphological evolvability are robustness, phenotypic plasticity, and modularity. Section 5 discusses how mathematical models are involved in explanations of robustness, and prepares my point that robustness, phenotypic plasticity, and modularity go beyond the philosopher's typical focus of a mechanism's structural aspects and its actual organization and operation. The concluding section describes the broader philosophical conception of mechanisms required, and makes relevant general observations about the nature of scientific explanation.

2   Evolutionary Developmental Biology: Integrative and Diverse

Though evo-devo's new molecular-experimental techniques have fueled its scientific promise and prominence, it is not these new techniques which best characterize the discipline, but the intellectual problems it addresses, problems which were neglected by neo-Darwinian evolutionary theory concerned with adaptation and speciation (Love and Raff 2003). Evo-devo does not just study the evolution of developmental processes, but it addresses evolutionary questions where development is essential to the explanation. The claim is that these questions cannot be answered using a traditional framework focused on the dynamics of genes within populations (Müller and Wagner 2003; Wagner 2000).[2] One core item on the evo-devo agenda is the evolutionary origin of morphological novelty (Brigandt and Love 2010, 2012a; Müller and Newman 2005). An *evolutionary novelty* (or innovation) is a morphological trait that is qualitatively different from traits of ancestral lineages, which is often expressed by the definition that a novelty is a trait that is not homologous to any ancestral feature. Examples are the origin of fins in fish and—to mention a trait on a lower level of organization—the evolution of vertebrate neural crest cells, which among other things form craniofacial bone, smooth muscle, and some types of neurons, so that after its origin the neural crest came to be involved in the evolutionary modification and generation of a variety of structures. Explaining the origin of novelty involves an account of how ancestral developmental mechanisms were so modified as to give rise to a new developmental system that produces the novelty in question (Brigandt 2010).

Another related issue that evo-devo attempts to explain is morphological *evolvability* (Brigandt 2015; Hendrikse et al. 2007). Evolvability is the ability of biological systems to evolve, and a core aspect of morphological evolvability is the generation of heritable phenotypic variation on which natural selection can subsequently act (Gerhart and Kirschner 2003; Kirschner and Gerhart 1998). A key question in the study of evolvability is how a sufficient amount of *viable and functional* morphological variation could have been created so as to permit the significant morphological change that has occurred in evolution (Gerhart and Kirschner 2007; Kirschner and Gerhart 2005). One contributing factor is that organismal structures are organized as trait complexes, where individual phenotypic traits in a complex tend to change together upon mutation, e.g., the particular covariation structure among the individual traits in the mammalian skull (Jamniczky

---

[2]" ... *mechanistic* models of how developmental systems produce phenotypes and how changes within these systems contribute to corresponding changes in phenotypes. This differs from the Modern Synthesis view that evolutionary processes are driven largely by (random) genetic changes, on the one hand, and by functional interactions of organisms with their environment, on the other hand, ... What the molecular analysis of developmental processes and regulatory gene networks provides is a *mechanistic* understanding of both the development and evolution of phenotypic characters." (Laubichler 2010, pp. 202 and 208, my emphasis)

and Hallgrímsson 2011). Such an integration of individual traits permits coordinated change of several phenotypic traits based on a few genetic changes. The reverse situation is that some traits are uncorrelated and thus one can be modified by natural selection without impacting other traits and diminishing their fitness contribution. Even traits on different levels of organismal organization, such as developmental processes and morphological structures, can evolve independently of each other (Brigandt 2007). It is the particular mechanism of development that explains how among organisms of a species functional morphological variation can be generated, how complex traits can change in an integrated fashion, and how some traits can vary and evolve independently of each other.

Evo-devo is an integrative approach that is currently making a lot of progress, yet its future disciplinary nature is not yet settled (Brigandt and Love 2010, 2012a). Typically, evo-devo is portrayed as an emerging synthesis of evolutionary biology and developmental biology (which were unrelated for most of the twentieth century), with developmental genetics creating the link. However, the label 'synthesis' suggests the merging of different fields into a single field. This is inconsistent with the plurality of partially independent disciplines and subdisciplines within contemporary biology. There are also open questions of how to relate evo-devo (and developmental biology) to more traditional approaches within evolutionary biology (Laubichler 2010; Wagner 2007), with some tending to describe evo-devo as an autonomous discipline that has its own questions, explanations, and methods (Hendrikse et al. 2007). In any case, one can capture evo-devo's integrative nature by highlighting that it is an *interdisciplinary* approach (Love 2013). The complex explanatory problems it addresses require the use of ideas from many different biological disciplines (Brigandt 2010; Brigandt and Love 2012a; Love 2008a, b). In addition to evolutionary genetics and developmental biology—which are explicitly noted by the notion of a synthesis of evolution and development—accounting for evolutionary novelty involves intellectual contributions from paleontology (fossil data on ancestral morphological change), phylogeny (trees of species to determine character polarity and phylogenetic junctures relevant to a character change), and morphology (composition of structures and performance of anatomical functions), among other fields. Explanatory frameworks in the context of evo-devo coordinate data, ideas, and explanatory models from a variety of fields, and evo-devo reveals its integrative potential by setting up new connections between such items of knowledge (Brigandt 2010).

Evo-devo is a diverse field including different methodological and theoretical perspectives (Brigandt 2012b; Love 2015). Though many experimentally minded evo-devo biologists may not recognize it under the 'evo-devo' label (Green et al. 2015), there is the mathematical modelling of phenomena studied by evo-devo, and my discussion will pay particular attention to such mathematical models due to their relevance for philosophical accounts of mechanistic explanation. Evo-devo's diversity holds even for its development component. Many studies of development focus on the regulation of individual genes (Prud'homme et al. 2011; Shigetani et al. 2002), or complete gene regulatory networks (Davidson 2006, 2010; Linksvayer et al. 2012), so that morphological evolution is conceived as change in gene regulation (Carroll 2008; Davidson and Erwin 2006; Erwin and Davidson 2009; Laubichler 2009). However, not all explanations are restricted to developmental genetics. *Epigenetic* processes in development, and their role in morphological evolution, are often taken into account (Forgacs and Newman 2005; Hallgrímsson and Hall 2011; Hallgrímsson et al. 2007; Müller and Newman 2003; Newman and Müller 2000, 2005; Schnell et al. 2008). Though causally enmeshed with gene activity, epigenetic processes are any influences on development that do not solely depend on the expression of genes, for example biophysical interactions among cells, mechanical influences on tissues, and physical and biochemical processes of self-organization. The environment can also influence epigenetic-developmental processes (Gonzalez et al. 2011) and this is particularly important in the case of phenotypic plasticity, the ability of organisms to develop several phenotypic outcomes depending on environmental factors (Gilbert 2001; Whitman and Agrawal 2009). Phenotypic plasticity can be significant for the evolution of novel morphological traits, and shows that sometimes morphological evolution is initiated by phenotypic change, with genetic change only subsequently taking place (Palmer 2004; West-Eberhard 2003, 2005). There are also contexts in which the active behavior of an organism during its development or its adult life-time is instrumental in the evolution of novelty (Müller 2003; Palmer 2012).

Overall, evo-devo studies development and its impact on evolution in terms of the relations and interactions among entities and processes on several levels, from the molecular and cellular to the organ and whole-organism, which (apart from being interdisciplinary) is an additional way in which evo-devo's explanations are integrative. The idea championed by many of its practitioners that evo-devo is a mechanistic approach obscures that beyond explaining morphological evolution in terms of changes to developmental mechanisms, explanatory contributions from several other disciplines than developmental biology are needed, requiring scientists to take a balanced approach that does not neglect considerations about historical patterns for questions about causal processes, and that addresses both empirical and theoretical issues (Brigandt and Love 2012a). While some evo-devo biologists contrast explanation in terms of developmental mechanisms with traditional evolutionary theory's explanation in terms of the dynamics of allele frequencies within populations, there are possible connections between developmental and population processes (Rice 2008, 2012; Wagner 2007).

3   Explanatory Relevance and How Mathematical Models Can Mechanistically Explain

Philosophical accounts of mechanistic explanation have been developed as an alternative to seeing explanation as the derivation from laws (Brigandt 2013a). In molecular and experimental biology, there are hardly laws, and instead research involves breaking a whole system down into its concrete structural parts. Rather than being able to logically deduce an explanandum from laws and other premises, explanatory understanding stems from mentally simulating how a mechanism's components are organized and interact so as to bring about the phenomenon to be explained (Bechtel and Abrahamsen 2005).

The availability of different philosophical models of explanation does not necessarily entail that explanation in terms of mechanisms and quantitative principles are incompatible. However, Rasmus Winther (2006) distinguishes between compositional biology (which produces explanations in terms of the parts of a whole) and formal biology (which explains using mathematical theories) as distinct styles of theorizing used in different fields (but see Winther 2011). As one of the main developers of accounts of mechanistic explanation, Carl Craver (2006, 2007, 2008) has gone so far as to claim that while mathematical models are widely used and indeed represent and predict, unlike mechanistic accounts they typically do not *explain*. At least, Craver contends for every mathematical model he has considered that it is merely a phenomenological model, which represents without explaining. He has illustrated his position in the case of the Hodgkin and Huxley model which describes how action potentials are generated and transmitted along the surface of neurons. As a characteristic change of a neuronal membrane's electric potential, there is a quantitative aspect to an action potential, and the original work by Hodgkin and Huxley modeled this phenomenon using equations. Yet Craver claims that the Hodgkin and Huxley model merely represented the phenomenon, and the explanation came with later research, in particular the discovery of the molecular structure and mechanistic operation of transmembrane ion channels (Craver 2006, pp. 364–367, 2007, pp. 54–58; for arguments that the Hodgkin and Huxley equations are explanatory see Levy 2014; Weber 2005, 2008).

Craver maintains that only causal-mechanistic accounts explain, and his vision of a mechanistic explanation involves entities, qualitative activities, and sufficient detail about their organization and physical interaction (as opposed to the representation of how a change in one entity quantitatively relates to some other, non-contiguous entity without a consideration of intermediates).

> Mechanistic models are *ideally complete* when they include all of the relevant features of the mechanism, its component entities and activities, their properties, and their organization. (Craver 2006, p. 367)
>
> Complete descriptions of mechanisms exhibit productive continuity without gaps from the set up to termination conditions. (Machamer et al. 2000, p. 3)

Craver's argument for his dichotomy between mechanisms and mathematical models is that since not every representation is an explanation, there have to be normative constraints on when a representation is explanatory (Craver 2006, pp. 357–358). But philosophical proponents of mathematical modeling cannot provide such constraints:

> My objection to the covering-law model … is that [it is] *too weak to capture the distinctions between description and explanation*, between explanation sketches and (more) complete explanations, and between how-possibly and how-actually explanations. (Craver 2008, p. 1024, my emphasis)

> ... the strong predictivist has difficulty expressing the explanatory limits of mere how-possibly models or theories ... that could produce the phenomenon in question but that, in fact, do not produce the phenomenon. (Kaplan and Craver 2011, pp. 608–609)

Craver's (2007) mechanistic approach is able to provide constraints for when an account is explanatory. Not any physical part of a whole qualifies as a component of a mechanism; instead, on his account something is a component if it is causally relevant to the mechanism's behavior, in that changing this component would lead to a change in the mechanism's activity. To spell this out, Craver relies on James Woodward's (2003) interventionist theory of causation.

I agree that not every representation is an explanation, in fact, some mathematical models are not even meant to be explanatory. The use of models and equations serves various epistemic purposes (Bogen 2005; Weisberg 2013). Some models aim at capturing data as precisely (and simply) as possible, so as to highlight statistical trends, without the assumption that the model reflects the causal origins of the represented features. A quantitative model may be needed to represent a phenomenon that is explained by a different representation. A model may also be set up for heuristic purposes and serve theoretical discovery by revealing what (surprising) phenomena follow from assumptions that need not be realistic.

However, some mathematical models are intended to be explanatory. A clear counterexample to Craver's mechanisms/models dichotomy is systems biology, an approach that, among other things, studies molecular and cellular processes by developing mathematical models based on experimentally acquired molecular data (Baetu 2015; Boogerd et al. 2007a; Brigandt 2013c; Fagan 2012a, b; Gross 2015; Isaad and Malaterre 2015; Mekios 2015; Théry 2015). Mathematical models can be explanatory by representing causally relevant factors. Craver's reservation about quantitative models ignores that Woodward's account of causation and causal explanation, on which Craver's mechanistic approach relies, is formulated for *quantitative generalizations*. Woodward (2003) represents putative causes as variables that may take quantitative values, and possible causal relations are equations involving variables (he includes examples from physics and economics). His central notion—to which Craver likewise appeals—is 'invariance under intervention,' as A is a cause of B only if the quantitative relation between A and B is invariant under some interventions on A. The idea is that while an intervention on A changes the value of B, the *relation* among A and B is thereby not broken. In fact, one can change B by manipulating A precisely because this causal relation is still intact. Invariant generalization is Woodward's proxy for laws of nature, as his goal is to develop an account of explanation for scientific domains where there are no laws. A universal law holds across an enormous range of conditions. A generalization between A and B may be invariant only under some small range of changes to A, but this generalization still serves the purpose of causal explanation.

Moreover, Woodward (2002) has applied his account specifically to *mechanisms*, laying out conditions "for a representation to be an acceptable model of a mechanism" (p. S375). One condition is 'modularity,' the situation where one quantitative generalization can be changed by intervention without modifying other generalizations. This condition means that different generalizations represent non-overlapping parts of an overall mechanism. While Craver (2008) complains that the proponents of mathematical models do not provide criteria for what constitutes an explanation, Woodward (2002) emphasizes that his account has normative impact. While many models in psychology (unlike in biology) may not reveal actual causes, his account specifies what would count as a cognitive mechanism.

In mathematical modeling contexts, there is a good deal of talk about making 'predictions' from models. The biological cases discussed in the following sections are no exception (e.g., Fisher et al. 2007; Manu et al. 2009b). While Craver tends to view prediction and explanation in opposition, in the below and many other cases where mathematical models are based on molecular data, the model and its predictions are to hold not only for the naturally occurring organism, but also for different experimental modifications to the organism, typically the phenotypes of various mutants. Since the model is meant to capture the effects of interventions on molecular components internal to the biological system modeled, the model is meant to get at causal features, so that the 'predictive model' is explanatory (assuming that it is fully realistic). In some cases a model-derived prediction about a novel modification to the organismal system motivates the experimental creation of a new mutant, so as to test the prediction about the intervention in turn (Herrgen et al. 2010; Manu et al. 2009b).

To Woodward's account of when equations represent causes, I add the following considerations about *explanatory relevance*, which I discuss in more detail in Brigandt (2013c). A biological system may contain a variety of causes, and my point is that which of them are explanatorily relevant depends on the *particular explanandum* considered, and thus on the epistemic goal.

> (*ER*) A component of an account representing causal features (including a mathematical equation) is explanatorily relevant, if omitting it or changing it results in an account from which the particular explanandum does not follow any longer. Features that are not explanatorily relevant for the explanandum at hand (and the criteria of explanatory adequacy) are to be excluded from the explanation.

By this criterion, not every quantitative detail is significant. If the explanandum can be derived not only given a mathematical model's precise parameter values (representing the actual quantities in nature) but still follows under a range of values, what explains is the fact that the parameter values are within a particular range, since being inside vs. outside of this range makes the difference to the explanandum (Strevens 2008). But molecular-mechanistic detail can likewise be explanatorily irrelevant. Mathematical models of molecular and cellular processes may represent the relations among genes and gene products as networks, without including the structure of these molecules (network nodes) and how one molecule mechanistically affects another through intermediate steps and structural interactions (see Footnote 4 below; Bechtel 2015; Levy and Bechtel 2013). This exclusion of detail is legitimate—and in fact mandatory—if from the structure of the network and the individual functional relations among the entities modeled the explanandum phenomenon follows.

An analogous point pertains to situations in which one does not have to reductively break down a causal factor into its lower-level components. Recently David Kaplan and Carl Craver (2011) have acknowledged that mathematical models can explain, if they represent entities and activities of mechanisms and the equations correspond to causal relation. This suggests that it is not their being mathematical that has motivated Craver's reservations about mathematical models, but that he deems most of them non-explanatory on the following (though invalid) ground:

> ... the variables [dynamical models] posit are not low level (e.g., neural firing rates) but, rather, macroscopic quantities at roughly the level of the cognitive performance itself ... If so, they are phenomenal models. They describe the phenomenon. They do not explain it any more than Snell's law explains refraction or the Boyle-Charles gas law explains why heat causes gases to expand. (Kaplan and Craver 2011, p. 616)

This appears to confuse being explanatory and being a *reductive* explanation.[3] The Boyle-Charles gas law does explain why the gas volume expanded by reference to an increase in temperature. If *A* causes *B*, then *B* can be explained in terms of *A*, regardless of them being on the same level. To be sure, explaining why this causal relation holds requires an appeal to lower-level entities (statistical thermodynamics in the case of why the temperature-volume relation of the Boyle-Charles law obtains). But this is a different explanandum, and if it was the explanandum Craver actually had in mind, the above quote would assert that the Boyle-Charles laws does not explain the Boyle-Charles law and Craver's challenge to mathematical models would be the trivial observation that they do not explain themselves. In the case of the (non-trivial) explanandum of why the gas expanded, according to criterion *ER*, the lower-level entities should not be included as they are not explanatorily relevant in the sense of making a difference to the given explanandum (see Footnote 7). Mathematical models can legitimately abstract away from some molecular-mechanistic detail, for instance by aggregating the effects of many individual molecular events (Levy 2014). Abstraction is explanatorily virtuous, not because it makes models simpler or more general and unified (by having more concrete instances), but because upon proper abstraction a model includes precisely those factors that are relevant to the explanandum (Putnam 1975; Strevens 2008).

The lesson is that one cannot categorically say that a model is non-explanatory or—to use Craver's label—phenomenological. Models are always developed for certain epistemic purposes, and a model's explanatory credentials depend on the particular explanandum considered (and on additional standards of explanatory adequacy). In Sects. 4 and 5 I use criterion *ER* to argue that for some explanations

---

[3]Likewise, in his argument that the Hodgkin and Huxley equations are merely phenomenological, Craver (2006) acknowledges that the equations "allow neuroscientists to predict how current will change *under various experimental interventions*" (p. 363, my emphasis)—which given Woodward's interventionist account of causation entails that the equations capture some causal factors and thus explain. Craver still rules them to be non-explanatory, apparently on the grounds that they do not provide an account of how the quantitative relation is brought about by lower-level constituents.

found in evo-devo equations are *indispensable*, in that without the use of any equations the explanandum phenomenon does not follow at all. As it may seem trivial that quantitative models are needed if the explanandum is the precise temporal dynamics of a system, I will point to *qualitative* explananda where quantitative equations are still needed as part of the explanans. For instance, when a qualitative phenomenon is an emergent property resulting from nonlinear interactions among the components of a system, the phenomenon cannot be anticipated by the component's qualitative interactions—on which philosophical accounts of mechanistic explanation have focused. Such a qualitative property can only be predicted and explained by citing the quantitative nonlinear interactions. (Further examples of qualitative explananda requiring equations are discussed by Brigandt 2013c in the context of systems biology.)

Given Craver's worry that too many mathematical models are merely phenomenological and black-box a system without representing its internal causal workings, I highlight that the quantitative models discussed below were developed based on molecular-mechanistic data and in many instances are tested against the properties of mutants and other experimental interventions. A mismatch between theoretical models and biological systems motivates changes of a model or further experimental investigation, so that there is interplay between molecular-experimental research and mathematical modeling (see also Baetu 2015).

4 Mathematical Models of the Origin of Morphological Structures

In addition to the experimental investigation of the development of morphological structures and its explanation in terms of the spatial organization and qualitative interactions among molecular and non-molecular entities, there are mathematical models of developmental processes (Morelli et al. 2012; Murray 2003; Schnell et al. 2008). Several such developmental models are relevant to evolutionary issues or explicitly meant to be evo-devo accounts, as they explain how developmental processes create phenotypic variation within a species, how developmental mechanisms can be changed so as to result in morphological change across species, and how a novel morphological structure and its underlying mode of development originated in evolution.

Many models of the temporal formation of patterns and morphological structures involve reaction-diffusion equations. These nonlinear equations are partial differential equations (representing change in both space and time) containing terms representing local chemical reactions and terms expressing the spatial diffusion of reaction products. A common version is *activator-inhibitor systems*, which—while having two molecular entities at its core—are compatible with the situation that the entities are the products of gene activation (rather than a simple chemical reaction) and are spatially transported across cells by active mechanisms (rather than passively diffusing in a non-cellular medium). By positive feedback the activator enhances its own production—locally increasing its concentration—and it also positively regulates the inhibitor, which because of its higher diffusion rate acts in surrounding regions and there inhibits the activator. Over time this process can lead to spatial waves of substance concentrations and the formation of stable patterns, such as stripe patterns or regularly spaced spots, as seen in sea shell coloration patterns (Meinhardt 2003, 2009). As the labels 'activator' and 'inhibitor' pertain to the mutual relations between these two components, both the activator and the inhibitor may either activate or suppress downstream developmental pathways, depending on the biological case. This means that the spatial distributions of the activator and inhibitor can cause the developmental formation of morphological features (Meinhardt 2012). As first proposed by Alan Turing (1952), over most of their history reaction-diffusion systems have offered speculative, how-possibly explanations of the biological development of patterns, as the underlying molecular substances and interaction properties were unknown. But nowadays experimental evidence for the presence of activator-inhibitor systems exists, for example the interaction of pigment cells in zebrafish (Nakamasu et al. 2009), the regeneration of hair stem cells in mice and rabbits (Plikus et al. 2011), and palate growth in mice (Economou et al. 2012).

Of evolutionary significance are mathematical models of the development of mammalian teeth. The cusp number and shape of the teeth of a mammal are quite distinctive, making them a criterion for taxonomically distinguishing species. Teeth develop based on the signaling among epithelial and mesenchymal tissues, where the tooth crowns form based on a limited number of epithelial signaling centers called the enamel knots. The model of Salazar-Ciudad and Jernvall (2002) proceeds from prior experimentally generated information about the relations among several molecular components, i.e., genes producing proteins which in turn affect the expression of other genes and their products (Fig. 7.1a). From this causal network, a basic activator-inhibitor-system is abstracted, where in this case the activator suppresses epithelial growth and induces epithelial cells to differentiate to form an enamel knot, while the inhibitor promotes growth and represses knot differentiation (Fig. 7.1b).[4] The computational model predicts the three-dimensional tooth shape and the activator and inhibitor distributions across developmental time (Fig. 7.1c). These predictions can be compared with empirical, in vivo data (Fig. 7.1d). Not only does the prediction align with the developing tooth shape in mice, but the predicted activator and inhibitor distributions roughly align with the expression patterns of p21 and Fgf4, respectively, seen at different developmental stages of mice and voles, which suggests that the mathematical model is realistic.

---

[4]Note that the model abstracts away from entities mediating the interaction of the activator and inhibitor, e.g., DAN (Fig. 7.1a, b). This omitting of molecular-mechanistic detail is licit assuming that it does not alter the functional interaction and dynamics of the activator and inhibitor. If so, by my criterion *ER* such (for the target phenomenon) explanatorily irrelevant detail ought to be excluded from the explanation. This shows that a mechanistic account of *how* an effect is produced (citing all intermediate steps and structural interactions) and an explanation of *why* it occurs are consistent, but not identical.

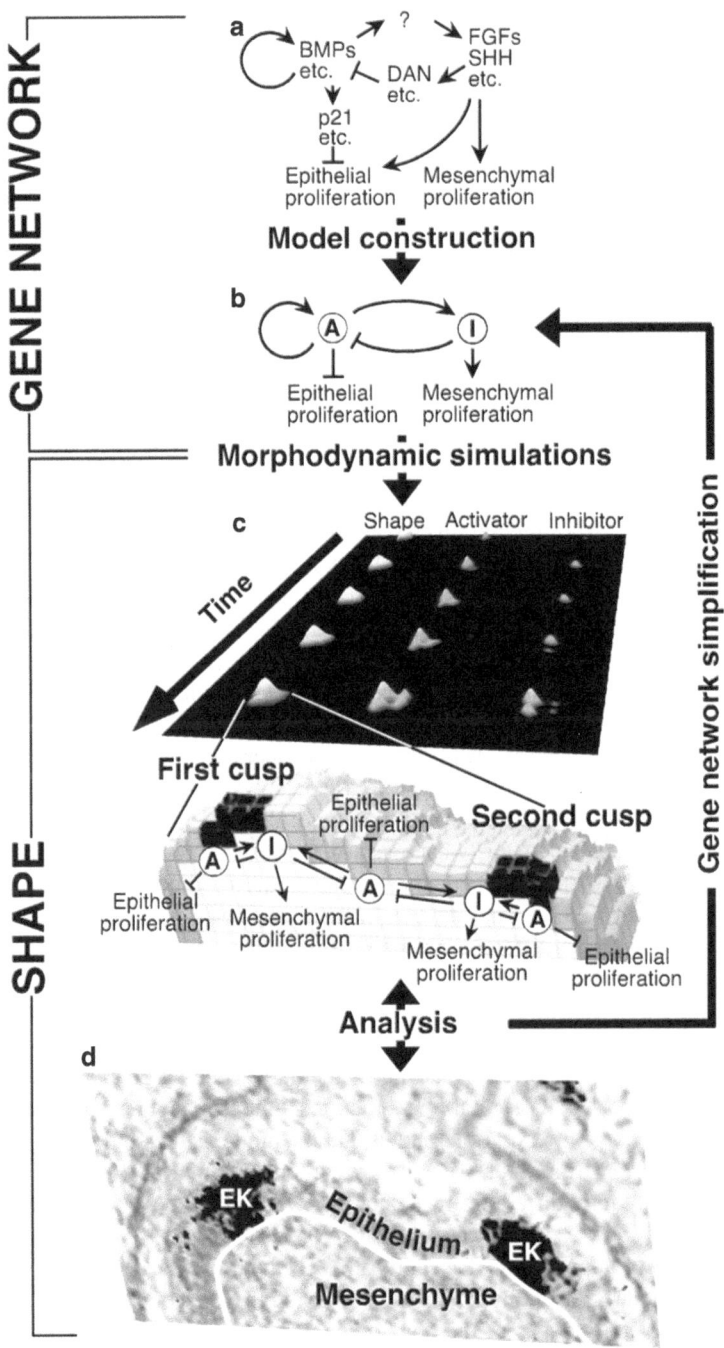

Fig. 7.1 Modeling of the development of mammalian teeth. An experimentally obtained causal network of molecular components (**a**) forms the basis of a simplified activator-inhibitor system (**b**). The mathematical model's prediction of the three-dimensional tooth shape and the distribution of the activator and inhibitor across developmental time (**c**) is compared to empirical information about mice teeth (**d**); *EK* enamel knot (From Salazar-Ciudad and Jernvall 2002)

Importantly, Salazar-Ciudad and Jernvall's (2002) model is able to generate teeth with different cusp numbers, cusp positions, and overall shapes by a variation of some of the parameters, yielding clues to the developmental basis of morphological evolution across species. Salazar-Ciudad and Jernvall (2010) present an improved model, that apart from gene activation by molecular signaling, takes into account the mechanical forces that tissues exert on cells. This model is compared to empirical data in ringed seals and used to account for the large variation in tooth shape found within this species. According to these mathematical models of tooth development, large morphological differences can often be achieved by small developmental changes. This shows that the correct *mechanistic* explanation of why a tooth with a certain cusp number and position develops requires the use of a *mathematical* model with precise parameter values.

One of the best-studied cases of vertebrate morphological development is the limb skeleton, which is essential to understanding the evolutionary origin of fins in fish and their subsequent evolutionary modification, in particular the fin-to-limb transition in land-living vertebrates (Hall 2006). In addition to a plethora of fossil studies and experimental investigations in extant species (e.g., chicken), the formation of the broadest aspects of the shape of different skeletal elements and their basic spatial position has been mathematically modeled using activator-inhibitor systems (Hentschel et al. 2004; Newman et al. 2008; Newman and Müller 2005). This basic skeletal pattern in the adult organism is a fairly *qualitative* phenomenon and its developmental explanation involves quantitative models. For it needs to be understood how relatively undifferentiated tissues give rise to a highly structured pattern. The spatial pattern resulting from the operation of biological processes involving nonlinear interactions (as in the case of activator-inhibitor systems) cannot be predicted from the qualitative organization of the system's components, so that some equations are explanatorily necessary (criterion *ER* from Sect. 3).

Zhu et al. (2010) present a mathematical model of limb development which not only replicates the normal development of the basic skeletal features of the chicken wing, but also different instances of modified development, including the experimental removal of the apical ectodermal ridge (a causally crucial zone at the tip of the growing limb bud) at different points in early development, and the expansion of the early limb bud either by tissue graft or as seen in two different genetic mutants. This indicates that the mathematical model gets at some causal-mechanistic aspects of the actual phenomenon studied. In any case, the attempt to capture the effects of interventions shows that the model is meant to be explanatory (see the discussion from Sect. 3). With their model, Zhu et al. (2010) are able to generate several quite different fin skeletal patterns known only from distinct taxa of fossil fish.

Mathematical approaches of a quite different type model *genetic oscillations*, which are regularly oscillating levels of gene activity and thus of gene products. This is the molecular basis of the development of segments in vertebrates, and of relevance to an evolutionary explanation of the origin of the vertebrate body plan and the diversity of segment number in different vertebrates (30 in some fish to hundreds in snakes). In early embryonic development, these segments originate as somites, which then guide the formation of vertebrae and ribs. Crucially, somites develop in a rhythmic pattern, where one somite forms temporally after the other, from the anterior to the posterior end of the embryo, until the species-specific somite number is reached. The basic explanation of this rhythmic development of somites of equal length is the clock and wavefront model (Dequéant and Pourquié 2008; Oates et al. 2012). It involves the interaction of two processes, a segmentation clock consisting of synchronized cellular oscillations in the tissues where somites form, and a wavefront of a molecular substance moving at constant speed from anterior to posterior end (where somites yet have to form). When the wavefront passes by oscillating cells, it arrests their clock at the present stage of the cycle, so the temporal pattern of the clock is transformed into a repeated spatial molecular pattern along the anterior-posterior axis. In a nutshell, the clock determines the timing of the formation of a new somite, and the constantly moving wavefront determines the position of the somite boundaries. The length of each somite is the speed of the wavefront multiplied by the period of the clock, which is 30 min in zebrafish embryos (the experimentally best-studied model organism in this context) but longer in different land-living vertebrates.

In the last decade, many of the molecular components and pathways making up the wavefront mechanism and the segmentation clock have been identified by the analysis of mutants with defective segmentation, among other techniques (Dequéant and Pourquié 2008; Oates et al. 2012). Figure 7.2 shows major components of the segmentation clock in mice. Apart from the mechanistic interaction between the wavefront and the segmentation clock, research efforts are devoted to investigating how oscillations are mechanistically generated within each individual cell and how the oscillations are synchronized across cells by means of cell-cell signaling. Different aspects of somite formation have also been mathematically modeled (Baker and Schnell 2009; Baker et al. 2008; Mazzitello et al. 2008; Santillán and Mackey 2008).

Mathematical models that focus on oscillations within an individual cell are gene regulatory network accounts, which by means of differential equations represent the interaction among the levels of mRNAs (transcribed from oscillating genes) and of proteins (some of which are regulating gene transcription). Even the explanation of the oscillation of one such gene's activity requires a mathematical account. Take the mouse gene *Hes1*, whose oscillating transcription has been directly shown by real-time imaging studies (Masamizu et al. 2006). A non-mathematical mechanistic account can lay out the various entities involved and how they qualitatively interact with each other, e.g., whether there is positive or negative causal interaction. But this alone does not yield the phenomenon to be explained. Figure 7.2 shows that the protein produced from *Hes1* in turn inhibits the transcription of *Hes1*, so as to create a negative feedback loop.[5] Negative feedback loops are known to yield oscillations. However, whether this results in stable oscillations or damped oscillations that

---

[5]Figure 7.2 schematically depicts the four oscillating genes *Hes1*, *Hes7*, *Hes5*, and *Hey2* (all of which engage in negative feedback) together.

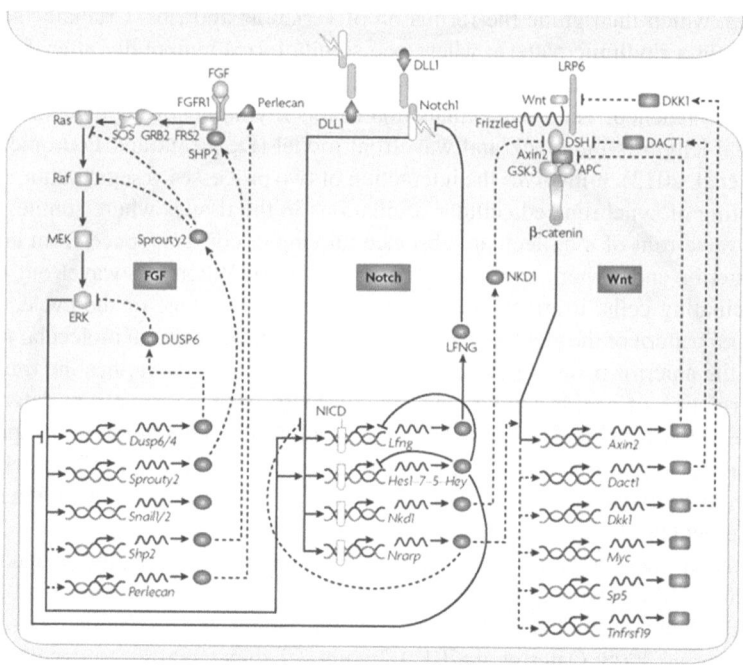

Fig. 7.2 The segmentation clock in mice. Apart from components involved in the signaling and synchronization with another cell (*top*), the figure shows the basic causal network of one cell's components underlying the generation of oscillations. Three major signaling pathways (FGF, Notch, and Wnt) are represented, where the area with light background highlights oscillatory genes, their mRNA transcripts, and their protein products (From Dequéant and Pourquié 2008; *dashed lines* are regulatory interactions that were inferred from other species or microarray data)

fade out after some while, depends on the quantitative interaction parameters, so the explanation has to include some of this quantitative detail.[6] Moreover, the explanation has to show why—beyond the presence of one negative feedback loop—*Hes1* stably oscillates in its actual mechanistic context which includes other components influencing its transcription. Overall, the various oscillating genes of the FGF pathway and of the Notch pathway oscillate in phase, while the genes of the Wnt pathway (see Fig. 7.2) are in antiphase to this (Dequéant et al. 2006). This fact likewise requires explanation. The mathematical model by Goldbeter and Pourquié (2008) addresses this synchronization among the FGF, Notch, and Wnt pathways by modeling the quantitative interactions among several of the components involved.

---

[6]Explaining why the oscillation has a period of 120 min (in mice) would definitely necessitate a quantitative account (see also Baetu 2015). In the related context of circadian rhythms (genetic oscillations with a period of about a day), for a philosophical account indicating the relevance of mathematical modeling see Bechtel and Abrahamsen (2010, 2011) and Bechtel (2013).

Another important aspect of somite formation is that the oscillations of different cells are in synchrony. Notch-Delta signaling among adjacent cells is one mechanistic component underlying this, but an explanation of why synchronized oscillations occur requires a mathematical account. Mathematical models of between-cell synchronization are typically phase oscillator models. These represent each cell as one oscillator with a certain phase, so as to abstract away from the complex gene regulatory mechanism generating oscillations within the cell (which has the advantage that the many interaction values within a cell need not be experimentally known). The models mathematically study how the phases of different cells influence each other with a time-delay, showing that the coupling between adjacent cells results in overall synchrony (Morelli et al. 2009). Such mathematical modeling can accompany experimental studies of manipulated synchrony behavior, by means of changes to the timing of Delta-Notch signaling across cells (Riedel-Kruse et al. 2007). Using a phase oscillator model approach, Herrgen et al. (2010) theoretically predicted how the segmentation clock period would change in a novel zebrafish mutant. Their prediction was borne out upon creation of the mutant. It is experimentally known that the individual oscillating cells move, changing their relative position, and that synchronization across cells is recovered upon perturbations that initially destroy synchrony. Uriu et al. (2010) present a quantitative model that shows that synchronized oscillations can be maintained under random cell movement, and that such random movement in fact reduces the time needed to reestablish synchrony upon perturbation.

The questions of why somites of identical length develop and why stable gene activity oscillations of a regular period occur (be it within a single cell, or in synchrony across cells) are *qualitative* explananda. Knowledge of the structure of the mechanism alone, including the molecular components and their qualitative (positive or negative) interactions, is insufficient to predict that regular oscillations will in fact occur. Thus, by my criterion *ER*, the mechanistic explanation of why somites of identical length develop and of why periodical oscillations occur (in the actual, quite complex system) requires the involvement of equations laying out the quantitative and dynamic influences among the components.

5   How Mechanisms Adaptively React to Modification: Robustness, Phenotypic Plasticity, and Modularity

Robustness, phenotypic plasticity, and modularity are developmental properties—an organism exhibits them because of its particular mode of development—but because of their evolutionary implications, they are highly important for evo-devo. All three properties pertain to how developmental systems adaptively react to modifications or how they permit modification while remaining functional, and thus yield morphological evolvability and provide the basis for the evolution of structural novelty (Sect. 2). In the following section, I will examine the implications of such dispositions to react to modification for philosophical conceptions of mechanisms, while the present discussion on robustness also continues the theme on the relevance of mathematical models.

Robustness is the ability of a cellular or developmental system to produce certain traits despite perturbations to the system. Robustness to non-genetic changes means that a phenotype develops regardless of certain environmental disturbances or internal developmental perturbations. Robustness to genetic changes is also possible if upon mutation the same phenotype is still present in other organisms possessing the mutation. The latter is often encountered in experimental contexts. Knockout studies, in which a particular gene is deactivated in a model organism, are conducted to trace the developmental effect of this gene. When there are good reasons to assume that a gene is part of a developmental pathway leading up to a phenotype, it comes as a surprise that the knockout hardly shows any phenotypic difference. But this is possible when the knockout organism adjusts the regulation of other genes so as to compensate for the deactivated gene.[7]

Robustness is of evolutionary significance for the following basic reasons (Kitano 2004; Wagner 2008). In the case of robustness to genetic modifications, some mutants will still have the same phenotype and not be removed by natural selection, so that this type of robustness leads to the accumulation of cryptic variation, i.e., genetic variation without phenotypic variation. Though it does not make any phenotypic difference for the time being, such increased genetic variation sets the stage for rapid future evolution, once the cryptic variation is uncovered by further genetic or environmental changes (Delattre and Félix 2009; Masel and Siegal 2009). Robustness to non-genetic modifications allows organisms to survive in changing environments, and is thus the product of evolution. But this ability to develop a functional phenotype in the case of environmental impact also has the side-effect that even in the case of a genetic change, the resulting phenotype is likely to be functional. The presence of such new genotypes yielding functional phenotypes—some of which may be preserved by natural selection—enables morphological evolution, in other words, evolvability (Gerhart and Kirschner 2007; Kirschner and Gerhart 2005; see also Merlin 2015).

Robustness can be found on various levels of organization, from the genetic code and the structure of RNAs and proteins, up to more complex organismal subsystems (Wagner 2005b; Huneman 2010; see also Breidenmoser and Wolkenhauer 2015). Individual metabolic pathways and complete metabolic networks can be robust in that the overall metabolic flux is maintained even if the reaction rate of individual

---

[7]At the end of Sect. 3, I pointed out that not every explanation requires the reductive decomposition of a mechanism's components. According to my criterion *ER*, if the component is explanatorily relevant—if changing it would lead to a change in the surrounding mechanism's features to be explained—but the component's lower-level constituents are not relevant to the particular explanandum, then the explanation should cite the component but not its constituents. The component exhibiting robustness is a clear way in which this can be the case, as a change in the component's constituents does not make a causal difference to the component's robust properties (which are relevant to the explanation).

enzymes is significantly decreased. With a robust gene regulatory network, the phenotypic trait (e.g., a spatial pattern of signaling molecules, eventually giving rise to anatomical structures) forms regardless of whether some the network's genes are altered or deactivated. Robustness can result from redundancy, where two copies of a structure (e.g., a duplicated gene) are present so that the loss of one structure does not have any impact (Dean et al. 2008). Even in such cases, there often have to be functional amendments, as an active mechanism has to turn on the second gene copy which is normally not expressed (Baggs et al. 2009; Kafri et al. 2005). Often robustness is not just due to structural redundancy, but is a *distributed process* in that the overall system undergoes various functional changes to compensate for the loss of one component (Ihmels et al. 2007; Wagner 2005a, b). A case in point is developmental regulatory networks which contain several feedback loops so as to buffer against perturbations (Li et al. 2009).

Another example is exploratory behavior, which yields robustness on several levels of organization through its ability to generate many, if not an unlimited number, of phenotypic outcome states, any of which can be physiologically stabilized if it is adaptive to the organism (Gerhart and Kirschner 2007; Kirschner and Gerhart 2005). Microtubules generate the shape of eukaryotic cells as each of the many microtubules grow and shrink in an exploratory fashion, until some of their lengths are stabilized by a signal from outside the cell. In this fashion, many cell shapes can be produced in an individual organism, permitting the remodeling of cells. The vertebrate limb consists of various skeletal elements, muscles, blood vessels, and nerves, which need to be arranged in a certain way to yield a functioning limb. This organization is not represented in some organismal blueprint; rather, it emerges by means of exploratory developmental processes, in which blood vessels and nerves grow from the body core toward the developing limb, guided by chemical signals and their surrounding milieu, with those nerves that do not find a target degenerating by cell death. Many perturbations to development will have a temporary impact, but not prevent the development of the final anatomy. Apart from illustrating that exploratory behavior robustly produces functional phenotypes by means of a distributed process, this kind of robustness to non-genetic modifications also enhances evolutionary change. The size and placement of limbs differs dramatically across vertebrates, but given the mode of limb development a *simple* genetic change to the placement of the limb is likely to yield a functioning limb with all its components properly connected, and thus a heritable, *complex* morphological change (Kirschner and Gerhart 2005).

Beyond the analysis of natural variation within a species, experimental studies offer a clear causal way to demonstrate a developmental mechanism's robustness (Baggs et al. 2009). Sometimes such experiments are very hard to conduct in higher organisms; and exhaustively showing that a mechanism is robust to changes in several components (each across a specific range) requires considering all possible component state combinations. While such a large number of modifications to a mechanism cannot be produced experimentally, a computational model of the system permits different factors to be independently varied to any quantitative degree. Apart from showing *that* a system is robust, the issue I want to emphasize is that sometimes a mathematical model is needed to explain *why* the system is robust—in line with my general thesis that equations may be needed as part of mechanistic explanations. A mechanism's robustness in some properties with respect to certain modifications can count as a *qualitative* phenomenon to be explained, at least the explanandum is not the temporal change in all of the mechanism's quantitative properties. Robustness as a distributed process can be explained only by accounting for the structural and functional organization of a larger organismal system. Some mathematical approaches attempt to infer robustness of a network from the topology of causal relations (Barabási and Oltvai 2004; Huneman 2010). In scale-free networks (where there are few nodes with many connections) the network topology is likely to be such that the system is robust to the elimination of individual nodes (Greenbury et al. 2010; Jeong et al. 2000). However, in general the biological impact of modifications cannot be inferred from gene network topology alone, so that an actual explanation of robustness involves the dynamical modeling of the perturbed mechanism's behavior based on experimental data about quantitative interactions and gene functions (Gross 2015; Siegal et al. 2007).

There are several robustness studies which integrate mathematical modeling and experimentally obtained molecular data. In various unicellular organisms whose genome has been sequenced and molecular functioning has been well-characterized, including the bacterium *Escherichia coli* and the eukaryotic yeast *Saccharomyces cerevisiae*, the robustness of metabolic networks has been mathematically modeled (Edwards and Palsson 1999, 2000a; Smart et al. 2008). Edwards and Palsson (2000b) show that the rate of two individual enzymatic reactions in *E. coli* can be reduced to 15 % and 19 %, respectively, of the optimal rate, without significantly diminishing the system's overall metabolic flux (if an individual reaction's rate goes below these values the system's flux drops rapidly). In contrast, for a third reaction, the threshold rate above which overall metabolic flux is largely unaffected is 70 %. A quantitative explanation is required as the system's response to change (with changes to different reactions having a different impact) depends, among other things, on the quantitative rates of the various reactions in the metabolic network.

The nematode *Caenorhabditis elegans* is among the six most prominent animal model organisms in developmental biology. There are many studies pertaining to the vulva (in hermaphrodites), and its development exhibits robustness. While originally the laser ablation of individual cells was one of the primary experimental methods for investigating the causes of development and the impacts of developmental perturbations, nowadays accounts of the robustness of vulva development can rely on experimental data about molecular pathways and signaling networks (Braendle and Félix 2008; Milloz et al. 2008). Based on information about gene interactions, Fisher et al. (2007) present a mathematical model of vulva development, which is additionally validated by the subsequent experimental verification of two model-derived predictions (one about the wild-type, the other about a mutant). The computational model sheds light on the mechanistic basis of stable cell fate patterns as an instance of robust development.

In the fruit fly *Drosophila melanogaster*, many studies attempt to uncover the molecular interactions and networks underlying the formation of the basic body axes and the different body segments in early embryonic development. One case of robustness is the segment polarity network, which determines in each body segment its anterior and posterior part. The classical study by von Dassow et al. (2000) quantitatively modeled this mechanism involving 48 interaction parameters representing such features as gene transcription rates, decay rates of gene products, and the degree of cooperative interaction among entities jointly affecting gene transcription. Their analysis shows that a functional network results in about 90 % of random interaction parameter value assignments, so that the mechanism is robust to a large number of modifications. More recent accounts incorporate additional molecular detail into their mathematical models to yield clues about different molecular aspects underlying the robustness of the segment polarity network (Albert and Othmer 2003; von Dassow and Odell 2002). The mathematical model of Ingolia (2004) entails that the final segment polarity gene expression pattern forms a stable steady state that is due to the presence of distinct expression states, where each individual cell can be in one such expression state, corresponding to different cell types. Positive feedback among components—which is a qualitative, topological property of a gene regulatory network—is a necessary condition for the existence of distinct stable states. Ingolia, however, argues that positive feedback is not sufficient, as different stable states are present only if the interaction parameters satisfy certain inequalities. Thus the explanation of robustness has to include *quantitative* information about the interactions of gene network components.

A different aspect of *Drosophila* segment development are gap genes, each of which is expressed in a specific continuous region of the early embryo, where a deactivation of a gap gene results in the loss of the corresponding body segment. The mathematical model of gap gene regulation by Manu et al. (2009a, b) is based on and tested by data from high precision gene expression studies. It explains the robustness of gap gene pattern formation by showing that gap gene expression patterns form dynamical attractors, i.e., quantitative states toward which the system evolves and that the system tends to occupy even if temporarily removed from such an attractor state by disturbance. (For a review of models pertaining to different aspects of *Drosophila* segmentation and spatial patterning see Umulis et al. 2008.)

Phenotypic plasticity, as the situation where different phenotypes develop in different environmental conditions for one genotype/organism, is in a sense the opposite of robustness. But in either case, the issue is the ability of a developmental system to develop a functional phenotype. Phenotypic plasticity is not just an environmental change of a passive developmental mechanism, but an adaptive response to external conditions, so that overall a developmental system can produce one phenotype despite environmental perturbations if this is the most adaptive phenotype (robustness), or it can produce different phenotypes in different conditions (plasticity). Of additional evolutionary significance is that phenotypic plasticity makes it possible for a novel phenotype to originate, not by means of genetic mutations, but through environmental changes and only subsequently being genetically stabilized. In such a case phenotypic change precedes genetic change in evolution (Palmer 2004; West-Eberhard 2003, 2005). Phenotypic changes in response to environmental circumstances concern not only physiological and behavioral traits, but even morphological structures. Vertebrate bone can change its size, shape, and density in response to frequency of use and intensity of load, so that a human's asymmetric arm use can lead to a different bone size and mineral density in the right versus left arm (as seen in tennis players), and the developed morphology of fish jaws can be contingent upon the hardness of the particular organism's diet (Müller 2003; Palmer 2012).

While these are instances of continuous phenotypic variation, there are many cases of plasticity consisting in the development of two (or more) qualitatively distinct alternative morphologies, so-called polyphenisms (Gilbert 2001; Whitman and Agrawal 2009). Daphnia (waterfleas) have two morphs. If they are in water containing chemicals indicating the presence of predators, juveniles develop with a large helmet-like extension of their head and an elongated tail which makes them less likely to be swallowed by a predator. The parasitic wasp *Trichogramma semblidis* develops either of two distinct morphologies, one with, the other without wings, depending on whether it grew up inside a butterfly or alderfly host. The genetically identical individuals of insect societies can have cast-specific morphologies (e.g., soldiers and small workers) depending on how they were reared (Whitman and Agrawal 2009). Most aphids are cyclically parthenogenetic and viviparous. During the summer, females reproduce asexually and offspring develops from an unfertilized oocyte inside the mother, who gives birth to live young. After several such asexual generations, in response to environmental cues during the fall, asexual females produce sexual males and females, where sexual females make frost-resistant eggs that are fertilized by male sperm. These eggs overwinter before asexual females hatch from them in the spring. Since in asexual females the oocytes produced do not have a half set of chromosomes, which would result from the reduction division of meiosis found in sexual reproduction, there are also major differences in chromosomal and cellular activities between the sexual and asexual morphs. An interesting question arises: how, depending on the environmental cues, can two such divergent developmental programs be executed by one organismal mechanism using one genome (Davis 2012)?

The final property to be discussed is *modularity*, which is the organization of a developmental system into partially dissociated modules (Braillard 2015). These modules are moderately independent component structures or processes, such that one component can change in evolution without a change in the others, making modularity a developmental property of evolutionary significance and thus of concern to evo-devo (Callebaut and Rasskin-Gutman 2005; Schlosser and Wagner 2004). Modularity can result if the degree of functional integration (number, strength, and complexity of causal relations) within a module is greater than between modules, so that natural selection can adaptively modify one module without diminishing the functionality of others (Wagner 1996). Modules can be present at various level of organization, from gene regulatory networks and signaling pathways, to developmental processes and morphological structures (Bolker 2000; Glass and Bolker 2003; Prum 2005; von Dassow and Munro 1999). Some evo-devo discussions focus on how different modules can be rearranged so as to generate novel phenotypic outcomes, more precisely—since it is not a spatial shuffling of structures but a change in the procedural relations among developmental processes operating at different points in time—how one module that was once causally connected with a second module becomes functionally detached from the latter and causally connected with a different module.

Some instances of modularity have a structural or simple functional basis. Multicellular organisms are spatially arranged into different cells, each of which structurally contains its own DNA, so that corresponding genes can be differentially activated in different cells, making different cell types and cellular behaviors possible. A protein may be structurally arranged so as to have two separate sites of functional interaction, so that one can be modified without the other. For example, in allostery, the protein's active site (and the reaction it catalyzes) is distinct from its allosteric site, where effector molecules can bind so as to enable or disable the operation of the active site. In a similar vein, though triggering a complex developmental and morphological outcome, many cellular signals do not contain the information for the complex response. The signal may lead to a simple response of a receptor, as in case when the receptor either activates or deactivates a downstream developmental process, which actually embodies the complexity. As a result, a highly integrated (and internally hard to modify) developmental process can, by a change of the receptor, become tied to and activated by a quite different signal (Kirschner and Gerhart 2005). Some gene regulatory networks are arranged into separable components, such as input/output switches and plug-in subcircuits, which can be deployed in different combinations (Davidson and Erwin 2006; Erwin and Davidson 2009).

However, for evo-devo modules are not the same as the spatial parts of organisms. Rather, something is a module to the extent to which it can change independently in evolution. Modularity is not just due to an organism's structural arrangement, but due to its functional-developmental organization, where a module's partial dissociation results from the larger developmental context in which it figures (Breuker et al. 2006; Gonzalez et al. 2011; Jamniczky and Hallgrímsson 2011). The different body segments of segmented animals are structurally distinct and can evolve independently, but many developmental pathways are involved in the formation of all segments. Thus, attention to the underlying developmental process is required to understand what makes segments separate modules.

It is instructive that sometimes traits on different levels are separate modules. A morphological structure is generated by a developmental process which is orchestrated by the activity of genes. Despite the presence of such close functional and developmental connections among levels, features on different levels can evolve independently of each other. There are many instances in which a gene is involved in different developmental pathways and the formation of different morphological structures in different species, and, conversely, where the same, homologous structure develops by means of different developmental processes, from different tissues, or by the involvement of different genes in different species (Brigandt 2007; Brigandt and Griffiths 2007; Wagner and Misof 1993). A case in point is digit identity in the bird forelimb. In the hand of typical land-living vertebrates, the bones of the five digits DI to DV develop from precartilage cell condensations CI to CV, respectively. In the evolution of birds, two digits have been lost, where paleontological evidence indicates that the remaining digits are DI, DII, and DIII. However, developmental evidence seems to suggest that the digits of extant birds are DII, DIII, and DIV. This conflict is resolved by the hypothesis that in extant birds there remain condensations CII, CIII, and CIV, but that digits of the identity DI, DII, and DIII develop out of them—the so-called frameshift hypothesis, which is strongly supported by current evidence (Wagner 2005c; Young and Wagner 2011). In other words, while in the ancestor condensation CII developed into digit DII, CII came to develop into DI during the evolution of birds due to a developmental frameshift, raising interesting questions of what features make such a dissociation of a developmental precursor and the structure it gives rise to mechanistically possible. Whether a developmental process and the resulting structure make up one module or are distinct modules, depends on the dynamics of an overall developmental system.

Apart from the point that mathematical models are needed to explain why a developmental mechanism is robust, this discussion has also dealt more generally with robustness, phenotypic plasticity, and modularity as they pertain to the ways developmental systems react to modifications and permit modification. In the concluding section I discuss how this goes beyond the philosophical focus on the actual organization and operation of mechanisms, and offer further considerations of what revised vision of mechanisms and mechanistic explanation is called for.

6 A Broader Philosophical Conception of Mechanisms and Mechanistic Explanation

Although some of the preceding discussion focused on the explanatory relevance of mathematical models, the examples from evo-devo I have analyzed suggest the need for a broader philosophical conception of mechanisms and mechanistic explanation more generally. The considerations I adduce may not be objectionable to philosophers, but they go beyond stereotypical philosophical portrayals of mechanisms. Apart from indicating where such stereotypical characterizations are wrong (at least for developmental processes), the aim here is to point to important aspects of biological mechanisms that philosophers have neglected, so as to lay out a more adequate account of how biologists explain using mechanisms.

While philosophers have focused on the qualitative structure of mechanism components and their qualitative interactions (e.g., binding, activating, opening), this neglects the *quantitative properties and quantitative changes* of these entities and activities. Apart from the specific reaction rates of enzymes and quantitative changes in the concentrations of various metabolites, of particular interest is that rather than just being switched on or off, a gene produces a certain copy number of its transcript per time unit, a quantitative amount that varies among cells and changes in a cell over time. Beyond the precision of cellular processes, these quantitative features are vital for the complexity of developmental processes, which is enhanced by the fact that the regulation of a single gene is often influenced by many transcription factors which interact in a cooperative fashion (Wang et al. 2009). Such synergistic interactions can transform a transcription factor concentration to gene expression rate curve that is shallow (for a single transcription factor binding) into a threshold-like curve where a quantitative difference in a cell-cell signal yields a qualitative cellular response and developmental effect. Quantitative changes of mechanism components are important to the mathematical models of structure formation and robustness discussed in Sects. 4 and 5.

Simplistic portrayals suggesting that a mechanism consists of a fixed stock of entities (that move around and interact) are erroneous in that there is the *disappearance of entities and generation of novel entities*. Many molecular entities of the cell are quite *transient*, with biochemical reactions rapidly transforming one molecule into a different kind of molecule, forming complexes of several proteins, and breaking down entities into smaller molecular components.[8] This also holds for entities on higher levels of organization. In neuronal pruning, individual synaptic connections and axons are removed. Whole cells disappear due to apoptosis, i.e., controlled cell death in which the cell systematically disassembles itself and its remaining fragments are removed by the immune system. Chunks of tissue can disappear based on apoptosis, a process which is instrumental for the formation of morphological structures in normal development. For example, in the limbs of land-living vertebrates, the different digits form by the removal of soft tissue in between the forming digits (Abud 2004). The pathological death of cells and tissues occurs in autoimmune and neurodegenerative diseases. Of particular relevance in our context is the generation of new entities, in fact, of new *types* of entities. New cell types are generated by means of differentiation. During ontogeny (developmental time), tissues, morphological structures and organs are formed that the developing organism did not previously possess. Evo-devo likewise studies the phylogenetic origin of novel structures (not present in ancestral species) in the course of evolutionary time. Although such higher-level entities originate by changes to lower-level entities, the generation of new types of entities has to be a solid ingredient of any philosophical conception of mechanisms, given that the formation of new structures in development and evolution is a major explanatory target for developmental biology and evo-devo, respectively.

When the operation of a mechanism is described as consisting of "regular changes from start or set-up to finish or termination conditions," where "what makes [the mechanism] regular is the productive continuity between stages ... represented schematically by A→B→C" (Machamer et al. 2000, p. 3), such a stereotypical characterization creates the impression that every mechanism consists in a single causal sequence. Yet causal pathways may branch or merge, and in fact

---

[8]Baetu (2015) points out that the functioning of a mechanism can be due not so much to stable entities, but to a stable concentration (of a type of entity), where individual entities are very short-lived and constantly replaced.

form *complex networks* of causal interactions among components. This is reflected by biologists stating that they study 'gene regulatory networks' and by the recent terminological shifts from 'metabolic pathway' to 'metabolic network' and from 'signaling pathway' to 'signaling network.' Researchers emphasize that since there is 'cross-talk' between what used to be considered separate pathways, the larger network needs to be studied (Barabási et al. 2011; Fraser and Germain 2009; Jørgensen and Linding 2010; Layek et al. 2011; Wing et al. 2011). Moreover, the organization of causal interactions is not unidirectional and acyclic, but mechanisms have *feedback loops* (Bechtel 2011). This complex causal structure of mechanisms (the topology of functional relations) needs to be taken into account because it is vital for the mechanism's actual behavior and the higher-level properties it generates. The case of the generation of synchronized oscillations across cells as the molecular basis of vertebrate segment formation (Sect. 4) and the robustness in gene regulatory network activity (Sect. 5) illustrate this point.

Related to the generation of novel entities, another relevant aspect of mechanisms is *emergence* (Bedau 2003; Boogerd et al. 2005; Huneman 2012; Mitchell 2012). I do not require a strong ontological type of emergence to be tied to the concept of a mechanism. Rather, I use emergence to refer to situations in which some of a system's qualitative properties can only be predicted if the full system dynamics are simulated, or more loosely, if qualitative properties cannot be foreseen from the system's components and their basic interactions. For example, bistability occurs when a subsystem is in either of two distinct states (though switching from one to the other state is possible), so that continuous processes and the components' quantitative interactions yield some discontinuous and qualitatively different properties (Eissing et al. 2004; Ferrell and Xiong 2001; Goldbeter et al. 2007). The presence of distinct *Drosophila* segment polarity gene expression patterns corresponding to different cell types, mentioned in Sect. 5, is an instance of bistability (Ingolia 2004). In Brigandt (2013c), I discuss spontaneous symmetry-breaking, which occurs when extremely small stochastic fluctuations in a system eventually lead to the crossing of a threshold that determines which of two branches of a bifurcation (in possible states of the system) is taken. Found also in molecular and cellular biology contexts, spontaneous symmetry-breaking makes it possible for a nearly homogeneous state to give rise to a distinct and stable structural pattern.

Emergent properties can result from nonlinear interactions among mechanism components combined with the presence of a complex organization with feedback loops (Bhalla and Iyengar 1999). Emergence even in the above weak sense is scientifically important because it entails the need to study the mechanism's complete organization, not just its structural, but also its functional organization. Moreover, while the operation and resulting features of some mechanisms represented by diagrams can be understood be means of mental simulation, complex mechanisms with feedback loops or nonlinear interactions require a mathematical treatment (Boogerd et al. 2007b; Brigandt 2013c; Ihekwaba et al. 2005; Noble 2002; Westerhoff and Kell 2007).

The discussion so far about the appropriate ontological construal of a mechanism has already relied on scientifically, and thus epistemically, important aspects of mechanisms. Now I comment explicitly on the epistemological issue of *mechanistic explanation*. Philosophers typically assume that a mechanistic account explains in terms of the mechanism's structural components, the component's spatial organization, and their qualitative interactions. Carl Craver (2006, 2007, 2008) has promoted the picture that unlike mechanistic accounts in this sense, mathematical models describe but typically do not explain. In contrast, in Sect. 3 I laid out a criterion for equations being indispensable components of an explanation and applied it in Sects. 4 and 5 to different cases from evo-devo, arguing that there are even *qualitative* explananda about developmental mechanisms where quantitative models are a necessary part of the explanans (see also Brigandt 2013c). Overall, mathematical modeling is needed for two related reasons. First, according to criterion *ER*, if omitting or changing a feature results in the explanandum not following any longer, this feature is explanatorily indispensable—and thus has to be *included in the explanatory model*. And some such features can only be mathematically represented or are even quantitative. In some cases the mathematical model may represent qualitative relations of the molecular entities involved, e.g., the topology of a complex network showing positive or negative regulatory influence among components. However, if the property of the system to be explained is sensitive to quantitative parameters, a quantitative model is explanatorily indispensable (Sects. 3 and 4 mentioned instances of this). Second, beyond a model representing different components of a mechanism and their organization, the explanation has to shows *how (or at least that) the explanandum results* from this. If the mechanism's operation cannot be understood by mental simulation, a mathematical analysis of the model or a computer simulation is needed.

My examples illustrate that some studies in developmental biology and evo-devo integrate concrete molecular knowledge gained from experiments with mathematical modeling. As a result, mechanistic explanations can—and sometimes must, depending on the explanandum—include *quantitative considerations and mathematical models* (see also Baetu 2015; though Gross 2015; Isaad and Malaterre 2015; Théry 2015 analyze the differences between mechanistic explanation and explanations using mathematical models). I have focused on the developmental prong of evo-devo, as developmental mechanisms pertain to molecular and experimental biology for which philosophical accounts of mechanisms are meant to hold. For a full picture of explanation in evo-devo (regardless of whether all aspects qualify as mechanistic explanation), one needs to bear in mind that developmental processes are not the only features of evo-devo. Other explanatory contributions are involved, including phylogenetic trees, historical patterns of change in morphological structures, considerations about natural selection (e.g., organism-environment and organism-organism interactions),[9] and the dynamics of genotype

---

[9]There is disagreement on whether philosophical accounts of mechanisms can capture natural selection (Barros 2008; Skipper and Millstein 2005). My view is that explanations in terms of natural selection (in particular when using mathematical models) abstract away from many concrete properties and activities of individual organism. But abstraction from mechanistic detail happens even in mathematical models in molecular and developmental biology (Sect. 3; Bechtel

and phenotype distributions within populations. In the latter case, mathematical models from population genetics, quantitative genetics, and evolutionary ecology can be components of evo-devo explanations (Rice 2008, 2012).

My discussion has highlighted the *functional aspects of mechanisms*, for instance in the contexts of robustness, phenotypic plasticity, and modularity. Such properties go beyond spatial organization on which many philosophical discussions have centered. The functional organization of a mechanism need not align with its structural organization, as shown by the case of modularity. It may be easy to recognize the structures involved in a developmental process, but something is a module to the extent to which it can be modified or rearranged in morphological evolution. While there are clear developmental-functional connections between genes and anatomical structures (as structures on different levels of organization), these functional relations can sometimes be rearranged so that genes and anatomical structures evolve independently of each other. As a result, modularity can be a very complex kind of organization determined by an organism's developmental-functional dynamics. Functionality can also be *distributed* across a mechanism, so that beyond local interactions the system has global causal properties.[10] A mechanism may be robust in maintaining one component while another component is modified, but this potential may not just reside in the relations among a few components but in a more system-wide response (Sect. 5; Mitchell 2009). Phenotypic plasticity is likewise due to the functioning of larger developmental processes. While modularity means that two components are sufficiently developmentally dissociated (to be able to be rearranged in evolution), a look at the larger functional context may be required to account for why this dissociation exists.

Philosophical accounts have emphasized the actual organization and the actual, regular operation of a mechanism, given that a how-possibly mechanism postulated does not in fact explain (Craver 2006, 2007). But robustness, phenotypic plasticity, and modularity pertain to the *mechanism's modified organization and modified operation*. For these are dispositional properties of how a developmental mechanism reacts to perturbations or permits modification. In evo-devo explanations of why a developmental process exhibits robustness, phenotypic plasticity, or modularity, the very explanandum is the response to a mechanism's modification—so that there are important scientific questions that are not just about the actual behavior of a mechanism, but also its dispositions.

2015; Brigandt 2013c; Levy 2014; Levy and Bechtel 2013), so that the broad conception of mechanistic explanation advocated here is more likely to accommodate natural selection. One difficulty is that natural selection is about fitness *differences* among phenotypes. Even if each of two phenotypes is part of a mechanism (by each phenotype being possessed by concrete organisms), what matters is how the phenotypes differ and the phenotypes' differential behavior across time, which is a complex and unusual aspect of a 'mechanism.'

[10]On related grounds, Baetu (2015) argues that molecular mechanisms are not neatly individuated objects.

Section 5 indicated why these dispositional aspects of developmental mechanisms are scientifically important by discussing how robustness, phenotypic plasticity, and modularity increase morphological evolvability and the potential for the generation of structural novelty. Now I emphasize this issue again by tying it to intelligent design ideas against evolution (Brigandt 2013b). According to Michael Behe (1996), a biological system is irreducibly complex when the removal of any part leads to the system ceasing to function (the alleged implication being that such a system cannot have evolved gradually but must have originated with all parts in place). This is actually a resurrection of Paley's (1802) watchmaker argument, though Behe claims irreducible complexity of molecular systems. But robustness is the very opposite of irreducible complexity. The prevalence of robustness shows that organisms are not like Paley's watch which breaks down upon modification (Kirschner and Gerhart 2005). In contrast to the machine and artifact metaphors that intelligent design creationists use to portray cells and organisms, developmental systems are highly flexible. The flexibility of developmental mechanisms is also of evolutionary significance, and since biologists have it in view, flexibility and active response to perturbations must be part of philosophical conceptions of mechanisms.

Let me conclude with some general remarks on scientific explanation (see also Brigandt 2013c). Evo-devo shows that not only is each explanatory account a work in progress with new contributions constantly being added, but explanatory accounts are so complex that they do not consist in and *cannot be captured by a single representation* (O'Malley et al. 2014). Accounts of morphological evolvability and the evolutionary origin of novelty coordinate a plethora of descriptions, explanatory ideas, and models. Such individual representations come from different biological fields, pertain to different levels of organization, focus on organismal structure or address function, consist in qualitative-mechanistic accounts or quantitative models, provide empirical data or theoretical models, and address change in developmental time or change in evolutionary time. To reflect this complexity, it is better to speak of an explanatory *account or framework* than one explanation. While past philosophical theories such as the deductive-nomological model (Hempel and Oppenheim 1948) attempted to characterize a scientific explanation by laying out conditions for what makes a set of statements an explanans, no such simple philosophical account is possible.

In the last three decades there has been a laudable trend in philosophy of science to not just studying the *content* and results of science, but also the *practice* and changing activities of scientists (Brigandt 2013a, b). In the present context, beyond analyzing explanatory theories, this involves philosophically studying how scientists develop and use explanations. Accounts of mechanistic explanation are already tied to the process of discovery by paying attention to the discovery of mechanisms and the shifts between reductive research episodes and integrative strategies (Bechtel 2006, 2010; Craver 2005; Craver and Darden 2013; Darden 2006). In my argument that some mathematical models are explanatory, I have heeded Carl Craver's admonition that not every representation is an explanation. However, one also has to point out that a single model can be used for both explanatory purposes and non-explanatory purposes (describing a phenomenon, predicting to test a hypothesis, exploring conceptual possibilities) depending on the context.[11] Likewise, in their research geared toward the generation of an explanatory account, scientists make use of many representations, some of which are non-explanatory. Since in scientific practice, explanations and other representations are jointly used (guided by epistemic aims and values; Brigandt 2012a, 2013a), philosophical theories of explanation have to be related to other epistemological notions, including description, prediction, model, standard, and method. Since the scientific activity of explaining is related to such other activities as predicting, confirming, modeling, and choosing theoretical and experimental strategies, isolated philosophical accounts of discovery, confirmation, and explanation are impossible.

Acknowledgments I am indebted to Pierre-Alain Braillard, Christophe Malaterre, and two anonymous referees for detailed comments on an earlier version of this paper. I thank Emma Kennedy for proofreading the manuscript and Arnon Levy, Bill Bechtel, Carl Craver, and Maureen O'Malley for discussions on mechanistic explanation and mathematical models. Figure 7.1 was reprinted from Salazar-Ciudad and Jernvall (2002) with the permission of the copyright holder, the National Academy of Sciences, USA. Figure 7.2 was reprinted from Dequéant and Pourquié (2008) by permission from Macmillan Publishers Ltd.

References

Abud, H. E. (2004). Shaping developing tissues by apoptosis. *Cell Death and Differentiation, 11*, 797–799.

Albert, R., & Othmer, H. G. (2003). The topology of the regulatory interactions predicts the expression pattern of the segment polarity genes in *Drosophila melanogaster*. *Journal of Theoretical Biology, 223*, 1–18.

Baetu, T. (2015). From mechanisms to mathematical models and back to mechanisms: Quantitative mechanistic explanations. In P.-A. Braillard & C. Malaterre (Eds.), *Explanation in biology. An enquiry into the diversity of explanatory patterns in the life sciences* (pp. 345–363). Dordrecht: Springer.

Baggs, J. E., Price, T. S., DiTacchio, L., Panda, S., FitzGerald, G. A., & Hogenesch, J. B. (2009). Network features of the mammalian circadian clock. *PLoS Biology, 7*, e1000052.

Baker, A. (2015). Mathematical explanation in biology. In P.-A. Braillard & C. Malaterre (Eds.), *Explanation in biology. An enquiry into the diversity of explanatory patterns in the life sciences* (pp. 229–247). . Dordrecht: Springer.

Baker, R. E., & Schnell, S. (2009). How can mathematics help us explore vertebrate segmentation? *HFSP Journal, 3*, 1–5.

Baker, R. E., Schnell, S., & Maini, P. K. (2008). Mathematical models for somite formation. In S. Schnell, P. K. Maini, S. A. Newman, & T. J. Newman (Eds.), *Multiscale modeling of developmental systems* (pp. 183–203). New York: Academic.

Barabási, A.-L., & Oltvai, Z. N. (2004). Network biology: Understanding the cell's functional organization. *Nature Reviews Genetics, 5*, 101–113.

---

[11]Baetu (2015) discusses how a mathematical model can reveal a previous molecular-mechanistic account to be explanatorily incomplete. This can prompt and guide further experimental discovery, so a mathematical model can be involved in both discovery and explanation.

Barabási, A.-L., Gulbahce, N., & Loscalzo, J. (2011). Network medicine: A network-based approach to human disease. *Nature Reviews Genetics, 12*, 56–68.

Barros, D. B. (2008). Natural selection as a mechanism. *Philosophy of Science, 75*, 306–322.

Bechtel, W. (2006). *Discovering cell mechanisms: The creation of modern cell biology.* Cambridge: Cambridge University Press.

Bechtel, W. (2010). The downs and ups of mechanistic research: Circadian rhythm research as an exemplar. *Erkenntnis, 73*, 313–328.

Bechtel, W. (2011). Mechanism and biological explanation. *Philosophy of Science, 78*, 533–557.

Bechtel, W. (2013). From molecules to behavior and the clinic: Integration in chronobiology. *Studies in History and Philosophy of Biological and Biomedical Sciences, 44*, 493–502.

Bechtel, W. (2015). Generalizing mechanistic explanations using graph-theoretic representations. In P.-A. Braillard & C. Malaterre (Eds.), *Explanation in biology. An enquiry into the diversity of explanatory patterns in the life sciences* (pp. 199–225). Dordrecht: Springer.

Bechtel, W., & Abrahamsen, A. (2005). Explanation: A mechanist alternative. *Studies in History and Philosophy of Biological and Biomedical Sciences, 36*, 421–441.

Bechtel, W., & Abrahamsen, A. (2010). Dynamic mechanistic explanation: Computational modeling of circadian rhythms as an exemplar for cognitive science. *Studies in History and Philosophy of Science Part A, 41*, 321–333.

Bechtel, W., & Abrahamsen, A. (2011). Complex biological mechanisms: Cyclic, oscillatory, and autonomous. In C. Hooker (Ed.), *Philosophy of complex systems* (pp. 257–285). Amsterdam: Elsevier.

Bedau, M. A. (2003). Downward causation and the autonomy of weak emergence. *Principia, 3*, 5–50.

Behe, M. J. (1996). *Darwin's black box: The biochemical challenge to evolution.* New York: Free Press.

Bhalla, U. S., & Iyengar, R. (1999). Emergent properties of networks of biological signaling pathways. *Science, 283*, 381–387.

Bogen, J. (2005). Regularities and causality; generalizations and causal explanations. *Studies in History and Philosophy of Biological and Biomedical Sciences, 36*, 397–420.

Bolker, J. A. (2000). Modularity in development and why it matters to evo-devo. *American Zoologist, 40*, 770–776.

Boogerd, F. C., Bruggeman, F., Richardson, R., Stephan, A., & Westerhoff, H. (2005). Emergence and its place in nature: A case study of biochemical networks. *Synthese, 145*, 131–164.

Boogerd, F. C., Bruggeman, F. J., Hofmeyr, J.-H. S., & Westerhoff, H. V. (Eds.). (2007a). *Systems biology: Philosophical foundations.* Amsterdam: Elsevier.

Boogerd, F. C., Bruggeman, F. J., Hofmeyr, J.-H. S., & Westerhoff, H. V. (2007b). Towards philosophical foundations of systems biology: Introduction. In F. C. Boogerd, F. J. Bruggeman, J.-H. S. Hofmeyr, & H. V. Westerhoff (Eds.), *Systems biology: Philosophical foundations* (pp. 3–19). Amsterdam: Elsevier.

Braendle, C., & Félix, M.-A. (2008). Plasticity and errors of a robust developmental system in different environments. *Developmental Cell, 15*, 714–724.

Braillard, P.-A. (2015). Prospect and limits of explaining biological systems in engineering terms. In P.-A. Braillard & C. Malaterre (Eds.), *Explanation in biology. An enquiry into the diversity of explanatory patterns in the life sciences* (pp. 319–344). Dordrecht: Springer.

Breidenmoser, T., & Wolkenhauer O. (2015). Explanation and organizing principles in systems biology. In P.-A. Braillard & C. Malaterre (Eds.), *Explanation in biology. An enquiry into the diversity of explanatory patterns in the life sciences* (pp. 249–264). Dordrecht: Springer.

Breuker, C. J., Debat, V., & Klingenberg, C. P. (2006). Functional evo-devo. *Trends in Ecology & Evolution, 21*, 488–492.

Brigandt, I. (2007). Typology now: Homology and developmental constraints explain evolvability. *Biology and Philosophy, 22*, 709–725.

Brigandt, I. (2010). Beyond reduction and pluralism: Toward an epistemology of explanatory integration in biology. *Erkenntnis, 73*, 295–311.

Brigandt, I. (2012a). The dynamics of scientific concepts: The relevance of epistemic aims and values. In U. Feest & F. Steinle (Eds.), *Scientific concepts and investigative practice*(pp. 75–103). Berlin: de Gruyter.

Brigandt, I. (Ed.). (2012b). *Perspectives on evolutionary novelty and evo-devo*. Special issue of the *Journal of Experimental Zoology Part B: Molecular and Developmental Evolution, 318*(6), 417–517.

Brigandt, I. (2013a). Explanation in biology. Reduction, pluralism, and explanatory aims. *Science & Education, 22,* 69–91.

Brigandt, I. (2013b). Intelligent design and the nature of science: Philosophical and pedagogical points. In K. Kampourakis (Ed.), *Philosophical issues in biology education* (pp. 205–238). Dordrecht: Springer.

Brigandt, I. (2013c). Systems biology and the integration of mechanistic explanation and mathematical explanation. *Studies in History and Philosophy of Biological and Biomedical Sciences, 44,* 477–492.

Brigandt, I. (2015). From developmental constraint to evolvability: How concepts figure in explanation and disciplinary identity. In A. C. Love (Ed.), *Conceptual change in biology: Scientific and philosophical perspectives on evolution and development* (pp. 305–325). Dordrecht: Springer.

Brigandt, I., & Griffiths, P. E. (2007). The importance of homology for biology and philosophy. *Biology and Philosophy, 22,* 633–641.

Brigandt, I., & Love, A. C. (2010). Evolutionary novelty and the evo-devo synthesis: Field notes. *Evolutionary Biology, 37,* 93–99.

Brigandt, I., & Love, A. C. (2012a). Conceptualizing evolutionary novelty: Moving beyond definitional debates. *Journal of Experimental Zoology Part B: Molecular and Developmental Evolution, 318,* 417–427.

Brigandt, I., & Love, A.C. (2012b). Reductionism in biology. In E. N. Zalta (Ed.), *The Stanford encyclopedia of philosophy*. http://plato.stanford.edu/entries/reduction-biology

Callebaut, W., & Rasskin-Gutman, D. (Eds.). (2005). *Modularity: Understanding the development and evolution of natural complex systems*. Cambridge, MA: MIT Press.

Cañestro, C., Yokoi, H., & Postlethwait, J. H. (2007). Evolutionary developmental biology and genomics. *Nature Reviews Genetics, 8,* 932–942.

Carroll, S. B. (2008). Evo-devo and an expanding evolutionary synthesis: A genetic theory of morphological evolution. *Cell, 134,* 25–36.

Craver, C. F. (2005). Beyond reduction: Mechanisms, multifield integration and the unity of neuroscience. *Studies in History and Philosophy of Biological and Biomedical Sciences, 36,* 373–395.

Craver, C. F. (2006). When mechanistic models explain. *Synthese, 153,* 355–376.

Craver, C. F. (2007). *Explaining the brain: Mechanisms and the mosaic unity of neuroscience*. Oxford: Oxford University Press.

Craver, C. F. (2008). Physical law and mechanistic explanation in the Hodgkin and Huxley model of the action potential. *Philosophy of Science, 75,* 1022–1033.

Craver, C. F., & Darden, L. (2013). *In search of mechanisms: Discoveries across the life sciences*. Chicago: University of Chicago Press.

Darden, L. (2005). Relations among fields: Mendelian, cytological and molecular mechanisms. *Studies in History and Philosophy of Biological and Biomedical Sciences, 36,* 349–371.

Darden, L. (2006). *Reasoning in biological discoveries: Essays on mechanisms, interfield relations, and anomaly resolution*. Cambridge: Cambridge University Press.

Davidson, E. H. (2006). *The regulatory genome: Gene regulatory networks in development and evolution*. London: Academic.

Davidson, E. H. (2010). Emerging properties of animal gene regulatory networks. *Nature, 468,* 911–920.

Davidson, E. H., & Erwin, D. H. (2006). Gene regulatory networks and the evolution of animal body plans. *Science, 311,* 796–800.

Davis, G. K. (2012). Cyclical parthenogenesis and viviparity in aphids as evolutionary novelties. *Journal of Experimental Zoology Part B: Molecular and Developmental Evolution, 318,* 448–459.

Dean, E. J., Davis, J. C., Davis, R. W., & Petrov, D. A. (2008). Pervasive and persistent redundancy among duplicated genes in yeast. *PLoS Genetics, 4,* e1000113.

Delattre, M., & Félix, M.-A. (2009). The evolutionary context of robust and redundant cell biological mechanisms. *BioEssays, 31,* 537–545.

Dequéant, M.-L., & Pourquié, O. (2008). Segmental patterning of the vertebrate embryonic axis. *Nature Reviews Genetics, 9,* 370–382.

Dequéant, M.-L., Glynn, E., Gaudenz, K., Wahl, M., Chen, J., Mushegian, A., & Pourquié, O. (2006). A complex oscillating network of signaling genes underlies the mouse segmentation clock. *Science, 314,* 1595–1598.

Economou, A. D., Ohazama, A., Porntaveetus, T., Sharpe, P. T., Kondo, S., Basson, M. A., Gritli-Linde, A., Cobourne, M. T., & Green, J. B. A. (2012). Periodic stripe formation by a Turing mechanism operating at growth zones in the mammalian palate. *Nature Genetics, 44,* 348–351.

Edwards, J. S., & Palsson, B. O. (1999). Systems properties of the *Haemophilus influenzae* Rd metabolic genotype. *Journal of Biological Chemistry, 274,* 17410–17416.

Edwards, J. S., & Palsson, B. O. (2000a). The *Escherichia coli* MG1655 in silico metabolic genotype: Its definition, characteristics, and capabilities. *Proceedings of the National Academy of Sciences of the United States of America, 97,* 5528–5533.

Edwards, J. S., & Palsson, B. O. (2000b). Robustness analysis of the *Escherichia coli* metabolic network. *Biotechnology Progress, 16,* 927–939.

Eissing, T., Conzelmann, H., Gilles, E. D., Allgöwer, F., Bullinger, E., & Scheurich, P. (2004). Bistability analyses of a caspase activation model for receptor-induced apoptosis. *Journal of Biological Chemistry, 279,* 36892–36897.

Erwin, D. H., & Davidson, E. H. (2009). The evolution of hierarchical gene regulatory networks. *Nature Reviews Genetics, 10,* 141–148.

Fagan, M. B. (2012a). Materia mathematica: Models in stem cell biology. *Journal of Experimental & Theoretical Artificial Intelligence, 24,* 315–327.

Fagan, M. B. (2012b). Waddington redux: Models and explanation in stem cell and systems biology. *Biology and Philosophy, 27,* 179–213.

Ferrell, J. E., & Xiong, W. (2001). Bistability in cell signaling: How to make continuous processes discontinuous, and reversible processes irreversible. *Chaos, 11,* 227–236.

Fisher, J., Piterman, N., Hajnal, A., & Henzinger, T. A. (2007). Predictive modeling of signaling crosstalk during *C. elegans* vulval development. *PLoS Computational Biology, 3,* e92.

Forgacs, G., & Newman, S. A. (2005). *Biological physics of the developing embryo.* Cambridge: Cambridge University Press.

Fraser, I. D. C., & Germain, R. N. (2009). Navigating the network: Signaling cross-talk in hematopoietic cells. *Nature Immunology, 10,* 327–331.

Gerhart, J. C., & Kirschner, M. W. (2003). Evolvability. In B. K. Hall & W. M. Olson (Eds.), *Keywords and concepts in evolutionary developmental biology* (pp. 133–137). Cambridge, MA: Harvard University Press.

Gerhart, J. C., & Kirschner, M. W. (2007). The theory of facilitated variation. *Proceedings of the National Academy of Sciences of the United States of America, 104,* 8582–8589.

Gilbert, S. F. (2001). Ecological developmental biology: Developmental biology meets the real world. *Developmental Biology, 233,* 1–12.

Glass, G. L., & Bolker, J. A. (2003). Modularity. In B. K. Hall & W. M. Olson (Eds.), *Keywords and concepts in evolutionary developmental biology* (pp. 260–267). Cambridge, MA: Harvard University Press.

Goldbeter, A., & Pourquié, O. (2008). Modeling the segmentation clock as a network of coupled oscillations in the Notch, Wnt and FGF signaling pathways. *Journal of Theoretical Biology, 252,* 574–585.

Goldbeter, A., Gonze, D., & Pourquié, O. (2007). Sharp developmental thresholds defined through bistability by antagonistic gradients of retinoic acid and FGF signaling. *Developmental Dynamics, 236*, 1495–1508.

Gonzalez, P. N., Oyhenart, E. E., & Hallgrímsson, B. (2011). Effects of environmental perturbations during postnatal development on the phenotypic integration of the skull. *Journal of Experimental Zoology Part B: Molecular and Developmental Evolution, 316B*, 547–561.

Green, S., Fagan, M., & Jaeger, J. (2015). Explanatory integration challenges in evolutionary systems biology. *Biological Theory, 10*, 18–35.

Greenbury, S. F., Johnston, I. G., Smith, M. A., Doye, J. P. K., & Louis, A. A. (2010). The effect of scale-free topology on the robustness and evolvability of genetic regulatory networks. *Journal of Theoretical Biology, 267*, 48–61.

Gross, F. (2015). The relevance of irrelevance: Explanation in systems biology. In P.-A. Braillard & C. Malaterre (Eds.), *Explanation in biology. An enquiry into the diversity of explanatory patterns in the life sciences* (pp. 175–198). Dordrecht: Springer.

Hall, B. K. (Ed.). (2006). *Fins into limbs: Evolution, development and transformation.* Chicago: University of Chicago Press.

Hallgrímsson, B., & Hall, B. K. (Eds.). (2011). *Epigenetics: Linking genotype and phenotype in development and evolution.* Berkeley: University of California Press.

Hallgrímsson, B., Lieberman, D. E., Liu, W., Ford-Hutchinson, A. F., & Jirik, F. R. (2007). Epigenetic interactions and the structure of phenotypic variation in the cranium. *Evolution & Development, 9*, 76–91.

Hempel, C. G., & Oppenheim, P. (1948). Studies in the logic of explanation. *Philosophy of Science, 15*, 135–175.

Hendrikse, J. L., Parsons, T. E., & Hallgrímsson, B. (2007). Evolvability as the proper focus of evolutionary developmental biology. *Evolution & Development, 9*, 393–401.

Hentschel, H. G. E., Glimm, T., Glazier, J. A., & Newman, S. A. (2004). Dynamical mechanisms for skeletal pattern formation in the vertebrate limb. *Proceedings of the Royal Society of London. Series B: Biological Sciences, 271*, 1713–1722.

Herrgen, L., Ares, S., Morelli, L. G., Schröter, C., Jülicher, F., & Oates, A. C. (2010). Intercellular coupling regulates the period of the segmentation clock. *Current Biology, 20*, 1244–1253.

Huneman, P. (2010). Topological explanations and robustness in biological sciences. *Synthese, 177*, 213–245.

Huneman, P. (2012). Determinism, predictability and open-ended evolution: Lessons from computational emergence. *Synthese, 185*, 195–214.

Ihekwaba, A. E. C., Broomhead, D. S., Grimley, R., Benson, N., White, M. R. H., & Kell, D. B. (2005). Synergistic control of oscillations in the nf-κb signalling pathway. *IEE Proceedings Systems Biology, 152*, 153–160.

Ihmels, J., Collins, S. R., Schuldiner, M., Krogan, N. J., & Weissman, J. S. (2007). Backup without redundancy: Genetic interactions reveal the cost of duplicate gene loss. *Molecular Systems Biology, 3*, 86.

Ingolia, N. T. (2004). Topology and robustness in the *Drosophila* segment polarity network. *PLoS Biology, 2*, e123.

Issad, T., & Malaterre, C. (2015). Are dynamic mechanistic explanations still mechanistic? In P.-A. Braillard & C. Malaterre (Eds.), *Explanation in biology. An enquiry into the diversity of explanatory patterns in the life sciences* (pp. 265–292). Dordrecht: Springer.

Jamniczky, H. A., & Hallgrímsson, B. (2011). Modularity in the skull and cranial vasculature of laboratory mice: Implications for the evolution of complex phenotypes. *Evolution & Development, 13*, 28–37.

Jeong, H., Tombor, B., Albert, R., Oltvai, Z. N., & Barabási, A. L. (2000). The large-scale organization of metabolic networks. *Nature, 407*, 651–654.

Jørgensen, C., & Linding, R. (2010). Simplistic pathways or complex networks? *Current Opinion in Genetics & Development, 20*, 15–22.

Kafri, R., Bar-Even, A., & Pilpel, Y. (2005). Transcription control reprogramming in genetic backup circuits. *Nature Genetics, 37*, 295–299.

Kaplan, D. M., & Craver, C. F. (2011). The explanatory force of dynamical and mathematical models in neuroscience: A mechanistic perspective. *Philosophy of Science, 78*, 601–627.

Kirschner, M. W., & Gerhart, J. C. (1998). Evolvability. *Proceedings of the National Academy of Sciences of the United States of America, 95*, 8420–8427.

Kirschner, M. W., & Gerhart, J. C. (2005). *The plausibility of life: Resolving Darwin's dilemma.* New Haven: Yale University Press.

Kitano, H. (2004). Biological robustness. *Nature Reviews Genetics, 5*, 826–837.

Laubichler, M. D. (2009). Form and function in Evo Devo: Historical and conceptual reflections. In M. D. Laubichler & J. Maienschein (Eds.), *Form and function in developmental evolution* (pp. 10–46). Cambridge: Cambridge University Press.

Laubichler, M. D. (2010). Evolutionary developmental biology offers a significant challenge to the neo-Darwinian paradigm. In F. Ayala & R. Arp (Eds.), *Contemporary debates in the philosophy of biology* (pp. 199–212). Malden: Wiley-Blackwell.

Layek, R. K., Datta, A., & Dougherty, E. R. (2011). From biological pathways to regulatory networks. *Molecular BioSystems, 7*, 843–851.

Levy, A. (2014). What was Hodgkin and Huxley's achievement? *British Journal for the Philosophy of Science, 65*, 469–492.

Levy, A., & Bechtel, W. (2013). Abstraction and the organization of mechanisms. *Philosophy of Science, 80*, 241–261.

Li, X., Cassidy, J. J., Reinke, C. A., Fischboeck, S., & Carthew, R. W. (2009). A microRNA imparts robustness against environmental fluctuation during development. *Cell, 137*, 273–282.

Linksvayer, T. A., Fewell, J. H., Gadau, J., & Laubichler, M. D. (2012). Developmental evolution in social insects: Regulatory networks from genes to societies. *Journal of Experimental Zoology Part B: Molecular and Developmental Evolution, 318*, 159–169.

Love, A. C. (2008a). Explaining evolutionary innovations and novelties: Criteria of explanatory adequacy and epistemological prerequisites. *Philosophy of Science, 75*, 874–886.

Love, A. C. (2008b). From philosophy to science (to natural philosophy): Evolutionary developmental perspectives. *The Quarterly Review of Biology, 83*, 65–76.

Love, A. C. (2013). Interdisciplinary lessons for the teaching of biology from the practice of evo-devo. *Science & Education, 22*, 255–278.

Love, A. C. (Ed.). (2015). *Conceptual change in biology: Scientific and philosophical perspectives on evolution and development.* Dordrecht: Springer.

Love, A. C., & Raff, R. A. (2003). Knowing your ancestors: Themes in the history of evo-devo. *Evolution & Development, 5*, 327–330.

Machamer, P., Darden, L., & Craver, C. F. (2000). Thinking about mechanisms. *Philosophy of Science, 67*, 1–25.

Manu, Surkova, S., Spirov, A. V., Gursky, V. V., Janssens, H., Kim, A.-R., Radulescu, O., et al. (2009a). Canalization of gene expression and domain shifts in the *Drosophila* blastoderm by dynamical attractors. *PLoS Computational Biology, 5*, e1000303.

Manu, Surkova, S., Spirov, A. V., Gursky, V. V., Janssens, H., Kim, A.-R., Radulescu, O., et al. (2009b). Canalization of gene expression in the *Drosophila* blastoderm by gap gene cross regulation. *PLoS Biology, 7*, e1000049.

Masamizu, Y., Ohtsuka, T., Takashima, Y., Nagahara, H., Takenaka, Y., Yoshikawa, K., Okamura, H., & Kageyama, R. (2006). Real-time imaging of the somite segmentation clock: Revelation of unstable oscillators in the individual presomitic mesoderm cells. *Proceedings of the National Academy of Sciences of the United States of America, 103*, 1313–1318.

Masel, J., & Siegal, M. L. (2009). Robustness: Mechanisms and consequences. *Trends in Genetics, 25*, 395–403.

Mazzitello, K. I., Arizmendi, C. M., & Hentschel, H. G. E. (2008). Converting genetic network oscillations into somite spatial patterns. *Physical Review E, 78*, 021906.

Meinhardt, H. (2003). Complex pattern formation by a self-destabilization of established patterns: Chemotactic orientation and phyllotaxis as examples. *Comptes Rendus Biologies, 326*, 223–237.

Meinhardt, H. (2009). *The algorithmic beauty of sea shells* (4th ed.). Berlin: Springer.

Meinhardt, H. (2012). Modeling pattern formation in hydra: A route to understanding essential steps in development. *International Journal of Developmental Biology, 56*, 447–462.

Mekios, C. (2015). Explanation in systems biology: Is it all about mechanisms? In P.-A. Braillard & C. Malaterre (Eds.), *Explanation in biology. An enquiry into the diversity of explanatory patterns in the life sciences* (pp. 47–72). Dordrecht: Springer.

Merlin, F. (2015). Developmental noise: Explaining the specific heterogeneity of individual organisms. In P.-A. Braillard & C. Malaterre (Eds.), *Explanation in biology. An enquiry into the diversity of explanatory patterns in the life sciences* (pp. 91–110). Dordrecht: Springer.

Milloz, J., Duveau, F., Nuez, I., & Félix, M.-A. (2008). Intraspecific evolution of the intercellular signaling network underlying a robust developmental system. *Genes & Development, 22*, 3064–3075.

Mitchell, S. D. (2009). *Unsimple truths: Science, complexity, and policy.* Chicago: University of Chicago Press.

Mitchell, S. D. (2012). Emergence: Logical, functional and dynamical. *Synthese, 185*, 171–186.

Morelli, L. G., Ares, S., Herrgen, L., Schröter, C., Jülicher, F., & Oates, A. C. (2009). Delayed coupling theory of vertebrate segmentation. *HFSP Journal, 3*, 55–66.

Morelli, L. G., Uriu, K., Ares, S., & Oates, A. C. (2012). Computational approaches to developmental patterning. *Science, 336*, 187–191.

Müller, G. B. (2003). Embryonic motility: Environmental influences and evolutionary innovation. *Evolution & Development, 5*, 56–60.

Müller, G. B., & Newman, S. A. (Eds.). (2003). *Origination of organismal form: Beyond the gene in developmental and evolutionary biology.* Cambridge, MA: MIT Press.

Müller, G. B., & Newman, S. A. (2005). The innovation triad: An EvoDevo agenda. *Journal of Experimental Zoology Part B: Molecular and Developmental Evolution, 304B*, 487–503.

Müller, G. B., & Wagner, G. P. (2003). Innovation. In B. K. Hall & W. M. Olson (Eds.), *Keywords and concepts in evolutionary developmental biology* (pp. 218–227). Cambridge, MA: Harvard University Press.

Murray, J. D. (2003). *Mathematical biology II: Spatial models and biomedical applications.* Berlin: Springer.

Nakamasu, A., Takahashi, G., Kanbe, A., & Kondo, S. (2009). Interactions between zebrafish pigment cells responsible for the generation of Turing patterns. *Proceedings of the National Academy of Sciences of the United States of America, 106*, 8429–8434.

Newman, S. A., & Müller, G. B. (2000). Epigenetic mechanisms of character origination. *Journal of Experimental Zoology Part B: Molecular and Developmental Evolution, 288*, 304–317.

Newman, S. A., & Müller, G. B. (2005). Origination and innovation in the vertebrate limb skeleton: An epigenetic perspective. *Journal of Experimental Zoology Part B: Molecular and Developmental Evolution, 304B*, 593–609.

Newman, S. A., Christley, S., Glimm, T., Hentschel, H. G. E., Kazmierczak, B., Zhang, Y.-T., Zhu, J., & Alber, M. S. (2008). Multiscale models for vertebrate limb development. In S. Schnell, P. K. Maini, S. A. Newman, & T. J. Newman (Eds.), *Multiscale modeling of developmental systems* (pp. 311–340). New York: Academic.

Noble, D. (2002). Modeling the heart—from genes to cells to the whole organ. *Science, 295*, 1678–1682.

O'Malley, M. A., Brigandt, I., Love, A. C., Crawford, J. W., Gilbert, J. A., Knight, R., Mitchell, S. D., & Rohwer, F. (2014). Multilevel research strategies and biological systems. *Philosophy of Science, 81*, 811–828.

Oates, A. C., Morelli, L. G., & Ares, S. (2012). Patterning embryos with oscillations: Structure, function and dynamics of the vertebrate segmentation clock. *Development, 139*, 625–639.

Paley, W. (1802). *Natural theology, or evidences of the existence and attributes of the deity, collected from the appearances of nature.* London: R. Faulder.

Palmer, A. R. (2004). Symmetry breaking and the evolution of development. *Science, 306*, 828–833.

Palmer, A. R. (2012). Developmental plasticity and the origin of novel forms: Unveiling of cryptic genetic variation via "use and disuse". *Journal of Experimental Zoology Part B: Molecular and Developmental Evolution, 318*, 466–479.

Plikus, M. V., Baker, R. E., Chen, C.-C., Fare, C., de la Cruz, D., Andl, T., Maini, P. K., Millar, S. E., Widelitz, R., & Chuong, C.-M. (2011). Self-organizing and stochastic behaviors during the regeneration of hair stem cells. *Science, 332*, 586–589.

Prud'homme, B., Minervino, C., Hocine, M., Cande, J. D., Aouane, A., Dufour, H. D., Kassner, V. A., & Gompel, N. (2011). Body plan innovation in treehoppers through the evolution of an extra wing-like appendage. *Nature, 473*, 83–86.

Prum, R. O. (2005). Evolution of the morphological innovations of feathers. *Journal of Experimental Zoology Part B: Molecular and Developmental Evolution, 304B*, 570–579.

Putnam, H. (1975). Philosophy and our mental life. In H. Putnam (Ed.), *Mind, language and reality* (Philosophical papers, Vol. 2, pp. 291–303). Cambridge: Cambridge University Press.

Rice, S. H. (2008). Theoretical approaches to the evolution of development and genetic architecture. *Annals of the New York Academy of Sciences, 1133*, 67–86.

Rice, S. H. (2012). The place of development in mathematical evolutionary theory. *Journal of Experimental Zoology Part B: Molecular and Developmental Evolution, 318*, 480–488.

Riedel-Kruse, I. H., Müller, C., & Oates, A. C. (2007). Synchrony dynamics during initiation, failure, and rescue of the segmentation clock. *Science, 317*, 1911–1915.

Salazar-Ciudad, I., & Jernvall, J. (2002). A gene network model accounting for development and evolution of mammalian teeth. *Proceedings of the National Academy of Sciences of the United States of America, 99*, 8116–8120.

Salazar-Ciudad, I., & Jernvall, J. (2010). A computational model of teeth and the developmental origins of morphological variation. *Nature, 464*, 583–586.

Santillán, M., & Mackey, M. C. (2008). A proposed mechanism for the interaction of the segmentation clock and the determination front in somitogenesis. *PLoS ONE, 3*, e1561.

Schlosser, G., & Wagner, G. P. (Eds.). (2004). *Modularity in development and evolution*. Chicago: University of Chicago Press.

Schnell, S., Maini, P. K., Newman, S. A., & Newman, T. J. (Eds.). (2008). *Multiscale modeling of developmental systems*. New York: Academic.

Shigetani, Y., Sugahara, F., Kawakami, Y., Murakami, Y., Hirano, S., & Kuratani, S. (2002). Heterotopic shift of epithelial-mesenchymal interactions in vertebrate jaw evolution. *Science, 296*, 1316–1319.

Siegal, M., Promislow, D., & Bergman, A. (2007). Functional and evolutionary inference in gene networks: Does topology matter? *Genetica, 129*, 83–103.

Skipper, R. A., Jr., & Millstein, R. L. (2005). Thinking about evolutionary mechanisms: Natural selection. *Studies in History and Philosophy of Biological and Biomedical Sciences, 36*, 327–347.

Smart, A. G., Amaral, L. A. N., & Ottino, J. M. (2008). Cascading failure and robustness in metabolic networks. *Proceedings of the National Academy of Sciences of the United States of America, 105*, 13223–13228.

Strevens, M. (2008). *Depth: An account of scientific explanation*. Harvard: Harvard University Press.

Théry, F. (2015). Explaining in contemporary molecular biology: Beyond mechanisms. In P.-A. Braillard & C. Malaterre (Eds.), *Explanation in biology. An enquiry into the diversity of explanatory patterns in the life sciences* (pp. 113–133). Dordrecht: Springer.

Turing, A. M. (1952). The chemical basis of morphogenesis. *Philosophical Transactions of the Royal Society of London, Series B: Biological Sciences, 237*, 37–72.

Umulis, D., O'Connor, M. B., & Othmer, H. G. (2008). Robustness of embryonic spatial patterning in *Drosophila melanogaster*. In S. Schnell, P. K. Maini, S. A. Newman, & T. J. Newman (Eds.), *Multiscale modeling of developmental systems* (pp. 65–111). New York: Academic.

Uriu, K., Morishita, Y., & Iwasa, Y. (2010). Random cell movement promotes synchronization of the segmentation clock. *Proceedings of the National Academy of Sciences of the United States of America, 107*, 4979–4984.

von Dassow, G., & Munro, E. (1999). Modularity in animal development and evolution: Elements for a conceptual framework for EvoDevo. *Journal of Experimental Zoology Part B: Molecular and Developmental Evolution, 285*, 307–325.

von Dassow, G., & Odell, G. M. (2002). Design and constraints of the Drosophila segment polarity module: Robust spatial patterning emerges from intertwined cell state switches. *Journal of Experimental Zoology Part B: Molecular and Developmental Evolution, 294*, 179–215.

von Dassow, G., Meir, E., Munro, E. M., & Odell, G. M. (2000). The segment polarity network is a robust developmental module. *Nature, 406*, 188–192.

Wagner, G. P. (1996). Homologues, natural kinds and the evolution of modularity. *American Zoologist, 36*, 36–43.

Wagner, G. P. (2000). What is the promise of developmental evolution? Part I: Why is developmental biology necessary to explain evolutionary innovations? *Journal of Experimental Zoology Part B: Molecular and Developmental Evolution, 288*, 95–98.

Wagner, A. (2005a). Distributed robustness versus redundancy as causes of mutational robustness. *BioEssays, 27*, 176–188.

Wagner, A. (2005b). *Robustness and evolvability in living systems*. Princeton: Princeton University Press.

Wagner, G. P. (2005c). The developmental evolution of avian digit homology: An update. *Theory in Biosciences, 124*, 165–183.

Wagner, G. P. (2007). How wide and how deep is the divide between population genetics and developmental evolution? *Biology and Philosophy, 22*, 145–153.

Wagner, A. (2008). Gene duplications, robustness and evolutionary innovations. *BioEssays, 30*, 367–373.

Wagner, G. P., & Misof, B. Y. (1993). How can a character be developmentally constrained despite variation in developmental pathways? *Journal of Evolutionary Biology, 6*, 449–455.

Wagner, G. P., Chiu, C.-H., & Laubichler, M. D. (2000). Developmental evolution as a mechanistic science: The inference from developmental mechanisms to evolutionary processes. *American Zoologist, 40*, 819–831.

Wang, Y., Zhang, X.-S., & Xia, Y. (2009). Predicting eukaryotic transcriptional cooperativity by Bayesian network integration of genome-wide data. *Nucleic Acids Research, 37*, 5943–5958.

Weber, M. (2005). *Philosophy of experimental biology*. Cambridge: Cambridge University Press.

Weber, M. (2008). Causes without mechanisms: Experimental regularities, physical laws, and neuroscientific explanation. *Philosophy of Science, 75*, 995–1007.

Weisberg, M. (2013). *Simulation and similarity: Using models to understand the world*. New York: Oxford University Press.

West-Eberhard, M. J. (2003). *Developmental plasticity and evolution*. Oxford: Oxford University Press.

West-Eberhard, M. J. (2005). Phenotypic accommodation: Adaptive innovation due to developmental plasticity. *Journal of Experimental Zoology Part B: Molecular and Developmental Evolution, 304B*, 610–618.

Westerhoff, H. V., & Kell, D. B. (2007). The methodologies of systems biology. In F. C. Boogerd, F. J. Bruggeman, J.-H. S. Hofmeyr, & H. V. Westerhoff (Eds.), *Systems biology: Philosophical foundations* (pp. 23–70). Amsterdam: Elsevier.

Whitman, D. W., & Agrawal, A. A. (2009). What is phenotypic plasticity and why is it important? In D. W. Whitman & T. N. Ananthakrishnan (Eds.), *Phenotypic plasticity of insects: Mechanisms and consequences* (pp. 1–63). Enfield: Science Publishers.

Wing, S. S., Lecker, S. H., & Jagoe, R. T. (2011). Proteolysis in illness-associated skeletal muscle atrophy: From pathways to networks. *Critical Reviews in Clinical Laboratory Science, 48*, 49–70.

Winther, R. G. (2006). Parts and theories in compositional biology. *Biology and Philosophy, 21*, 471–499.

Winther, R. G. (2011). Part-whole science. *Synthese, 178*, 397–427.

Woodward, J. (2002). What is a mechanism? A counterfactual account. *Philosophy of Science, 69*, S366–S377.

Woodward, J. (2003). *Making things happen: A theory of causal explanation.* Oxford: Oxford University Press.

Young, R. L., & Wagner, G. P. (2011). Why ontogenetic homology criteria can be misleading: Lessons from digit identity transformations. *Journal of Experimental Zoology Part B: Molecular and Developmental Evolution, 316B*, 165–170.

Zhu, J., Zhang, Y.-T., Alber, M. S., & Newman, S. A. (2010). Bare bones pattern formation: A core regulatory network in varying geometries reproduces major features of vertebrate limb development and evolution. *PLoS ONE, 5*, e10892.

Chapter 8
The Relevance of Irrelevance: Explanation in Systems Biology

Fridolin Gross

Abstract In this chapter I investigate explanations in systems biology that rely on dynamical models of biological systems. I argue that accounts of mechanistic explanation cannot easily make sense of certain features of dynamical patterns if they restrict themselves to change-relating relationships. When investigating the use of such models, one has to distinguish between the concepts of causal or constitutive relevance on the one hand, and explanatorily relevant information on the other. I show that an important explanatory function of mathematical models consists in elucidating relationships of non-dependence. Notably, the fundamental concept of robustness can often be accounted for in this way, and not by invoking separate mechanistic features. Drawing on examples from the scientific literature, I suggest that an important aspect of explaining the behavior of a biological mechanism consists in elucidating how in the systemic context components are not, or only weakly, dependent on each other.

Keywords Systems biology • Mechanism • Dynamical system • Explanatory relevance • Robustness • Invariance

1 Introduction

The molecules inside a living cell do not behave like the molecules in a gas. In a gas the single particles freely move around and interact randomly (if at all), while showing no apparent organization. A gas seems to be the perfect example of a simple aggregate (Wimsatt, 1997, 2007) whose macro-level properties are invariant under various kinds of changes at the micro-level. Obviously, the cell is not such a simple aggregate. However, a cell does not appear to behave like a mechanical clock either. The mechanism of a clock almost certainly breaks down if we remove one part or try to exchange two different components, whereas living

F. Gross (✉)
Institute for Theoretical Biology, Charité Universitätsmedizin, Philippstraße 13, 10115 Berlin, Germany
e-mail: fridolin.gross@hu-berlin.de

© Springer Science+Business Media Dordrecht 2015
P.-A. Braillard, C. Malaterre, *Explanation in Biology*, History, Philosophy and Theory of the Life Sciences 11, DOI 10.1007/978-94-017-9822-8_8

systems are often surprisingly stable under a wide range of perturbations.[1] From this one may infer that on the spectrum of organizational complexity cells assume a position somewhere between gases and clocks. But arguably many people would strongly object to this classification. Aren't biological systems much *more* complex and organized than the artifacts of mechanical engineering? And aren't there very specific kinds of perturbations, such as mutations, to which living systems react in very sensitive ways? Isn't it more plausible to assume that living systems are very complex mechanisms, but differently from clocks they have additional features that account for their particular ways of resisting perturbations? The coexistence of extreme complexity and robustness is undoubtedly one of the most fascinating features of life. As the physiologist Walter Cannon remarked:

> When we consider the extreme instability of our bodily structure, its readiness for disturbance by the slightest external forces and the rapid onset of its decomposition as soon as favoring circumstances are withdrawn, its persistence through many decades seems almost miraculous. (Cannon, 1932, 20)

Cannon and many others have shown that the stability of physiological processes can often be explained with reference to particular mechanisms. For example, the homeostasis of blood sugar levels can be explained with reference to a simple feedback mechanism involving the hormones insulin and glucagon.

Systems biology has recently started to investigate similar phenomena at a more fine-grained level, such as the robustness of genetic or metabolic networks. Is it obvious that the robustness of such fundamental structures of living systems can be understood in the same way? In what follows I want to show how some of the insights gained by systems biologists challenge widely held intuitions about mechanistic explanations in the life sciences. By doing so, I do not want to suggest that systems biology will eventually give rise to an alternative, non-mechanistic paradigm of explanation in biology. I merely want to draw attention to certain issues that mechanistic accounts will need to address in order to capture the explanatory ambitions of systems biologists. The relationship of systems biology to the framework of mechanistic explanation and the particular importance of the concept of robustness have already been discussed elsewhere (e.g. Braillard, 2010; Breidenmoser and Wolkenhauer, 2015; Brigandt, 2015). Here, I want to pursue some of the already developed ideas further and address some of the issues from a different angle by focusing on the idea of "non-dependence."

The currently most refined accounts of mechanistic explanation link the understanding of systemic behavior to manipulationist criteria of causal or constitutive relevance (e.g. Craver, 2007; Woodward, 2010). The claim I want to defend in this chapter is that these accounts rely on a clock-like picture of biological mechanisms by assigning importance solely to *change-relating* relationships. I argue instead for

---

[1] An impressive example of such robustness is revealed, for instance, in the experiments of rewiring the *E. coli* gene regulatory network by Isalan et al. (2008).

the explanatory value of relationships that are not change-relating,[2] especially when it comes to explanations of the behaviors of very complex mechanisms. Examples of explaining biological robustness by means of dynamical modeling will provide the right kind of illustrations for this point.

The chapter is organized as follows. In Sect. 2, I introduce the manipulationist account of mechanistic explanation and discuss how biological robustness might be accounted for from this perspective. Section 3 takes a closer look at Woodward's concepts of *invariance* and *stability*, which seem to be closely related to robustness, and investigates their role in manipulationist explanations. For illustrative purposes, a toy example of gene expression is presented in Sect. 4. Here my aim is to show that the significance of non change-relating relationships arises especially in the context of dynamic modeling. Dynamic (or steady-state) equilibrium is arguably the simplest case of dynamic stability, and it reveals some typical features of explanations that are given in systems biology. Then, in Sect. 5, I will present a real case study from systems biology that illuminates how information about non-dependence plays an essential role in our understanding of biological mechanisms. After this more formal discussion, these issues are connected in Sect. 6 to a general perspective on robustness and the global architecture of living systems.

2   Mechanistic Explanation

It seems natural to think that real understanding of a system implies the ability to predict how it will respond to various kinds of interventions. To understand a phenomenon means to know how changes in it can be brought about, and this idea seems, at least implicitly, to underlie many of the recent conceptions of mechanistic explanation in the philosophy of science. The relationship between intervention and explanation has been made most explicit by James Woodward (2002, 2003) in his manipulationist account of causation and explanation which was subsequently adopted and further developed by some of the main proponents of mechanistic explanation (e.g. Craver, 2007; Glennan, 2002). On Woodward's account a causal relationship holds between two variables or events, roughly speaking, if it is possible (at least in principle) to systematically bring about changes in one by intervening on the other. The importance of these relationships for explanation, according to Woodward, lies in the fact that they allow us to answer a range of counterfactual *what-if-things-had-been-different* questions about the explanandum. We may, for instance, explain why a particular person has contracted lung cancer by referring to the fact that the person was a heavy smoker. The causal knowledge that the occurrence of cancer can be influenced by intervening on smoking behavior

---

[2]I will synonymously speak of "non change-relating relationships" and "relationships of non-dependence".

increases our understanding since it allows us to infer the counterfactual claim that the person (probably) wouldn't have gotten the cancer if she hadn't smoked.

However, mechanistic explanations are not necessarily explanations of effects in terms of their causes, but usually are understood as explanations of the activity of a whole in terms of the properties of its parts. This distinction can be further illuminated by considering the different types of questions that different explanations are supposed to answer. While explanations of effects in terms of their causes are directed towards *why-questions*, such as 'why did this person get lung cancer?,' the description of the mechanism underlying a behavior may be understood most intuitively as answering a *how-question*, such as 'how does the heart pump blood?'. In this context, Wesley Salmon referred to explanations in terms of underlying structure as *constitutive* and distinguished them from *etiological* explanations that cite the causal history of an event or phenomenon (Salmon, 1984, 275). More recently, Craver (2007) has argued that constitutive dependencies between mechanisms and their components are metaphysically distinct from causal dependencies holding between objects at the same level. He argues, however, that usually both causal and constitutive relationships are employed in mechanistic descriptions, and that both can be understood within Woodward's general manipulationist perspective.

Mechanistic accounts are often presented as an alternative to nomological models of explanation. According to the latter, explanation consists in logically deriving the explanandum from premises that include law-like generalizations. Mechanistic accounts, by contrast, hold that the explanation does not lie in a derivation, but is directly given by objective relationships, causal or constitutive, that account for the explanandum phenomenon. The core idea of applying the manipulationist framework to mechanistic explanations can be illustrated with the following quote by Carl Craver:

> One need not be able to derive the phenomenon from a description of the mechanism. Rather, one needs to know how the phenomenon is situated within the causal structure of the world. That is, one needs to know how the phenomenon changes under a variety of interventions into the parts and how the parts change when one intervenes to change the phenomenon. When one possesses explanations of this sort, one is in a position to make predictions about how the system will behave under a variety of conditions. Furthermore when one possesses explanations of this sort, one knows how to intervene into the mechanism in order to produce regular changes in the phenomenon. (Craver, 2007, 160)

Craver thus thinks of *explanatory relevance* (which components and relationships should figure in a mechanistic explanation) in terms of *manipulationist relevance* (which factors can be manipulated to change the phenomenon). The proponents of the manipulationist view argue that, restricting oneself to change-relating relationships one obtains a criterion for explanatory relevance that avoids many of the problems afflicting both nomological accounts and traditional versions of causal-mechanical explanation (e.g. Railton, 1981; Salmon, 1984).

If one conceives of biological systems as clock-like, it is plausible to equate manipulationist relevance with explanatory relevance. In a clock it seems that exactly those interventions that bring about changes in the overall behavior are the ones that reveal the relationships one needs to know in order to grasp the underlying mechanism. For instance, if the balance spring in a clock is replaced by an otherwise similar spring with greater stiffness, the balance wheel will oscillate with increased frequency, and, as a consequence, the hands of the clock will move faster. Generalizing from this example, one may say that in a certain sense machines like clocks are extremely fragile because changes in the components are rigidly connected to changes in systemic behavior. The observation that clocks do not easily fall apart, in spite of this fragility, is explained by the fact that the properties of the parts are not easily changed in the first place. Consider the effect of temperature on a clock. A clock made of metal can work reliably both in most climates because the temperatures that could significantly deform its components lie outside the typically encountered range. Similarly, most mechanical devices owe their robustness to the fact that properties of their parts are insensitive to a wide range of external perturbations or changes in external conditions. 'Mechanical robustness' therefore can be taken to mean that the components of a system are insensitive to interventions on external conditions.

It seems that biological systems are not fragile in the same sense. Biological robustness means roughly "that some property of the system remains the same under perturbation" (Gunawardena, 2010, 35). However, in this context 'perturbation' is understood not as a change in external conditions, but as a change in the components or the organization of the system itself. Thus biological robustness implies that certain interventions on the components do *not* bring about changes in a phenomenon, which in turn means that there are relationships between properties of the system and its components that are *not* change-relating. What role do such relationships play in our attempts to understand biological systems? And how could they be interpreted within a manipulationist account of mechanistic explanation?

There seem to be two strategies of dealing with such relationships from within the manipulationist framework. On the one hand, one may argue that they simply fail to meet the criterion for explanatory relevance. For example, a clock's behavior will not be altered by changing the color of the balance spring. Consequently, the color of the spring is considered irrelevant when it comes to explaining how the clock works. On the other hand, there might be occasions where the observation that something *doesn't change* itself is of explanatory interest. The manipulationist will then set out to look for an explanation of this behavior in terms of underlying relationships that actually *are* change-relating, such as in the case of blood sugar homeostasis (see Issad and Malaterre, 2015). Insofar as robustness is a somehow "surprising" or "almost miraculous" (Cannon, 1932) property of living systems, she will attempt to explain it by looking for specific mechanisms that are responsible for the resistance to change. In any case, change-relating relationships are the fundamental building blocks of mechanistic explanations for the manipulationist. Relationships that are not change-relating either are irrelevant or themselves have to be explained in terms of change-relating relationships.

By investigating examples of dynamical modeling in systems biology, I will show that relationships that are not change-relating (relationships of non-dependence) point to something deeper and draw our attention to complementary aspects of scientific understanding that have been neglected in recent discussions on mechanistic explanation. Before turning to these examples, however, I will have to say a bit more about the connection between change-relating relationships and explanation according to the manipulationist picture and the problems it raises for explaining biological robustness.

3   Explanation and Invariance

To illustrate his counterfactual account of causal explanation, James Woodward (2003, 187) makes use of a simple example from physics that probably can be found in any textbook on electrostatics (Fig. 8.1). A very long straight wire carries a uniformly distributed electric charge with density λ. The explanandum in this example is the force of the electric field on a test charge at position P at a perpendicular distance r from the wire. Woodward describes a derivation that is based on Coulomb's law and determines the strength of the field at P by summing up the contributions dq from all the infinitesimal sections dx of the wire. As a result he obtains the following expression:

$$E = \frac{1}{2\pi\epsilon_0} \frac{\lambda}{r} . \tag{8.1}$$

Woodward argues that this relationship, together with its derivation, explains the field intensity at P because it allows us to predict how the value of E changes if we intervene on the system in various ways. For instance, if we increase the distance between the wire and the test charge, the formula tells us that the field intensity decreases proportionally to the reciprocal of the distance. Changing the relative charge λ, on the other hand, results in a proportional change in intensity.

Generalizing from this example, Woodward proposes that explanation amounts to exhibiting the systematic patterns of counterfactual dependence that can be expressed as functional relationships between variables. In doing so, however, he restricts himself to relationships that are change-relating, that is, to relationships in which an intervention on one variable brings about a change in the other. If we consider the derivation of (8.1), however, we notice that it also elucidates relationships that are not change-relating. For instance, we learn that moving the test

Fig. 8.1 Woodward's example of the charged wire. Reprinted by kind permission of Oxford University Press, Oxford, from Woodward (2003, 188)

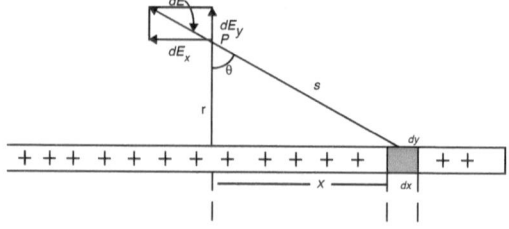

charge to a new position P' at the same distance from the wire will not change the value of the field intensity because, as can easily be shown, such a transformation would not affect the result of the calculation. Likewise, we can infer that the x-component E_x of the field vector will never change to a value different from zero, no matter how we intervene, provided that we do not destroy the symmetry of the geometrical setup.

In Woodward's picture information about such non change-relating relations plays no direct role in explanation. However, the notion of *invariance*, that figures prominently in his account, seems closely related. He argues that causal claims are always associated with claims about invariant relationships:

> Invariance under at least one testing intervention (on variables figuring in the generalization) is necessary and sufficient for a generalization to represent a causal relationship or to figure in explanations. (Woodward, 2003, 250)

The idea is roughly the following. A generalization on which a causal claim is based can be described as a functional relationship between two variables of the type $Y = G(X)$. It is not necessary that such a generalization holds under all circumstances, instead it is required only that there are *some* possible changes of X under which it continues to hold. Invariance obviously comes in degrees, but as long as there is a minimum of invariance, a relationship is causal and, therefore, potentially explanatory. Highly invariant generalizations, such as the fundamental laws of physics, do not necessarily give rise to better explanations, even though they might have other desirable features.

In a more recent article Woodward uses the terms of *invariance* and *stability* interchangeably, and gives a slightly different characterization in terms of background circumstances. He argues that in order to qualify as causal, it is sufficient that a relationship of counterfactual dependence holds in some set of circumstances B_i. He then states:

> The *stability* of this relationship of counterfactual dependence has to do with whether it would continue to hold in a range of other background circumstances B_k different from the circumstances B_i. (Woodward, 2010, 291–292, emphasis in original)

According to this characterization, a claim of invariance or stability can be formally expressed as:

$$Y = G(X, B_i) = G(X, B_k) \text{for all } k \text{ in some set } K, \tag{8.2}$$

which obviously implies the existence of a relationship $F(B) = G(X, B)$ of non-dependence, that is, $F(B_k) = F(B_l)$ even if $B_k \neq B_l$. Thus Woodward's account actually relies on *both* change-relating and non change-relating relationships. However, the two kinds seem to play very different roles. Change-relating relationships on Woodward's view are the crucial elements; they provide the *content* of the explanation, so to speak, and elucidate the features of the explanandum phenomenon by providing patterns of counterfactual dependence. Relationships of invariance, on the other hand, largely keep in the background; they are necessary for specifying the range of application, or generality, of an explanatory claim, but strictly speaking do not provide any explanatory information.

There is thus a clear conceptual separation between the two types of functional relationships reflecting Woodward's distinction between causal explanatory claims, on the one hand, and claims about invariance, on the other hand. However, if we think again of the derivation of the field strength in the wire example, we recall that it also provides information about interesting relationships of non-dependence. If we keep track of both the x- and y-components of the field strength, we notice that all the infinitesimal contributions to E_x exactly cancel each other out, independently of the position P at which the field is evaluated. The relationship

$$E_x(P) = 0 \text{ for all } P, \tag{8.3}$$

however, does not play the role of an invariance condition, and it does not seem to be irrelevant in the same way as, for instance, the color of the wire. At least for mathematical structures it seems that both change-relating and non change-relating relationships are potentially important for our understanding. It is not obvious why one type of relationship should be more interesting or more informative than the other. The reason why non-change-relating relationships are often neglected might be due to the following feature: they can be represented in a very compressed way, or simply be left implicit. As Herbert Simon put it: "Mother Hubbard did not have to check off the list of possible contents to say that her cupboard was bare" (1962, 478). But the fact that non-change-relating relationships lend themselves to descriptive economy does not imply that they are explanatorily irrelevant.

Woodward's account suggests that change-relating relationships exhaust all that is needed for explaining the behavior of a system. But if relationships of non-dependence can contribute to our understanding of mathematical models, why shouldn't they be taken as contributing to our understanding of phenomena that are explained by means of such models?[3] As will be further illustrated later in this chapter, the functional relationships that play a role in the models of systems biology often are change-relating in some particular range of values while being non-change relating in a different range. I will show that usually both types of information are crucial for an understanding of complex behavior, without one necessarily being reducible to the other.

By that I do not want to deny the important role that change-relating relationships play in determining the causal or constitutive links within a mechanism. There is no doubt that these relationships provide explanations by allowing us to answer to why-questions of a particular type. But this alone does not entail the equivalence of information about manipulationist relevance and explanatory information when it comes to more complex mechanistic explanations.

A related issue, that Woodward's account leaves unclear, is how invariance or stability itself is explained. As Robert Batterman notices:

---

[3]For a more sophisticated analysis of the role of mathematics in scientific explanation see Baker (2015).

> Woodward stresses the importance for explanation of a kind of invariance and robustness
> that may be present in a given regularity to some degree or other. Thus, he discusses how
> "nonlaw-like" regularities may, because of their robustness, play crucial explanatory roles.
> Woodward is not concerned to answer why-questions about the universality or degree of
> universality of the regularities that he discusses. That is, he does not, as far as I can tell, ask
> the question why the regularity has the robustness that it has or has it to the degree that it
> has. (Batterman, 2002, 59)

Batterman argues that in the explanation of a phenomenon one has to distinguish
between two different kinds of why-questions:

> A type (i) why-question asks for an explanation of why a given instance of a pattern
> obtained. A type (ii) why-question asks why, in general, patterns of a given type can be
> expected to obtain. Thus, a request to explain an instance of universality is a request to
> provide an answer to a type (ii) why-question. (Batterman, 2002, 23)

Batterman's ambition to explain universality and Woodward's efforts to elucidate
explanation in terms of contingent causal generalizations point to different but
possibly complementary aspects of scientific curiosity. These may be seen as loosely
related to the different types of questions that are typically asked in the physical and
the biological sciences, respectively. It is a philosophically interesting question how
the new field of systems biology locates itself on this spectrum since, with regard to
its methodological and explanatory resources, it has often been perceived as pushing
biology more towards a physics attitude (see e.g. Poon, 2011). A closer look at some
examples may help to shed light on this issue.

4   Explaining Equilibria

Let us start with a very simple case and consider the following minimal model
of gene expression. The system consists of a protein with concentration X that is
synthesized at a constant rate $S = \sigma$, while its degradation rate, $D = \delta \cdot X$, is
proportional to the concentration. Figure 8.2 graphically represents the qualitative
features of this model. The dynamics of X is captured by the following differential
equation:

$$\frac{dX}{dt} = S - D(X) = \sigma - \delta \cdot X . \tag{8.4}$$

Solving this equation allows us to obtain the temporal behavior of X depending on a
given initial concentration X_0 at time $t = 0$. As can be checked, the explicit solution
is given by:

$$X(t, X_0) = \left(X_0 - \frac{\sigma}{\delta} \right) \exp(-\delta t) + \frac{\sigma}{\delta} . \tag{8.5}$$

After sufficient time, the value of the exponential will become very small and the
first part of the right hand side of (8.5) can be neglected. Formally,

Fig. 8.2 Rate balance plot for the simple gene expression model. Synthesis is constant, while degradation depends linearly on protein concentration. Both can be represented as *straight lines*. The intersection of the two lines corresponds to the equilibrium state. The stability of the equilibrium can be inferred from the sign of the resulting rate when subtracting degradation from synthesis

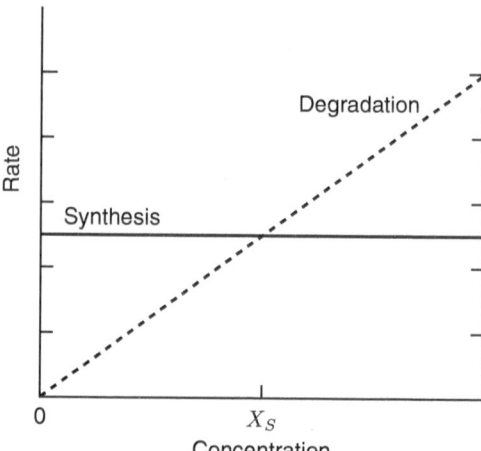

$$X(t, X_0) \rightarrow \frac{\sigma}{\delta} \quad \text{for} \quad t \rightarrow \infty. \tag{8.6}$$

Note that the expression to which X converges does not contain X_0. This means that the protein concentration in the long run does not depend on its initial value, but assumes an equilibrium (or steady state) value $X_S = \sigma/\delta$ that depends only on the rates of synthesis and degradation. A further consequence is that, whenever the system is perturbed by changing the concentration to some value $X \neq X_S$, it will always return to X_S eventually. At first sight this derivation seems to provide a perfectly satisfactory explanation of equilibrium.

The model just described is very similar to an example that Elliott Sober (1983) has used to raise some questions about causal-mechanical approaches to explanation. He refers to an explanation given by R. A. Fisher for the 1:1 sex ratio observed in many sexually reproducing species.[4] Instead of providing a particular causal history for the occurrence of the ratio, Fisher points out why the long run ratio in many sexually reproducing populations does *not* depend on particular causal details. As Sober reports:

> Fisher's account shows why the actual initial conditions and the actual selective forces don't matter; whatever the actual initial sex ratio had been, the selection pressures that would have resulted would have moved the population to its equilibrium state. Where causal explanation shows how the event to be explained was in fact produced, equilibrium explanation shows how the event would have occurred regardless of which of a variety of causal scenarios actually transpired. (Sober, 1983, 202)

Sober concludes that equilibrium explanations are not causal explanations of the etiological type:

---

[4]Baker (2015) briefly mentions the same example and identifies equilibrium explanation as one of three general classes of mathematical explanation in science.

> The causal explanation focuses exclusively on the actual trajectory of the population; the equilibrium explanation situates that actual trajectory (whatever it may have been) in a more encompassing structure. It is in this way that equilibrium explanations can be more explanatory than causal explanations even though they provide less information about what the actual cause was. This difference arises from the fact that explanations provide understanding, and understanding can be enhanced without providing more details about what the cause was. Equilibrium explanations are made possible by theories that describe the dynamics of systems in certain ways. (Sober, 1983, 207)

Sober thus hints at a discrepancy between information about particular causal events and information that is relevant for explanation. He suggests that explaining equilibrium means showing why particular causal facts do not make a difference to the outcome. The question thus arises how this idea relates to Woodward's account according to which such facts are simply explanatorily irrelevant. Is it straightforward to capture equilibrium explanations within the manipulationist framework?

Before trying to determine what kinds of explanations they are—or aren't, we should clarify what it is that equilibrium explanations are supposed to explain. Regarding the sex ratio, the general question is 'Why is there an equilibrium at a sex ratio of 1:1 in so many species?.' However, this question can be interpreted as actually including three different calls for explanation, depending on where we put the stress in the sentence.[5] First, it may be read as the question of why it is one and the same ratio that is observed across a wide range of sexually reproducing species. In other words, why does the rate not assume different values for different species? Second, as the question of why that ratio has the particular numerical value of $r = \#males/\#females \approx 1$, and not some other number in the interval $(0, \infty)$. And third, as the question of why the observed ratio represents an equilibrium point, that is, why it is stable and adjusts itself after perturbations.

Each way of interpreting the initial question calls for an account that makes use of different explanatory resources. The first interpretation, even though interesting in its own right, is not relevant for the current discussion since it seems to mainly depend on empirical facts that are specific to evolutionary biology. For this reason, my focus will be on the differences between the second and the third interpretation that more directly pertain to the phenomenon of equilibrium in general, and roughly correspond to Batterman's type (i) and type (ii) why-questions. I will discuss these differences in more detail using the particularly clear example of the gene expression model.

Let us therefore look at the explanation-seeking question and the two relevant interpretations when transferred to this example. The question is, 'Why is there an equilibrium at a concentration $X = X_S$?', and it can be interpreted as expressing an interest either in the particular numerical value or in the fact that there is an equilibrium. Responding to the first, the derivation of (8.6) can be taken to show why the protein concentration at steady state is given by the particular ratio σ/δ.

---

[5]See Morange (2015) for a general discussion of how the ambiguity of questions can lead to explanatory diversity.

This seems to represent a paradigmatic case of a Woodwardian explanation since the steady state concentration is explained in terms of the dependency relations characterizing the system. It clearly allows us to answer a range of counterfactual *what-if-things-had-been-different* questions. For instance, we can predict how the steady state value would change if we were to intervene in ways that change the rates of synthesis or degradation. Differently from the type of causal explanation that are the target of Sober's argument, however, this explanation refers to structural features of the model rather than to causal history. In the terminology introduced earlier, this explanation might, therefore, best be understood as constitutive. This is the way in which Kuorikoski (2007) interprets equilibrium explanations within a manipulationist framework:

> If explanations indeed track dependencies instead of persistence, the interesting explanatory relationship cannot be the one between the initial conditions and the equilibrium state, as might first be surmised, and indeed as seems to have been Sober's view. Instead, what the equilibrium state does depend on are the structural features of the system. *Equilibrium explanations are not causal explanations of events but structural or constitutive explanations of system-level properties.* (Kuorikoski, 2007, 154, emphasis in original)

However, stating the dependency relations between parameters and steady state value alone does not give an answer to Batterman's type (ii) question of why the pattern, in this case equilibrium, obtains in the first place. Instead, as we have seen, equilibrium seems to be explained precisely by deriving a relation of *non-dependence* between the initial conditions and the long-run concentration. Is there another way in which we can understand this aspect of equilibrium within a manipulationist framework of causation while avoiding Sober's puzzle about the irrelevance of particular causal facts?

To maintain a contrastive focus, one may try to interpret the existence of a single stable equilibrium as a property that systems either do or do not possess, and determine exactly what this property depends on. It turns out that in the present example this property depends only on the structure of the model.[6] This dependency may be expressed in terms of a binary variable $P \in \{0, 1\}$ in the following way:

$$P(\{S, D\}, \{X\}, \{\sigma, \delta\}) = 1, \tag{8.7}$$

where $S = \sigma$ and $D = \delta \cdot X$ represent the particular types of functions used to express the dynamic relationships, while $\{X\}$ and $\{\sigma, \delta\}$ stand for the sets of variables and parameters that appear in the model. By modifying this structure in particular ways, one may obtain a different model for which $P = 0$, that is, a model without an equilibrium state, or perhaps with more than one. An example of such a modification is the complete disruption of degradation, i.e. setting $D = 0$, or the addition of a more complex dependency $S(X)$ of synthesis on the concentration.

---

[6]Since the equilibrium is global it does, for instance, not depend on the initial concentration being within a particular range. However, similar arguments can be made for cases of non-global equilibrium.

This reasoning suggests that in principle it might be possible to find a representation of the (potentially very complicated) dependency relation between P and the structural properties of the model. Subsequently, one could make use of this relation to explain why a particular instance of the model does or does not possess the equilibrium property P. Furthermore, one may argue that P's structural dependency explains equilibrium by showing how it appears when the structural parameters are changed in particular ways. But have we thereby really explained equilibrium? It seems that by using the complex dependency relation, we have at best been able to give a more sophisticated answer to a type (i) why-question. We have explained that a particular system shows equilibrium because it belongs to a particular structural class. If we intervene on the structure of the system in such a way that it no longer belongs to this class, it will exhibit qualitatively different behavior.[7] We have not explained the behavior itself. It seems that by using a manipulationist strategy, we do not reach beyond the explanation of instances of equilibrium.

To summarize, a satisfactory explanation of equilibrium in causal-etiological terms fails for the reasons discussed in Sober's paper. In order to explain equilibrium constitutively, the manipulationist may invoke relationships that relate quantitative or qualitative changes in behavior to changes in structural features of the system, but thereby fails to give an account of how the behavior is produced in the first place. As I argue, and as Sober suggests, equilibrium is best explained by referring to a relationship of non-dependence. The case of equilibrium thus shows how knowledge about such relationships can be explanatorily relevant.

5   Dissecting a Dynamical Switch

After these initial considerations about relationships of non-dependence, one may ask whether they are of any importance for the description of actual scientific explanations. For this reason I will now turn to a real example taken from the scientific literature. The biological phenomenon I will discuss is an instance of so-called *bistable switching* which plays a role in many important biological processes, for instance in the control of gene expression, in cellular differentiation, cell-cycle progression, and in neural signaling. It is thus representative for a class of phenomena that are biologically relevant and widely discussed among theoretically minded molecular biologists (see e.g. Bhalla and Iyengar, 1999; Ferrell and Xiong, 2001; Novak et al., 2007; Savageau, 2001). My aim is to show how in the explanatory practice of systems biology manipulationist reasoning about causal mechanisms is integrated with dynamical modeling. Notably, it will become clear

---

[7]In the theory of dynamical systems the investigation of equilibrium states when varying the parameter values is known as bifurcation analysis. This type of analysis is carried out to investigate the circumstances under which a system shifts between qualitatively different behaviors, not to explain the behaviors themselves.

that relationships of non-dependence are crucial for an understanding of systemic behavior, and not only used to establish the invariance of the causal or constitutive relationships. Conveniently, the philosophically interesting features of this example can be elucidated without going into the mathematical details.

At a particular stage during the process of egg formation in the frog *Xenopus laevis*, oocytes are arrested in an immature state. When exposed to the hormone progesterone, they undergo maturation and complete the first meiotic division. This maturation has been observed to occur in a switch-like manner: cells are either in the immature or in the mature state, but apparently cannot be in intermediate states for extended periods of time (Ferrell and Machleder, 1998). A crucial step in triggering maturation, and a convenient read-out, is the phosphorylation of the protein kinase p42 MAPK. When treating individual oocytes with intermediate doses of progesterone, Ferrell and Machleder observed either very high ($>90\%$) or very low levels ($<10\%$) of phosphorylated p42 MAPK. In the following they were interested in understanding how "a continuously variable stimulus—the progesterone concentration—is converted into an all-or-none biological response" (Ferrell and Machleder, 1998, 895). They hypothesized that the all-or-none character of the phenomenon is due to bistability, that is, the system can shift between two alternative stable equilibrium states. In what follows I will present the way in which these and other scientists have explained the switching behavior in oocyte maturation.

Bistability can arise in certain types of dynamical systems that involve nonlinear relationships.[8] It is easy to see how the existence of multiple equilibria is possible when we consider a rate balance plot which, in contrast with the case of simple equilibrium discussed in Sect. 4, is not restricted to straight lines. Figure 8.3 shows the balance of phosphorylation (forward reaction) and dephosphorylation (back reaction) of the kinase p42 MAPK in the oocyte system. The two curves indicate how the rates of these reactions depend on the fraction of phosphorylated kinase (denoted by A^*/A_{tot}). While the back reaction curve is simply a straight line, the forward reaction curve is essentially flat in the left portion of the plot and has a pronounced maximum when about half of the total amount of kinase is phosphorylated. The particular nonlinear behavior of the forward reaction curve is due to an underlying positive feedback: the phosophorylation reaction is slow unless there is a considerable fraction of kinase that is already phosphorylated.[9] Since the slope of this curve is initially less steep than the slope of the back reaction curve, there are three intersections of the two curves and hence three equilibrium points.

---

[8]Nonlinear relationships are not necessarily change-relating. It is perhaps as a result of the hype around chaos theory that 'nonlinearity' is usually associated with the idea that small changes can have large effects. However, in nonlinear dynamical systems the converse is also possible: large changes with negligible or small effects.

[9]Ferrell and Xiong (2001) suggest that several mechanisms are jointly responsible for this effect. Notably, p42 MAPK is involved in a positive feedback loop by contributing to the accumulation of Mos, its upstream activating kinase.

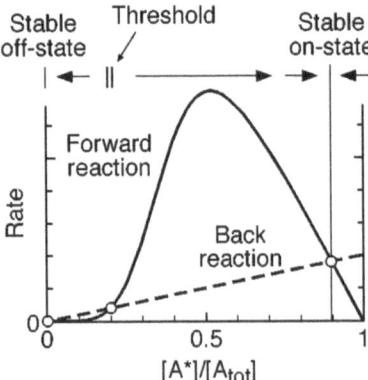

Fig. 8.3 Rate balance plot for the oocyte maturation model. Due to non-linearities the forward reaction is not a straight line. The three intersections correspond to three equilibrium points. The one in the middle is unstable. Note that, instead of balance of degradation and synthesis, equilibrium in this case requires equal rates of forward and back reaction. Reprinted by kind permission of AIP Publishing LCC, from Ferrell and Xiong (2001, 232)

The middle one is unstable, however, since in its vicinity the system will always be driven away from it, towards one of the two outer equilibria, which are both stable.

In this way we have established that there are two stable equilibria at low and high concentrations of phosphorylated kinase, respectively. These can be interpreted as *off* and *on*-states of a switch; but how can this maturation switch actually be effectuated? It turns out that such a shift from *off* to *on* occurs at a critical level of progesterone concentration because the *basal rate* of the forward reaction is proportional to the level of the activating progesterone stimulus. The basal rate is the rate at which the reaction would proceed in the absence of the positive feedback mechanism, and its dependence on the stimulus affects the shape of the total forward reaction curve as shown in Fig. 8.4. With increasing stimulus the *off*-state and the unstable equilibrium point move closer together until the two points finally coalesce. The curves corresponding to even higher levels of stimulus each have only one intersection with the back reaction curve. Therefore, if the system is initially in the *off*-state, it will at some critical level of stimulus jump to the *on*-state which then is the only remaining equilibrium. This explains the observed all-or-none behavior in the maturation of oocytes.

The behavior of the switch can be further illustrated by representing the position of the stable equilibria as a function of the stimulus (Fig. 8.5). This plot elucidates another important property of the switch: After the system has been driven from the *off*-state to the *on*-state by continuously increasing the stimulus, it will remain in the *on*-state even if the stimulus is subsequently withdrawn. Thus, once an oocyte has received a hormonal stimulus of sufficient size, it is irreversibly committed to maturation and does not shift back and forth between the two states. Obviously, this irreversibility is crucial for the reliability of developmental pathways.

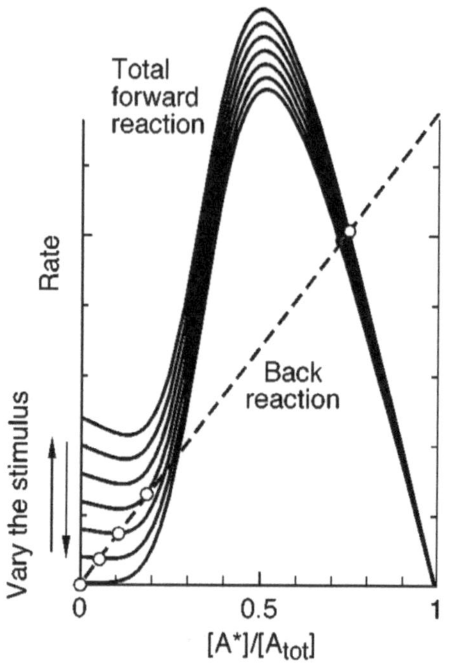

Fig. 8.4 Rate balance plot
for varying progesterone
concentration. Above a
critical level the *off*-state
disappears. Reprinted by kind
permission of AIP Publishing
LCC, from Ferrell and Xiong
(2001, 233)

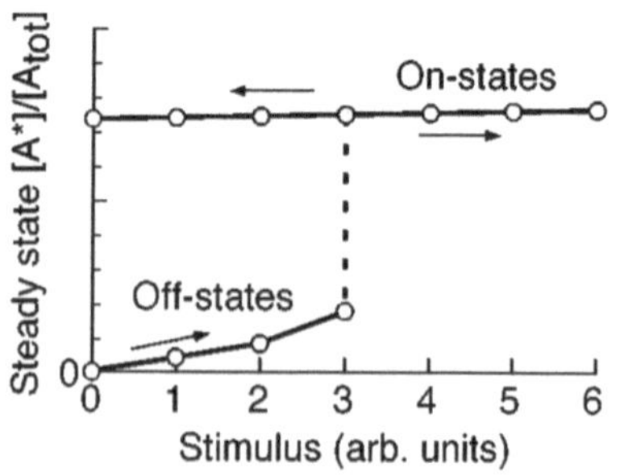

Fig. 8.5 Stimulus response
curve. Once the stimulus
reaches the threshold level,
the system is locked in the
on-state in which the
concentration of the
phosphorylated kinase A^* is
always high. Reprinted by
kind permission of AIP
Publishing LCC, from Ferrell
and Xiong (2001, 233)

Let us now try to understand in more detail how the given account explains
the initiation of maturation. If we first consider only the "switching on" part of
the story, we can represent the mechanism in terms of a simple causal relationship
between two binary variables: A stimulus variable that can take on the values 'below
threshold' or 'above threshold', and a kinase activity variable that accordingly
assumes either of the values 'on' or 'off'. Obviously, this is exactly what we expect
from a simple switch. Note, however, that this behavior is exhibited by a system
with a high number of degrees of freedom. The simplicity of the behavior, as will
be shown, arises from the fact that possible dependencies among the variables are
removed or attenuated.

Let us go back to Fig. 8.4 from which we can infer how the total forward reaction
rate curve changes as the stimulus is varied. The first thing to notice is that the

important changes concern only the lower left portion of the plot. Which are the relevant features for the behavior of the switch? First of all, it is necessary, as we have seen, that there exists a threshold level for the stimulus above which the curves do not intersect in this region of the graph. The value of this threshold is biologically important since it determines the sensitivity of the switch. A very low threshold would cause the system to shift already at small levels of hormone, which might lead to premature differentiation. Second, it is crucial that the highest value of phosphorylated kinase in the *off*-state, is not so high as to already activate the maturation process. As long as these conditions are met, however, the details of the relationship between stimulus and *off*-state concentration do not matter. The organization of the mechanism, notably the positive feedback, ensures that there is a range within which the level of phosphorylated kinase depends only weakly on the stimulus (corresponding to the branch of *off*-states in Fig. 8.5) and ensures that the system remains in the *off*-state even if the progesterone level is varied significantly.[10] Only around the threshold level there is sensitive dependence on the stimulus. But as soon as the system has switched, the level of phosphorylated kinase becomes virtually independent from the level of hormone (the branch of *on*-states in Fig. 8.5). The reason is that the stimulus does not significantly affect the reaction rates at high levels of phosphorylated kinase and cannot destabilize the equilibrium.

Thus in order to explain the phenomenon, we have to invoke both the non-dependence of the *off*- and *on*-states on the stimulus as well as the very sensitive dependence around the threshold. What the example shows is that the explanation of complex dynamical behaviors requires information both about relations of dependence and of non-dependence. In order to understand features of persistence, such as robustness or memory, we have to illuminate how some variables in certain ranges do not or only weakly depend on others. Moreover, this kind of knowledge allows us to explain how systems built of many parts may show behaviors that can be described in comparatively simple terms. The simplicity of the behavior at the level of the whole mechanism is due to the fact that many changes at the level of the components are not constitutively relevant, in the sense of not being change-relating. We cannot fully comprehend how this behavior is brought about if we restrict ourselves to information about manipulationist relevance. This suggests that the manipulationist conception of mechanistic explanation is insufficient to account for many aspects of phenomena that involve dynamical patterns.

As already noted, the description of the switching behavior itself represents a change-relating generalization. Therefore, it can be used as a basis for further explanation. For example, one may explain why one particular oocyte did not initiate maturation by referring to this generalization plus the fact that the given hormonal stimulus was not sufficient. Relationships of non-dependence partly account for the invariance of this generalization. They illuminate, for instance, why different

---

[10]Note however, that the role of feedback is not merely to confer robustness to the system. Instead, it is an integral part of the switching mechanism since in its absence the system would not show bistability in the first place.

oocytes initiate maturation even when given slightly different doses of stimulus. It might be argued, therefore, that information about change-relating generalizations is sufficient to explain the phenomenon of interest, and that information about non-dependence comes into play only if we want to generalize for further purposes of explanation. But as I hope to have shown, both kinds of relationships are in fact already used in the explanation of the basic features of the switch. The mechanistic explanation that shows *how* the system brings about the behavior contains answers to both types of Batterman's *why*-questions. Systems biologists want to understand the factors on which changes in observed dynamical patterns depend, but they also want to explain why these patterns obtain in the first place.

The discussion of this particular mechanism has touched upon the concept of robustness on several occasions. In the following section I will return to the idea of robustness as a fundamental property of living systems and show that relationships of non-dependence play an important explanatory role here as well.

6   Robustness and the Architecture of Living Systems

Investigating robustness is often invoked as one of the key motivations for research in systems biology. Hiroaki Kitano, for instance, holds that "[it] is one of the fundamental and ubiquitously observed systems-level phenomena that cannot be understood by looking at the individual components" (Kitano, 2004, 826). How does this idea of robustness as a fundamental property of living systems connect to the discussion about relations of non-dependence in the preceding sections? We have seen in the example of the switch that the particular dynamical organization of a mechanism can lead to weak relationships between variables or components, which in turn confers reliability and robustness to the system as a whole. However, in the just cited article Kitano notes:

> Robustness is often misunderstood to mean staying unchanged regardless of stimuli or mutations, so that the structure and components of the system, and therefore the mode of operation, is unaffected. In fact, robustness is the maintenance of specific functionalities of the system against perturbations, and it often requires the system to change its mode of operation in a flexible way. (Kitano, 2004, 827)

This seems to imply that it would be overly simplistic to explain robustness by referring to the causal or the constitutive irrelevance of particular factors under certain conditions. Instead, the quote suggests that the reliable performance of a system requires sophisticated underlying structures. A more refined view of the mechanistic structure of living systems would, therefore, consist in holding that robustness can be explained by invoking particular 'robustness mechanisms.' Indeed, Kitano mentions four different features that could play the role of such mechanisms: system control, redundancy, modularity, and decoupling (Kitano, 2004, 827). Even though it may not have been his intention, the fact that Kitano is speaking of "mechanisms that insure the robustness of a system" (Kitano, 2004,

827) suggests a particular picture: A living system may at its core be clock-like, but its reliable functioning in environments that are characterized by uncertainty and noise is guaranteed by an intricate machinery of additional features that has evolved around this core. If this picture were accurate, the general strategy of understanding mechanisms in terms of change-relating generalizations alone would be justified after all. Robustness would not be a fundamental property of the mechanisms themselves, but rather a separate phenomenon that could be explained by referring to independent mechanistic features. Yet, we have seen in the previous section that there are at least some cases where it is not possible to separate the explanation of a behavior from an explanation of its robustness. Moreover, a closer look at Kitano's alleged robustness mechanisms suggests that this conceptual separation might in general not be obvious. What he means, for example, by 'systems control' is the use of certain control strategies in the building of biological circuits, something that also fits the example discussed in the previous section. So his robustness 'mechanisms' are probably better understood as 'design features' of biological mechanisms. Just as in the case of oocyte maturation, robustness is often *in-built* and not in any obvious way *added* to the mechanism. In the remainder of this section, I will illustrate how mathematical modeling has recently been applied to elucidate the features underlying biological robustness of this kind.

In the attempt to simulate the interactions among the genes responsible for segmentation in *Drosophila*, von Dassow et al. (2000) developed a dynamical model and systematically investigated its behavior under changes in parameters. The segment polarity network described by this model generates a periodic expression pattern across cells early in development. Initially, von Dassow et al. had hoped that the requirement to reproduce the behavior of the target system would impose sufficient constraints on the model to obtain reasonable estimates for the nearly 50 parameters of the model. Consequently, they expected that only a relatively small subset among all the states in the high-dimensional parameter space would lead to biologically meaningful versions of their model. Strikingly, however, they found that solutions in this space were not rare at all:

> Among 240,000 randomly-chosen parameter sets we found 1,192 solutions ($\sim$ 1 in 200). This is very frequent; as this search involved 48 parameters, on average a random choice of parameter value has roughly a 90% chance of being compatible with the desired behaviour. (von Dassow et al., 2000, 189)

Apart from their abundance, solutions were apparently not isolated in parameter space. For many of them the model was found to be tolerant to variation of individual parameters over several orders of magnitude. Thus the scientists concluded that the model's ability to reproduce the target behavior was "intrinsic to its topology rather than to a specific quantitative tuning" (von Dassow et al., 2000, 189).

The case of the segment polarity network, therefore, supports the idea that robust behavior is not always achieved by adding structural components to an otherwise fragile mechanism. Robustness, therefore, cannot necessarily be analyzed as a separate feature, but instead appears to be entangled with a system's overall functionality. We have also seen this clearly in the example of the bistable switch in Sect. 5, where the positive feedback is necessary both for the robustness and for the basic behavior of the mechanism. In general it seems that the molecular organization of biological mechanisms can lead to a weak dependence of system behavior on the behavior of the components. The following quote nicely illustrates how the scientists' initial assumptions about the robustness of the segment polarity network were overturned by their detailed investigation of the mathematical model:

> We originally expected the core topology to be frail and easily perturbed, and expected to achieve robustness only by adding additional complexity; we expected the reconstitution approach to tell us which architectural features confer robustness. Confounding that expectation, the simplest model that works at all emerged complete with unexpected robustness to variation in parameters and initial conditions. (von Dassow et al., 2000, 191)

Robustness of this kind does not seem to be restricted to the generation of developmental patterns in *Drosophila*. Gutenkunst et al. (2007) investigated 17 different systems biology models and systematically examined the sensitivity of their behavior to parameter changes. Their set of models covers a wide range of different biological mechanisms and, aside from von Dassow et al.'s network, includes models of circadian rhythm, metabolism, and signaling. In all of them they found what they call 'sloppy parameter spectra:' the behavior of the model is sensitive to variation along a few 'stiff' directions in parameter space, but insensitive along a large number of 'sloppy' directions. It is important to emphasize that these directions (or 'eigenvectors,' mathematically speaking) do not correspond to individual model parameters but rather to combinations thereof:

> Naively, one might expect the stiff eigenvectors to embody the most important parameters and the sloppy directions to embody parameter correlations that might suggest removable degrees of freedom, simplifying the model. Empirically, we have found that the eigenvectors often tend to involve significant components of many different parameters. (Gutenkunst et al., 2007, 1873)

This essentially means that the investigated systems do not react in a clock-like fashion to most perturbations on individual components. In order to bring about significant changes in systemic behavior, it is necessary to intervene on multiple components simultaneously.

The observed 'sloppiness' of biological systems provides resilience towards many disturbances at the molecular level and therefore appears to be beneficial for the survival of the organism, but it is not necessarily an evolved feature of living systems. Daniels et al. (2008), for example, conjecture that sloppiness might be a universal property of a particular class of dynamical models which naturally accounts for many types of robust behavior. With regard to von Dassow et al.'s model of the segment polarity network they note:

> The model is robust in these [sloppy] directions not because of evolution and fitness, but because of the mathematical behavior of chemical reaction networks, which are naturally weakly dependent on all but a few combinations of reaction parameters. (Daniels et al., 2008, 393)

In general, however, it is clear that the investigation of biological robustness must pay attention both to evolved robustness mechanisms and to generic features of biological organization.

The results of Gutenkunst et al. suggest that the relation between components and system behavior is often not as straightforward as analogies to machine-like mechanisms would make us believe. Information about non-dependence is highly relevant for the explanation of the behavior of many biological mechanisms, and their robustness is not just an interesting feature that requires separate explanation. Thinking about robustness may have a profound influence on the way in which we conceive of mechanistic explanations in the life sciences.

7   Conclusion

In this chapter I have tried to assess particular accounts of mechanistic explanation, according to which explanatory relevance relies on change-relating generalizations, by looking at dynamical modeling in systems biology. The motivating question was whether this manipulationist framework can adequately account for what we know about the robustness of living systems, a property that is extensively studied by systems biologists. I have argued that certain aspects of the explanation of dynamical patterns, first and foremost simple dynamical equilibrium, are not captured by approaches that solely focus on change-relating relationships. Instead, the explanation of such features relies on information about relationships of non-dependence, that is, on information about factors or relationships that are irrelevant from a manipulationist standpoint. By presenting the example of a bistable switch in the maturation of *Xenopus* oocytes, I have shown that this kind of reasoning is actually applied in mechanistic explanations as they are found in the scientific literature. I have tried to illuminate how in the case of this mechanism the discussion of robustness cannot be separated from its functional behavior. I have then turned to a more general discussion of the concept of robustness. There is evidence that at least some of the robustness we find in biological systems cannot be accounted for by invoking separate 'robustness mechanisms.' Instead, it can often be explained by the fact that interactions among components of a system are reducible to only a few significantly sensitive dependencies. Therefore, if we want to mechanistically explain how these systems work, we have to understand how biological organization results in weak dependencies and leads to coherent behavior and robustness at the systemic level.

Robustness is often presented as one of the paradigmatic examples of an emergent property; at least systems biologists frequently describe it as such. The ideas discussed in this chapter may shed some light on the reasons for this usage. Philosophical accounts of emergence have mostly focused on system properties that somehow 'exceed' the capacities of the components, or are unpredictable based on information about individual parts (e.g. Bedau, 1997; Boogerd et al., 2005; Kim, 1999). However, as the French philosopher Edgar Morin has noticed, a system is not only more than the sum of its parts, it is also *less* than the sum of its parts in certain respects (Morin, 2008). The behavior of the components is constrained in various ways by the structure and organization of the system, which keeps them from exhibiting many of the properties that they might show in isolation or in different contexts. Robustness is thus a striking example of a restriction of component potential. What scientists mean by 'emergence' might often simply be the idea that the system is *different* from the sum of its parts. Restricting ourselves to change-relating relationships may prevent us from understanding how such kinds of emergent behavior are brought about.

I briefly discussed in Sect. 3 that the manipulationist picture might be defended by maintaining that a relationship of non-dependence simply expresses the fact that some element is explanatorily irrelevant. For instance, even though physics tells us that the current positions of remote stars exert a non-vanishing gravitational force on objects on the earth, we do not mention them in biological explanations because they are not considered to make significant differences to biological phenomena. However, many phenomena sensitively depend on factors that we would not want to include in their explanations either. The croaking of a frog, for instance, depends on whether or not the frog has just been run over by a car, yet we do not cite facts about cars when we explain how frogs croak. Acknowledging this problem of extravagant causes, Carl Craver resorts to a pragmatic solution by restricting himself to changes that *typically* occur in a system. But extravagant causes are a threat only if we insist on equating the notions of manipulationist and explanatory relevance. If it is true, as I have tried to argue, that information about causal or constitutive irrelevance can be explanatory, then the line between what is relevant for explanation and what is not must be drawn elsewhere anyway. I propose that there are interesting 'non-dependencies' just as there are uninteresting dependencies. The question of what makes a relationship interesting, however, may not be answered so easily.

Acknowledgements Apart from the editors, I would like to thank Pierre-Luc Germain, Mark Bedau, Jan Baedke, Sara Green, Veli-Pekka Parkkinen, and Robert Meunier for their helpful suggestions.

References

Baker, A. (2015). Mathematical explanation in biology. In P.-A. Braillard & C. Malaterre (Eds.), *Explanation in biology. An enquiry into the diversity of explanatory patterns in the life sciences* (pp. 229–247). Dordrecht: Springer.

Batterman, R. W. (2002). *The devil in the details: asymptotic reasoning in explanation, reduction, and emergence.* Oxford: Oxford University Press.

Bedau, M. A. (1997). Weak emergence. *Noûs, 31*(Suppl s11), 375–399.

Bhalla, U. S., & Iyengar, R. (1999). Emergent properties of networks of biological signaling pathways. *Science, 283*(January), 381–387.

Boogerd, F. C., Bruggeman, F. J., Richardson, R. C., Stephan, A., & Westerhoff, H. V. (2005). Emergence and its place in nature: a case study of biochemical networks. *Synthese, 145*(1), 131–164.

Braillard, P.-A. (2010). Systems biology and the mechanistic framework. *History and Philosophy of the Life Sciences, 32*, 43–62.

Breidenmoser, T., & Wolkenhauer, O. (2015). Explanation and organizing principles in systems biology. In P.-A. Braillard & C. Malaterre (Eds.), *Explanation in biology. An enquiry into the diversity of explanatory patterns in the life sciences* (pp. 249–264). Dordrecht: Springer.

Brigandt, I. (2015). Evolutionary developmental biology and the limits of philosophical accounts of mechanistic explanation. In P.-A. Braillard & C. Malaterre (Eds.), *Explanation in biology. An enquiry into the diversity of explanatory patterns in the life sciences* (pp. 135–173). Dordrecht: Springer.

Cannon, W. B. (1932). *The wisdom of the body*. New York: W. W. Norton.

Craver, C. F. (2007). *Explaining the brain*. New York: Oxford University Press.

Daniels, B. C., Chen, Y. J., Sethna, J. P., Gutenkunst, R. N., & Myers, C. R. (2008). Sloppiness, robustness, and evolvability in systems biology. *Current Opinion in Biotechnology, 19*(4), 389–95.

von Dassow, G., Meir, E., Munro, E. M., & Odell, G. M. (2000). The segment polarity network is a robust developmental module. *Nature, 406*(6792), 188–92.

Ferrell, J. E., & Machleder, E. M. (1998). The biochemical basis of an all-or-none cell fate switch in Xenopus oocytes. *Science, 280*, 895–898.

Ferrell, J. E., & Xiong, W. (2001). Bistability in cell signaling: how to make continuous processes discontinuous, and reversible processes irreversible. *Chaos, 11*(1), 227–236.

Glennan, S. (2002). Rethinking mechanistic explanation. *Philosophy of Science, 69*(3), S342–S353.

Gunawardena, J. (2010). Models in systems biology: The parameter problem and the meanings of robustness, chap 2. In H. M. Lodhi & S. H. Muggleton (Eds.), *Elements of computational systems biology* (pp. 21–47). Hoboken: Wiley.

Gutenkunst, R. N., Waterfall, J. J., Casey, F. P., Brown, K. S., Myers, C. R. & Sethna, J. P. (2007). Universally sloppy parameter sensitivities in systems biology models. *PLoS Computational Biology, 3*(10), 1871–1878.

Isalan, M., Lemerle, C., Michalodimitrakis, K., Horn, C., Beltrao, P., Raineri, E., Garriga-Canut, M., & Serrano, L. (2008). Evolvability and hierarchy in rewired bacterial gene networks. *Nature, 452*(April), 840–846. doi:10.1038/nature06847

Issad, T., & Malaterre, C. (2015). Are dynamic mechanistic explanations still mechanistic? In P.-A. Braillard & C. Malaterre (Eds.), *Explanation in biology. An enquiry into the diversity of explanatory patterns in the life sciences* (pp. 265–292). Dordrecht: Springer.

Kim, J. (1999). Making sense of emergence. *Philosophical Studies, 95*, 3–36.

Kitano, H. (2004). Biological robustness. *Nature Reviews Genetics, 5*(11), 826–37.

Kuorikoski, J. (2007). Explaining with equilibria. In J. Persson & P. Ylikoski (Eds.), *Rethinking explanation* (pp. 149–162). Dordrecht: Springer.

Morange, M. (2015). Is there an explanation for … the diversity of explanations in biological studies? In P.-A. Braillard & C. Malaterre (Eds.), *Explanation in biology. An enquiry into the diversity of explanatory patterns in the life sciences* (pp. 31–46). Dordrecht: Springer.

Morin, E. (2008). *On complexity*. New York: Hampton Press.

Novak, B., Tyson, J. J., Gyorffy, B., Csikasz-Nagy, A. (2007). Irreversible cell-cycle transitions are due to systems-level feedback. *Nature Cell Biology, 9*(7), 724–728.

Poon, W. C. K. (2011). Interdisciplinary reflections: The case of physics and biology. *Studies in History and Philosophy of Biological and Biomedical Sciences, 42*(2), 115–118.

Railton, P. (1981). Probability, explanation, and information. *Synthese, 48*(2), 233–256.

Salmon, W. C. (1984). *Scientific explanation and the causal structure of the world*. Princeton: Princeton University Press

Savageau, M. A. (2001). Design principles for elementary gene circuits: Elements, methods, and examples. *Chaos, 11*(1), 142–159.

Simon, H. A. (1962). The architecture of complexity. *Proceedings of the American Philosophical Society, 106*(6), 467–482.

Sober, E. (1983). Equilibrium explanation. *Philosophical Studies, 43*, 201–210.

Wimsatt, W. C. (1997). Aggregativity: Reductive heuristics for finding emergence. In *Proceedings of the 1996 biennial meetings of the philosophy of science association*, Cleveland (Vol. 64, pp. S372–S384).

Wimsatt, W. C. (2007). *Re-engineering philosophy for limited beings: Piecewise approximations to reality*. Cambridge: Harvard University Press.

Woodward, J. (2002). What is a mechanism? A counterfactual account. *Philosophy of Science, 69*, S366–S378.

Woodward, J. (2003). *Making things happen: A theory of causal explanation*. Oxford: Oxford University Press.

Woodward, J. (2010). Causation in biology: Stability, specificity, and the choice of levels of explanation. *Biology & Philosophy, 25*(3), 287–318.

Chapter 9
Generalizing Mechanistic Explanations Using Graph-Theoretic Representations

William Bechtel

Abstract Mechanistic explanations appeal to the parts, operations, and organizations of mechanisms to explain the phenomena for which they are responsible. Scientists have developed accounts of myriads of mechanisms thought to be operative in biology, each involving distinctive parts and operations organized in idiosyncratic ways. The focus on specific mechanisms (e.g., those found in a particular cell type in a given model organism) appears opposed to the idea that explanations ought to be generalizable to new instances. Some generalizability can arise from the fact that many biological mechanisms inherit their parts and operations from mechanisms found in ancestral species and one can often identify commonalities in these parts and the operations they perform. But organization seems to be idiosyncratic to specific mechanisms, thwarting attempts to develop generalizations about how mechanisms organized in a specific way will behave. For example, as a result of different genes being expressed, the organizational pattern of interactions between proteins varies in different tissues or in the same tissue in different strains of a species. This poses an even greater problem when it is recognized that many biological mechanisms exhibit non-sequential organization of non-linear operations, making it difficult to use mental simulation to determine the behavior of the mechanism. Instead researchers resort to computational models, resulting in dynamic mechanistic explanations that integrate mathematical modeling with empirically ascertained details of parts and operations. These models, though, appear to be even more idiosyncratic, revealing only the behavior of the specific organization employed in a specific mechanism.

In recent years, however powerful tools have been developed for abstracting from the details of individual networks, providing a basis for informative generalizations about how networks employing the same abstract design will behave. These involve developing graph-theoretic representations of mechanisms and analyzing the properties of classes of graphs. Watts and Strogatz, Barabási, and Sporns have shown that large networks (e.g., neural circuits or gene networks) often exhibit

W. Bechtel (✉)
Department of Philosophy, Center for Circadian Biology, and Interdisciplinary Program in Cognitive Science, University of California, 9500 Gilman Drive, La Jolla, San Diego, CA, 92093-0119 USA
e-mail: bill@mechanism.ucsd.edu

© Springer Science+Business Media Dordrecht 2015
P.-A. Braillard, C. Malaterre, *Explanation in Biology*, History, Philosophy and Theory of the Life Sciences 11, DOI 10.1007/978-94-017-9822-8_9

a scale-free, small-world organization capable of efficient, flexible coordination of operations and appeal to these properties to explain behaviors of specific mechanisms. Alon and Tyson, focusing on sub-networks with just two to four nodes, have identified different *motifs* (distinctive micro-architectures such as feedforward loops and double negative feedback loops) that are specialized for particular types of processing. These tools offer abstract organizational principles to which researchers appeal in their efforts to explain the behaviors generated by the mechanisms in which they are implemented. In this paper I show how these projects provide a basis for developing generalizable accounts of complex mechanisms and their dynamical behavior.

Keywords Dynamic mechanistic explanations • Generalizing explanations • Graph-theory formalizations • Motifs • Scale-free small worlds

1   Introduction: Formalizing Mechanisms in Graphs

In a broad range of biological disciplines the explanations researchers have pursued over the past two centuries have taken the form of characterizing the mechanisms responsible for various phenomena and demonstrating how they are able to generate the phenomenon. For example, metabolic research has described mechanisms for phenomena such as glycolysis or fatty acid metabolism, molecular biology has proposed mechanisms for cell division and protein synthesis, and neuroscience has investigated mechanisms for generating action potentials in neurons and for spatial navigation by animals. Although during much of the past two centuries philosophers had little to say about such explanations, over the last two decades several philosophers have advanced accounts of what mechanistic explanations are, how they are discovered and revised, and how they are evaluated (Bechtel and Richardson 1993, 2010; Bechtel 2011; Craver 2007; Glennan 1996, 2002; Machamer et al. 2000).[1]

A fundamental step in developing a mechanistic account is to characterize the phenomenon to be explained and to identify a mechanism taken to be responsible for its occurrence in the situations in which it occurs. This is often a non-trivial endeavor requiring experimental manipulation and quantitative analysis of the results. Research then seeks to decompose this mechanism into its components and operations. Biologists have developed powerful tools and inference strategies for decomposing mechanisms into their component parts and determining the operations for which these parts are responsible. But explanation also requires recomposing the mechanism to show that, in a given situation, it is able to produce the phenomenon (Bechtel 2011; Bechtel and Abrahamsen 2009). Recomposing the

---

[1] See Théry (2015, this volume) for a critical assessment of mechanistic explanations. See also Baetu (2015, this volume) and Brigandt (2013) about how mathematical models and mechanistic explanations may mutually support each other.

mechanism requires determining how the mechanism is organized and demonstrating that when so organized the operations identified are capable of jointly producing the phenomenon.

Researchers often approach the task of recomposition by assuming that the operations are organized sequentially. In such cases researchers can simulate the behavior of the mechanism mentally by representing the effects of each operation in the mechanism successively. In many cases, however, researchers come to recognize that sequential organization is insufficient—multiple operations may be performed simultaneously and individual components may have effects on other components responsible for other operations. The result is that individual components function differently depending on the context within the mechanism or beyond it. As various operations modulate the execution of other operations, the behavior of the whole mechanism may begin to produce a complex dynamic pattern such as oscillatory behavior or synchronization of independent oscillators. Mental simulation, even when assisted by a diagrammatic representation of the mechanism, commonly fails to provide accurate accounts of how such mechanisms will behave. In these circumstances it often proves highly useful to represent the mechanism in a computational model and to run computer simulations to determine how the mechanism will behave. I refer to explanations that rely on computational models to understand the dynamic behavior of a mechanism as *dynamic mechanistic explanations* (Bechtel and Abrahamsen 2010, 2011).[2] In the case of complicated mechanisms, these models may involve dozens or hundreds of differential equations. If appropriate parameter values are employed, these equations may together accurately simulate the behavior of the target mechanism. But often this accuracy comes at the expense of reducing the intelligibility of the explanation.[3]

A further problem with mechanistic or dynamic mechanistic explanations of mechanisms is that they are often developed for specific instances of the mechanism that have been investigated. Typically in biology mechanistic research is pursued on specific model systems or model organisms, which may not be the target of interest. A model organism may be chosen because of what is already known about the model organism, tools that have been developed for intervening on it, and the hope that the mechanisms in it will be simpler (although not simple) and so more tractable than

---

[2]While recognizing that one might integrate them, Woodward (2013) emphasizes the differences between more prototypical mechanistic approaches and those that adopt a more holistic perspective in, for example, explaining the robustness of a biological system or the dynamical behavior of a complex network. Woodward is right to emphasize the difference in research strategies and assumptions of the two approaches. More traditional mechanistic approaches assume modularity of the mechanism and dependence on the fine-tuned details of the system. In developing dynamical models of integrated systems researchers typically abstract away from these details. My reason for emphasizing their integration in dynamic mechanistic explanations is that many contemporary biological explanations draw on both details about the components of the mechanism and the dynamics of the integrated system, attempting to integrate the approaches that on their own make different assumptions and pursue different research strategies.

[3]See Issad and Malaterre (2015, this volume) for a discussion of the explanatory force of dynamic mechanistic explanations.

those in the target. This raises the question of whether the mechanistic or dynamic mechanistic account developed on the basis of research on the model system will generalize to the target system. Typically there will be differences between the model system and the target system, and researchers must assess whether these differences matter. Researchers working with computational models involving non-linearities recognize that often small changes lead to unpredicted major changes in the behavior of the model. This points to the challenge in generalizing from the model to the target.

To support generalizing results from the model to the target, researchers often appeal to the conservation of the mechanism through phylogeny—the mechanisms in contemporary are treated as variants that evolved through descent with modification from a common ancestor. When the model organism has evolved fewer additional components (as a result, for example, of gene duplications), it is expected to provide a less complicated version of the mechanism of interest in the target organism. The fact that the model is less complicated, however, points to differences between the model and the target, making it important to independently test the applicability of the explanation in the target system. The important issues in generalizing from model organisms are increasingly the focus of philosophical inquiry (Ankeny 2001; Ankeny and Leonelli 2011; Leonelli et al. 2014; Meunier 2012).

Another mode of generalization, which will be my focus here, is to abstract from the specific details of the mechanism to identify the pattern of organization (Levy and Bechtel 2013; Overton 2011). This has proven particularly valuable when the non-sequential organization of a mechanism results in dynamical behavior and researchers can explain that dynamical pattern as an expected consequence of a particular mode of organization. As I will demonstrate, computational modeling of the abstract mode of organization can contribute to the explanation of why mechanisms implementing the form of organization exhibit the behavior they do. But the key step is abstraction—leaving out details about the component parts and operations so as to focus only on the organization and determine its implications. Generalization then applies the information acquired through analyzing the abstract pattern of organization to different mechanisms that implement that pattern.

A particularly powerful means of identifying the abstract organization of a mechanism is to formalize the organization of the mechanism in graph-theoretical terms.[4] A graph consists of a collection of nodes connected by edges. The nodes represent entities, which in the case of a mechanism will include the component parts but may also include any other entities with which the mechanism or its parts interact. By treating different types of entities as nodes, a graph abstracts from the

---

[4]Jones (2014) discusses graph representations of the topology of a system that are not intended to be mechanistic but to identify such things as vulnerabilities of a system to disruption. While the diagrams he discusses and those discussed below often abstract from the details of the mechanism and thus provide a different type of understanding than do accounts of the parts and operations of a mechanism, they nonetheless play an important role in understanding how, as a result of its organization, a mechanism behaves either normally or when disrupted.

specific nature of the parts. Edges represent relations between nodes, which in many cases will be causal, although they may reflect correlations or even just the presence of physical connections. When the edges are causal, they are directional and in the case of a mechanism reflect the fact that an operation performed by one part or entity has effects on another. As in the case of nodes, edges abstract from the specific character of the operation.

As a basis for the discussion in subsequent sections, I present two examples in which, in the pursuit of research on a mechanism, researchers formalized the organization graph-theoretically. My goal at this stage is not to discuss the ways in which such graphs are analyzed, but simply to show contexts in which they are introduced. Working from slices examined with the confocal electron microscope, White et al. (1986) mapped the entire nervous system of the nematode worm *Caenorhabditis elegans*. The researchers identified 302 neurons and over 7,000 synapses. In the course of presenting their analysis, they employed many figures, including microscopic images, showing individual neurons and their projections in context. The last set of figures, however, abstracts from this detail and offers six circuit diagrams of the nervous system, of which the one relating amphids involved in chemoreception is shown on the top in Fig. 9.1. This circuit diagram does maintain some detail of the actual nervous system (chemical synapses are distinguished by arrows whereas gap junctions are represented by bar-ended lines, sensory neurons are shown as triangles and interneurons by hexagons, and the prominence of synapses is indicated by the number of cross-hatches on arrows using a scale from 1 to 4), much of the detail has been removed so as to focus on the pattern of connections.

Graph representations have also been developed to represent interactions between genes and proteins in a number of organisms (for pioneering work in this project, see Kauffman 1969, 1974). In developing the graph on the bottom in Fig. 9.1, Jeong et al. (2001) drew upon four datasets of possible interactions between proteins in Baker's yeast *Saccharomyces cerevisiae*. The datasets mostly relied on two-hybrid analyses to determine which proteins could interact. The nodes in this graph represent 1,870 individual proteins and the edges show 2,240 interactions (the coloring of the nodes will be discussed below).

In the sections that follow I will focus first on graphs such as these representing whole mechanisms or even the larger systems in which mechanisms of interest might be situated. In a later section I will turn to sub-graphs that show relations between components within the larger graphs. In both cases, the focus will be on how graph-theoretic tools for analyzing network organization reveal how mechanisms that instantiate the organization exhibited in the graph will behave. Although such analyses often employ mathematical models and simulations using those models, I focus on the qualitative conclusions that are reached. In the final section I show how the tools for sub-graphs can be employed after performing suitable manipulations of whole-graphs of a mechanism.

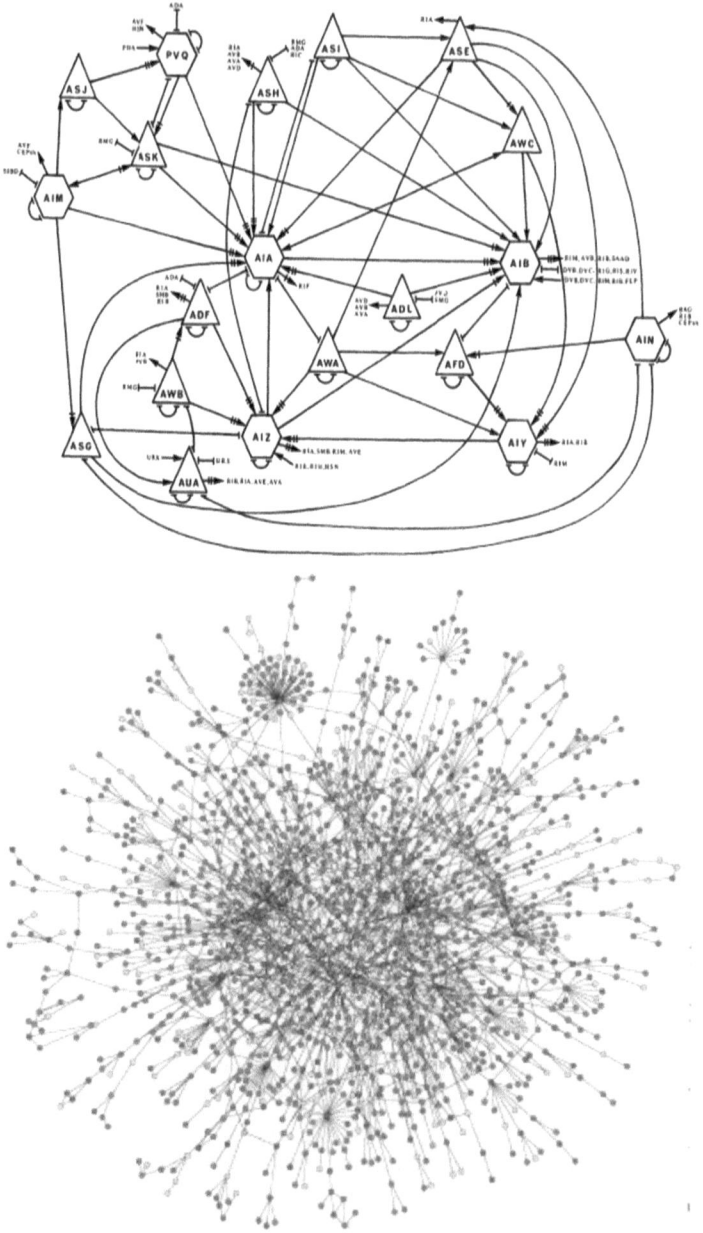

Fig. 9.1 On the *top* White et al. (1986) constructed a graph representation of that part of the nervous system of *C. elegans* involved in chemoreception. chemoreception (Reprinted from White et al., The Structure of the Nervous System of the Nematode *Caenorhabditis elegans*, *Proceedings of the Royal Society B, Biological Sciences*, 314, Figure 21, by permission of the Royal Society). On the *bottom* Jeong et al. (2001) proposed a graph representation of the largest cluster in the protein interaction network of *Saccharomyces cerevisiae*, containing approximately 78 % of all proteins (Reprinted by permission from Macmillan Publishers Ltd: Nature, copyright (2001))

2 Whole-Graph Organization: Scale-Free Small Worlds

Since graphs such as those shown in Fig. 9.1 are based on evidence about a particular set of entities and their connections or interactions, each is different. At first it might not be apparent that one could develop generalizable insights into how different networks will behave. Instead, one might expect that when mechanisms have sufficient parts, the graph showing the interactions will be different for each mechanism. Researchers might take advantage of the graph to develop a computational model of how a mechanism instantiating that graph might behave (Jones and Wolkenhauer 2012, demonstrate how scientists utilize graph representations as guides in developing computational models), but not be able to use them to generalize results from one mechanism to another. But pioneering graph theorists in the mid-twentieth century developed ways of analyzing the behavior of a few idealized forms of network organization that have proven very useful. A common concern for much of this work was to understand how individual entities (neurons or fireflies) that exhibit oscillatory behavior are able to synchronize their oscillations depending on how the nodes are connected. Winfree (1967) developed one of the first analyses by assuming that each of the nodes in the network is an oscillator with its own intrinsic frequency and that each is weakly connected to all the others. In his mean field model, each oscillatory node is treated as being affected by the mean activity of the other nodes. He investigated what happened as the strength of the connections between nodes is increased and demonstrated two global transitions. At first some of the oscillators begin to exhibit synchronization in their oscillation while later all of the oscillators become phase and amplitude locked. Kuramoto (1984) further advanced the analysis of mean field models.

Although the mean field approach is often used as a first approximation, a connection between each pair of nodes is not common in actual biological systems. Connections are more likely to develop between some pairs of nodes than others. Accordingly, Erdös and Rényi (1960) investigated networks in which connections are established randomly and showed that if the number of connections is much less than the number of nodes, only small, disconnected clusters of connected nodes appear in which oscillators would synchronize. However when the number of connections increases to approximately half the number of nodes, a phase transition occurs and a single giant cluster emerges in which nodes all synchronize. Even though most biological networks do not appear to exhibit random connectivity, the idea that a giant cluster will emerge at such a phase transition has been applied in analyzing networks such as protein interaction networks in yeast; the network shown on the right in Fig. 9.1 shows such a giant cluster and leaves out nodes not included in the giant cluster.[5]

---

[5]A special case on which extensive research has been conducted are Boolean networks (networks in which the behavior of a node is a Boolean function of the state of other nodes). See Derrida and Pomeau (1986), Flyvbjerg (1988), Kauffman (1993), Luque and Solé (1997), Rohlf and Bornholdt (2002) for analyses of Boolean networks.

Yet other graph theorists focused on networks organized as regular lattices in which nodes are connected only to their neighbors around a ring. Ermentrout and Kopell (1984) demonstrated that such networks could generate traveling waves in which oscillatory nodes around the ring would reach their maximum in succession. This approach provided a powerful tool for explaining the traveling waves exhibited in neuromuscular rhythms in the mammalian intestine as well as in the motor behavior of species such as the lamprey (Kopell and Ermentrout 1988).

Not only do random networks and regular lattices exhibit different types of behavior (synchronized activity versus traveling waves), they also exhibit patterns of connectivity that have enabled subsequent researchers to place them along a continuum. In the giant cluster of a random network, the number of edges that must be traversed on average to get from one node to another (referred to as the *characteristic path length*) is quite small whereas in regular lattices it is much greater. On the other hand, around each node in a regular lattice there is a cluster of highly interconnected nodes, whereas in a random network these clusters are much smaller. The *clustering coefficient* measures the size of these clusters. The short characteristic path length of random networks entails that activity across such a network is integrated over a relatively short period of time but the low clustering coefficient entails that there are no concentrated sets of nodes able to work together. With the high clustering in regular lattices, nodes can collaborate on a given process, but it takes much longer to relate the activity in one cluster to that occurring in others.

Watts and Strogatz (1998) explored the region between these endpoints. Starting from a regular lattice such as is shown on the left of Fig. 9.2 they investigated what happened as they replaced local connections with long-range connections and discovered that the characteristic path length dropped rapidly as the percentage of connections were replaced but that the clustering coefficient remained high until a much higher percentage was replaced. Watts and Strogatz referred to networks in this intermediate range as *small-world* networks. They argued that a host of real world networks exhibited small-world properties including not only the neural

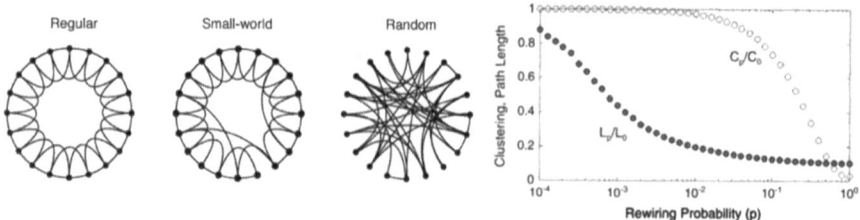

Fig. 9.2 The graphs on the *left* illustrate how small world networks are intermediate between regular lattices and random networks. One can view small-world networks as resulting from rewiring a few edges in a regular lattice. As illustrated on the *right*, as the probability of rewiring increases, the characteristic path length drops rapidly but the clustering coefficient remains high. This intermediate region is where small worlds are found. (Reprinted by permission from Macmillan Publishers Ltd: Nature, copyright (1998))

network in *C. elegans* shown in Fig. 9.1, but also the electric power grid of the Western U.S. and networks of co-appearance of actors in films. They also addressed the functional significance of small-world organization, showing for example that infectious diseases would spread much more rapidly in a small-world network than in a regular lattice. In addition, they highlighted the computational power of such networks. The key idea is that nodes that are highly clustered can be organized into appropriate sub-networks for executing particular information-processing tasks but that, as a result of the short characteristic path length, the nodes can be modulated by activity occurring elsewhere.

The two measures Watts and Strogatz introduced—characteristic path length and clustering coefficient—provide potent general measures for assessing the behavior of any systems, including mechanisms, organized in such networks. Subsequently Barabási and his colleagues addressed another measure of network architecture—node degree. This is a measure of the number of edges linking a given node. Specifically, Barabási focused on how node degree is distributed in a network. Investigations of random networks by researchers such as Erdös and Rényi assumed that node degree was distributed in a Gaussian fashion over a fairly narrow range. In such cases there is a characteristic scale at which the network can be analyzed—that is, at which component patterns can be identified. But Barabási and Albert (1999) discovered that in many real world networks, including the protein interaction network shown on the bottom in Fig. 9.1, node degree is distributed according to a power-law function so that most nodes have only a limited number of connections to other nodes while a very few have a very large number. These networks are referred to as scale-free since there is no characteristic scale at which the pattern of network activity can be identified. Barabási's initial interest seemed to be in the robustness of scale-free networks to random damage. A random removal of a node or edge would typically have little effect on the overall network. For example, mutations in the proteins shown in yellow and green in the protein interaction network in Fig. 9.1 have little or no detectable effects on the life of the yeast. These are by far most of the proteins. Those shown in red, on the other hand, are required for the cell to live while eliminating those shown in orange significantly affect growth. The red and orange nodes typically have a far higher node degree.

When node degree is distributed according to a power-law, networks tend to have modules organized around hubs. Hubs are highly connected nodes through which a large amount of the activity in a system flows. Two different types of hubs can be distinguished. Those with high clustering coefficients serve to coordinate activity within a module and are often referred to as *provincial* hub. Others, however, serve primary to link between modules, and are referred to as *connector* hubs. In many networks, including brain networks that I discuss below, there are not sufficient numbers of units to demonstrate true scale-freeness. But as long as some units have many more connections than others, hubs will likely result and will be important for understanding the behavior of the network. These networks can thus be regarded as having scale-free characteristics even if they are too small to show true scale-freeness.

I will illustrate the use of these different tools for understanding the functioning of mechanisms by focusing on research on the primate brain. Beginning with the pioneering research of Hubel and Wiesel (1962, 1968), who recorded from individual neurons in the region now known as V1 while presenting visual stimuli and demonstrated that neurons in this region responded to edges in visual scenes, researchers have identified a large number of brain regions important to visual processing, extending from the occipital lobe into the parietal and temporal cortex, and including a few frontal areas. Felleman and van Essen (1991) synthesized information about 32 cortical regions in the *Macaque* monkey involved in visual processing and the anatomical connections between them into the matrix shown on the left in Fig. 9.3. Pluses indicated the existence of a connection between the area listed on the left of each row and the area indicated at the top of each column. Visual inspection reveals clusters of areas that are interconnected. Although Felleman and van Essen did not themselves apply the measures discussed above, Sporns and Zwi (2004) calculated path length and clustering and showed the network described met the conditions for a small world. They applied similar measures and reached a similar conclusion about the matrix they developed for the whole *Macaque* brain based on data from Young (1993), shown on the right in Fig. 9.3. They also focused on sub-regions in these matrices and showed that some regions have different properties. Thus, V4 exhibits both a short path length and a low clustering coefficient, characteristic of random networks. This suggests it is a connector hub. Area A3a in the somatosensory cortex, on the other hand, is only connected to A1 and A2, which are themselves connected, giving it a long path length but high clustering. It is a provincial hub.

Techniques such as diffusion imaging and tractography have made it possible to identify in non-invasive ways interconnections of brain regions, enabling the extension of the approach to humans. Analyses of the networks identified through these techniques indicate again the existence of a small-world architecture with scale-free properties. Sporns et al. (2005) introduced the term *connectome* for these efforts at identifying and analyzing connectivity of the brain. One particularly intriguing finding from these studies which exhibits the usefulness of analyzing brain connectivity in a graph-theoretical manner was the discovery by Hagmann et al. (2008) of a set of hub regions along the anterior-posterior medial axis of the human brain which includes the rostal and caudal anterior cingulate, the paracentral lobule, and the precuneus. These regions are connected to each other and taken together connect to virtually all the other cortical areas in both hemispheres. This suggests the existence of a central network that is important for directing and coordinating information processing throughout the brain. Focusing on these areas promises to reveal how the brain coordinates specialized processing mechanisms located in different areas of the brain, enabling them to work together as needed to perform specific tasks.

The connectome project has focused primarily on anatomical connections, but a major reason for identifying anatomical connections is the assumption that these connections are likely to be important in coordinating functional activities in the brain. Bullmore and Sporns (2009) have proposed a strategy for developing

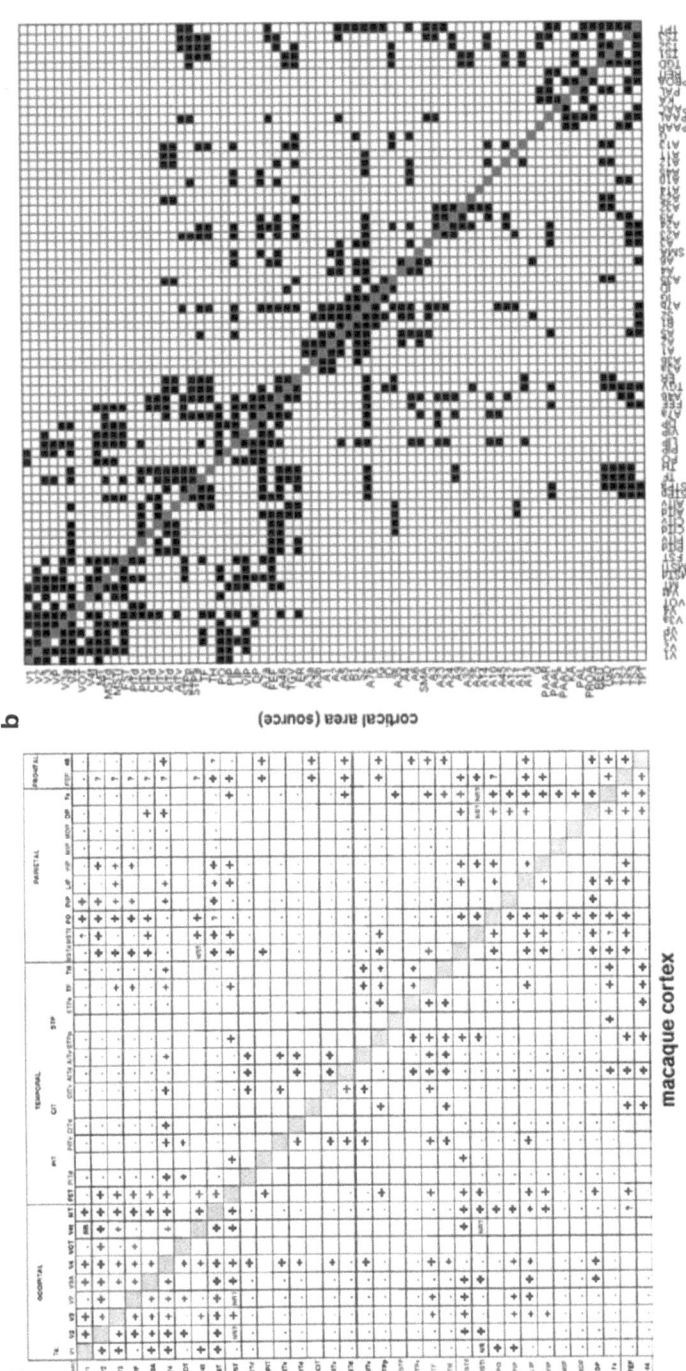

Fig. 9.3 On the *left* a matrix indicating where connections have been found between brain regions in the visual system of the *Macaque* (Felleman and van Essen, Distributed Hierarchical Processing in the Primate Cerebral Cortex, Cerebral Cortex, 1991, vol. 1, issue 1, Table 3, by permission of Oxford University Press). On the *right* is a matrix showing connectivity in the whole *Macaque* brain (Sporns and Zwi 2004, Figure 1b, with kind permission from Springer Science and Business Media)

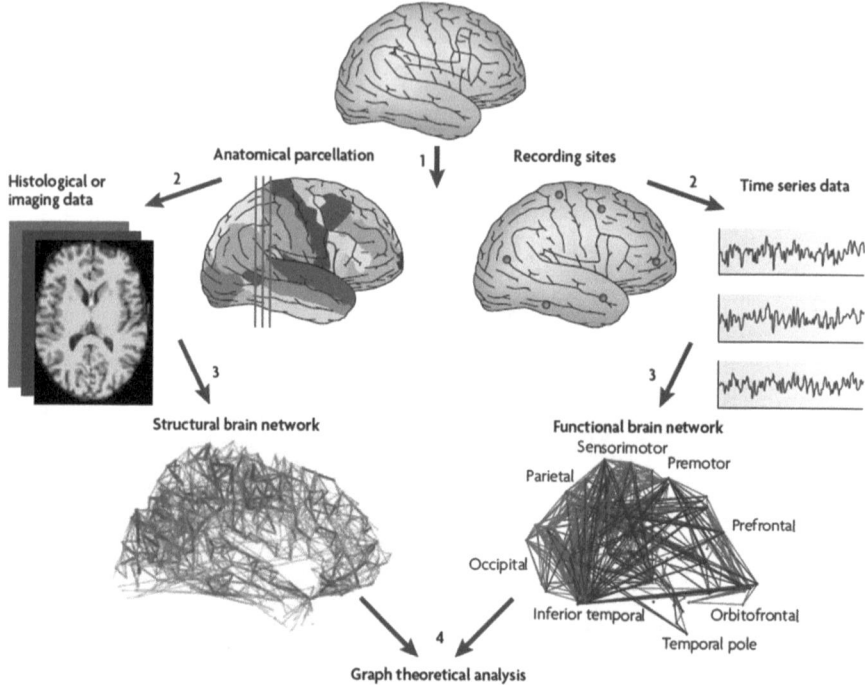

Fig. 9.4 Bullmore and Sporns (2009) proposal of a method for integrating structural and functional connectivity analyses. From each approach a graphical representation is constructed whose similarities can then be accessed. Reprinted by permission from Macmillan Publishers Ltd: Nature Reviews Neuroscience, copyright (1991)

comparable graphs of functional relations that can then be correlated to the structural graphs (see Fig. 9.4). The key to doing so is to employ techniques that provide time-series data of brain activity. EEG, which detects oscillations in electrical potentials in the 1–100 Hz range, offers one such measure. However, it is difficult to link EEG activity to specific brain regions. fMRI provides far greater localization of activity in the brain, but in initial applications was not used to support time-series analysis. When Biswal et al. (1995) did perform time series analyses on successive scans, they discovered ultraslow oscillations (<0.1 Hz) in sensorimotor areas that were correlated across hemispheres. At about the same time Raichle and his colleagues began studying people lying quietly in the scanner not performing a task (a condition referred to as the *resting state*). Shulman et al. (1997) identified several regions that exhibited greater BOLD signal in the resting state than in task conditions: the junction of precuneus and posterior cingulate cortex, the inferior parietal cortex, the left dorsolateral prefrontal cortex, a medial frontal strip that continued through the inferior anterior cingulate cortex, the left inferior frontal cortex, the left inferior frontal gyrus, and the amygdala. Since these areas were more active in the absence of a task, they interpreted them as constituting the brain's default mode network (the network of brain regions active in non-task conditions).[6] Cordes et al. (2000) developed the technique of functional connectivity MRI (fcMRI) analysis in which they identified correlations in the oscillations in resting state BOLD signal between regions and interpreted those areas whose activity was correlated as constituting functional networks. Greicius et al. (2003) applied this approach to the default mode network and demonstrated correlated resting state activity between the regions identified as part of the default mode network. In a subsequently study Greicius et al. (2009) integrated fmMRI studies to diffusion tensor imaging of structures to establish structural connections between regions identified as parts of the default mode network, thereby implementing Bullmore and Sporns' proposed approach.

Other researchers have identified several additional networks whose activity is correlated in the resting state (Mantini et al. 2007). van den Heuvel et al. (2009) directly compared nine networks found in the resting state with diffusion tensor imaging scans and found that "well-known anatomical white matter tracks interconnect at least eight of the nine commonly found resting-state networks, including the default mode network, the core network, primary motor and visual network, and two lateralized parietal-frontal networks." Analyzing these networks graph-theoretically, He et al. (2009) found that the combined networks exhibited a small-world architecture with modules characteristic of scale-free networks, but that the individual modules exhibited different architectures that might be appropriate for the computations each performs.

In this section I have focused on graph-theoretical formalization of whole networks that represent mechanisms either individually or as constituents of larger systems such as the brain. Starting with tools to analyze random networks or regular lattices, graph theorists have developed measures such as the characteristic path length, the clustering coefficient, and the distribution of node degree. Using these measures, they have revealed small-world networks with scale-free characteristics such as nodes and hubs in many biological systems. Appeals to such features support generalizing conclusions about how mechanisms behave across different mechanisms that instantiate the same graph structure with different component parts and operations.

3   Sub-graph Organization: Motifs

In addition to ways of characterizing the organization found in whole graphs and showing the consequences different organizations have for the functioning of mechanisms exhibiting such organization, researchers have also begun to focus on sub-graphs embedded in these larger graphs that can be shown to yield specific

---

[6]The one type of task that yields comparable activation in these regions are episodic memory tasks, resulting in the suggestion that while resting quietly in the scanning individuals are remembering events in their lives or planning future events (see Buckner et al. 2008).

types of behavior. Until near the end of the twentieth century, only one such pattern, negative feedback, in which an edge from a subsequent node in a pathway has an inhibitory effect on one earlier in the pathway, had been subject to systematic analysis. Pioneering cyberneticists saw this as a mode of organization that could keep a process at a target level and Rosenblueth et al. (1943) argued that it was a design capable of explaining the apparent teleology or goal directness of various biological or engineered systems. While negative feedback was soon recognized as a common mode of organization in many biological systems, few biologists attended to another feature of negative feedback that physicists and engineers had identified—its capacity to support oscillations. Among the exceptions was Goodwin (1965), who analyzed negative feedback with a particular emphasis on conditions, such as non-linear interactions, that would generate sustained oscillations.

A major impetus for research on additional sub-graph structures with characteristic effects on mechanism behavior was a mode of analysis developed by Alon in his investigation of gene transcription and protein interaction networks in bacteria and yeast (for a comprehensive overview, see Alon 2007a). Figure 9.5 shows a network that represents about 20 % of the transcription interactions in *E. coli*; the nodes are genes and edges are included when the product of one gene regulates the transcription of another. Alon and his collaborators identified a number

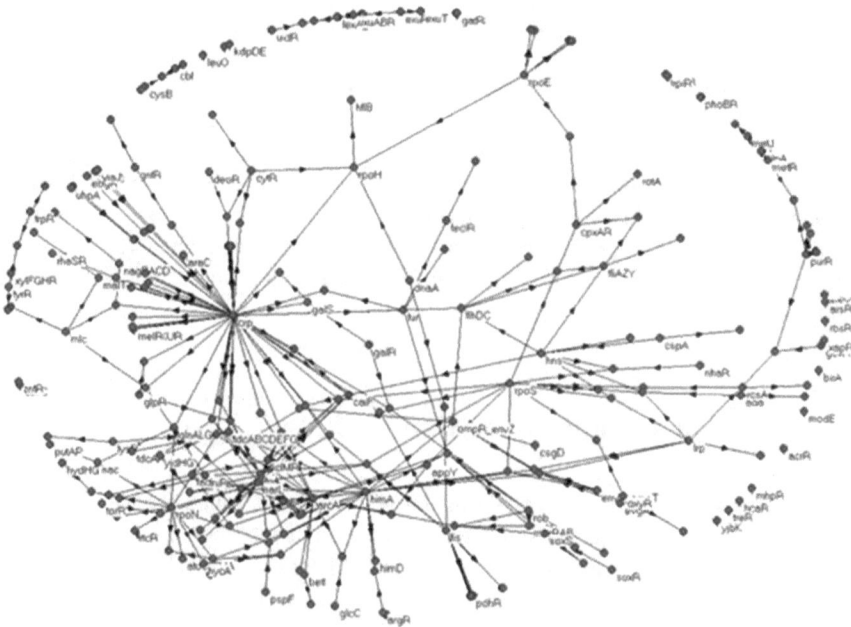

Fig. 9.5 Network representation of approximately 20 % of the transcription interactions in *E. coli*. The nodes represent genes and an edge indicates that the product of one gene exhibits regulatory effects on those of another. (Figure from Professor Uri Alon, who kindly has provided permission to reproduce it here)

of "recurring, significant patterns of interconnections" within sub-graphs (of 1–4 nodes) that appeared far more frequently than would be expected by chance (Milo et al. 2002, p. 824). He calculated the frequency expected by chance by counting how often these patterns occurred in randomly constructed networks of the same degree of node connectivity and devised an algorithm for searching databases specifying network connectivity for unusually frequently occurring subgraphs. Shen-Orr et al. (2002, p. 64) applied the term *motif* to these sub-graphs: "We generalize the notion of motifs, widely used for sequence analysis, to the level of networks. We define 'network motifs' as patterns of interconnections that recur in many different parts of a network at frequencies much higher than those found in randomized networks."[7]

Figure 9.6 illustrates three variations of a three-node sub-graph known as the *feedforward loop* that Mangan et al. (2003) reported as occurring in "hundreds of non-homologous gene systems" in the full transcription network of *E. coli*. This motif consists of three nodes representing operons in which the first (X) responds to an input signal (S) by producing a transcription factor that both regulates an operon for producing an output protein (Z) and an intermediate operation (Y) which also produces a transcription factor that regulates the output. The edges between operons can represent situations in which the first activates or inhibits the second. Making

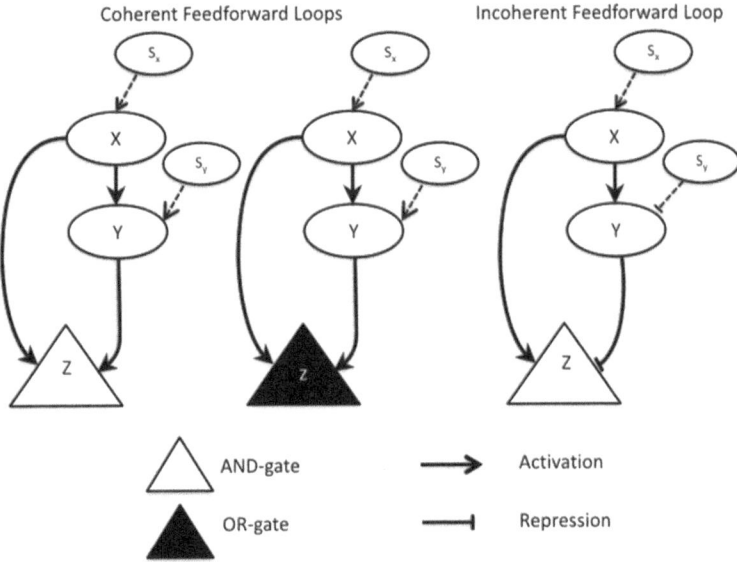

Fig. 9.6 Three versions of the feedforward loop motif investigated by Alon and his collaborators

---

[7]See also Braillard (2015, this volume) for a discussion of the heuristic and explanatory role of such *motifs* and more generally of modules in biology.

only very general assumptions about the component parts and operations (e.g., that each node represents a molecule that is synthesized proportional to the input that is received on the edges coming into it, that it takes time for the concentration of the molecule to build up, and that the triangular nodes operate as either AND or OR gates), Shen-Orr et al. (2002) demonstrated with a Boolean analysis that the motif on the left operates as a persistence detector. That is, it ensures that synthesis of product Z does not begin unless the input to X endures for some time. Only then will Y also accumulate so that both inputs to the AND-gate regulating Z are present. If the input to X is transient, by the time Y responds sufficiently to provide input to Z the input from X will have degraded.

By changing the AND-gate to an OR-gate, as illustrated in the middle of Fig. 9.6, the motif provides a buffer against interruption of the input to X. Keeping an AND-gate for the output operon but making the effect of Y on Z repression results in a motif that generates pulses.[8] These motifs are simple enough that one can simulate their functioning mentally. But Alon and his collaborators also provide computational simulations to support their assessment. They further identified specific instances in which the motif occurs in *E. coli* and showed how the imputed function is appropriate for controlling the particular operon occurring at Z. Thus, Mangan et al. (2003) identify the network on the left in Fig. 9.6 as occurring in the operon regulating the arabinose operon in which it is important not to commence the synthesis of the enzymes required to metabolize arabinose unless the absence of glucose, that plays the role of X, is sustained. Kalir et al. (2005) report that the motif in the middle occurs in the system regulating the construction of the flagellum where it ensures that the construction will continue until completion even if the initiating signal is interrupted. Finally, Mangan et al. (2006) identify the motif on the right as occurring in the galactose utilization system to generate a pulse of the needed enzyme galE when glucose is absent and galactose is present.

Alon and his colleagues have explored various ways in which motifs can be extended (Alon 2007b). For example, X could be replaced by two units, each of which sends inputs to Y and Z. This pattern occurs frequently in the *E. coli* transcription network. Or Z could be replaced by two or more outputs. This occurs frequently in the *C. elegans* neural network. A variant occurs in the mechanism for creating the flagellum in bacteria such as *E. coli*. The flagellum is a complex structure of nearly three-dozen parts that is built, for the most part, from the membrane out (Berg 2004; Macnab 2003). Each part must be added in sequence, which Shen-Orr et al. (2002) demonstrate can be achieved by a multi-output feedforward loop where the different outputs are regulated by parameters that set thresholds for response (see Fig. 9.7). By increasing threshold parameters K_1 to

---

[8]In his analysis of the neural network in *C. elegans* that he helped identify (discussed above), White (1985) identified the frequent occurrence of the feedforward loop and considered the case in which the connection from Y to Z was inhibitory. He proposed X would initially elicit a response from Z, but this would be soon be suppressed as a result of the negative connection from Y to Z. He suggests: "The whole system would therefore act as a differentiator, the output from [Z] being proportional to the rate of change of stimulus. As the animal is constantly moving, this reformation is probably of more value to it than an absolute measure of the stimulus."

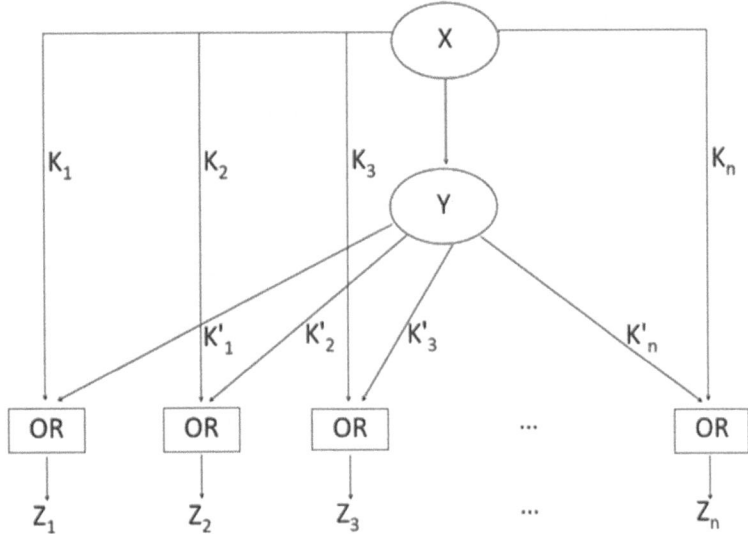

Fig. 9.7 Multi-output feedforward loop. Threshold parameters K_1 to K_n determine the order in which operons Z_1 to Z_n are activated while threshold parameters K_1' to K_n' determine the order in which they are turned off

Fig. 9.8 Five different 3-node subgraphs that occur especially frequently in different networks and so qualify as motifs on Alon's approach

K_n on the connection between X and each of Z_1 to Z_n it is possible to begin each synthesis in order as the input from X increases over time (each successive Z becomes active only when the input from X has exceeded its threshold). If K_1 to K_n were the only thresholds employed, when X declined, Z_n would be the first to stop. But it is also important to stop the activity required to build the first part when it is completed. This is achieved by decreasing the parameters K_1' to K_n' on the connection between Y and each of Z_1 to Z_n. Now the first Z_n to fall below threshold will be Z_1. The resulting arrangement Shen-Orr et al. refer to refer to as *first in, first out*.

A key point of Alon's analysis is that different sub-graphs meet the frequency condition for being a motif in different systems. Of the 13 possible sub-graphs connecting 3 nodes, in the *E. coli* transcription network only variants of the feedforward loop occur at a frequency more than 10 standard deviations above the mean frequency found in randomly constructed networks (and just 1 of 199 possible sub-graphs connecting 4 nodes, the bi-fan, meets this condition). But in food web networks, a simple chain (shown on the top in Fig. 9.8) meets this standard, and in electronic circuits the feedback loop (second from the left) qualifies. In their analysis of the connectivity of brain regions in the cat and monkey cortex, Sporns and Kötter (2004) report that the dual dyad (middle) occurs frequently, although its occurrence is most frequent when the apex node is a hub region.[9] In the *C. elegans* nervous system the two sub-graphs shown on the right are especially frequent.

In differentiating sub-graphs as motifs, Alon required that they occur in a given graph far more frequently then would be expected by chance. The motivation for this appears to be to provide the basis for proposing that motifs are adaptations promoted by natural selection. Several commentators have challenged use of random networks as a basis for establishing that motifs are adaptations (Artzy-Randrup et al. 2004; Solé and Valverde 2006).[10] However, one need not be an adaptationist to look to the analysis of sub-graphs to understand the design principles found in biological mechanisms (see Green et al. 2015). The analysis that Alon provides as to how sub-graphs function does not depend on motifs being adaptations or even occurring frequently. Even if a sub-graph occurs infrequently in a network, it can still perform the function revealed by mental or computational simulation of it. Accordingly, several researchers have generalized the concept of a motif to apply to small sub-graphs independently of their frequency of occurrence. Tyson and Novák (2010), for example, have developed differential equation models to explore the behavior of all two and three unit sub-graphs over a range of parameter values. A particularly interesting example from their analysis is the double-negative feedback sub-graph shown on the left in Fig. 9.9. For a range of parameter values, as illustrated on the right, this motif yields a bi-stable toggle switch—it has two stable configurations between which it can flip between but where the activation of S needed to turn on unit R is much higher than that which turns it off. In the case shown, S must be increased above $Q_{activate}$ to turn R on, but it must decrease below $Q_{inactivate}$ to turn R off. Such a sub-graph is extremely useful in situations in which it is important to maintain directionality in a process that will eventually repeat (such as the eukaryotic cell cycle) as it prevents temporary drops in the input from reversing the output but allows substantial drops to do so.

Beginning with Alon's work, researchers have analyzed sub-networks of 2–4 units and shown distinctive ways each functions by means of mental simulations and computational analyses. This provides a powerful tool for analyzing how different modes of organization found in mechanisms will affect the behavior of these mechanisms. As with whole-graph analyses, these analyses abstract from the details of the components found in the network and so, subject to certain minimal conditions, can be applied to any networks in which the sub-graphs are found.

---

[9]They report that computational analyses show that dual dyads are particularly effective in promoting zero phase-lag synchrony over long distances, which is important if hubs are to coordinate synchronous activity between brain regions.

[10]Ward and Thornton (2007) propose a possible origin of feedforward loops from another motif the authors term the bi-fan array to genome-wide duplication followed by random rewiring. This provides an explanation for their occurrence that doesn't depend on selection, although Ward and Thornton do not deny that selection may have also favored these motifs.

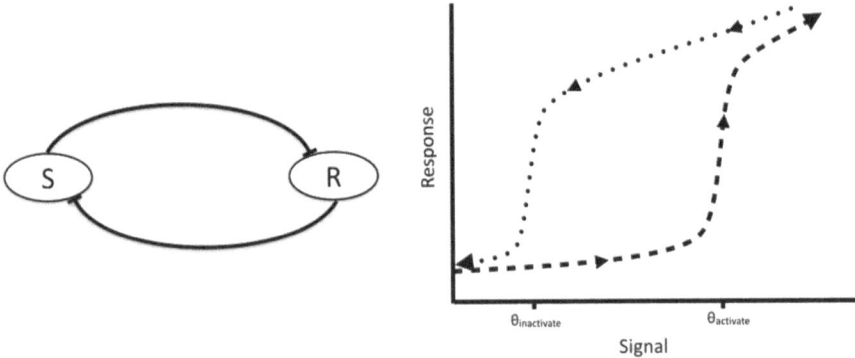

Fig. 9.9 The double negative feedback loop. The signal S and the response R each inhibit the other. With appropriate parameter values, this motif gives rise to a bi-stable switch in which the value of the signal must rise above $Q_{activate}$ for R to turn on, but once on it must decrease below $Q_{inactivate}$ before R turns off

4   Manipulating Whole-Graph Representations to Extract Motifs

In the examples discussed above, motifs were identified directly by analyzing the graphical representation of the mechanism. But in some cases investigators need to perform a series of operations on an initial graphical representation to identify motifs. I offer an example from circadian rhythm research in which such manipulations of a graphical representation have provided new insights into the working of the mechanism.

Circadian rhythms are endogenously generated oscillations of approximately 24-h that can be entrained to local cues such as light exposure and that regulate a wide range of physiological activities and behaviors. The critical mechanism is located within individual cells. As a result of the discovery of a gene (*period* or *per*) in which mutations resulted in aberrant circadian behavior in fruit flies (Konopka and Benzer 1971) and subsequent discovery that the mRNA and proteins synthesized from this gene oscillated with an approximately 24-h period, researchers hypothesized a transcriptional-translational feedback mechanism (Hardin et al. 1990). On this proposal, the protein PER synthesized from *per* mRNA feeds back to inhibit the transcription of *per* by interfering with activities at its promoter site (see the left side of Fig. 9.10). When PER finally breaks down, the inhibition is relaxed and transcription begins again. With appropriate time delays and non-linear interactions, Goldbeter (1995) showed that this mechanism could generate sustained circadian rhythms. However, subsequent research has revealed many more components to the mechanism involving multiple interacting feedback loops. The feedback loop involving *Per* is shown in the upper quadrant of the diagram on the right in Fig. 9.10 where it is now accompanied by, among other things, a positive feedback loop that produces the CLOCK:BMAL1 dimer that binds to the promoter on *Per*.

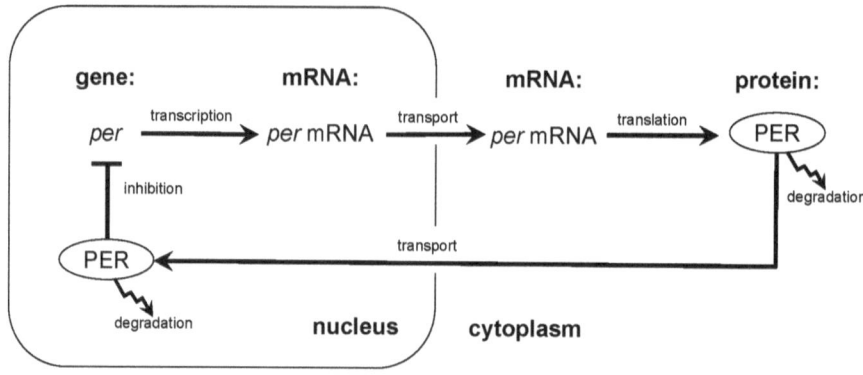

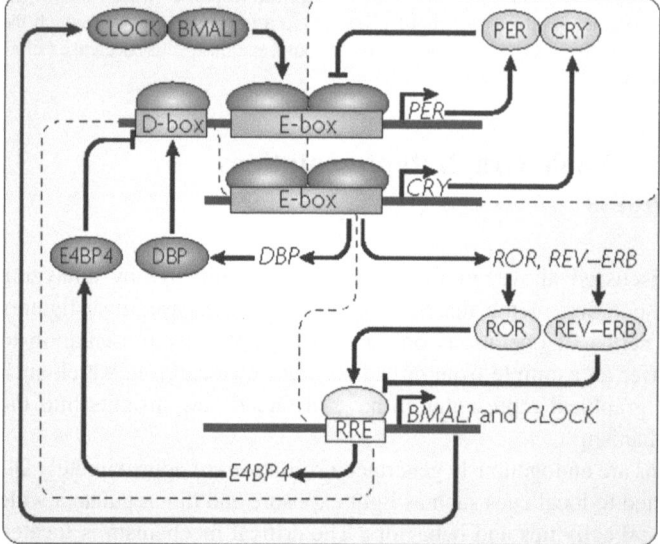

Fig. 9.10 The simple transcription-translation feedback loop as initially proposed by Hardin et al. (1990) is shown on the *top*. On the *bottom* is a representation by Zhang and Kay (2010) of the mammalian clock mechanism as it was understood circa 2005. (Reprinted by permission from Macmillan Publishers Ltd: Nature Reviews Molecular Cell Biology, copyright (2010))

A notable feature of both representations in Fig. 9.10 is that time is not explicitly represented. Rather, the viewer must follow through the sequence of operations to understand how the processes are carried out over a 24-h period. With one feedback loop, as on the *top* in Fig. 9.10, this is not too difficult, but with multiple feedback loops, as shown on the *bottom*, this becomes increasingly challenging. Activities are occurring in several feedback loops at the same time, and the figure provides no support for determining when each occurs in relation to the other. As a result, it becomes increasingly hard to understand what about the mechanism generates oscillation. In an attempt to understand the systemic organization of this mechanism that generates oscillation, Ueda and his colleagues developed a different technique for representing the mechanism. Instead of focusing on the feedback loops themselves, they focused on what they call *clock controlled elements* (CCEs): the short DNA sequences shown near the promoter region of clock genes on the right in Fig. 9.10 and labeled E-box, D-box, and RRE. Each of these adjoins the promoter of a number of clock genes and in turn is regulated positively or negatively by another, partially overlapping set of clock genes. Using a cell-culture system in which a firefly luciferase reporter was inserted downstream of the clock-controlled promoters, Ueda et al. (2005) determined the circadian time of peak activity for each CCE: the E/E′-box regulates gene expression in the morning, the D-box regulates gene expression 5 h later in the evening, and the RRE regulates expression 8 h later during the night.

Given the distinctive timing at which each CCE regulates transcription, Ueda and his colleagues put them at the center of their graphical formalization of the mechanism (see the left side of Fig. 9.11). Time can be visualized as moving clockwise around the triangle formed by the three CCEs. The various transcription factors that regulate a CCE are shown around it—those that serve to activate a particular CCE are shown in green ovals and linked to it with green arrows while those that repress it are shown in red ovals and linked to it with bar-ended lines. There is a grey line from the node for the CCE to the transcription factors of the genes with which the CCE is associated. As an illustration, there are grey lines between the E/E′-box and the D-box to Per1 (indicating that both boxes figure in its regulation). There is also a red arrow back to the E/E′-box, indicating that Per1 represses the E/E′-boxes. This graphical formalization renders the clock genes/proteins simply as intermediaries between the three CCEs. This is further

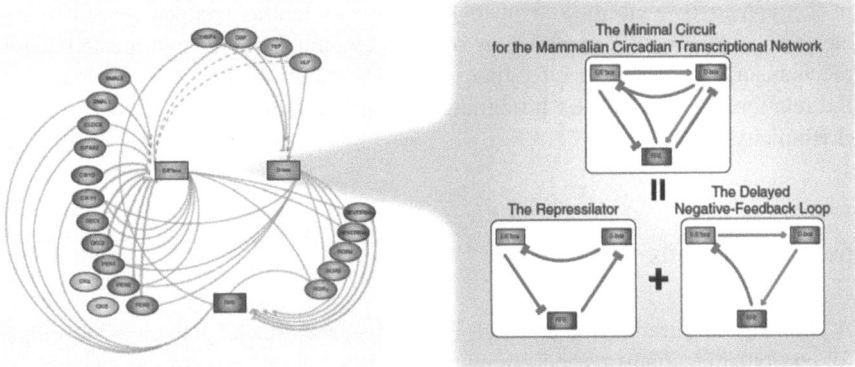

Fig. 9.11 On the *left* is Ueda's re-representation of the mammalian circadian clock mechanism in which the clock controlled elements (CCEs) are shown as *rectangles* in the center with *grey lines* linking the CCEs to the transcription factors whose genes they regulate and *green* and *red arrows* represent the effect of these transcription factors on other CCEs. On the *right* the details of the intermediary transcription factors are removed to reveal the minimal circuit shown at the *top*, which is then decomposed into two component circuits (Reprinted from Minami et al. 2013, With kind permission from Springer Science and Business Media)

illustrated by the minimal circuit diagram shown on the right in which there are excitatory and inhibitory arrows between two CCEs if there is a gene/protein associated with one CCE that has such an effect on the other CCE. For example, there is a green arrow between the E/E'-box and the D-box since the E/E-box is associated with *Dpb* and DPB is an activator of the D-box.

The point of this graphical formalization of the clock mechanism is to advance an explanation of how this mechanism is able to generate oscillatory behavior. Ueda shows that the minimal circuit can be treated as a composite of the two component circuits shown on the bottom right in Fig. 9.11—the repressilator and delayed negative feedback. Although these circuits are intended to be simplifications of complex networks, they have the same form as motifs for specific circuits. In both computational models and synthetic models, each has been shown to give rise to sustained oscillations (Elowitz and Leibler 2000; Stricker et al. 2008). When their analysis reveals the same circuit structure in these larger networks, Ueda and his colleagues conclude that the occurrence of the circuit structure in those networks explains their oscillatory behavior.

As this case indicates, researchers sometimes find it useful to modify the initial graphical formalization of a mechanism to reveal the underlying graph structural that explains the behavior of the mechanism. Thus, in this case, researchers started with a graph formalization in which all of the transcription factors were indicated (although this formalization already simplified by not distinguishing between the genes and the transcription factors). This was already an innovative formalization since it put the CCEs at the center. But the researchers then abstracted further, collapsing all the transcription factors linking one CCE to another into a single edge. This resulted in a *minimal circuit* that the researchers could further decompose into two motifs, already known from work on sub-graphs to be able to generate oscillations. The fact that they were able to reduce the whole network as formalized in a graph to two motifs known to generate oscillations enabled researchers to advance an distinctive explanation of how the much more complicated whole mechanism operates as a circadian clock. On their account, it is the fact that the relations between CCEs instantiate appropriate motifs that explains circadian rhythmicity.

5   Conclusions

A challenge in mechanistic research is how to generalize explanations developed for particular mechanisms occurring in specific tissues or organisms to mechanisms that employ similar modes of organization. Each type of mechanism (e.g., the mechanism for generating circadian rhythms in mouse SCN cells) employs its own mode of organization and this would seem to require an idiosyncratic explanation for each specific mechanism. When mechanisms employ parts that behave similarly, this knowledge can be generalized, but it has not been clear how one could do this for organization. Graph-theoretic approaches, however, are providing powerful tools for formalizing organization patterns. By formalizing a given mechanism in a graph in which the parts are nodes and the interactions are edges, investigators abstract from the details of the specific parts and operations. Making minimal assumptions about the operations represented in edges, researchers can often draw inferences about how any mechanism exhibiting that organization will behave. Moreover, they can generalize across graphs by showing, for example, that despite differences, a set of properties is shared by the graphs that result in the same pattern of behavior. Once researchers have analyzed how mechanisms instantiating a graph structure or related graph structures behave, they can generalize that explanation to other mechanisms that instantiate the same abstract structure.

I have focused on graph theoretic analyses both of graphs representing whole mechanisms or networks of mechanisms and of sub-graphs found within larger graphs. By employing measures such as the characteristic path length, the clustering coefficient, and degree distribution, researchers have been able to classify whole-graphs and demonstrate some of the expected properties of these graphs when realized in different systems. In particular, research on scale-free small-world networks provides ways to understand both the specialized processing capacities within a larger system and the ways in which these modulate each other to achieve coordinated function. By focusing on sub-graphs of just a few nodes, researchers have demonstrated patterns of behavior different sub-graphs or motifs are expected to produce such as buffering against noise or executing operations in sequence. These can be applied to any network in which the motif is realized. I concluded with an example that showed how researchers have found ways to manipulate whole-graphs to reveal patterns of organization whose behavior is well understood. They then appealed to these organizational patterns to explain the behavior of the whole mechanism.

As biologists have discovered more and more complicated mechanisms incorporating non-linear interactions of components, the challenge of explaining their behavior has increased. Computational modeling has resulted in dynamical mechanistic analyses that attempt to integrate details about parts and operations and information about how these are organized. Graph theoretic approaches provides additional resources, enabling researchers to identify common abstract graph structures in different mechanisms and generalize the results developed for one mechanism to other mechanisms sharing that structure.

Acknowledgment Initial research on this project began when I was a Fellow at the Institute for Advanced Studies at Hebrew University. I thank the members of the group for productive discussions and especially Arnon Levy for introducing me to the work of Uri Alon and facilitating a meeting with him. Subsequently I have benefited from many further discussions with Arnon and with Sara Green. I presented much of the material here at colloquia at the University of California, Irvine and the University of Cincinnati, at workshop at the University of Wollongong, and to the reunion conference of the research group at the Institute for Advanced Studies. I thank the audiences at these various forums for very helpful comments. I also thank members of the WORGODS research group (Adele Abrahamsen, Daniel Burnston, and Benjamin Sheredos) at the University of California, San Diego for valuable discussion of the diagrammatic representations of circadian clock mechanisms. Thanks as well to Marta Halina and to the editors of this volume, Pierre-Alain Braillard and Christophe Malaterre, for valuable comments and suggestions.

References

Alon, U. (2007a). *An introduction to systems biology: Design principles of biological circuits.* Boca Raton: Chapman & Hall/CRC.

Alon, U. (2007b). Network motifs: Theory and experimental approaches. *Nature Reviews Genetics, 8,* 450–461.

Ankeny, R. A. (2001). Model organisms as models: Understanding the 'Lingua Franca' of the human genome project. *Philosophy of Science, 68,* S251–S261.

Ankeny, R. A., & Leonelli, S. (2011). What's so special about model organisms? *Studies in History and Philosophy of Science Part A, 42,* 313–323.

Artzy-Randrup, Y., Fleishman, S. J., Ben-Tal, N., & Stone, L. (2004). Comment on 'network motifs: Simple building blocks of complex networks' and 'superfamilies of evolved and designed networks'. *Science, 305,* 1107.

Baetu, T. (2015). From mechanisms to mathematical models and back to mechanisms: Quantitative mechanistic explanations. In P.-A. Braillard & C. Malaterre (Eds.), *Explanation in biology. An enquiry into the diversity of explanatory patterns in the life sciences* (pp. 345–363). Dordrecht: Springer.

Barabási, A.-L., & Albert, R. (1999). Emergence of scaling in random networks. *Science, 286,* 509–512.

Bechtel, W. (2011). Mechanism and biological explanation. *Philosophy of Science, 78,* 533–557.

Bechtel, W., & Abrahamsen, A. (2009). Decomposing, recomposing, and situating circadian mechanisms: Three tasks in developing mechanistic explanations. In H. Leitgeb & A. Hieke (Eds.), *Reduction and elimination in philosophy of mind and philosophy of neuroscience* (pp. 173–186). Frankfurt: Ontos Verlag.

Bechtel, W., & Abrahamsen, A. (2010). Dynamic mechanistic explanation: Computational modeling of circadian rhythms as an exemplar for cognitive science. *Studies in History and Philosophy of Science Part A, 41,* 321–333.

Bechtel, W., & Abrahamsen, A. (2011). Complex biological mechanisms: Cyclic, oscillatory, and autonomous. In C. A. Hooker (Ed.), *Philosophy of complex systems. Handbook of the philosophy of science* (Vol. 10, pp. 257–285). New York: Elsevier.

Bechtel, W., & Richardson, R. C. (1993/2010). *Discovering complexity: Decomposition and localization as strategies in scientific research.* Cambridge, MA: MIT Press. 1993 edition published by Princeton University Press.

Berg, H. C. (2004). *E. coli in motion.* New York: Springer.

Biswal, B., Yetkin, F. Z., Haughton, V. M., & Hyde, J. S. (1995). Functional connectivity in the motor cortex of resting human brain using echo-planar MRI. *Magnetic Resonance in Medicine, 34,* 537–541.

Braillard, P.-A. (2015). Prospect and limits of explaining biological systems in engineering terms. In P.-A. Braillard & C. Malaterre (Eds.), *Explanation in biology. An enquiry into the diversity of explanatory patterns in the life sciences* (pp. 319–344). Dordrecht: Springer.

Brigandt, I. (2013). Systems biology and the integration of mechanistic explanation and mathematical explanation. *Studies in History and Philosophy of Biological and Biomedical Sciences, 44,* 477–492.

Buckner, R. L., Andrews-Hanna, J. R., & Schacter, D. L. (2008). The brain's default network: Anatomy, function, and relevance to disease. *Annals of the New York Academy of Sciences, 1124,* 1–38.

Bullmore, E., & Sporns, O. (2009). Complex brain networks: Graph theoretical analysis of structural and functional systems. *Nature Reviews Neuroscience, 10,* 186–198.

Cordes, D., Haughton, V. M., Arfanakis, K., Wendt, G. J., Turski, P. A., Moritz, C. H., Quigley, M. A., & Meyerand, M. E. (2000). Mapping functionally related regions of brain with functional connectivity MR imaging. *American Journal of Neuroradiology, 21,* 1636–1644.

Craver, C. F. (2007). *Explaining the brain: Mechanisms and the mosaic unity of neuroscience.* New York: Oxford University Press.

Derrida, B., & Pomeau, Y. (1986). Random networks of automata: A simple annealed approximation. *Europhysics Letters, 1*, 45–49.

Elowitz, M. B., & Leibler, S. (2000). A synthetic oscillatory network of transcriptional regulators.
Nature, 403, 335–338.

Erdös, P., & Rényi, A. (1960). On the evolution of random graphs. *Proceedings of the Mathematical
Institute of the Hungarian Academy of Sciences, 5*, 17–61.

Ermentrout, G. B., & Kopell, N. (1984). Frequency plateaus in a chain of weakly coupled
oscillators. 1. *Siam Journal on Mathematical Analysis, 15*, 215–237.

Felleman, D. J., & van Essen, D. C. (1991). Distributed hierarchical processing in the primate
cerebral cortex. *Cerebral Cortex, 1*, 1–47.

Flyvbjerg, H. (1988). An order parameter for networks of automata. *Journal of Physics A:
Mathematical and General, 21*, L955–L960.

Glennan, S. (1996). Mechanisms and the nature of causation. *Erkenntnis, 44*, 50–71.

Glennan, S. (2002). Rethinking mechanistic explanation. *Philosophy of Science, 69*, S342–S353.

Goldbeter, A. (1995). A model for circadian oscillations in the *Drosophila* period protein (PER).
Proceedings of the Royal Society of London B: Biological Sciences, 261, 319–324.

Goodwin, B. C. (1965). Oscillatory behavior in enzymatic control processes. *Advances in Enzyme
Regulation, 3*, 425–428.

Green, S., Levy, A., & Bechtel, W. (2015). Design sans adaptation. *European Journal for the
Philosophy of Science, 5*, 15–29.

Greicius, M. D., Krasnow, B., Reiss, A. L., & Menon, V. (2003). Functional connectivity in the
resting brain: A network analysis of the default mode hypothesis. *Proceedings of the National
Academy of Sciences of the United States of America, 100*, 253–258.

Greicius, M. D., Supekar, K., Menon, V., & Dougherty, R. F. (2009). Resting-state functional
connectivity reflects structural connectivity in the default mode network. *Cerebral Cortex, 19*,
72–78.

Hagmann, P., Cammoun, L., Gigandet, X., Meuli, R., Honey, C. J., Wedeen, V. J., & Sporns, O.
(2008). Mapping the structural core of human cerebral cortex. *PLoS Biology, 6*, e159.

Hardin, P. E., Hall, J. C., & Rosbash, M. (1990). Feedback of the *Drosophila period* gene product
on circadian cycling of its messenger RNA levels. *Nature, 343*, 536–540.

He, Y., Wang, J., Wang, L., Chen, Z. J., Yan, C., Yang, H., Tang, H., Zhu, C., Gong, Q., Zang, Y., &
Evans, A. C. (2009). Uncovering intrinsic modular organization of spontaneous brain activity
in humans. *PLoS One, 4*, e5226.

Hubel, D. H., & Wiesel, T. N. (1962). Receptive fields, binocular interaction and functional
architecture in the cat's visual cortex. *Journal of Physiology, 160*, 106–154.

Hubel, D. H., & Wiesel, T. N. (1968). Receptive fields and functional architecture of monkey striate
cortex. *Journal of Physiology, 195*, 215–243.

Issad, T., & Malaterre, C. (2015). Are dynamic mechanistic explanations still mechanistic? In
P.-A. Braillard & C. Malaterre (Eds.), *Explanation in biology. An enquiry into the diversity of
explanatory patterns in the life sciences* (pp. 265–292). Dordrecht: Springer.

Jeong, H., Mason, S. P., Barabasi, A. L., & Oltvai, Z. N. (2001). Lethality and centrality in protein
networks. *Nature, 411*, 41–42.

Jones, N. (2014). Bowtie structures, pathway diagrams, and topological explanation. *Erkenntnis,
79*, 1135–1155.

Jones, N., & Wolkenhauer, O. (2012). Diagrams as locality aids for explanation and model
construction in cell biology. *Biology and Philosophy, 27*, 705–721.

Kalir, S., Mangan, S., & Alon, U. (2005). A coherent feed-forward loop with a SUM input function
prolongs flagella expression in *Escherichia coli. Molecular Systems Biology, 1*, 2005.0006.

Kauffman, S. A. (1969). Metabolic stability and epigenesis in randomly constructed genetic nets.
Journal of Theoretical Biology, 22(3), 437–467.

Kauffman, S. A. (1974). The large scale structure and dynamics of gene control circuits: An
ensemble approach. *Journal of Theoretical Biology, 44*(1), 167–190.

Kauffman, S. A. (1993). *The origins of order: Self-organization and selection in evolution.* New
York: Oxford University Press.

Konopka, R. J., & Benzer, S. (1971). Clock mutants of *Drosophila melanogaster. Proceedings of the National Academy of Sciences of the United States of America, 89*, 2112–2116.

Kopell, N., & Ermentrout, G. B. (1988). Coupled oscillators and the design of central pattern generators. *Mathematical Biosciences, 90*, 87–109.

Kuramoto, Y. (1984). *Chemical oscillations, waves, and turbulence*. Berlin: Springer.

Leonelli, S., Ramsden, E., Nelson, N., & Ankeny, R. A. (2014). Making organisms model humans: Situated models in alcohol research. *Science in Context, 27*(3), 485–509.

Levy, A., & Bechtel, W. (2013). Abstraction and the organization of mechanisms. *Philosophy of Science, 80*, 241–261.

Luque, B., & Solé, R. (1997). Phase transitions in random networks: Simple analytic determination of critical points. *Physical Review E, 55*, 257–260.

Machamer, P., Darden, L., & Craver, C. F. (2000). Thinking about mechanisms. *Philosophy of Science, 67*, 1–25.

Macnab, R. M. (2003). How bacteria assemble flagella. *Annual Review of Microbiology, 57*, 77–100.

Mangan, S., Zaslaver, A., & Alon, U. (2003). The coherent feedforward loop serves as a sign-sensitive delay element in transcription networks. *Journal of Molecular Biology, 334*, 197–204.

Mangan, S., Itzkovitz, S., Zaslaver, A., & Alon, U. (2006). The incoherent feed-forward loop accelerates the response-time of the gal system of *Escherichia coli. Journal of Molecular Biology, 356*, 1073–1081.

Mantini, D., Perrucci, M. G., Del Gratta, C., Romani, G. L., & Corbetta, M. (2007). Electrophysiological signatures of resting state networks in the human brain. *Proceedings of the National Academy of Sciences, 104*, 13170–13175.

Meunier, R. (2012). Stages in the development of a model organism as a platform for mechanistic models in developmental biology: Zebrafish, 1970–2000. *Studies in History and Philosophy of Science Part C: Studies in History and Philosophy of Biological and Biomedical Sciences, 43*, 522–531.

Milo, R., Shen-Orr, S., Itzkovitz, S., Kashtan, N., Chklovskii, D., & Alon, U. (2002). Network motifs: Simple building blocks of complex networks. *Science, 298*, 824–827.

Minami, Y., Ode, K. L., & Ueda, H. R. (2013). Mammalian circadian clock: The roles of transcriptional repression and delay. *Handbook of Experimental Pharmacology, 217*, 359–377.

Overton, J. A. (2011). Mechanisms, types, and abstractions. *Philosophy of Science, 78*, 941–954.

Rohlf, T., & Bornholdt, S. (2002). Criticality in random threshold networks: Annealed approximation and beyond. *Physica A: Statistical Mechanics and its Applications, 310*, 245–259.

Rosenblueth, A., Wiener, N., & Bigelow, J. (1943). Behavior, purpose, and teleology. *Philosophy of Science, 10*, 18–24.

Shen-Orr, S. S., Milo, R., Mangan, S., & Alon, U. (2002). Network motifs in the transcriptional regulation network of Escherichia coli. *Nature Genetics, 31*, 64–68.

Shulman, G. L., Corbetta, M., Buckner, R. L., Fiez, J. A., Miezin, F. M., Raichle, M. E., & Petersen, S. E. (1997). Common blood flow changes across visual tasks: I. increases in subcortical structures and cerebellum but not in nonvisual cortex. *Journal of Cognitive Neuroscience, 9*, 624–647.

Solé, R. V., & Valverde, S. (2006). Are network motifs the spandrels of cellular complexity? *Trends in Ecology and Evolution, 21*, 419–422.

Sporns, O., & Kötter, R. (2004). Motifs in brain networks. *PLoS Biology, 2*, e369.

Sporns, O., & Zwi, J. D. (2004). The small world of the cerebral cortex. *Neuroinformatics, 2*, 145–162.

Sporns, O., Tononi, G., & Kötter, R. (2005). The human connectome: A structural description of the human brain. *PLoS Computational Biology, 1*, e42.

Stricker, J., Cookson, S., Bennett, M. R., Mather, W. H., Tsimring, L. S., & Hasty, J. (2008). A fast, robust and tunable synthetic gene oscillator. *Nature, 456*, 516–519.

Théry, F. (2015). Explaining in contemporary molecular biology: Beyond mechanisms. In P.-A. Braillard & C. Malaterre (Eds.), *Explanation in biology. An enquiry into the diversity of explanatory patterns in the life sciences* (pp. 113–133). Dordrecht: Springer.

Tyson, J. J., & Novák, B. (2010). Functional motifs in biochemical reaction networks. *Annual Review of Physical Chemistry, 61*, 219–240.

Ueda, H. R., Hayashi, S., Chen, W., Sano, M., Machida, M., Shigeyoshi, Y., Iino, M., & Hashimoto, S. (2005). System-level identification of transcriptional circuits underlying mammalian circadian clocks. *Nature Genetics, 37*, 187–192.

van den Heuvel, M. P., Mandl, R. C. W., Kahn, R. S., & Pol, H. E. H. (2009). Functionally linked resting-state networks reflect the underlying structural connectivity architecture of the human brain. *Human Brain Mapping, 30*, 3127–3141.

Ward, J. J., & Thornton, J. M. (2007). Evolutionary models for formation of metwork motifs and modularity in the *Saccharomyces* transcription factor network. *PLoS Computational Biology, 3*, e198.

Watts, D., & Strogratz, S. (1998). Collective dynamics of small worlds. *Nature, 393*, 440–442.

White, J. G. (1985). Neuronal connectivity in Caenorhabditis elegans. *Trends in Neurosciences, 8*, 277–283.

White, J. G., Southgate, E., Thomson, J. N., & Brenner, S. (1986). The structure of the nervous system of the nematode *Caenorhabditis elegans. Philosophical Transactions of the Royal Society of London. B Biological Sciences, 314*, 1–340.

Winfree, A. T. (1967). Biological rhythms and the behavior of populations of coupled oscillators. *Journal of Theoretical Biology, 16*, 15–42.

Woodward, J. (2013). II—Mechanistic explanation: Its scope and limits. *Aristotelian Society Supplementary Volume, 87*, 39–65.

Young, M. P. (1993). The organization of neural systems in the primate cerebral cortex. *Proceedings of the Royal Society of London. Series B: Biological Sciences, 252*, 13–18.

Zhang, E. E., & Kay, S. A. (2010). Clocks not winding down: Unravelling circadian networks. *Nature Reviews Molecular and Cell Biology, 11*, 764–776.

Part III
The Role of Mathematics
in Biological Explanations

Chapter 10
Mathematical Explanation in Biology

Alan Baker

Abstract Biology has proved to be a rich source of examples in which mathematics plays a role in explaining some physical phenomena. In this paper, two examples from evolutionary biology, one involving periodical cicadas and one involving bee honeycomb, are examined in detail. I discuss the use of such examples to defend platonism about mathematical objects, and then go on to distinguish several different varieties of mathematical explanation in biology. I also connect these discussions to issues concerning generality in biological explanation, and to the question of how to pick out which mathematical properties are explanatorily relevant.

Keywords Mathematical explanation • Biology • Indispensability • Optimization

1 Introduction

As is the case for most areas of contemporary science, biology is permeated with mathematics. This situation has potentially interesting implications both for biology and for mathematics. My focus in this paper is on explanation, and in particular on the explanatory role that mathematics sometimes plays in the context of biological theorizing. Philosophers interested in scientific explanation have started to look in more detail at biology as a potential source of test cases for more general theories of scientific explanation. And recent philosophical debates over the nature and role of scientific models have engaged with the issue of the explanatory role of models, in biology and elsewhere.

Philosophical investigation of the role of mathematics in empirical science has the potential to cast light not only on scientific methodology but also on epistemological and metaphysical issues concerning mathematics itself. Interest among philosophers in the topic of applied mathematics has been steadily increasing over the past several decades, spurred on initially by Quine and Putnam's so-called

A. Baker (✉)
Department of Philosophy, Swarthmore College, 500 College Ave. Swarthmore, PA, 19081 USA
e-mail: abaker1@swarthmore.edu

© Springer Science+Business Media Dordrecht 2015
P.-A. Braillard, C. Malaterre, *Explanation in Biology*, History, Philosophy
and Theory of the Life Sciences 11, DOI 10.1007/978-94-017-9822-8_10

"Indispensability Argument" for platonism about mathematics.[1] In its contemporary guise, platonism is the view that abstract mathematical objects exist in some objective sense, and that the goal of mathematics is to give a true description of the realm of mathematical abstracta. According to the Indispensability Argument (IA), we ought rationally to believe in the existence of mathematical objects if we are scientific realists, because quantification over such objects is indispensable to our best scientific theories. Put another way, our grounds for believing in numbers are analogous to our grounds for believing in electrons. Each kind of object plays an indispensable role in science.

Early reaction to IA focused on the further reaches of theoretical physics, because it is here that the most complex and sophisticated mathematical apparatus tends to be employed. Philosophers unsympathetic to platonism (often referred to as nominalists, or – more recently – as fictionalists) tried to show how quantification over mathematical objects might be avoided. Perhaps the most well-known example of this kind of project is that undertaken by Hartry Field in his book, *Science Without Numbers*, where he showed – among other things – how Newtonian gravitational theory might be formulated in a manner that was free of ontological commitment to mathematics.[2] Defenders of IA pointed out that physics has become much more entangled in mathematics since Newton's time. What about the curved geometries of general relativity, or the infinite-dimensional Hilbert spaces of quantum mechanics?

Rather than meet this challenge head on, a second wave of nominalists took a different tack, arguing that more attention needs to be paid to what role mathematics is actually playing in science. Returning to the earlier analogy between numbers and electrons, electrons are not merely indispensable but they also play an *explanatory* role in science. For example, we can explain the track in the cloud chamber by positing that an electron was emitted by the source material. Quine's holism led him to largely ignore the details of which role was played by which theoretical posits in science. For the second-wave nominalists, this was a crucial mistake: we can legitimately resist commitment to abstract mathematical objects unless they play an indispensable *explanatory* role in science. According to such nominalists, there are no genuine mathematical explanations in science. Hence nominalism can be upheld without any need to reformulate our scientific theories.[3]

With this new twist on IA, attention switched away from theoretical physics to other areas of science, and to biology in particular. I shall return in Sect. 4 to consider the question of why biology is such a fertile ground for putative mathematical explanations. For the moment, however, I simply note that the most widely discussed examples of mathematical explanation in science have biological subject matter, as

---

[1]Colyvan (2003) gives a good overview of the Indispensability Argument.
[2]Field (1980).
[3]Melia (2000, 2002).

we shall see shortly.[4] As a consequence, the nature of biological explanation has become a topic of considerable interest to many philosophers of mathematics.

The plan for this paper is as follows. In Sects. 2 and 3, I present and briefly discuss two cases of mathematical explanation in biology that have been prominent in the recent philosophical literature. The first concerns the life-cycles of periodical cicadas, and the second concerns the honeycomb-building behaviour of bees. In Sect. 4, I discuss the role of generality in explanation, and say a bit more about the different types of mathematical explanation within biology. In Sect. 5, I take up the question of why biology is so rich in mathematical explanations. In Sect. 6, I look at how to determine which mathematical features of a given biological phenomenon are genuine targets of explanation. I conclude, in Sect. 7, with some remarks about future directions for philosophical investigation into the role of mathematics in biological explanation.

2 Periodical Cicadas

The first case study, featuring a genuinely mathematical explanation of a biological phenomenon, is drawn from evolutionary biology. Its subject is the life-cycle of the North American 'periodical' cicada. Three species of cicada of the genus *Magicicada* share the same unusual life-cycle. In each species the nymphal stage remains in the soil for a lengthy period, then the adult cicada emerges after either 13 or 17 years depending on the geographical area. Even more strikingly, this emergence is synchronized among all members of a cicada species in any given area. The adults all emerge within the same few days, they mate, die a few weeks later and then the cycle repeats itself.

Biologists have long found features of the life-cycle of periodical cicadas mysterious, and this is reflected both in the substantial literature devoted to this topic and in biologists' specific remarks.[5] There are at least five distinct features of this life-cycle for which explanations have been sought by biologists;

(i)   The great duration of the cicada life-cycle.
(ii) The presence of two separate life-cycle durations (within each cicada species) in different regions.
(iii) The periodic emergence of adult cicadas.
(iv) The synchronized emergence of adult cicadas.
(v)   The prime-numbered-year cicada life-cycle lengths.

---

[4]See for instance Breidenmoser and Wolkenhauer (2015, this volume) who stress the importance of theorems such as the "robustness theorem" in biology, and Issad and Malaterre (2015, this volume) who emphasize the explanatory force of mathematical derivation in dynamic mechanistic explanations.

[5]For example, that "periodical cicadas are among the most unusual insects in the world" (Yoshimura 1997, p. 112).

Features (i) and (ii) concern the temporal range of the life-cycle. Biologists have argued that the long life-cycle of *Magicicada* is due both to the poor availability of nutrients for nymphs, and to the low soil temperatures for much of the year. Together these environmental stresses force nymphs to spend several years maturing into adults. Thus both (i) and (ii) seem explicable in terms of specific ecological constraints.

Features (iii) and (iv) concern coordination of the life-cycles of different individuals. Given that cicada nymphs require several years to develop into adults, and that the adult stage is very brief, having a fixed periodic emergence is advantageous in terms of maximizing mating opportunities. It ensures that the offspring of a particular mating generation will all appear at the same time, several years down the line. Synchronization makes sense for the same reason. Especially in areas which can support only a sparse population of cicadas, staggering different subpopulations to emerge at different times may produce so few adults at any one time that it is difficult to find mates. These explanations of (iii) and (iv) rely on (evolutionary) biological 'laws' which potentially apply to any organism with a long life-cycle and brief adult stage.

This leaves feature (v) to be explained, and with it one key question to be answered: why are the life-cycle periods *prime*? In other words, given a synchronized, periodic life-cycle, is there some evolutionary advantage to having a period that is prime? In seeking to answer this question, biologists have come up with two basic alternative theories.

An explanation of the advantage of prime cycle periods has been offered by Goles, Schulz and Markus (henceforth, GSM) based on avoiding predators. GSM hypothesize a period in the evolutionary past of *Magicicada* when it was attacked by predators that were themselves periodic, with lower cycle periods. Clearly it is advantageous – other things being equal – for the cicada species to intersect as rarely as possible with such predators. GSM's claim is that the frequency of intersection is minimized when the cicada's period is prime;

> For example, a prey with a 12-year cycle will meet – every time it appears – properly synchronized predators appearing every 1, 2, 3, 4, 6 or 12 years, whereas a mutant with a 13-year period has the advantage of being subject to fewer predators.[6]

A second explanation, proposed by Cox and Carlton and by Yoshimura, concerns the avoidance not of predators but of hybridization with similar subspecies.[7] A crucial factor for periodical insects is to have sufficient mating opportunities during their brief adult stage. Almost as important, however, is to avoid mating with subspecies that have different cycle periods to their own. For example, if some of a (hypothetical) population of synchronized 10-year cicadas were to mate with some 15-year cicadas then their offspring would likely have a period of around 12 or

---

[6]Goles et al. (2001, p. 33).

[7]Cox and Carlton (1988, 1998), Yoshimura (1997).

13 years.[8] These hybrid offspring would emerge well after the next cycle of the 10-year cicadas and hence their mating opportunities would be severely curtailed.

The mathematical underpinnings of both the predation and the hybridization explanations lie in number theory. The mathematical link between primeness and minimizing the intersection of periods involves the notion of *lowest common multiple* (lcm). The lcm of two natural numbers, m and n, is the smallest number into which both m and n divide exactly; for example, the lcm of 4 and 10 is 20. Taking the predation explanation first, let us assume that m is the life-cycle period (in years) of a given cicada species, C_m, and n is the period of a periodical predator, P_n. If C_m and P_n intersect in a particular year, then the year of their next intersection is given by the lcm of m and n. In other words, the lcm is the number of years between successive intersections.

In fact the fundamental property in this context is not primeness but *coprimeness*; two numbers, m and n, are coprime if they have no common factors other than 1 (i.e., neither number is divisible by the other). All that is needed to underpin the above explanations are the following two number-theoretic results;

Lemma 1 *The lowest common multiple of* m *and* n *is maximal if and only if* m *and* n *are coprime.*[9]

Lemma 1 implies that the intersection frequency of two periods of length m and n is maximized when m and n are coprime. We get from coprimeness to primeness *simpliciter* with a second result;

Lemma 2 *A number,* m, *is coprime with each number* $n < 2\,m$, $n \nmid m$ *if and only if* m *is prime.*

The mathematics for the predation explanation is already contained in the above two Lemmas. Predators are assumed to have relatively low cycle periods.[10] It therefore suffices to show that prime numbers maximize their lcm relative to all lower numbers. More formally, we need to show that for a given prime, p, and for any pair of numbers, m and n, both less than p, the lcm of p and m is greater than the lcm of n and m. But this follows directly from Lemmas 1 and 2. From Lemma 2, p is coprime with m, since $m < p$. So, from Lemma 1, the lcm of m and p is $m.p$. The highest the lcm of m and n can be is $m.n$, but $n < p$, by assumption, so $m.n < m.p$. Furthermore, only prime numbers maximize their lcm's in this way, so in this respect primes are uniquely optimal.

---

[8]Presuming that period length is a heritable trait, which is a presupposition of both candidate explanations.

[9]For proofs of these lemmas, see Landau (1958).

[10]Note that we are, by assumption, restricting attention to *periodical* predators, i.e., predators that have life-cycles that are greater than 1 year. Prime periods remain optimal even if annual predators are included. However, they are no better (or worse) than non-prime periods with respect to annual predators, since the lcm is n in both cases.

The hybridization explanation proceeds along very similar lines. The main difference is that, instead of a cicada species and a periodical predator, we assume that m and n are the life-cycle periods (in years) of two subspecies of cicada, C_m and C_n. A second difference, which does not affect the number-theoretic framework but may effect the way in which period lengths coevolve, is that there is a mutual benefit in the hybridization case for the two subspecies to have periods that are coprime. This makes it more likely that coprime periods may evolve which are not both prime. For example, if there are just two cicada subspecies in a given area, and they evolve to periods of 9 and 10 years respectively, then this is a local optimum. Nonetheless, primes remain the most robust solution, hence the optimization explanation still goes through.

The basic structure common to the predation and hybridization explanations is as follows:

1. Having a life-cycle period which minimizes intersection with other (nearby/lower) periods is evolutionary advantageous.   [**biological 'law'**]

2. Prime periods minimize intersection (compared to non-prime periods).
                                                        [**number theoretic theorem**]

3. Hence organisms with periodic life-cycles are likely to evolve periods that are prime.   [**'mixed' biological/mathematical law**]
   When the law expressed in (3) is combined with

4. Cicadas in ecosystem-type, E, are limited by biological constraints to periods from 14 to 18 years.[11]   [**ecological constraint**]
   it yields the specific prediction

5. Hence cicadas in ecosystem-type, E, are likely to evolve 17-year periods.

This explanation makes use of specific ecological facts, general biological laws, and number theoretic results. My claim is that the mathematical component, (2), is both essential to the overall explanation and genuinely explanatory in its own right. In particular, it explains *why* prime periods are evolutionarily advantageous in this case.

Following its initial presentation in Baker (2005), there have been several responses from philosophers with nominalist sympathies attacking various aspects of this example. Some have tried to show that the mathematical component is in fact dispensable.[12] Others have argued that there is a problematic circularity involved, since the fact to be explained – namely the primeness of the periods – itself involves mathematics.[13] Others have accepted the indispensability of the mathematics for the overall explanation, but claimed that the mathematical component is not itself

---

[11]Clearly a parallel constraint may be formulated for 13-year cicadas, in which the ecosystem limits potential periods to the range from 12 to 15 years.

[12]Saatsi (2007), Daly and Langford (2009), Rizza (2011).

[13]Bangu (2008). I examine this objection in more detail in Sect. 6, below.

explanatory.[14] Responses to these various objections can be found in Baker (2009), and the debate here is ongoing.

An important point to keep in mind is the way in which the above explanation works at two different levels of abstraction. The same mathematical framework can be used to explain both the general fact that periodic organisms are likely to evolve periods that are prime, and the specific fact that a given cicada sub-species has a 17-year period. One way of conceptualizing this is in terms of the interaction between 'top-down' mathematical constraints and 'bottom-up' physical and biological constraints. General mathematical considerations are enough to derive the conclusion that the cicada periods are prime. When combined with the biological constraints specific to cicada physiology and their environment, which gives a range of viable period lengths between 12 and 18, this entails that the only periods that are both mathematically optimal and biologically possible are 13 and 17. The potential of mathematics to allow different levels of abstraction – and thereby generalization – is a characteristic feature of mathematical explanation in biology, and of mathematical explanation in science more generally. There are of course other, non-mathematical means of obtaining abstract explanations in biology, whether based on functional descriptions of phenomena, or on general principles such as the principle of transcription. However, none of these alternative routes to abstraction has the complete topic-neutrality of mathematics, and hence cannot achieve the level of generality of a mathematical explanation. The mathematics of the cicada explanation applies, in principle, to *any* interacting periodical phenomena, both biological and non-biological.

3   Bees and Honeycombs

For the second example of mathematical explanation in biology, we move from number theory to geometry. Why do honeybees build the cells of their honeycombs in the shape of hexagons? Biologists have long hypothesized that the answer has to do with economizing on the amount of wax per unit area. Wax is energetically costly to produce, so it makes sense for bees to use as little as possible when building their combs. As it turns out, it can be proved that the hexagonal tiling of the plane into unit areas is optimal in terms of minimizing the perimeter of the individual cells. This explains why honeybees build hexagonal cells.[15] I set out one possible regimentation of the explanation below:

[14]Melia (2002), Yablo (2012).

[15]The first discussion of this example in the philosophical literature, as far as I am aware, is in Lyon and Colyvan (2008), although their remarks on it are relatively brief.

6. Building cells which minimize perimeter per unit area is evolutionarily advantageous (under constraints b_i)[16]
7. Regular hexagons minimize perimeter per unit area (among all tilings of the plane) [geometrical theorem]
8. Hence cell-building organisms are likely to evolve building techniques which produce hexagonal cells
9. Honeybees are limited by biological constraints b_i
10. Hence honeybees are likely to produce hexagonal cells

Biologists generally take this to be the best explanation of why honeybees build their cells in the shape of hexagons, and it clearly makes nontrivial use of mathematics.

One issue that has been the subject of recent debate among philosophers of applied mathematics concerns the link – if any – between mathematical explanation in science (including biology) and the 'internal' explanatoriness or otherwise of mathematical proofs. Mathematicians do seem to make distinctions between the relative explanatoriness of proofs, including between different proofs of the same result. The implication is that there may be proofs which are rigorous and which demonstrate *that* a result holds but which do not, in some important sense, show *why* it holds. However, there is little consensus, either among mathematicians or philosophers, concerning criteria for what makes a proof explanatory. Among the putative features cited are simplicity, non-redundancy, purity, and non-disjunctiveness. In an old paper, Mark Steiner asserts that a genuine mathematical explanation in science (MES) must feature an explanatory proof of whatever mathematical theorem lies at its core.[17] I think that Steiner is wrong about this, and that the honeycomb case study provides a particularly clear counterexample.

The issue turns on whether the core mathematical result has an explanatory proof. In the honeycomb example, we have the following theorem:

(Honeycomb Theorem) Any partition of the plane into regions of equal area has perimeter at least that of the regular hexagonal honeycomb tiling.

Despite the fact that the general conjecture dates back to antiquity, the above theorem was only proved in full generality by Thomas Hales in 2001. I shall begin by sketching some of the key ideas involved in Hales's proof.[18] As already mentioned, the basic problem is one of optimization, namely to minimize perimeter for a collection of cells of unit area. One of the major challenges in proving the Honeycomb Theorem is that it is not true locally. In other words, if the challenge is to enclose a single unit of area with the minimum perimeter then the optimal shape

---

[16]The constraints here might include conditions such as it being energetically costly to produce the material to build the walls of the cells, that the cells be contiguous, and that the cells be of uniform area.

[17]Steiner (1978).

[18]Hales (2001). The full proof runs to 18 pages, so this will be of necessity no more than a brief overview.

is not a hexagon but a circle.[19] Once we shift to consider multiple cells, however, it becomes clear that using circles is disadvantageous because they cannot be fitted together without leaving gaps between the cells. When we are dealing with regular polygons, therefore, local performance (with respect to area to perimeter ratio) can be improved by increasing the number of sides, so that the polygon more closely approximates a circle. However, if we want the polygons to fit together without gaps, then it is a direct consequence of Euler's formula ($v-e + f = 2$) that the average number of sides cannot be more than 6.[20] Hence any polygon with more than six sides must be counterbalanced by some other polygon in the tiling that has fewer than six sides. Hales's approach is to encapsulate the above insight by introducing into the key optimization equation a penalty term that quantifies the global 'cost' of a polygon having more than six sides.

A second way of locally enlarging the area enclosed is by using shapes with curved sides rather than straight sides. Thus, for a given polygon, replacing a given straight side with a convex curved side (i.e., one that bulges out) increases its area to perimeter ratio. However this means that one of its neighbouring polygons must have a corresponding side that is concave, thus reducing its own area to perimeter ratio. Hales therefore adds a second penalty term into the governing equation that represents the global cost of a polygon having curved sides. The proof then proceeds by verifying that the penalty terms in the optimization equation correctly characterize the effects of changing the shapes of the constituent polygons in various ways, and then by deriving that the regular hexagon is optimal with respect to these penalty terms.

What I want to argue is that Hales's proof does not provide an *explanation* of the truth of the theorem. There are at least four reasons for thinking that Hales's proof, ingenious though it is, does not explain *why* the hexagonal tiling of the plane is optimal. I shall present and discuss these reasons in increasing order of philosophical abstraction.

The first reason (purely circumstantial, but important nonetheless) is that mathematicians working in this area appear not to find Hales's proof especially explanatory. This is manifested both in comments made about the proof, and also in attempts to 'improve' various aspects of it. Some of these improvements have been in minor technical details (for example Frank Morgan has recently shown that one of the lower bounds in Hales's Chordal Isoperimetric Inequality can be raised from $\pi/8$ to $\pi/4$). But there have also been attempts, unsuccessful thus far, to find "a simpler, more geometric version of [Hales's] proof."[21]

---

[19]This in known in the mathematics literature as the *isoperimetric problem*.

[20]Since every vertex in a finite graph corresponds to at least three half edges, $e \geq (3/2)\, v$, so (by substitution into Euler's formula), $(2/3)\, e - e + f = 2$. Hence $f = (1/3)\, e + 2$, from which it follows that $e < 3f$. Since an edge borders two faces, the average number of edges per face cannot be greater than 6.

[21]Carroll et al. (2006, p. 1). Note the implication that both simplicity and purity will tend to enhance the explanatoriness of a proof.

Secondly, while there seem to be good motivating reasons for including penalty terms of the sort described in the previous section, these reasons do not show why these penalty terms take the particular values that they do. Consider the full statement of Hales's crucial *Hexagonal Isoperimetric Inequality*:

> Consider a curvilinear planar polygon of N edges, area A, and perimeter P. Let $P*$ denote the perimeter of a regular hexagon of area 1. For each edge, i, let a_i denote how much more area is enclosed than by a straight line. Then
>
> $$P/P* \geq \min\{A,\ 1\} - .5\sum a_i - .0505/24\sqrt{12\,(N-6)},$$
>
> with equality only for the regular hexagon of unit area (Morgan 2000, pp. 161–2)

Even ignoring the technicalities of the background definitions, it is possible to pick out from the above equation the penalty terms for a polygon having curved sides (as a function of a_i) and for having extra edges (as a function of N). Focusing on the second of these, the specific value of the penalty term is $.0505/24\sqrt{12\,(N-6)}$. The presence of the $(N-6)$ term can be explained by reference to the earlier remarks concerning Euler's formula: we know from this that the average number of sides of polygons tiling the plane cannot be greater than 6, hence it is only at this point that the overall penalty term becomes positive. But what about the coefficient $.0505/24\sqrt{12}$ on the front of the penalty term? Nothing in the general argument motivating Hales's approach gives any guidance about the specific value of this coefficient.[22]

A third reason for questioning the explanatoriness of Hales's proof is foreshadowed in the remark quoted earlier about searching for a 'simpler, more geometric version' of the proof. The fact is that the bulk of Hales's proof of the Honeycomb Theorem involves not geometry but other quite distinct areas of mathematics such as measure theory and numerical analysis. Implicit in the quoted remark is the assumption that a more geometric proof would also be more explanatory. The point behind this is presumably not that there is anything explanatorily distinctive about geometry per se, but rather that the theorem in question is geometrical and hence it makes sense to favour geometrical reasoning in its proof. The presumption is that 'purity' of this sort will, other things being equal, make for a more explanatory proof.

While I am hesitant to emphasize impurity as a reason in its own right for questioning the explanatoriness of Hales's proof, the above discussion does pave the way to a fourth and final reason which has considerably more force. In a nutshell, the problem with Hales's proof lies not with its impurity but rather with the specific non-geometrical apparatus that it utilizes. Recall that one of the two approaches mentioned as featuring in the proof is numerical analysis. The overall goal of this subfield of mathematics is the design and evaluation of techniques to give

---

[22]To be clear, Hales's proof is fully rigorous. In other words, the 'mysterious' coefficient works to establish the theorem that regular hexagons are optimal. What is not clear is *why* this coefficient works.

approximate but accurate solutions to 'hard' problems in continuous mathematics, in other words problems for which no closed-form solution is available.[23] In his proof of the Honeycomb Theorem, Hales uses numerical analysis to verify – through brute computation – some of the approximations he makes concerning upper and lower bounds. Indeed the central part of the proof involves various subdivisions into arbitrary-looking special cases. As Carroll puts it, Hales's proof "becomes a long, arduous case analysis using five separate intermediate lower bounds."[24] The two key features here are disjunctiveness and the role of computations. These are both features that philosophers have argued tend to weaken the explanatory power of mathematical proofs.[25] It also seems to fit with our intuitions. Typically, if a conjecture is broken down into a large number of separate subcases, each of which is then verified by a distinct computation, this in itself does not give a sense of *why* the conjecture is true. Instead it has more of the feeling of a 'brute force' verification, as when we check the truth of some claim about a finite domain by going through each case one by one.

I conclude that the Hale's proof does not explain why the Honeycomb Theorem holds, although it certainly establishes that it holds. It is also clear that the Honeycomb Theorem is an essential part of our best explanation of why bees build hexagonal cells. This is enough to show that Steiner's thesis is mistaken. *Contra* Steiner, it is not necessary for a mathematical result to have an explanatory proof in order for it to feature in a mathematical explanation of some biological (or other physical) phenomenon.

4 Aspects of Generality

As has already been mentioned, a key function of mathematics in the context of explanation is to facilitate generalization, by piggybacking on the inherent capacity of mathematics for abstraction. Thus the number-theoretic apparatus in the cicada example can potentially be applied to any periodical organism, and is not restricted to cicadas, or even just to insects. Similarly, the results in numerical analysis that underpin the honeycomb example can also be applied to any physical situation that involves perimeter minimization.

Philosophical discussions of scientific explanation often emphasize the connection between explanatoriness and generality, and this holds equally for philosophical discussions of explanation in mathematics. The basic idea – which gets cashed out differently in different analyses – is that an important way of explaining a given

---

[23]A distinctive feature of numerical analysis is the use of algorithms, and other methods of numerical approximation. This is in contrast to the symbol manipulation characteristic of purely analytic approaches. Hence there is a greater likelihood of numerical analysis producing methods and results that 'work', in some specified domain, but are such that it is not clear *why* they work.

[24]*ibid.*, p. 7.

[25]See e.g., Baker (2008).

phenomenon is to show how it is deducible from, or caused by, or is a special case of some more general phenomenon, law, or pattern. If there is something to this idea, then it would seem to suggest that the generality that contributes to the explanatoriness of a mathematical result in a MES is something that should also make the proof of the result more explanatory in purely mathematical terms.

To see why the above argument is too quick, we need to consider how generality plays out in some actual examples. So let us return to our two favourite case studies from biology. In the cicada case, the explanation is general in the sense of showing why any periodical organism with periodical predators is likely to evolve a life-cycle period that is prime. In the honeycomb case, the explanatory argument generalizes to cover any situation in which it is evolutionarily advantageous to enclose large numbers of equal areas using a minimum of materials. So these explanations are indeed general, in virtue of potentially applying to a wide range of organisms under a wide range of ecological conditions. However, the generality in question is restricted in various ways – to organisms acted upon by natural selection, given the actual laws of chemistry and physics – and thus, I shall argue, falls short of the kind of generality that is explanatorily relevant in mathematics.

Consider once again the honeycomb case. Faced with the core result of the Honeycomb Theorem, mathematicians are interested in whether – and, if so, how – Hales's proof generalizes to scenarios in which various key assumptions are altered or eliminated altogether. For example, are hexagons optimal for tiling other kinds of surface such as the sphere, or the torus, or the Möbius strip? What if we allow a mixture of two different sizes of cells? What if we assume that the walls of the cells have non-negligible width? And so on. None of these various questions are directly answered by Hales's proof, nor by any simple transformation of it, which is another important reason why mathematicians consider it to be relatively unexplanatory.

Notice, however, that there is no contradiction – nor even any real tension – in the honeycomb example featuring a biologically general result with a mathematically non-general proof. The point is that mathematicians are typically interested in a level of generality that is significantly greater than what is relevant to scientific applications. Hence lacking this degree of generality is no handicap to a given MES being a good scientific explanation, even though it may rule it out as being straightforwardly transformable into a good pure mathematical explanation.[26]

The distinction between mathematical and biological generality can also be couched in modal terms. A familiar distinction is between different strengths of possibility (or necessity). A weak form of possibility is logical possibility: any consistent state of affairs is logically possible. Whether mathematical possibility is a distinct kind of possibility is open to debate. In any case, a stronger form of possibility is physical possibility: roughly, a state of affairs is physically possible if it is consistent with the (actual) physical laws. Can sense be made of other more

---

[26]Note that nothing I have said here denies the importance of mathematical explanatoriness to scientific explanatoriness. I am simply arguing that the latter is not a sufficient condition for the former.

specific forms of possibility such as chemical possibility, biological possibility, and so on? Intuitively it does seem as if certain physically possible outcomes are not 'biologically possible'. For example, it seems physically possible for there to be a fish that dissolves when placed in water. Yet it is hard to see how such a creature could have actually evolved.

Reflection on the modal aspects of explanation may also be helpful in classifying different kinds of mathematical explanation in science. To date, philosophical discussion of MES has focused on specific case studies, and little attempt has been made to construct any broader classification scheme. As an initial step towards this goal, I suggest that the examples of MES that appear in the literature can be usefully grouped under three general headings: Constraint MES, Equilibrium MES, and Optimization MES.

A *Constraint MES* explains why some physical outcome is impossible by showing that (in some sense) it is mathematically impossible. A classic example of Constraint MES can be found in the Bridges of Königsberg problem, which Euler answered in the early eighteenth century. The question was why no-one could cross the seven bridges of Königsberg without crossing some bridge more than once. Euler's answer was couched in terms of graph theory, and showed that the corresponding pure mathematical graph could not be traversed by traveling along each edge exactly once. The modal link in such cases is very clear: mathematical impossibility entails physical impossibility.[27] There are also further entailments from physical impossibility to other forms of impossibility that are especially relevant to biological phenomena: for example, chemical impossibility, physiological impossibility, and (perhaps) evolutionary impossibility.

An *Equilibrium MES* explains why some physical outcome occurs by showing that it is mathematically inevitable (or almost inevitable) across a wide range of starting conditions. For example, the eventual resting point of a marble that is rolled around the inside of a bowl is insensitive to the initial position, angle, and velocity when it is released. A more sophisticated example of an Equilibrium MES is discussed by Colyvan, and concerns the presence of gaps in the asteroid belt between Mars and Jupiter.[28] Such explanations are often statistical in nature, hence the modal link between mathematical model and physical situation is less straightforward. A well-known example of an Equilibrium MES in biology is Fisher's principle concerning sex ratios. The mathematical argument shows that a 1:1 ratio of males to females is an evolutionarily stable strategy, and that – under a wide range of conditions – any deviation from this ratio will provide an incentive for parents to produce more of the minority sex and thus push the ratio back to 1:1.[29]

An *Optimization MES* explains why some physical outcome occurs by showing that it is mathematically optimal, in some relevant sense. Both the cicada and honeycomb cases lie within this category of MES. The modal element of Optimiza-

---

[27]For further discussion and references, see Pincock (2012, 51–54).

[28]Colyvan (2010).

[29]See Gould (2002, pp. 648–9).

tion MES's is easy enough to see. If a certain outcome is mathematically optimal then it is (mathematically) impossible for some distinct, better outcome to occur. However, the mathematically optimal solution may not be physically possible, or biologically feasible, or it may simply not have been 'found' (evolutionarily speaking). Although the cicada case involves a mathematically optimal solution, since the life-cycle periods are prime, and perfect hexagons are mathematically optimal in the honeycomb case, there is a second, related problem related to honeycombs where such mathematical optimality is not present. For there is also a three-dimensional version of the honeycomb problem, in which the cells are thought of as hexagonal 'tubes' and we consider how they should be fitted back-to-back so as to maximize volume to surface area ratio. In 1964, László Tóth discovered that the trihedral pyramidal shape used by bees, which is composed of three rhombi, is not mathematically optimal.[30] A cell end made up of two hexagons and two smaller rhombi would in fact be .035 % more efficient. Presumably the extra complexity of this second solution makes it inferior from an evolutionary point of view. This shows that sometimes the most general mathematical solution to an optimization problem may be undermined by considerations that were inadvertently excluded in the assumptions and constraints used to define the mathematical problem.

In the next section I shall take up the question of why so many of the prominent examples of MES are to be found in biology. One crucial part of the answer turns on the way in which Optimization MES's find a natural home within the biological sciences.

5   Why Biology?

As has already been noted, the history of the debate over the merits of indispensability-style arguments for mathematical platonism has seen a sharp switch of focus away from the higher reaches of theoretical physics and towards other areas of science such as biology and meteorology. Why has this switch taken place? I think that several interrelated reasons can be identified.

Firstly, the fact that explanation is a key component of recent versions of the Indispensability Argument introduces a countervailing pressure on the complexity of examples of applied mathematics. If all that matters is indispensability, then it makes sense to choose examples in which the mathematics (and often the science, also) is as sophisticated as possible. Hence the early predominance of examples from theoretical physics, with complex geometries and esoteric algebraic structures in abundance. But if what we care about is not indispensability per se, but rather indispensability for explanation, then it is better to choose examples in which our intuitions concerning explanatoriness are relatively clear. Such examples tend to be less complex from a mathematical point of view. In general, the special sciences

---

[30]Tóth (1964).

are a good place to look for examples of indispensable mathematics playing an explanatory role. Typically, the mathematical component is complex enough not to be trivially dispensable, but not so complex as to muddy the issue concerning explanatoriness. In this respect, the special sciences have a 'Goldilocks' aspect: the mathematical apparatus that is employed is neither too complex nor too simple.

A second reason that philosophers have moved away from the most technical examples of applied mathematics in physics is that, because of their highly theoretical nature, the divide between the mathematical and physical components of such examples is often unclear. This is an important drawback in the context of debates over IA, since the ultimate target of IA is to defend platonism, which asserts the existence of abstract objects. If the distinction between abstract and concrete is controversial, or open to interpretation, as is arguably the case in contemporary quantum mechanics, then the force of the examples is thereby undermined. Note that this second reason has nothing directly to do with explanation. It is also worth noting that the issue of where mathematics stops and the physical sciences start can also be an issue in some of the putative examples of MES in the special sciences. Thus in the honeycomb example, the mathematical explanation concerns geometrical objects such as perfect hexagons. But are these geometrical objects abstract or concrete?

The two reasons given so far help explain the shift of attention, in debates concerning IA, towards examples from the special sciences. But there is a further question of why biology in particular is so well-represented in discussions of MES. Here I think the crucial feature is the role of optimization explanations in biology. Optimization MES is one of the three main categories of MES, and there is no doubt that this style of explanation is a distinctive feature of theorizing in biology, especially in evolutionary biology. The use of such explanations is not without controversy, and there is a growing philosophical literature looking both at the pros and cons of optimization explanations and on more fully analysing what, if anything, makes such explanations distinctive.[31] Since my emphasis here is on the import of such examples on debates in the philosophy of mathematics, I will not spend more time here on this particular question. It is worth noting, however, that optimization explanations are not the exclusive preserve of evolutionary biology. Engineering, for example, often addresses itself to problems that involve optimization. In some cases such problems are analogous to ones that are found in evolutionary biology. The honeycomb example, at its core, concerns the problem of achieving a goal using the minimum amount of material, and as such is an optimization problem of the sort that is routinely faced in various human-devised manufacturing processes.[32]

---

[31]See e.g., Potochnik (2007). There are also important references in the philosophy of biology literature, for example Orzack and Sober (2001).

[32]More controversially, some physicists have argued for optimization explanations in cosmology, in which the values of the basic physical constants are 'explained' in terms of producing universes that are more conducive to the formation of black holes, which in turn (according to certain theories) spawn further universes as 'offspring.' For more on such explanations, see Smolin (1999).

6 Salient Mathematical Properties

The final question I shall address in connection with mathematical explanation
in biology concerns the choice of mathematical properties that feature in such
explanations. Especially if MES is supposed to carry ontological burdens, as is
the case in recent versions of the Indispensability Argument, it is important that
the mathematical properties involved are legitimate targets of scientific explanation.
Otherwise we run the risk of taking on ontological commitments to mathematical
entities that may in fact be avoidable. The problem arises because there are a
proliferation of mathematical properties that may happen to apply in a given
physical situation. Cicadas have periods of 13 or 17 years. These numbers are prime,
but they also share other mathematical properties, such as being odd, and being the
sum of two perfect squares ($13 = 4 + 9; 17 = 1 + 16$). Does this mean that it is
legitimate to ask *why* cicadas have periods that are the sum of two perfect squares?

A more general version of this worry has been articulated by Sorin Bangu as an
objection to explanatory versions of IA.[33] Take the cicada example. Bangu points
out that the question, "Why is the duration (in years) of the life-cycle of periodical
cicadas a prime number?", itself makes reference to mathematical objects. He
argues that the indispensability of mathematics (and in particular, number theory)
to explaining this fact does nothing to establish the platonist position because it
begs the question against the nominalist. The nominalist can simply reject the
explanandum, arguing that it is not true that the duration of cicada life cycles is
a prime number, because numbers do not exist! I have argued elsewhere that there
are various moves that the platonist can make in response to Bangu here, and hence
that the circularity objection per se is not fatal to the explanatory Indispensability
Argument.

The more specific worry I want to examine here concerns how to distinguish
genuine questions concerning mathematical properties of biological phenomena
from spurious questions. For example, "Why do cicadas have prime periods?",
seems like a genuine question., whereas "Why do human beings have a prime
number of legs?" does not. A simple way to draw the distinction would be just to
appeal to scientific practice. A question is a genuine scientific question if it is taken
seriously by a significant number of experts in the relevant scientific field. If enough
biologists take the question of the primeness of cicada periods seriously, then *ipso
facto* it is a genuine scientific question.[34]

But can more be said about the distinction, beyond appeal to scientific practice?
I think that it can, and I shall sketch an account that aims to draw the distinction in
a principled way. The question I am concerned with is the following: what makes

---

[33]Bangu (2008).

[34]This approach is explored further in Baker (2012), where the concept of *science-driven
mathematical explanation* is introduced.

a mathematical property, M, that applies to a physical phenomenon, P, a legitimate target of scientific explanation? For ease of exposition, I shall introduce the term *salience*, and say that a mathematical property is salient (in a given physical context) if it is a legitimate target of scientific explanation in that context. My claim is as follows:

> Salience Thesis   A mathematical property, M, is not salient with respect to a physical phenomenon, P, if there is some other mathematical property, M^*, such that the best explanation of why M applies to P also explains why M^* applies to P, but not vice-versa.

The above Thesis is easier to grasp in the context of an actual example. As mentioned above, the durations of cicada life-cycles have various mathematical properties in addition to primeness. Let M be the property of being the sum of two perfect squares. According to the Salience Thesis (ST), this is not a salient property in the cicada example. Why not? Because there is another mathematical property, primeness, that also applies. Moreover, the best explanation of why cicada periods are prime (as presented in Sect. 2), shows why the periods are prime and why they lie in the range $11 < n < 19$. These two facts jointly entail that the periods must be the sum of two perfect squares (since the only primes in this range are 13 and 17) and hence explain it. Conversely, there seems to be no explanation of this latter fact that does not go via the property of primeness. Perhaps there is some such explanation to be found. But it is hard to conceive of how being the sum of two squares could make a difference to the evolutionary fitness of a given period length.

A similar argument applies to the question of why human beings have a prime number of legs. Intuitively, the only reason we have a prime number of legs is that we have 2 legs, and 2 is prime. The salient property here is 2-ness. The property of primeness is not doing any independent explanatory work in this particular context. Or, for a slightly more biologically sophisticated example, consider the fact that *C. Elegans*, a favourite 'model organism' in biology, has 1,031 cells, and that 1,031 is a prime number. Why does *C. Elegans* have a prime number of cells? As with the human legs example, if there is an explanation of this fact that it will presumably proceed via an explanation of why *C. Elegans* has 1,031 cells in particular.

It should also be noted that the salience or non-salience of a given mathematical property in a given biological context may itself be an open scientific question, and a question that may end up being answered one way or the other either by further empirical research or by the development of a convincing explanation. Arguably this was initially the case in the cicada example. Until the predation and hybridization explanations were formulated, there were many biologists who assumed that primeness in this context was not a salient property, but merely a coincidental by-product of the fact that the life-cycles had periods 13 and 17. Being the target of scientific explanation is thus neither necessary nor sufficient for salience. In this sense, salience has a normative dimension and does not simply reduce to being coextensive with the web of mathematical properties that feature in actual scientific practice.

7 Conclusion

There are various interesting potential directions for further philosophical work on mathematical explanation in biology. Firstly, there are other areas of mathematics that may play an explanatory role in biology, such as statistics and topology, but which have not yet been discussed in any detail in the context of debates concerning indispensability. Secondly, there are interesting questions concerning the role of laws in biology – whether mathematical or otherwise – and how this impacts on different models of explanation, in particular Hempel's deductive-nomological account of explanation. Thirdly, some recent and scientifically controversial examples of explanation in biology have crucial mathematical components. One example is the formulation of scaling laws, such as laws relating the total length of an organism's vascular system to its body mass, and attempts to explain such laws by appeal to results from fractal geometry.[35] Finally, there is definitely further work to be done in the classification of different types of MES within biology, both to augment the tripartite scheme suggested in Sect. 4, and to get clearer about the philosophically important differences between the different types. There is little doubt, therefore, that biology will continue to be centre stage in debates concerning the explanatory role of mathematics in science.

References

Baker, A. (2005). Are there genuine mathematical explanations of physical phenomena? *Mind, 114*, 223–238.

Baker, A. (2008). Experimental mathematics. *Erkenntnis, 68*, 331–344.

Baker, A. (2009). Mathematical explanation in science. *British Journal for the Philosophy of Science, 60*, 611–633.

Baker, A. (2012). Science-driven mathematical explanation. *Mind, 121*, 243–267.

Bangu, S. (2008). Inference to the best explanation and mathematical realism. *Synthese, 160*, 13–20.

Breidenmoser, T., & Wolkenhauer, O. (2015). Explanation and organizing principles in systems biology. In P.-A. Braillard & C. Malaterre (Eds.), *Explanation in biology. An enquiry into the diversity of explanatory patterns in the life sciences* (pp. 249–264). Dordrecht: Springer.

Carroll, C., et al. (2006). *On generalizing the honeycomb theorem to compact hyperbolic manifolds and the sphere* (SMALL Geometry Group report). Williamstown: Williams College.

Colyvan, M. (2003). *The indispensability of mathematics*. New York: Oxford University Press.

Colyvan, M. (2010). There is no easy road to nominalism. *Mind, 119*, 285–306.

Cox, R., & Carlton, C. (1988). Paleoclimatic influences in the evolution of periodical cicadas. *American Midland Naturalist, 120*, 183–193.

Cox, R., & Carlton, C. (1998). A commentary on prime numbers and life cycles of periodical cicadas. *American Naturalist, 152*(1), 162–164.

Daly, C., & Langford, S. (2009). Mathematical explanation and indispensability arguments. *Philosophical Quarterly, 59*, 641–658.

---

[35] West and Brown (2004).

Field, H. (1980). *Science without numbers*. Princeton: Princeton University Press.

Goles, E., Schulz, O., & Markus, M. (2001). Prime number selection of cycles in a predator-prey model. *Complexity, 6*, 33–38.

Gould, S. (2002). *The structure of evolutionary theory*. Cambridge, MA: Belknap Press.

Hales, T. (2001). The honeycomb conjecture. *Discrete and Computational Geometry, 25*, 1–22.

Issad, T., & Malaterre, C. (2015). Are dynamic mechanistic explanations still mechanistic? In P.-A. Braillard & C. Malaterre (Eds.), *Explanation in biology. An enquiry into the diversity of explanatory patterns in the life sciences* (pp. 265–292). Dordrecht: Springer.

Landau, E. (1958). *Elementary number theory*. New York: Chelsea Publishing Company.

Lyon, A., & Colyvan, M. (2008). The explanatory power of phase spaces. *Philosophia Mathematica, 16*, 227–243.

Melia, J. (2000). Weaseling away the indispensability argument. *Mind, 109*, 458–479.

Morgan, F. (2000). *Geometric measure theory: a beginner's guide*. Academic Press.

Melia, J. (2002). Response to Colyvan. *Mind, 111*, 75–79.

Orzack, S., & Sober, E. (2001). *Adaptationism and optimality*. Cambridge: Cambridge University Press.

Pincock, C. (2012). *Mathematics and scientific representation*. Oxford: Oxford University Press.

Potochnik, A. (2007). Optimality modeling and explanatory generality. *Philosophy of Science, 74*, 680–691.

Rizza, D. (2011). Magicicada, mathematical explanation and mathematical realism. *Erkenntnis, 74*, 101–114.

Saatsi, J. (2007). Living in harmony: Nominalism and the explanationist argument for realism. *International Studies in the Philosophy of Science, 21*, 19–33.

Smolin, L. (1999). *The life of the cosmos*. Oxford: Oxford University Press.

Steiner, M. (1978). Mathematics, explanation, and scientific knowledge. *Noûs, 12*, 17–28.

Tóth, L. (1964). What the bees know and what they do not know. *Bulletin of the American Mathematical Society, 70*, 468–481.

West, G., & Brown, J. (2004, September). Life's universal scaling laws. *Physics Today, 57*, 36–42.

Yablo, S. (2012). Explanation, extrapolation, and existence. *Mind, 121*, 1007–1029.

Yoshimura, J. (1997). The evolutionary origins of periodical cicadas during ice ages. *American Naturalist, 149*(1), 112–124.

Chapter 11
Explanation and Organizing Principles in Systems Biology

Tobias Breidenmoser and Olaf Wolkenhauer

Abstract While explanation in biology is a well-established topic in philosophy of science, since the rise of new mechanistic approaches little has been said about explanation in Systems Biology. In this contribution, we analyze whether contemporary conceptualizations of explanation fit with the scientific practice of systems approaches in molecular and cell biology. We discuss how current views on mechanistic explanation can be applied to the use of mathematical models in system biology. To this end it is important to distinguish different kinds of models: While some of them merely save the phenomena, not being explanatory, others make mechanistic claims and can be embedded in the framework of mechanistic explanations. We will make a conceptual distinction and discuss the roles different kinds of models have in Systems Biology. In our view, the current mechanistic framework is not sufficient to capture all types of explanations occurring in Systems Biology. A multicellular system is not merely defined by its structural components, including genes, proteins and their interconnections, but has a functional organization, that is the system's behavior, which emerges from its structural organization. Such emergent properties cannot be understood or explained by merely describing the mechanisms of the underlying molecular and cellular processes. We argue that organizing principles in Systems Biology provides a way of explaining such high-level system properties. With the help of a case study we show that organizing principles explain system-properties complementary to mechanistic explanations. They cannot replace them but they also cannot be reduced to them.

Keywords Systems biology • Models in science • Mechanistic models • Mechanistic explanation • Organizing principles • Robustness

T. Breidenmoser (✉)
Systems Biology and Bioinformatics Group, University of Rostock, 18051 Rostock, Germany
e-mail: breidenmoser@googlemail.com

O. Wolkenhauer
Department of Systems Biology & Bioinformatics, University of Rostock, 18051 Rostock, Germany

Stellenbosch Institute for Advanced Study, Wallenberg Research Centre at Stellenbosch University, Stellenbosch, South Africa
e-mail: olaf.wolkenhauer@uni-rostock.de; www.sbi.uni-rostock.de

© Springer Science+Business Media Dordrecht 2015
P.-A. Braillard, C. Malaterre, *Explanation in Biology*, History, Philosophy and Theory of the Life Sciences 11, DOI 10.1007/978-94-017-9822-8_11

1   Introduction

Systems Biology is a relatively young approach within biology, focussing of the use of mathematical modelling and computer simulations. In contemporary philosophy of science, explanation in biology is usually described by mechanistic approaches. The aim of this paper is to analyse whether these conceptions are applicable to the kinds of models used in Systems Biology. We will first give an overview of the different kinds of models in Systems Biology, followed by a review of mechanistic approaches in philosophy of science. We distinguish non-explanatory phenomenological models from explanatory mechanistic models. Models in Systems Biology cover a wide range of model types, from simple curve fitting, to mechanistic models and abstract representations. It can be argued that the mechanistic approach is limited in explaining system properties, as we will show with the example of robustness. Organizing principles, on the other hand, are abstract conceptualizations which provide a basic framework to understand the behaviour of systems, on an abstract and general level similar to laws in physics. Organizing principles provide a different kind of explanation in biology, with some parallels to unifying accounts of explanation, and we will emphasise the epistemic need of such organizing principles.

2   Model Building in Systems Biology

Systems Biology combines mathematical modeling with experiments to understand the behavior of cellular systems. A cellular system is defined by two aspects. First its structural/material organization, i.e. the molecules, organelles, cells and compartments that provide the biophysical environment in which genes metabolites and proteins can biochemically interact. Second its functional organization, i.e. the system's (nonlinear dynamical) behavior emerging from the interactions of molecules and cells.

We consider Systems Biology an approach, aiming at understanding the spatio-temporal interactions among components of a cell, among cells, and their interaction with the environment (Wolkenhauer et al. 2009). Systems Biology is not a complementary sub-domain in biology next to molecular and cell biology, genetics and others, but another method, stance or paradigm to do biology. In our view, Systems Biology is therefore not a discipline as such, and hence there is also not really something like a "systems biologist". Systems Biology is an interdisciplinary approach to answer biological questions, requiring a range of expertise for the design of experiments, the analysis of data, the construction of mathematical models, realizing computer simulations and developing theoretical concepts and computational tools. The common ground is, or should however be, a shared biological research problem or question that is answered using a range of technologies and methodologies.

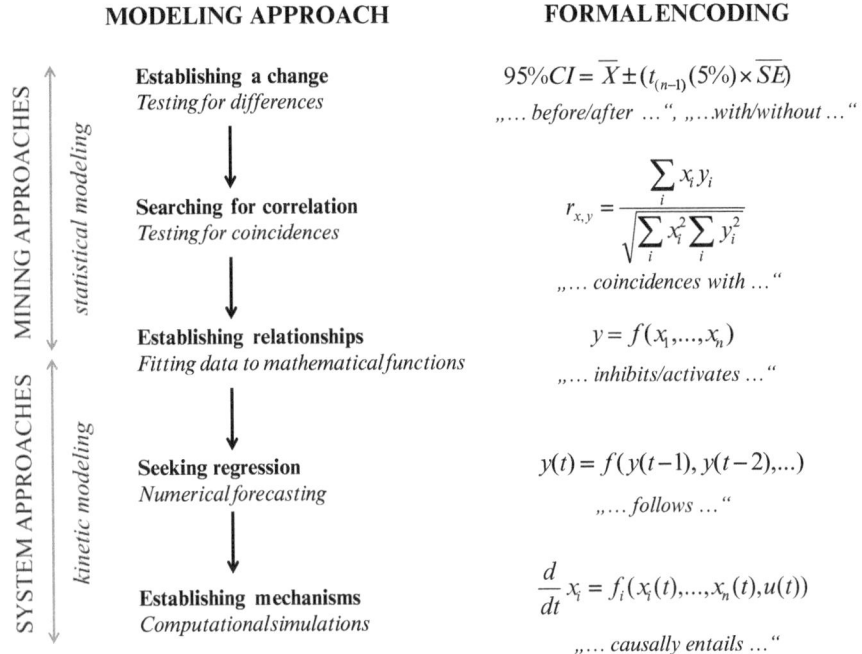

Fig. 11.1 Model building approaches and their formal encodings

Figure 11.1 gives a (necessarily incomplete) overview of widely used models in Systems Biology. The most common question asked in the context of experiments is one that asks for differences to establish whether or not different scenarios or changes imply different experimental observations, or not. Typical examples are experiments before and after a perturbation (e.g. knock-out, mutation, RNA interference) or the comparison of wild-type and mutant. The design of experiments is simple and the data are easily generated. To support or validate a claim for a change/difference, significance testing, based on a simple statistical model, is frequently employed. The mathematical model is here the assumption of certain statistical distributions.

The next level of sophistication in data generation is to measure levels and, for two or more variables, test for coincidence in levels (high/low). Statistical correlation analysis is the tool of choice here. A frequent and often discussed error is to infer causal relationships on the basis of correlations. Correlation analysis is thus limited to explain coincidences.

Regression analysis allows us to establish relationships between variables, encoded by mathematical expressions, by fitting these through data points. Here the intention is frequently to associate, with observations, purposeful relationships or the "function" of biological variables (e.g. efforts to establish whether a component is activated, inhibited etc.). Regression models can, in principle, also be the first step towards modeling dynamical phenomena. However, autoregressive modeling for time series analysis and numerical forecasting does not play any important role in Systems Biology. The reason is that these models only provide predictions about numerical values of a variable, without providing, or supporting an explanation about the mechanism by which the observations are generated.

The most frequently used approach for mathematical modeling in Systems Biology is based on differential equations. From the model types discussed here, differential equations have the greatest explanatory power in the sense that they encode a (hypothesized) mechanism by which observations are generated. Typical examples of such models are representations of biochemical reaction networks (e.g. gene regulatory networks, signal transduction and metabolic pathways) or spatio-temporal simulations in which typically biophysical and biochemical details of the processes are encoded by the rate equations.

As powerful as differential equation models are in deepening our understanding of mechanisms underlying our observations, the flipside is that their construction and validation puts the greatest demands on efforts to generate appropriate data in experiments. The parameter values (e.g. rate constants in biochemical reactions) can either be determined from biochemical binding experiments (characterizing interactions between individual molecules) or the values of parameters are estimated from time course data. In either case, when it comes to the validation of such models, matching simulations of the model with data, quantitative time series are required. This, for most practical cases will be time consuming, expensive and technically challenging (e.g. limited to only a few selected proteins and possibly not distinguishing (in)active states). True quantification, to generate absolute values and concentration measures is not only challenging for sufficiently rich time courses, multiple stimulation pattern and initial conditions, but also because in most cases it would be necessary to replicate the experiments to allow a quantification of the variability/uncertainty in the data. However, the nature and quality of data will also strongly influence the modeling approach employed. Mechanistic models, using representations from the theory of dynamical systems, are usually constructed to fit the particular experimental context.

Taken together, in the current practice of Systems Biology, the data available largely depend on the technologies and resources available. The choice of the experimental setup, the choice of the model organism, cell line and the choice of technologies used to generate data, usually define a narrow context for which an appropriate modeling approach has to be chosen.

3   Mechanistic Models and Mechanistic Explanations

Explanations in biology are frequently crafted within the framework of mechanisms. Definitions for mechanisms include the following:

> Mechanisms are entities and activities organized such that they are productive of regular changes from start or set-up to finish or termination conditions. (Machamer et al. 2000: 3)

> A mechanism for a behavior is a complex system that produces that behavior by the interaction of a number of parts, where the interactions between parts can be characterized by direct, invariant, change-relating generalizations. (Glennan 2002: 445)

> A mechanism is a structure performing a function in virtue of its component parts, component operations, and their organization. The orchestrated functioning of the mechanism is responsible for one or more phenomena. (Bechtel and Abrahamsen 2005: 423)

However, there are some divergences about these characterizations of a mechanism. Machamer, Darden and Craver describe mechanisms as sequential processes with start- and finish-conditions. In their view, a description of mechanisms contains a description of relevant entities and properties before the phenomenon is produced as well as "a privileged endpoint, such as rest, equilibrium, neutralization of charge, a repressed or activated state, elimination of something, or the production of a product" (Machamer et al. 2000: 11–12). However, not all mechanisms can be described as sequential processes. A very simple counter-example is the heart as a structure whose function is to pump blood (see Bechtel and Abrahamsen 2005: 424–425). Blood pumping is a cyclic event; it would be arbitrary to describe one phase of the mechanism as a starting or endpoint.

Bechtel subsequently calls the characterization of Machamer, Darden and Craver the *basic mechanistic account* (Bechtel 2011: 535). It remains qualitatively and fails to recognize quantitative organization in cells. The basic mechanistic account is insufficient because in Systems Biology, mechanisms are representations of networks, web-like structures including feedback and feedforward loops. This implies a dynamical organization of interactions, whose mathematical formulation involves concepts from the theory of (nonlinear) dynamical systems. Organelles, cells and tissues are autonomous systems and can only be described accurately by nonsequential or cyclic organization (ibid: 544). Referring to starting and finishing conditions, in the sense described above, should thus be avoided.

Mechanisms consist of parts (components) on a level lower than the mechanism itself. Components and the mechanism do not share all properties. The mechanism, as a whole, can have system properties such as robustness and self-organization that none of its components has. Thus, identification of components and of their mutual interactions plays a fundamental role in explaining the behavior of a mechanism.

Mechanisms are responsible for phenomena in the way that the interactions of components produce the phenomena. A mechanistic explanation is a description of the mechanism to understand how the phenomenon is produced by the mechanism. Phenomena are regular and stable events in the world (Bogen and Woodward 1988). In mechanistic explanations, phenomena have the role of the explanandum and mechanism the role of the explanans. Moreover, mechanisms are organized hierarchically: A component of a mechanism which explains the production of the phenomena is often a mechanism itself. Therefore, on a more fundamental level the function of the component itself might need an explanation.

Stuart Glennan defines mechanistic models as follows:

> A mechanical model consists of (i) a description of the mechanism's behavior (the behavioral description); and (ii) a description of the mechanism that accounts for that behavior (the mechanical description). (Glennan 2002: 446)

The behavioral description is the description of the phenomenon, in Systems Biology often realized with differential equations or Boolean networks. The mechanistic description goes further, it not only "saves the phenomena" mathematically but provides physical and biochemical claims, whether or not these claims can be experimentally justified.

Usually scientists cannot describe the underlying processes of a biological mechanism in reality completely (in contrast to designed mechanisms in engineering). Scientist starts with working hypotheses and mechanistic sketches. Machamer, Darden and Craver have introduced the concepts of *how-possibly*, *how-plausibly* and *how-actually* mechanistic models (Machamer et al. 2000: 21). They are described as follows:

> *How-possibly models* have explanatory purport, but they are only loosely constrained conjectures about the sort of mechanism that might suffice to produce the *explanandum phenomenon*. [...] *How-actually models*, in contrast, describe real components, activities, and organizational features of the mechanism that in fact produces the phenomenon. They show how a mechanism works, not merely how it might work. Between the extremes is a range of *how-plausibly models* that are more or less consistent with the known constraints on the components, their activities, and their organization. (Craver 2007: 112–113; see also Craver 2006)

How-actually models are rare in scientific practice. Often mechanistic models contain unobserved and undetected entities, black boxes and filler terms. Especially in Systems Biology, models frequently contain idealizations or counterfactual assumptions. Some models are not realized in nature even in principle, like models of three-sex organisms (Weisberg 2007: 223). Hence, we have mechanistic models on the one side, which involve approximations, idealizations, speculations and omissions; and mechanistic explanations on the other side, which describe biological mechanisms that are realized in a biological sense. We will now investigate whether mechanistic explanation can be made compatible with the modeling approach.

4 Phenomenological and Explanatory Models in Systems Biology

Not every model can provide an explanation. There are also mathematical models which merely summarize data ("save the phenomena"), especially regression models. We want to argue that models only have explanatory value if they are also accompanied with a mechanistic description.

To take an example, cyclic changes are frequently observed in biology (oscillations, rhythms). To answer the question to why there are oscillations is to ask for

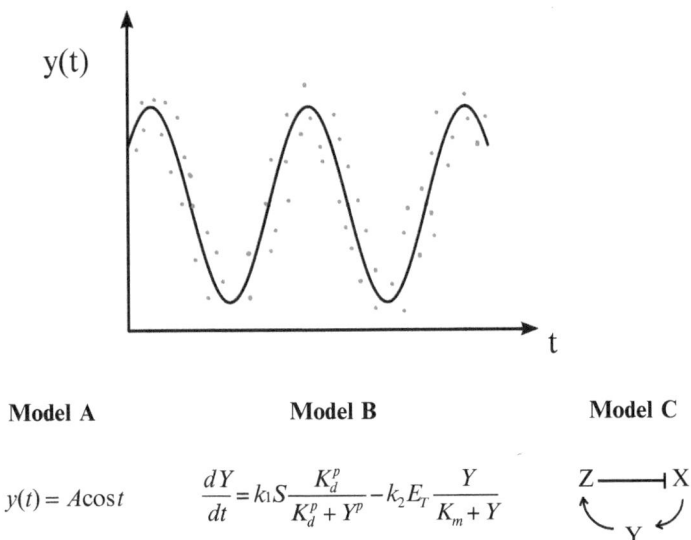

Model A   **Model B**   **Model C**

$$y(t) = A\cos t \qquad \frac{dY}{dt} = k_1 S \frac{K_d^p}{K_d^p + Y^p} - k_2 E_T \frac{Y}{K_m + Y}$$

Fig. 11.2 Different models of oscillation/cyclic changes (Model B and Model C are taken from Novác and Tyson 2008: 982 + 985)

how they are generated, which mechanisms generate them and to understand under which conditions they occur. In Fig. 11.2, Model A is nothing but curve fitting. It allows us to represent the data and also to make numerical predictions about further experimental outcomes, based on a history of data. In contrast, Model B is a kinetic model, here using differential equations, which provides the same simulation result but includes a biological interpretation. A formal analysis of the model or simulation experiments can be conducted to investigate the conditions under which oscillations arise and how the properties of oscillations depend on the structure of the model and values of parameters. Various systems-theoretic analyses for sensitivity, stability, robustness and bifurcations can be used to explain the occurrence and properties of oscillations (including frequency, amplitude and the consequences or stability in case of perturbations). Model C provides yet another level of abstraction. While Model B could be a mathematical representation of the diagrammatic Model C, the diagram includes a negative feedback loop as a dynamic motif, characteristic or necessary for oscillations. In some cases the structure of the network encapsulates information about the possible behaviors, quite independent of the parameter values or even the specific modeling framework used to encode the diagram for simulations (see Novák and Tyson 2008).

Model A does little more than mimicking observations, contributing nothing towards an explanation about mechanisms and conditions that generate the experimentally observed phenomenon. It is a description of the mechanisms behavior but there is no evidence on why the oscillation occurs. Therefore it is not explanatory. In contrast, Model B is not merely an equation, it is embedded in a scientific paradigm of how to generate and interpret kinetic models. Model B has more variables than Model A and each variable can be interpreted in a way that it corresponds to a biological feature. Hence, it is not merely a mathematical description of the behavior of the oscillator but describes the mechanism underlying this behavior. For that reason, Model B becomes explanatory. Model C neglect the details of the mathematical description of the oscillation and focus on general circumstances in which harmonic oscillations occur in biology. Therefore, it is a more general explanation with a wide scope on how mechanisms are linked to oscillation.

As Peter Lipton states, there is a "distinction between understanding why a phenomenon occurs and merely knowing that it does", in the former case we have "simply more knowledge: knowledge of causes" (see Lipton 2004: 30). If we want to distinguish models that merely save the phenomena and explanatory models, we need to know which models can provide causal knowledge, which is a precondition for describing the underlying mechanism which is responsible for the phenomenon.

We will here call models that merely describe the mechanism's behavior, i.e. describe the phenomenon which is produced by the mechanism a *phenomenological model*. Model A in Fig. 11.2 is an example of a phenomenological model. Phenomenological models *fit the curves* or describe some regularities between variables (if the value of A changes, then the value of B changes and vice versa). They do not offer causal knowledge and merely provide insufficient answers to what-if-things-has-been-different-questions (Woodward 2003). Craver argues that phenomenological models have only limited explanatory power. "Because phenomenal models summarize the phenomenon to be explained, they typically allow one to answer some w-questions. But an explanation shows why the relations are as they are in the phenomenal model, and so reveals conditions under which those relations might change or fail to hold altogether" (Craver 2006: 358). Hence, phenomenological models fail to provide causal knowledge, which is required for mechanistic explanations.

Although in biology there are a few law-like generalizations (or invariant change-relating generalizations), they have a quite different role than explaining phenomena. Simple formulae or basic functional relationships, like Model A above, simply describe and subsume data. The equation of Model A merely fits the curve and subsumes the behavior of the phenomena in a simple, unifying way. This subsumption is not able to tell us anything of how the phenomenon occurs.

Cummins (2000) concludes that laws in biology, more specifically in neurobiology and molecular and cell biology, are explananda, not explanans. Law-like generalizations are thus necessary to describe the behavior of the mechanism, but not sufficient to provide background knowledge or answers to what-if-things-has-been-different-questions. In Systems Biology, first you have to find regularities to extract stable phenomena out of the noise of data. Often this is done by law-like generalizations. But this is just the first step in generating biological knowledge. The detected phenomena need to be embedded in the complete picture of the target systems, where interconnections between phenomena and functional behavior have to be discovered. Only if this goal is reached, understanding and interventions can be successful. Therefore, we agree with Craver that for an explanation in molecular- and cell biology, at least a mechanistic sketch, which interprets the regularities as causal relations and describes how they are produced, is necessary.

Summarizing, we will define phenomenological models in the following way:

> A phenomenological model is a mathematical description of a phenomenon with the help of experimental data. It describes what happens, extracting the phenomenon out of the noise of data and can also describe correlations. Phenomenological models are a necessary starting point to summarize the behavior of what has to be explained.

Likewise, we can define explanatory models:

> An explanatory model links the description of a phenomenon with a description of an underlying mechanism. The description of the mechanistic behavior is responsible for the explanatory power of the model. There can either be a description of a specific mechanism or a more abstract description of how a type of mechanism is responsible for a type of phenomena.

We will now analyze robustness as an example of a system property and investigate whether explanatory models are limited to explain properties like that.

5   The Systemic Property of Robustness

The concept of robustness in Systems Biology does not have a single exact definition, but the main idea is the following:

> Robustness means the persistence of a system's characteristic behavior under perturbations or conditions of uncertainty. (Stelling et al. 2004: 675)

> Robustness is a property that allows a system to maintain its function despite external and internal perturbations. (Kitano 2004: 826)

Robustness is an important feature of many systems in biology. A system can be a single cell, a tissue or the whole organism. Note that robustness of a cell is not the same as robustness of the organism: Apoptosis of cells can be healthy for the organism, whereas cancer cells are very robust but damaging for the organism (Stelling et al. 2004: 677). Moreover, robustness is a relative property because systems can be highly robust to perturbations in common environments yet very fragile to uncommon environments. Therefore, systems can be "robust yet fragile" (Carlson and Doyle 2002: 2539).

There are many ways a system can become robust. First and foremost, *redundancy* is a fundamental feature of robust systems, i.e. alternative ways to realize a function. This can be realized by the duplication of identical genes, but also by structurally different entities or mechanisms performing similar functions (Stelling et al. 2004: 677). If one gene or structural entity is damaged, the function can still be realized due to the alternative. This is also called *alternative* or *fail-safe* mechanism (Kitano 2004: 828).

One example is the redundancy of the B-type cyclin genes CLB5 and CLB6. Both genes are regulating the initiation of S-phase in the cell cycle. The knockout of CLB6 has little effect and the knockout of CLB5 extends, but not prohibits the S-phase. Only if both genes at once are knocked out, the initiation of DNA replication is no longer possible (Schwob and Nasmyth 1993). CLB6 therefore has not an important function for itself, but is a fail-save gene if CLB5 is damaged.

A second feature of robustness is *systems control*, which is realized by positive and negative feedback loops in a network of interacting components. Negative feedback is signaling a gap between actual output and a given set point so autoregulation mechanisms can be activated. Positive feedback enhances sensitivity (Stelling et al. 2004: 677).

Chemotaxis of *E. coli* is a common example of robust negative feedback control. The bacterium is able to direct its motion towards attractors or away from repellents by modifying its tumbling frequency (Barkai and Leibler 1997). Experiments as well as theoretical models indicate that steady-state behavior and adaption time can be resolved under a great variety of protein concentrations (Alon et al. 1999). Hence the bacterium is robust to various changes in its environment.

Another feature of robustness is *modularity*. In a system that contains many modules, if one of them is damaged, the others still work. A cell is a module itself and is also composed of organelles which are modules of the cell (Kitano 2004: 828). While modularity itself cannot ensure robustness, a system is not affected as a whole if one module is damaged and does not work anymore. Other modules are independent and will not be affected by the damaged module. A very simple example is a single cell within a multi-cellular system. If one cell is damaged, the organism as a whole still works.

Kitano suggests that "the mechanisms that underlie robustness can be understood using an example of a sophisticated engineering object, such as an aeroplane" (ibid.: 829). While a system could work with fewer components, additional mechanisms secure the functionality even in uncertain environments. Therefore, robust systems are favored in evolutionary processes.

6   Mechanistic Explanations and Robustness

The example of robustness proves our suggestion of the insufficiency of mechanistic explanations for system properties. As we have seen, redundancy is a key element for robustness. No alternative mechanism or fail-save mechanism is making the system robust for itself, only the sum of all concrete mechanisms realize the system property of robustness. Hence, we cannot determine whether the system is robust by looking at one specific underlying mechanism.

Moreover, the concept of robustness does not refer to single mechanisms. A mechanism which is able to establish robustness on a system level does not have to be robust for itself. Even if this would be the case, we would regard such a mechanism as a robust module of the system which has its own underlying sub-mechanisms that enable robustness for the module. However, the concept of robustness refers to a system as a whole; if a system contains robust modules, then the concept of robustness also refers to the modules as a whole. Mechanistic explanations cannot predict or explain system properties sufficiently.

The concept of robustness can be compared to Cartwright's example on the concept of work:

> What did I do this morning? I worked. More specifically, I washed the dishes, then I wrote a grant proposal, and just before lunch I negotiated with the dean for a new position in our department. A well-known philosophical joke makes clear what is at stake: 'Yes, but when did you work?'. It is true that I worked; but it is not true that I did four things in the morning rather then three. 'Working' is a more abstract description of the same activities I have already described. (Cartwright and Le Poidevin 1991: 60)

The concept of robustness, like the concept of work, does not refer to a specific mechanism, component part or activity. It is an abstract concept which can only refer to the system as a whole. Therefore, no specific object makes the system robust, despite many specific structures are needed for a robust system.

A feedback loop, for example, can describe the origin of robustness in a simple network but the quantitative description of the mechanism makes it necessary to specify the experimental content of the system. In other words, the mechanistic model in mathematical terms is limited to a specific content and cannot be generalized. If we want to make general claims about system properties like robustness, we need a new framework that is able to give abstract and global explanation of system properties. We will analyze how organizing principles are doing this job in the next section.

7 The Robustness Theorem as an Example of Organizing Principles

Shinar and Feinberg (2011) have characterized the system property of robustness in a way that they were able to prove a theorem of robustness:

> Consider a mass-action system in which the underlying reaction network is conservative and has a deficiency of zero. Then, no matter what values the rate constants take, there is no species relative to which the system exhibits absolute concentration robustness. (Shinar and Feinberg 2011: 44)

A mass-action system is a reaction network taken together with an assignment of a positive rate constant to each reaction and can be described with first order polynomial differential equations. If a reaction network respects the conservation of mass, which is usually the case, it is called a conservative network. The deficiency of a reaction network is a non-negative integer index to classify reaction networks; the simple formula for deficiency is *complexes minus linkage classes minus rank*. (ibid: 42). A mass-action system has absolute concentration robustness "if the system admits at least one positive steady state and if the concentration of that species is the same in all of the positive steady states that the system might admit, regardless of the overall supplies of the various network constituents." (ibid.: 39). The difference to a conventional model is that the result is valid for a class of systems and thus a range of contexts.

The robustness theorem is a special case of a formal sensitivity analysis made by Shinar et al. (2009). The main goal was to understand the relationship between general features of chemical reaction networks and the sensitivity of its equilibria to changes in the overall supply of reactants. Their analysis is independent of specific parameters or even the particular equilibrium state (ibid.: 977), resulting in a purely mathematical nature of the paper.

Jeremy Gunawardena emphasizes the difference between models and theorems. Whereas models "make assumptions about a specific experimental setting", theorems "can apply to a setting of arbitrary molecular complexity" (Gunawardena 2010: 581). Hence, theorems are not connected directly to experiments. On contrary, the robustness theorem is not based on an experimental induction but is an analytical theorem. The sensitivity analysis that provides the basis for the robustness theorem is developed in a purely mathematical-analytical way; i.e. via definitions, lemmata and mathematical proofs. Obviously the mathematical calculations are afterwards interpreted biologically, but they are in no way based or derived from experience (See Baker 2015 this volume on distinctively mathematical explanations). Therefore, it is hard to claim that they are true or false; they are a mathematical framework that fits well together with biological phenomena. Biological experiments could not falsify the robustness theorem but only illustrate that the mathematical definitions are inadequate to catch biological phenomena.

Theorems in Systems Biology are not applied and do not explain single phenomena. Instead they are used on a more abstract level and define properties of the whole system by stating that every model with a specific structure necessarily has specific properties. If we want to understand a system, we need to acknowledge that they are organized. Mesarovic, a pioneer in the search for organizing principles, pointed out that "biological systems are organised, not necessarily in terms of mathematical/computer models but in terms of concepts. Multilevelness, concepts of feedbacks—negative, positive, multiple—feed-forward regulators, etc. are such principles" (Mesarovic et al. 2004: 20). The concept of robustness is another essential organization principle in Systems Biology, on a "higher level of perception", which can "be based on observations but not restricted to numerical data" (ibid.). Hence, organizing principles provide the conceptional framework to understand complex systems:

> The organising principles play a role akin to the laws in physics in the sense that they provide a starting framework to unravel the understanding of systems from observations and data. However, they are not numerical but 'relational', i.e. they indicate the functions of biological systems and/or their components. They provide the 'architecture' of the model—as an image of reality, i.e. projection of the reality on the focus/problem of interest. (Mesarovic et al. 2004: 20–21)

As we have seen, the search for organizing principles is the search for abstract conceptions, often connected to mathematical formalisms and theorems, which help to understand the biological systems by providing a conceptual framework. Obviously, this is completely different then describing specific mechanisms in nature. We will now sketch how organizing principles might be another kind of biological explanation.

8   Organizing Principles and Explanation

If organizing principles and theorems in Systems Biology are not connected directly to experiments, they cannot explain biological phenomena in a mechanistic way. One could argue that they are not explanatory at all. But we want to claim that organizing principles have explanatory power despite their lack of connection to the phenomena (see also Issad and Malaterre 2015 this volume on the explanatory power of mathematics in mechanistic explanations). Organizing principles *unify* mathematical models in a way that every model with a specific structure necessarily has specific properties. While causal and mechanistic models of explanation are very common in discussions about the nature of biological explanation, we believe that the unificatory model of explanation better fits the nature of explanation in Systems Biology. As Michael Friedman states, "the essence of scientific explanation [is that] science increases our understanding of the world by reducing the number of independent phenomena that we have to accept as ultimate or given" (Friedman 1974: 15). This is exactly what organizing principles do: System properties are not defined by many single mechanisms or many single phenomena, but in a more global and abstract way. The idiosyncratic mechanisms which establish system properties of specific systems are reduced to a general framework; hence we can understand the system property by merely looking at one single definition. This fits well to Friedman's statement that "we replace one phenomenon with a more comprehensive phenomenon, and thereby reducing the total number of accepted phenomena. We thus genuinely increase our understanding of the world" (ibid.: 19).

Philosophical accounts of unificatory explanation (Kitcher 1989 and others) are complex and sometimes idiosyncratic. We cannot go into details here as to which accounts may best fit our case studies and our general claim. However, as far as these accounts involve structuralism or set-theoretic reconstructions of scientific theories, we doubt that they are fully applicable to organizing principles. Knowledge in biology, especially in Systems Biology, is often presented by diagrams, which are hard to reduce to mathematical formalisms.

An organizing principle describes a property or properties of a class of systems, while a mechanistic model describes a specific subclass (e.g. defined by the structure of the equations or a range of parameter values). For a given set of specific parameter values, a mechanistic model describes a particular system or an instantiation of a more general. Therefore, organizing principles are linked with phenomena indirectly. As Morgan (1999) has argued, models are mediating between the phenomena and abstract laws. There are phenomenological models on the one side, which only save the phenomena, and abstract organizing principles on the other side, which provide an abstract framework for system properties. Between the two extremes are mechanistic models which explain the saved phenomena and are structured in a way that the results of organizing principles can be applied to them.

9   The Need for Organizing Principles

Contemporary research in the life sciences is largely driven by reductive approaches, with a tendency to neglect the larger picture. Although some general goals are always in mind (e.g. curing cancer) those targets do not play a big role in the everyday planning of research directions. While one would naively imagine that a research field would identify or hypothesize principles, laws and general rules and then work out the details, cancer research is an example where we know many details but where there is lack of theory—hardly any effort to generalize, to integrate details into a larger picture, not with more details but as an abstraction of the essence of a system, its organization and function, independent of the actual or specific biophysical/biochemical realization. Early molecular and cell biology produced some general theoretical models and theorems, like the Hodgkin-Huxley-model or the central dogma of molecular biology. On contrary, many domains in contemporary biology lack broad generalizations and the only place where such search for organizing principles is visible is in editorial notes in journals or outstanding high-level review articles, where the authors try to look down onto the forest that emerges as we walk up, out of the trees.

Systems Biology should not merely be about modeling of gene regulatory networks, signaling and metabolic pathways. There is a need to go beyond the current focus on pathway-centric mechanistic modeling to capture the complexity of biological systems and to identify organizing principles. Systems Biology needs general theorems which have a similar role as laws in physics. While mechanistic models analyze specific cell processes, theorems in Systems Biology are proving systems properties, often not only for a specific system but for a whole class of systems.

10   Summary and Outlook

Summarizing the results of our paper, we have argued that models in Systems Biology and the philosophical framework of mechanisms of the last 20 years are not opposed to each other. Not every model is explanatory, some models merely save the phenomena. This is not trivial, especially in big data sciences where sophisticated heuristic strategies are necessary to separate stable phenomena in the world from noise in the data. As we have seen, a major difference between explanatory and non-explanatory models is that only the former provides knowledge of the causal structure of the world, whereas the latter merely recognize correlations. Explanatory models needs to give at least a sketch of a mechanism. However, system biology is different from conventional molecular and cell biology: it is an approach that emphasizes the functional organization of systems and sheds light on system properties that cannot fully be explained within a mechanistic framework. We have argued that organizing principles are used for a different kind of explanation, complementary to mechanistic explanations, which unifies models with a specific structure and provides a framework that makes it possible to understand system properties. Moreover, organizing principles are similar to natural laws in physics, not directly connected to single phenomena, so mechanistic models need to mediate between organizing principles and experiments.

References

Alon, U., Surette, M. G., Barkai, N., & Leibler, S. (1999). Robustness in bacterial chemotaxis. *Nature, 397*, 168–171.

Baker, A. (2015). Mathematical explanation in biology. In P.-A. Braillard & C. Malaterre (Eds.), *Explanation in biology. An enquiry into the diversity of explanatory patterns in the life sciences* (pp. 229–247). Dordrecht: Springer.

Barkai, N., & Leibler, S. (1997). Robustness in simple biochemical networks. *Nature, 387*, 913–917.

Bechtel, W. (2011). Mechanism and biological explanation. *Philosophy of Science, 78*, 533–557.

Bechtel, W., & Abrahamsen, A. (2005). Explanation: A mechanistic alternative. *Studies in the History and Philosophy of Biology and the Biomedical Sciences, 36*, 421–441.

Bogen, J., & Woodward, J. (1988). Saving the phenomena. *The Philosophical Review, 97*, 303–352.

Carlson, J. M., & Doyle, J. (2002). Complexity and robustness. *Proceedings of the National Academy of Sciences, 99*, 2538–2545.

Cartwright, N., & Le Poidevin, R. (1991). Fables and models. *Proceedings of the Aristotelian Society, 65*, 55–82.

Craver, C. (2006). When mechanistic models explain. *Synthese, 153*, 355–376.

Craver, C. (2007). *Explaining the brain: Mechanisms and the mosaic unity of neuroscience.* Oxford: Oxford University Press.

Cummins, R. C. (2000). 'How does it work' versus 'what are the laws?': Two conceptions of psychological explanation. In F. Keil & R. A. Wilson (Eds.), *Explanation and cognition* (pp. 117–145). Cambridge, MA: MIT Press.

Friedman, M. (1974). Explanation and scientific understanding. *Journal of Philosophy, 71*, 5–19.

Glennan, S. (2002). Rethinking mechanistic explanation. *Philosophy of Science, 69*, S342–S353.

Gunawardena, J. (2010). Biological systems theory. *Science, 328*, 581.

Issad, T., & Malaterre, C. (2015). Are dynamic mechanistic explanations still mechanistic? In P.-A. Braillard & C. Malaterre (Eds.), *Explanation in biology. An enquiry into the diversity of explanatory patterns in the life sciences* (pp. 265–292). Dordrecht: Springer.

Kitano, H. (2004). Biological robustness. *Nature Reviews Genetics, 5*, 826–837.

Kitcher, P. (1989). Explanatory unification and the causal structure of the world. In P. Kitcher & W. Salmon (Eds.), *Scientific explanation* (pp. 410–505). Minneapolis: University of Minnesota Press.

Lipton, P. (2004). *Inference to the best explanation* (2nd ed.). London: Routledge.

Machamer, P., Darden, L., & Craver, C. (2000). Thinking about mechanisms. *Philosophy of Science, 67*, 1–25.

Mesarovic, M. D., Sreenath, S. N., & Keene, J. D. (2004). Searching for organizing principle: Understanding in systems biology. *Systems Biology, 1*, 19–27.

Morgan, M. (1999). Models as mediating instruments. In M. Morgan & M. Morrison (Eds.), *Models as mediators: Perspectives on natural and social science* (pp. 10–37). Cambridge: Cambridge University Press.

Novác, B., & Tyson, J. J. (2008). Design principles of biochemical oscillators. *Nature Reviews Molecular Cell Biology, 9*, 981–991.

Schwob, E., & Nasmyth, K. (1993). CLB5 and CLB6, a new pair of B cyclins involved in DNA replication in Saccharomyces cerevisiae. *Genes and Development, 7*, 1160–1175.

Shinar, G., & Feinberg, M. (2011). Design principles for robust biochemical reaction networks: What works, what cannot work, and what might almost work. *Mathematical Biosciences, 231*(2011), 39–48.

Shinar, G., Alon, U., & Feinberg, M. (2009). Sensitivity and robustness in chemical reaction networks. *SIAM Journal of Applied Mathematics, 69*, 977–998.

Stelling, J., Sauer, W., Szallasi, Z., Doyle, F. J., & Doyle, J. (2004). Robustness of cellular functions. *Cell, 118*, 675–685.

Weisberg, M. (2007). Who is a modeler? *British Journal for the Philosophy of Science, 58*, 207–233.

Wolkenhauer, O., et al. (2009). Advancing systems biology for medical applications. *IET Systems Biology, 3*, 131–136.

Woodward, J. (2003). *Making things happen: A theory of causal explanation*. Oxford: Oxford University Press.

Chapter 12
Are Dynamic Mechanistic Explanations Still Mechanistic?

Tarik Issad and Christophe Malaterre

Abstract Mechanistic explanations are one of the major types of explanation in biology. The explanatory force of mechanisms is apparent in such typical cases as the functioning of an ion channel or the molecular activation of a receptor: it includes the specification of a *model of mechanism* and the rehearsing of a *causal story* that tells how the explanandum phenomenon is produced by the mechanism. It is however much less clear how mechanisms explain in the case of complex and non-linear biomolecular networks such as those that underlie the action of hormones and the regulation of genes. While dynamic mechanistic explanations have been proposed as an extension of mechanistic explanations, we argue that the former depart from the latter in that they do not draw their explanatory force from a causal story but from the mathematical warrants they give that the explanandum phenomenon follows from a *mathematical model*. By analyzing the explanatory force of mechanistic explanation and of dynamic mechanistic explanation, we show that the two types of explanations can be construed as limit cases of a more general pattern of explanation – Causally Interpreted Model Explanations – that draws its explanatory force from a *model*, a *causal interpretation* that links the model to biological reality, and a *mathematical derivation* that links the model to the explanandum phenomenon.

Keywords Mechanisms • Models • Explanatory force • Mechanistic explanation • Dynamic mechanistic explanation • Mathematical explanation

T. Issad
Inserm, U1016, Institut Cochin, Paris, France

Cnrs, UMR8104, Paris, France

Université Paris Descartes, Sorbonne Paris Cité, Paris, France

Institut Cochin, Department of Endocrinology, Metabolism and Diabetes, 22 rue Méchain, 75014 Paris, France
e-mail: tarik.issad@inserm.fr

C. Malaterre (✉)
Département de philosophie & CIRST, Chaire de recherche UQAM en Philosophie des sciences, Université du Québec à Montréal (UQAM), Case postale 8888, Succursale Centre-Ville, Montréal, QC, H3C 3P8, Canada
e-mail: malaterre.christophe@uqam.ca

© Springer Science+Business Media Dordrecht 2015
P.-A. Braillard, C. Malaterre, *Explanation in Biology*, History, Philosophy and Theory of the Life Sciences 11, DOI 10.1007/978-94-017-9822-8_12

265

1 Introduction

It is now often taken for granted that mechanistic explanations (ME's) are one of the major types of explanation in biology. It is indeed often impossible to read a current scientific paper in biology without stumbling upon the concept of mechanism: mechanisms are found in articles dealing with subjects as diverse as regulation of food intake, transmission of nerve influx, cell division, cell growth, programmed cell death, production by mitochondria and chloroplasts of energy-rich molecules and so forth. It has been argued that the explanatory force of a mechanistic explanation – that is to say, what makes a mechanistic explanation explanatory – comes from displaying a mechanism while showing how this mechanism produces the phenomenon of interest. In this context, mechanisms are generally construed as particular types of models that include entities carrying out certain sets of activities and organized in such a way that they produce the phenomenon of interest (e.g. Machamer et al. 2000; Glennan 2002; Kaplan and Craver 2011). The explanatory force of ME's is claimed to be apparent in such canonical examples as the functioning of an ion channel involved in neuron firing or the molecular activation of a receptor. In this contribution, we question the explanatory force of ME when targeted at complex biological systems. More specifically, we argue that if the explanatory force of ME's is to be equated with displaying a mechanism and showing how this mechanism produces the phenomenon of interest, then such ME's are only possible with quite simple biological mechanisms. In the case of complex mechanisms – which are mechanisms that typically involve feedback loops and non-linear interactions (e.g. Bechtel and Abrahamsen 2010) – the explanatory force of proper explanations of phenomena produced by these mechanisms does not come from displaying the mechanisms and showing how these mechanisms produce the phenomena of interest, but rather from the use of dynamic mathematical models, thereby leading to specific dynamic mechanistic explanations (DME's). While it is a distinguishing feature of such mathematical models to be modally stronger than mechanisms or laws (Lange 2012), their use in complex biological systems highlights the very limited explanatory force of ME's. If there is anything distinctive to ME's, then their scope of application is actually quite restricted to fairly simple mechanisms, and their extension to complex biological systems (e.g. Bechtel and Abrahamsen 2010) is far from obvious when it comes to specifying the source of their explanatory force. We argue that, in the case of such complex systems, explanations are not mechanistic but follow a more general pattern of "Causally-Interpreted Model Explanations" (CIME, as we propose to call them). In the following section, we review what is usually defined as a mechanistic explanation (ME) and we highlight what makes mechanistic explanations explanatory: a model of mechanism and a causal story. We then show, in Sect. 3, how these elements of explanatory force are effective in the case of simple biological mechanisms, and how they are ineffective in the case of complex biological mechanisms. We illustrate our argument by looking at the case of the action of insulin on glucose homeostasis. In Sect. 4, we show that the explanatory force of explanations of phenomena produced by complex biological mechanisms (DME's) does not come from a causal-story but from their mathematical component and its stronger modality. As a consequence, we argue in Sect. 5, that, while ME's are perfectly explanatory in the case of simple biological systems, DME's are no longer mechanistic since they do not draw their explanatory force from the same types of elements. In Sect. 6, we show how the different elements of explanatory force of ME's and DME's can be reconstructed into three core elements that consist of a model, a causal interpretation and a mathematical derivation. We develop the notion of "Causally Interpreted Model Explanation" around those three elements and argue that ME's and DME's can be construed, within a continuum, as limit cases of CIME's.

2   Mechanistic Explanation and Its Explanatory Force

An explanation presupposes that there is something to be explained – the *explanandum* – and something that does the explaining – the *explanans*. In the case of mechanistic explanation (ME), the explanandum is usually taken to be a phenomenon produced by a mechanism, while the explanans usually consists in displaying the mechanism and in showing how this mechanism produces the phenomenon in question. Let us be more specific.

A *mechanism* is usually defined as a set of parts and activities organized in such a way that they produce the phenomenon in question. For instance, Craver proposes to define mechanisms as "entities and activities organized such that they exhibit the explanandum phenomenon" (2007, p. 6). And for Glennan, "a mechanism for a behavior is a complex system that produces that behavior by the interaction of a number of parts, where the interactions between parts can be characterized by direct, invariant, change-relating generalizations" (2002, p. S344). Mechanisms so construed are real things: they are composed of real parts that engage in real activities or change-relating generalizations. The way we formalize our knowledge about a given mechanism is typically by defining a model of this mechanism: a *model of mechanism*. Indeed, Glennan defines the notion of "mechanical model" which is a description of a mechanism (2002, p. S347), and Craver proposes to distinguish a range of "models of mechanisms", ranging from "how possibly models" to "how actually models". Note that, in the scientific and philosophical literature, mechanisms are not always distinguished from their representational counterparts. Indeed, mechanistic explanations are often directly framed in terms of mechanisms. Nevertheless, the distinction matters if we wish to be precise about what a mechanistic explanation consists of. We will therefore use "model of mechanism" to mean a representation of a "mechanism".

Similarly, when one speaks about explaining a phenomenon, one presupposes a *description* of this phenomenon, this description typically being given under the form of a proposition (or a set of propositions). In the case of mechanistic explanation, the phenomenon that constitutes the explanandum is supposed to be produced by a mechanism.

A mechanistic explanation somehow consists in linking a description of the phenomenon (produced by a mechanism) to a representational model of this mechanism. Of course, much hinges on what is meant by "linking" a description of a phenomenon to a model of mechanism. It has been argued that in a mechanistic explanation,

> intelligibility arises [...] from an elucidative relation between the explanans (the set-up conditions and intermediate entities and activities) and the explanandum (the termination condition or the phenomenon to be explained) [...]. Descriptions of mechanisms render the end stage intelligible by showing how it is produced by bottom out entities and activities. To explain is not merely to redescribe one regularity as a series of several. Rather, explanation involves revealing the *productive* relation. (Machamer et al. 2000, p. 22)

In other words, providing a mechanistic explanation is a matter of showing how the entities of the mechanism produce the phenomenon through carrying out their respective activities. This "productive" relation is in turn often understood in a causal sense, with causation understood in a manipulationist sense (e.g. Craver 2007). A mechanistic explanation therefore explains a phenomenon (produced by a mechanism) by showing how the phenomenon is causally brought about by the entities and activities of the mechanism.

In practice, this is done in two steps, first by exhibiting a model of the mechanism, second by showing how this model of mechanism causally accounts for the explanandum phenomenon. This model includes entities and activities that, ideally, refer to the real entities and activities of the real mechanism (and less ideally to plausible or even to possible ones – see (Craver 2007, p. 139)), while also specifying their spatial and temporal organization, all of this being relevant to the phenomenon to be explained. Along these lines, Craver details five normative elements that a model of a mechanism must necessarily fulfill in order to contribute to a mechanistic explanation: (i) the model of mechanism must account for all aspects of the explanandum phenomenon, (ii) the model of mechanism must be based on components and activities that are real, (iii) these activities must be causal (in a manipulationist sense), (iv) the entities and activities are spatially and temporarily organized, (v) the entities and activities are all relevant with respect to the explanandum phenomenon (for more details, see (Craver 2007, pp. 161–162)). These normative elements summarize what we expect of a model of a mechanism that does some explaining. Indeed, we expect that the entities and activities mentioned in the model of mechanism relate to some real entities and activities, and that they thereby somehow tell us something about the real world out there. We expect that these entities and activities are indeed relevant – and not superfluous – when it comes to explaining the phenomenon in question. We expect that the model of mechanism can account fully for the explanandum. And that the entities and activities are organized and causally meaningful is also something that makes sense if explaining entails showing how things work. Yet even if it fulfills all these normative elements, a model of a mechanism *alone* falls short of being an explanation. Indeed, a list of entities and activities, even supplemented with their spatio-temporal organization, is rarely explanatory in itself. What needs to be shown is how the mechanism *produces* the phenomenon of interest.

Hence a second step in a mechanistic explanation that consists in showing how the entities and activities that are represented in the model causally interact so as to *produce* the very phenomenon to be explained.[1] In other words, one needs to show how the explanans (the model of the mechanism) is linked to the explanandum (the phenomenon produced by the mechanism).[2] What appears to matter in this second step is to somehow animate the model of the mechanism so as to reveal the ways in which the represented entities interact with one another through their respective activities and in function of their organization so as to reproduce the phenomenon that is the target of the explanation. When we say "animate", we mean that a mechanistic explanation appears to involve some kind of rehearsing of the unfolding of the activities of the different entities that compose the mechanism. It is as if we were to run through each of the different causal links in the right sequential order. The explanatory force of a mechanistic explanation thereby stems from the intelligibility that arises when one rehearses this sequence of causal interactions and realizes that this sequence ends up in the phenomenon to be explained. For instance, an explanation of the movement of Na+ ions through the neuronal membrane during the phenomenon of the action potential – a canonical example – may involve the following passage:

> At V_{rest} [the membrane potential at rest], a positive extracellular potential holds the α-helix [a subunit of a trans-membrane molecular complex known as the Na^+ channel] in place. Weakening that potential, which happens when the cell is depolarized, allows the helix to rotate out toward the extracellular side (carrying a "gating charge" as positively charged amino acids move outward). This rotation, which occurs in each of the Na^+ channel's subunits, destabilizes the balance of forces holding the channel in its closed state and bends the pore-lining S6 region in such a way as to open a channel through the membrane. Another consequence of these conformation changes is that the pore through the channel is lined with hairpin turn structures, the charge distribution along which accounts for the channel's selectivity to Na^+. (Craver 2007, p. 119)

This excerpt clearly aims at showing how things work inside the mechanism and how this results in the production of the explanandum phenomenon. It mentions entities ("α-helix", "hairpin turn structures" etc.) that engage in a range of activities

---

[1] It could be argued that the causal account or "causal story" of how the entities and activities bring about the explanandum phenomenon is an integral part of the model of the mechanism, as proposed for instance by Glennan (2005, p. 446). We argue here that distinguishing the causal story from the model reveals two distinct dimensions of the explanatory force of mechanistic explanation, as we specify at the end of this Sect. 2.

[2] In the covering law model of explanation, the explanation takes the form of an argument whose premiss is the explanans and conclusion is the explanandum. The explanans is thereby linked to the explanandum by means of a deductive inference. It could be argued that mechanistic explanations are explanatory in just the same argumentative way, the description of the explanandum phenomenon being deduced from the description of the mechanism. Yet mechanistic philosophers specifically insist that mechanistic explanation is not an argument for "the explanation lies not in the logical relationship between these descriptions but in the causal relationships between the parts of the mechanism that produce the behavior described" (Glennan 2002, p. S348). What matters therefore in a mechanistic explanation is to somehow show how the explanans causally produces the explanandum.

("holding in place", "rotating", "bending" etc.) according to a specific spatial and temporal organization (which appears in the way the account runs). And it is typically this type of causal account that is expected to supplement a model of mechanism in order to constitute a truly explanatory mechanistic explanation. It has even been proposed that the way such causal mechanistic accounts are explanatory is by relating the interactions of the mechanism's components to familiar experiences "by an extension of sensory experience with ways of working" (Machamer et al. 2000, p. 22).

The explanatory force of a mechanistic explanation therefore appears to come in two steps: (1) the exhibition of a *model of mechanism*, that refers to real and relevant entities and activities, and whose behaviour reproduces the explanandum phenomenon; and (2) a *causal story*, that provides intelligibility to the model of mechanism by describing how the entities and activities of that mechanism work together so as to produce the explanandum phenomenon. More formally, the explanatory force of mechanistic explanations can be made more specific in the following way:

> A mechanistic explanation (ME) explains a phenomenon (P) produced by a system (S) in virtue of fulfilling the following two necessary and sufficient conditions:
>
> (*MM*)   Displaying a model of mechanism that represents a real mechanism in S with its entities, activities and spatio-temporal organization
>
> (*CS*)   Rehearsing a causal story that enumerates the cause-effect relationships taking place in the mechanism up to the production of the phenomenon P.

These conditions explicate how mechanistic explanations gain their explanatory force. Such explanations are deemed explanatory in virtue of displaying a model of mechanism (that tells something about the world), and of rehearsing how this model produces the explanandum through a sequence of causal links.[3] In the next section, we discuss the explanatory force of mechanistic explanation and show that this explanatory force, while being effective in the case of simple mechanisms, becomes ineffective in the case of complex mechanisms.

3   Mechanistic Explanation in the Face of Increasingly Complex Biological Mechanisms

There are many examples of mechanistic explanations that have already been discussed in the philosophical literature. The neuronal action potential, the chemical

---

[3]The two conditions of explanatory force we specify here do not specifically commit to the details of what constitutes a model of mechanism. For reasons of convenience, we use a terminology inspired from (Machamer et al. 2000), yet the argument could be adapted to others such as (Glennan 2005) for instance. Similarly, the two conditions are not linked to any specific account of causation, even if manipulationism as per (Woodward 2003) is often used in the mechanistic literature.

transmission at synapses, and the synthesis of proteins are all canonical examples of mechanistic explanations. In what follows, we focus on another biochemical mechanism: the *mechanism of insulin action*. As will become apparent, this mechanism is interesting because it lends itself well to a mechanistic explanation, but does so only to a certain limit.

Insulin is a hormone, that is to say a molecule secreted in the blood circulation by an organ or a tissue, and capable of acting on another tissue. Discovered during the first half of the twentieth century, insulin is produced by the pancreas during ingestion of a meal. It has long been known that one of the main function of insulin is to permit the utilization and/or storage of the nutrients ingested during meals (See Fig. 12.1). Insulin plays a crucial role in the regulation of glucose homeostasis, and more specifically in keeping concentration of glucose in the blood (glycaemia) within a narrow range, around a basal value of 0.8–1 g per liter. Dysfunctions in the regulation of glucose homeostasis can lead to chronic hyperglycaemia, which deleterious effects will eventually lead to the development of overt diabetes. One of the main effects of insulin is to stimulate glucose uptake by a number of insulin target cells, which then metabolize glucose or store it either as glycogen (a glucose polymer) in liver and muscle cells, or in the form of fat in adipose tissue. This permits it to rapidly bring back glycaemia at its basal level, and to avoid the development of chronic hyperglycaemia. In this paper, we will examine more specifically the action of insulin on one of its main target tissues, the adipose tissue. One of the main functions of the adipose cells is to store nutrients in the form of lipids, that will be released in the blood circulation during periods of starvation.

At the beginning of the 1980s, a first step in the understanding of the mechanism of insulin action was taken, when Cushman and Wardzala proposed that insulin

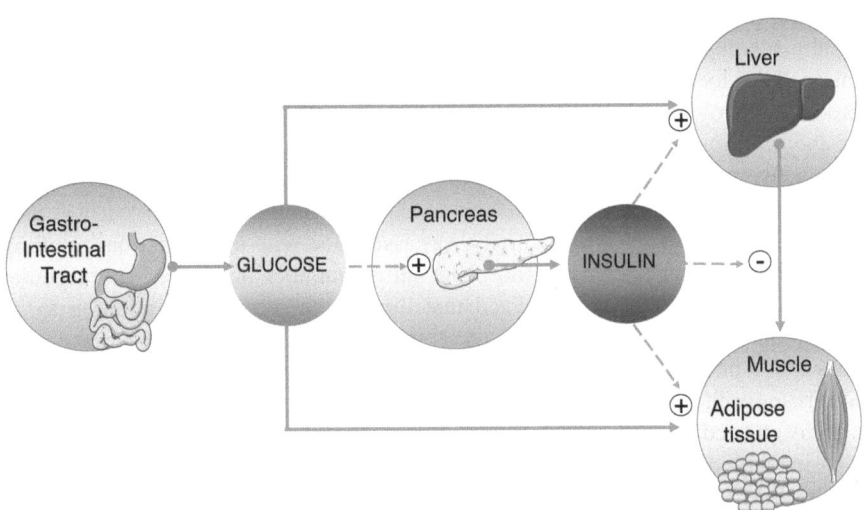

Fig. 12.1 High level representation of the glucose-insulin control system

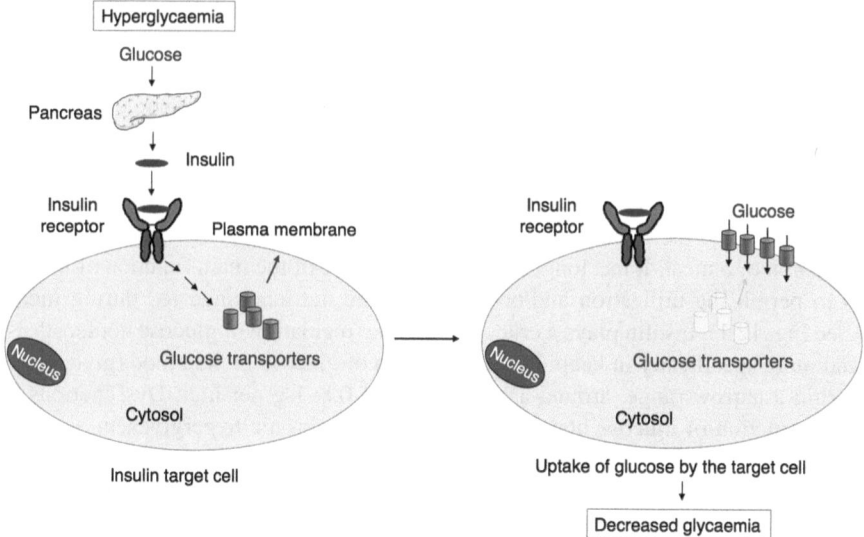

Fig. 12.2 Insulin-induced relocalization of glucose transport systems

stimulated glucose uptake by the adipocyte by inducing the relocalisation of the "glucose transport system" from an intracellular compartment to the plasma membrane (Cushman and Wardzala 1980). This relocalisation provided an explanation of how insulin increased the activity of glucose transport at the surface of the cell, and thereby increased glucose uptake (see Fig. 12.2).

Further investigations revealed the entities bearing these activities (insulin receptor, glucose transporter) and the intermediary entities and activities that link insulin binding on its receptor to the movement of glucose transporters from an intracellular location towards the plasma membrane. Because numerous biological processes involve phosphorylation of proteins (a reversible modification consisting of the addition of phosphate groups on proteins), several investigators, using radioactive phosphate, studied the phosphorylation state of intracellular proteins following insulin stimulation (Belsham et al. 1980; Smith et al. 1980). It was thus shown that a phosphorylating activity was associated with insulin action, resulting in increased incorporation of phosphate into a number of entities (intracellular proteins). It was then soon realized that the insulin receptor, in addition to its insulin binding activity located on the extra-cellular side of the cell surface, also possesses a phosphorylating activity, located inside the cell (Kasuga et al. 1982a, b). Surprisingly enough, the first target of this phosphorylating activity (tyrosine kinase activity) turned-out to be the insulin receptor itself, which autophosphorylates on tyrosine amino-acids upon insulin stimulation. Once phosphorylated on tyrosines, the insulin receptor is fully active and can phosphorylate intracellular proteins on tyrosines, such as IRS1 (See Fig. 12.3).

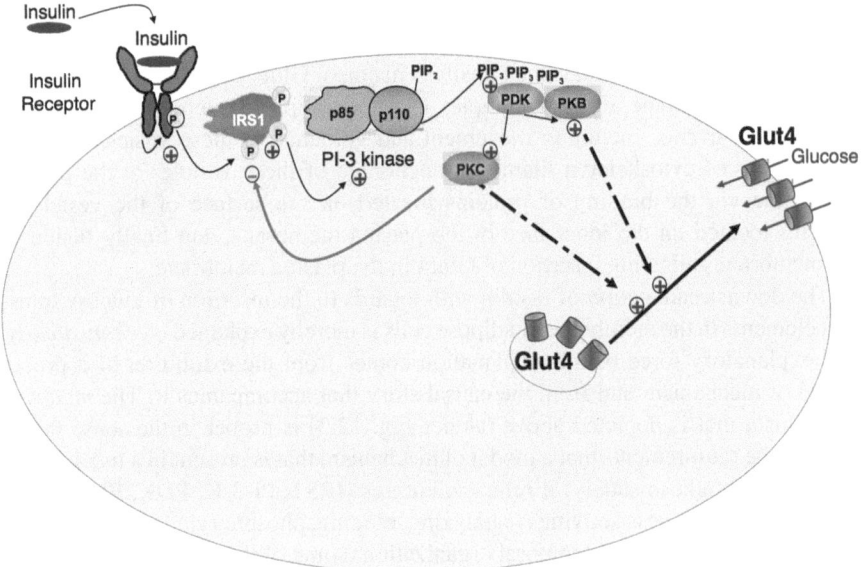

Fig. 12.3 Mechanism of insulin action with regards to the activation of glucose transporters

During the 1980s and 1990s, the generalization of the tools of molecular biology (determination of DNA and corresponding protein sequences) permitted the identification of a large number of entities involved in insulin action downstream of the receptor (Combettes-Souverain and Issad 1998). In addition, parts of these proteins (e.g., the domain bearing the tyrosine kinase activity of the insulin receptor in its inactive (Hubbard et al. 1994) and active forms (Hubbard 1997)) were crystallized, thereby bringing important insight into the tri-dimensional structure of these entities.

One of the proteins that bind phosphorylated IRS-1 is the Phosphatidyl Inositol-3 kinase constituting of a regulatory (p85) and a catalytic (p110) (See Fig. 12.3). The regulatory subunit binds to phosphorylated tyrosines of IRS-1, and this induces a stimulation of the catalytic activity of p110. We therefore have again a proper enzymatic activity and an interaction between IRS-1 and p85, which can also be defined as a binding activity. The enzymatic activity of p110 is a lipid-kinase activity, which phosphorylates lipids (Phosphatidyl Inositol-2 phosphate or PIP2) located on the inner face of the cell membrane. Once phosphorylated, these lipids (Phosphatidyl Inositol-3 Phosphate or PIP3) constitute anchorage platforms for membrane proteins with enzymatic activities (the serine/threonine kinases PDK, PKB and PKC ζ/λ). Relocalisation of these protein kinases at the plasma membrane permit phosphorylation of AKT and PKC by PDK, resulting in stimulation of their activities (Combettes-Souverain and Issad 1998). Activated PKB and PKC then return into the cytosol and phosphorylate intracellular proteins, thereby releasing membrane vesicles containing glucose transporters (now identified at the molecular level as membrane channels denominated Glut4) from an intracellular location and their translocation to the plasma membrane (Hoffman and Elmendorf 2011). Insertion of Glut4 in the membrane then involves fusion of Glut4 containing vesicles with the plasma membrane, through complex and not completely defined "geometrico-mechanical" events, including movement and guidance of these vesicles through remodelling of cytoskeleton filaments, anchorage of these vesicles at the plasma membrane via the binding of proteins located on the surface of the vesicle to proteins located on the inner face of the plasma membrane, and finally fusion of the membranes allowing insertion of Glut4 in the plasma membrane.

The downstream action of insulin with regards to the insertion of glucose transport elements in the membrane of adipose cells is thereby explained *mechanistically*. The explanatory force of this explanation comes from the exhibition of a proper model of mechanism and from the causal story that accompanies it. The *model* of mechanism that is depicted above (as per Fig. 12.3) is proper in the sense that it satisfies the requirements that a model of mechanism that is present in a mechanistic explanation ought to satisfy: it refers to entities (IRS1, PI-3 K, PDK, PKB, PKC, Glut4 etc.) that possess activities (catalyzing, binding, phosphorylating etc.) and that display both a spatial and temporal organization (some of these entities are located at the membrane of adipose cells, others within the cells, still others move from one place to another so as to enter into contacts with one another; the activities are carried out in a specific sequence of events). Simple negative feedback loops (e.g. phosphorylation of IRS1 by PKC) can also be included in the model without loosing its intelligibility. The entities and activities that the model refers to are taken to be real entities and activities, that are all causally relevant to accounting for the glucose intake of adipose cells upon stimulation by insulin. Furthermore, the model fully accounts for the phenomenon in question, including for its normal functioning under normal conditions as well as its dysfunctioning under a set of degraded or inhibitory conditions.

The explanatory force of this mechanistic explanation also comes from the *causal story* that accompanies the model of the corresponding mechanism (most of the above paragraphs of the present Section). It is this causal story that shows how the mechanism produces the explanandum phenomenon. The causal story thereby brings intelligibility to the model by revealing how the explanandum phenomenon causally follows from the entities and activities (and their spatio-temporal organization) that figure in the explanans model.

It is therefore fair to say that in this case, the mechanistic explanation works fine by virtue of fulfilling the two conditions of explanatory force that we have specified above for mechanistic explanations (ME): the condition of displaying a model of mechanism (*MM*) and the condition of rehearsing a causal story (*CS*). Like in many other examples of mechanistic explanation, the model of the mechanism and the causal story that accompanies it deliver the right intelligibility when it comes to explaining the explanandum phenomenon. However, a most interesting feature of this mechanism with respect to a discussion of the explanatory force of mechanistic explanation is that many of its entities and activities happen also to be linked to numerous other entities and activities located within the same adipose cells.

Indeed, careful investigations at each step of the mechanism resulted in the discovery of an important number of *new entities*: IRS-2, 3, 4, 5 and 6, different isoformes of PI-3 kinase displaying different functions, a considerable number of regulatory proteins bearing enzymatic activities capable of dephosphorylating proteins as well as PIP3 (protein- or lipid-phosphatases), protein kinases capable of phosphorylating and modulating the activities of different entities, as well as other proteins devoid of enzymatic activity but capable of binding to active sites of kinases and thereby blocking their catalytic activity (Carracedo and Pandolfi 2008; Lacasa et al. 2005; Nouaille et al. 2006; Virkamäki et al. 1999) (See Fig. 12.4). These entities are linked together through stimulatory or inhibitory interactions, resulting in a much more complex model of mechanism than it had been previously.

This complexity is further increased due to the discovery of *new activities* that are carried either by the original entities that are part of the simple mechanism of insulin action (as in Fig. 12.3) or by the new entities that complete this model (as in Fig. 12.4). Indeed, it is a known fact that a single entity can display different activities, because it can bear different domains with different functions (e.g., the insulin receptor, which has an insulin binding domain and a tyrosine kinase domain). But furthermore, the same domain within an entity can also display different

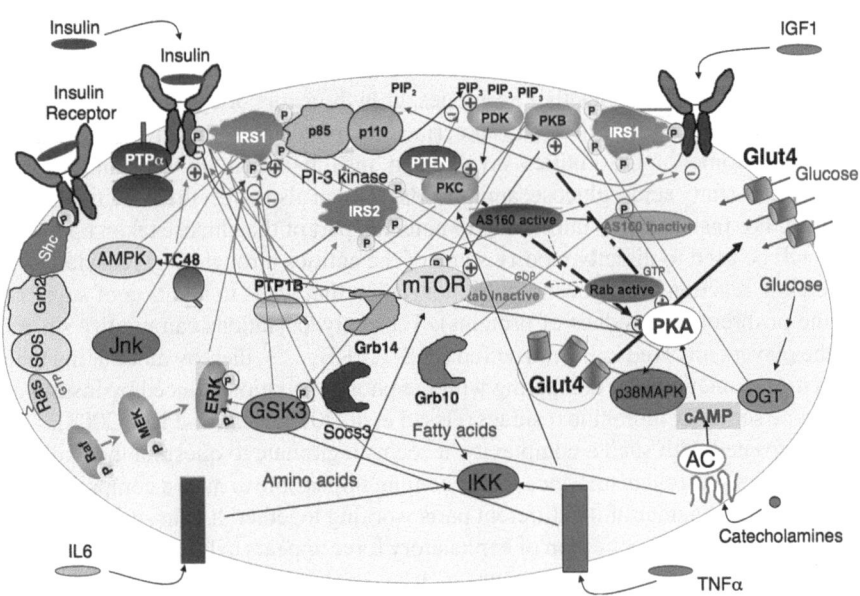

Fig. 12.4 Mechanism of insulin action completed with some of the known interacting entities and activities within adipose cells (this model still represents a very light version of the information available to date)

activities, depending on the partner it interacts with. For instance, a lipid-kinase such as PI-3 Kinase (which phosphorylates PIP2 into PIP3) turns-out to be also a protein kinase (Hunter 1995), whereas a lipid-phosphatase such as PTEN (which dephosphorylates PIP3 into PIP2) is also a protein phosphatase (Vazquez and Sellers 2000). More recently, it was shown that OGT, which glycosylates proteins according to glucose availability, is also a protease, capable of proteolytic cleavage of its substrates (Capotosti et al. 2011).

The complexity of the model of mechanism depicted in Fig. 12.4 is further increased by the fact that insulin triggers at the same time other mechanisms involving some of the entities described here as well as other entities not mentioned in this scheme, and resulting in other biological effects (protein synthesis, gene expression, etc. . . .).

In addition, because adipose cells in living organisms are not isolated as they are in experimental set-ups, they are continuously impacted by a blood stream that contains not only insulin but also numerous hormones, which act upon the considered cell through numerous receptors, capable of triggering different mechanism that may interfere with the mechanism elicited by insulin. The concentrations of these hormones in the blood vary during the day (circadian rhythms) and according to the physiological state (nutritional state, physical activity, stress, etc.). These hormones affect the activities of some of the entities involved in the mechanism of insulin action. For instance, catecholamines (produced during a stress or under fasting conditions) bind to a family of hormone receptors (adrenergic receptors) that trigger different mechanisms capable of inhibiting several steps of insulin action in adipocytes (Issad et al. 1995), while activating other steps (Gerhardt et al. 1999; Moule et al. 1997). Other hormones, growth factors or cytokines, sometimes secreted by the adipose cell itself (autocrine factors), can also, at any time, activate or inhibit some of the entities involved in insulin action. In addition, several metabolites (fatty acids, glucose, aminoacids . . .) can also affect the activity of these entities. For instance, it is interesting to note that part of the glucose taken up by the cell will be used to directly modify some of the entities through O-glycosylation (a reversible reaction analogous to phosphorylation that leads to addition of sugar on serine or threonine residues of proteins). These glycosylations can modify several of the players involved in insulin action (IRS-1, PKB, . . .), thereby modulating their activities (sometimes by competing with the phosphorylation induced by insulin on the same serine or threonine residues) (Issad et al. 2010; Issad and Kuo 2008).

Confronted with such a complexity, it seems legitimate to question the explanatory value of a mechanism, since it is becoming impossible to have a comprehensive view of the behaviour of the different parts working together. It is in such cases that the causal story (CS) condition of explanatory force appears to break down: a causal rehearsal that would take into account such a complexity is no longer possible.

4   When Mechanistic Explanations Become Dynamic

To deal with such complex bio-systems, scientists may recur to different dynamic modelling techniques and simulation tools, including, for instance, ordinary differential equations (ODEs), process calculi, boolean networks, Bayesian networks, Petri nets, bipartite graphs, stochastic equations (see, e.g., Baldan et al. 2010; de Jong 2002; Deville et al. 2003; Mandel et al. 2004; Wiekert 2002 for some surveys). These different modelling techniques may be quantitative or qualitative, discrete or continuous, deterministic or probabilistic, and are frequently put to use in the context of imprecise and often incomplete knowledge. The choice of modelling techniques depends on the extent of the available data, but also on the nature of the molecular processes at stake (e.g. chemical reaction, complex formation, binding process, transcriptional regulation, etc.), and on the chosen level of abstraction (for instance, a model of gene activation may include all different steps from transcription to translation, while another one may simply account for activation state). When much is known about the bio-system under scrutiny, and in particular when most kinetic data are available, scientists often turn to ODE-models as a means of gaining insight into the dynamic behaviour of the system. These models are then analyzed using bifurcation and phase plane analysis tools, and through numerical analyses run on computer simulation softwares (e.g., MATLAB, MATHEMATICA, SCILAB etc.).

Acknowledging the importance of such dynamic modelling tools in biology, Bechtel, Abrahamsen and Kaplan have proposed to extend the concept of mechanistic explanation so as to include what they call "dynamic mechanistic explanations" (Bechtel and Abrahamsen 2010; Kaplan and Bechtel 2011). A dynamic mechanistic explanation is an explanation that supplements a mechanism with a dynamic modelling. For instance, in the case of circadian rhythms (Bechtel and Abrahamsen 2010), it has been argued that a proper explanation of this phenomenon includes both a mechanism (whose components are specific molecular compounds interacting with each other in specific ways, such as promoting or inhibiting the synthesis of each other) and a dynamic modelling of this mechanism (that makes it possible to assess the time-related variations of key molecular compounds and therefore to account for the phenomenon at stake). So construed, dynamic mechanistic explanations involve both a mechanistic model ("mechanistic decomposition of the system into parts and operations"), and a mathematical model (typically from "computational modelling or dynamical systems theory") which are tightly coupled together in the sense that "some or all of the variables and terms in the mathematical model correspond to properties of identified parts and operations" (Bechtel and Abrahamsen 2012).

As Bechtel and colleagues have stressed, the display of the mechanism itself is not sufficient to provide an explanation of why there are circadian rhythms: one must appeal to dynamic modelling. Indeed, the time-related behaviour of the concentrations of the different molecules that figure in the mechanism can be accounted for by a set of three differential equations. Solving this set of equations numerically, that is to say typically by running a computer simulation, provides estimates of the time-related variations of each one of the different molecular concentrations. And, in this particular case, numerical simulations have shown that concentrations display oscillatory patterns with a period relatively close to 24 h, hence an explanation of circadian rhythms.[4] This has led Bechtel and colleagues to propose an *extension* of the mechanistic view of explanation: "Mentally rehearsing operations sequentially is not sufficient to determine how such a mechanism will behave, and the basic mechanistic account must be *extended* in the direction of dynamic mechanistic explanation in which computational modelling and dynamic systems analysis is invoked to understand the dynamic behaviour of biological mechanisms" (Bechtel 2011, p. 554, our italics). A key question is how one construes such an extension.

One way to tackle this question is to look at how mechanistic explanations (ME's) on the one hand and dynamic mechanistic explanations (DME's) on the other gain their *explanatory force*. Recall that the explanatory force of a ME comes from the fulfillment of the two conditions (*MM*) of displaying a model of mechanism, and (*CS*) of rehearsing a causal story of the production of the explanandum phenomenon (see Sect. 2). On the other hand, if we follow Bechtel and colleagues, it seems that the explanatory force of a DME comes from (1) displaying a model of mechanism, and (2) providing a mathematical dynamic modelling of this mechanism that accounts for the explanandum phenomenon, and which is such that some or all of its variables and terms correspond to properties of identified parts and operations of the mechanism (Bechtel and Abrahamsen 2012). While the first element corresponds to the (*MM*) condition above, the second explanatory element is of a different kind: it consists in specifying a mathematical model, in solving it – typically by means of numerical methods – and in showing how the solutions fit with the explanandum phenomenon. The explanatory force of DME's can therefore be made more specific in the following way:

> A dynamic mechanistic explanation (DME) explains a phenomenon (P) produced by a system (S) in virtue of fulfilling the following two necessary and sufficient conditions:
>
> (*MM*)   Displaying a model of mechanism that represents a real mechanism in S, with its entities, activities and spatio-temporal organization
> (*MA*)   Displaying a mathematical model such that some or all of its variables and terms correspond to properties of identified parts and operations of the mechanism, and providing mathematical warrants that its resolution/simulation fits P.

The explanatory force of a DME is therefore not coming from the very same elements as the explanatory force of a ME. While a DME and a ME share the (*MM*) condition of displaying a model of mechanism, the ME gains explanatory force by rehearsing a causal story (condition *CS*), while the DME gains explanatory force by providing and solving a mathematical modelling (condition *MA*).

---

[4]For the sake of brevity, we over-simplify this example. See (Bechtel and Abrahamsen 2010, 2012) for more details.

Somewhat contrary to what Bechtel and colleague propose, it is difficult to see how DME's could constitute an extension of ME's since the explanatory force of ME's include an element that is not even present in the case of DME's, namely the condition (CS) of rehearsing a causal story. How can one make sense of such an "extension"? It seems that we have here two possibilities. The first one would be to extend the concept of "mechanistic explanation" such that this concept should include DME's in addition to traditional ME's. If we follow this route, the explanatory force of a mechanistic explanation* would come either from the explanatory elements of a traditional ME *or* from the explanatory elements of a DME.[5] Put more formally, we would have the following explication of the explanatory force of a mechanistic explanation*:

> A mechanistic explanation* explains a phenomenon (P) produced by a system (S) in virtue of fulfilling the following condition: $(MM) \bigwedge [(CS) \bigvee (MA)]$.[6]

Such an option leads to defining a mechanistic explanation as a disjunction of two types of explanations that only share one common explanatory feature: the (MM) condition. It would therefore seem that the key explanatory element of a mechanistic explanation comes from being based on the display of a mechanistic model. Yet this option, we argue, strongly weakens the traditional construal of a mechanistic explanation, since it removes an essential part of it that consists precisely in showing how the phenomenon is causally produced by the mechanism. As such, this would run contrary to the idea that a key feature of a mechanistic explanation is to exhibit causal relevance (e.g. Craver 2007, Chap. 2).

The second possibility to make sense of an extension of mechanistic explanation in the direction of dynamic mechanistic explanation could be to criticize our account of what constitutes the explanatory force of a DME, and argue that the causal story condition (CS) ought also to be included. In this case, ME's and DME's would share the two core explanatory elements (MM) and (CS). Note that, as a result, DME would not truly be an extension of ME, but rather a special type of ME whose explanatory force would come from the adjunction of a further condition, namely a mathematical modelling as described in (MA). Yet, is this a viable option? Do DME's indeed lend themselves to causal story telling and do they gain explanatory force that way? We think not, and will argue the case with the help of an example that builds on the explanation of the insulin action we have started to describe above (Sect. 3). In a nutshell, we argue that, in the case of DME's, causal stories fall short of showing what is precisely expected of a causal story, namely how the system under study indeed causally produces the explanandum phenomenon.

---

[5]We have added a * to "mechanistic explanation" to distinguish this new extended concept from the traditional and narrower concept of mechanistic explanation that we have initialed (ME) everywhere else in the paper.

[6]With the conditions (MM), (CS) and (MA) as previously defined.

5 Explanatory Force in the Case of Complex Biological Systems

We have seen earlier that the translocation of glucose transporters from the cytoplasm to the cell membrane, thereby activating glucose uptake, is properly explained by a specific mechanistic explanation based on a model of mechanism and an associated causal story within which insulin plays a triggering causal role. Yet, the activation of glucose transporters is but one of the numerous mechanisms that happen to be at work in the broader phenomenon of glucose homeostasis. Recall that glucose homeostasis is the capacity of the human organism to keep its concentration of glucose in the blood (glycaemia) within a narrow range, around a basal value of 1 g per litre. Understanding how this is achieved is important as it may shed light on major dysfunctions such as chronic hyperglycaemia and diabetes. Explaining glucose homeostasis is however no trivial matter, as this phenomenon involves several organs, tissues and numerous molecular compounds and interactions across the whole body. Body-level models have been developed that link the concentration of glucose in the blood to the concentration of insulin, the latter inducing storage or release of glucose by the liver and its uptake by muscle and adipose tissue (e.g. Dalla Man et al. 2006). Such body-level models have also been linked to tissue-level models, and in particular to models of the adipose tissue with regards to the action of insulin onto the activation of glucose transporters and glucose storage (e.g. Nyman et al. 2011). These tissue-level models are based on the mechanism of insulin action that we have seen earlier, but have also been detailed further so as to take into account non-linear interactions between insulin molecules and insulin receptors at the cell surface (e.g. Brännmark et al. 2010; Kiselyov et al. 2009). This has led to the study of multi-level models as a means of explaining the action of insulin onto glucose homeostasis by unraveling a cascade of interactions, from the binding on insulin to its receptor, to the activation of glucose transporters, and finally the absorption of glucose by the adipose tissue and its effect on the whole glucose-insulin control system (Nyman et al. 2011). Such a model is depicted in Fig. 12.5.

It can be argued that what we have here is nothing but a classic "hierarchically organized mechanism", that is to say a mechanism involving several levels of finer-grained mechanisms (e.g. Craver 2007, Chap. 5). And we grant that such a hierarchical multi-level model can be said to be a model of mechanism in the sense that it includes entities, activities and some form of spatio-temporal organization. An explanation that would be based on such model would therefore draw explanatory force from meeting the (*MM*) condition. Recall, however, that the key question is whether such explanation would be explanatory by means of rehearsing a causal story, hence fulfilling also the (*CS*) condition.

Before examining this question, let us analyze in more detail the purported explanation that scientists put forward here (Nyman et al. 2011). The claim made by Cedersund, Strålfors and their colleagues is that such multi-level model explains the behaviour of the human organism with respect to glucose homeostasis as controlled by the adipose tissue. Following meal ingestion, the glucose concentration in the

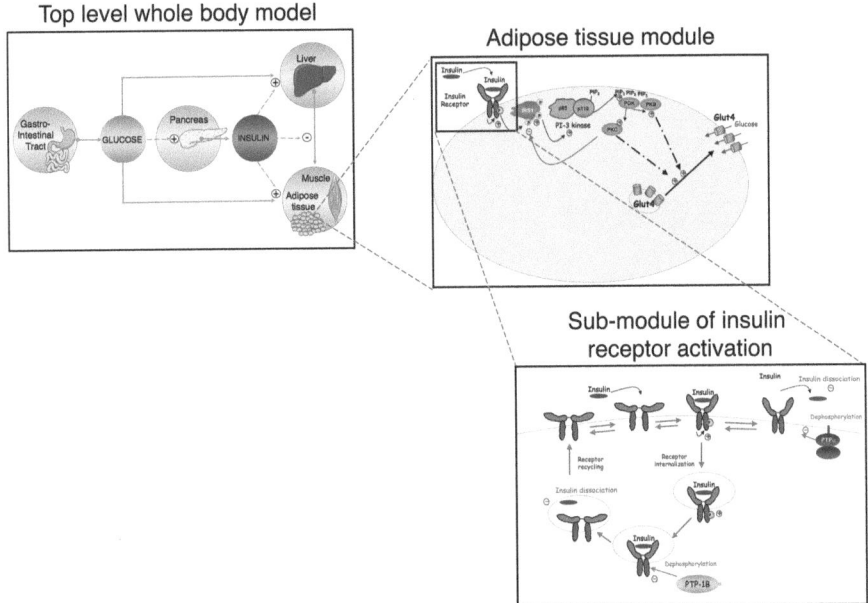

Fig. 12.5 Multi-level dynamic modelling of the action of insulin on glucose homeostasis (Adapted from Nyman et al. 2011)

blood increases.[7] This triggers the secretion of insulin by beta-cells in the pancreas and produces an "insulin signal." This insulin signal is then decoded by the liver and by the adipose tissue. It is this latter adipose tissue that is the focus of the model. The presence of insulin in the adipose tissue triggers a cascade of events, from the binding of insulin to insulin receptors, to the translocation of glucose transporters and the uptake of sugar by adipose cells. In turn, this uptake of sugar decreases the concentration of sugar in the blood, enabling the concentration of sugar to return to normal. The explanandum is why does the blood concentration return to normal after an increase due to meal ingestion? More precisely, what needs to be explained is the time-dependent concentration of glucose. The explanandum therefore is a dynamic feature of the human organism: the plotted curve of sugar uptake that has been measured experimentally as a function of time (See Fig. 12.6). And the scientists' claim is that their multi-level model explains precisely that: why glucose uptake has the time-dependent curve it has.[8]

---

[7]Glucose homeostasis is not just involved during meal ingestion but also during fasting. For the sake of simplicity, we do not cover this aspect of the phenomenon, as this is not central to our philosophical argument.

[8]To be accurate, the model aims at accounting not only for the time-variation of the concentration of glucose, but also for the time-variations of other compounds that can be measured experimentally, such as the % of insulin receptors and of the % of IRS1 that are phosphorylated in response to different insulin concentration variations.

Fig. 12.6 Plotted curve of experimentally measured sugar uptake as a function of time (Adapted from Nyman et al. 2011)

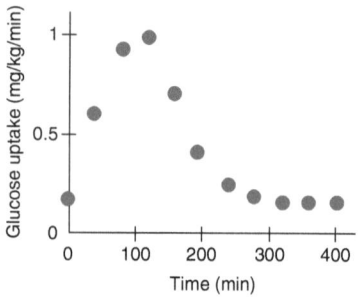

$$\dot{IR} = k1b \cdot IRp - k1f \cdot IR \cdot insulin - k1basal \cdot IR$$

$$\dot{IRp} = -k1b \cdot IRp + k1f \cdot IR \cdot insulin + k1basal \cdot IR$$

$$\dot{IRS} = k2b \cdot IRSp - k2f \cdot IRS \cdot IRp$$

$$\dot{IRSp} = -k2b \cdot IRSp + k2f \cdot IRS \cdot IRp$$

$$\dot{PKB} = k3b \cdot PKBp - k3f \cdot PKB \cdot IRSp$$

$$\dot{PKBp} = -k3b \cdot PKBp + k3f \cdot PKB \cdot IRSp$$

$$\dot{GLUT4} = k4b \cdot GLUT4pm - k4f \cdot GLUT4 \cdot PKBp$$

$$\dot{GLUT4pm} = -k4b \cdot GLUT4pm + k4f \cdot GLUT4 \cdot PKBp$$

$$glucose\ uptake = k_{glut1} \cdot \frac{glucose}{Km_{G1} + glucose} + k_{glut4} \cdot \frac{glucose}{Km_{G4} + glucose} \cdot GLUT4pm$$

Fig. 12.7 Example of a set of ODE's used in one of the models of (Nyman et al. 2011)

The multi-level model tackles this question by exhibiting a hierarchy of models of mechanisms at the levels of the organism, the adipose tissue and the insulin receptor. All of these models are looked at from a time-dependent perspective: what matters is what happens as a function of time. And they are all non-linear, involving numerous feedback mechanisms, including some that run across levels.[9] The way the different entities and activities of these models are represented is by means of a large set of ordinary differential equations (ODE's) that account for the variation over time of the concentrations of the different compounds as a function of their interactions with each other. For instance, the time-derivative of the concentration of insulin receptors is a function of the concentration of phosphorylated insulin receptors, of the concentration of insulin receptors and of the concentration of insulin (see Fig. 12.7 for a larger set of such ODE's).

---

[9]For instance, Cedersund and Strålfors have included a "blood flow effect" linking the whole-body model and the adipose tissue model, so as to account for the fact that blood flow may affect the availability of glucose and insulin in the local interstitial tissue surrounding the adipocytes, which in turn may effect glucose uptake by the adipocytes. For more details, see (Nyman et al. 2011).

Using ODE's is a mathematical way of encapsulating the causal knowledge that the variation of the concentration of certain entities is produced by the presence of other entities, at specific rates (reflected by kinetic constants). In a way, each ODE tells a causal story. Yet, these causal stories, even when patched together into a larger causal story do not lead to the production of the phenomenon to be explained. Recall that the explanandum phenomenon is the variation of sugar concentration as a function of time. At best, the causal story relates the variation of sugar uptake as a function of the concentration of other entities and of the variations of these concentrations or, more generally speaking, of other variables. The only way to untangle the many functional dependencies is to either analytically solve the set of ODE's when this is mathematically tractable, or to numerically estimate the solution functions by running computer simulations of the set of ODE's with the help of a computing software such as MATLAB, MATHEMATICA, SCILAB, or others. The output of these computer simulations is a set of functions that represent the values of each variable as a function of time. Among these, one will have the simulated values of sugar uptake as a function of time. And if this curve matches – with certain margins of error – the experimentally measured one, one will claim to have reproduced the explanandum phenomenon, and thereby to have explained it (see Fig. 12.8). Yet, where is the causal story in this account?

If causal stories are indeed possible when it comes to interpreting the individual ODE's that link the different variables with each other, such causal stories stop as soon as one steps into the process of mathematically solving the set of ODE's. Indeed solving the set of ODE's involve numerous technical steps that make it extremely difficult to see how any causal story could be rehearsed at all. For instance, in the very simple case of a set of first order linear differential equations, finding solutions typically involves writing the system of equations under a matrix

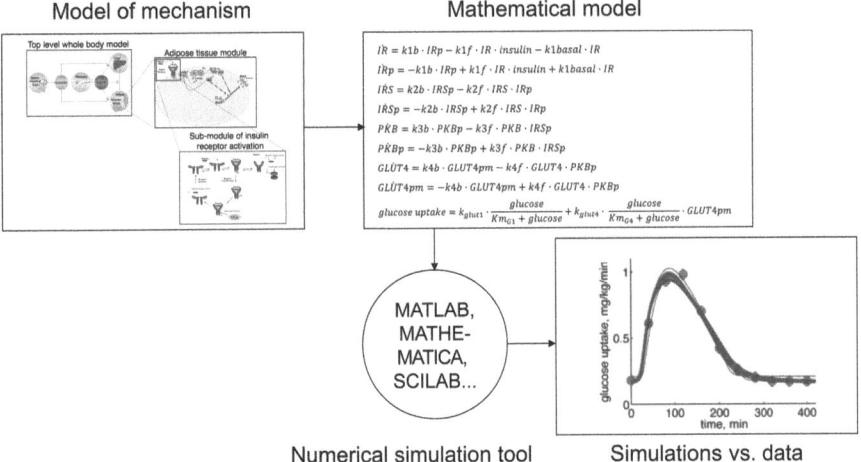

Fig. 12.8 Explanatory steps in the explanation of glucose homeostasis

form and finding the eigenvalues and eigenvectors of that very matrix. These different mathematical steps, many of which are very intricate and technical, relying in turn on sophisticated mathematical theorems, make it difficult to see how any causal story could make sense there. And of course, sets of differential equations can be much more tricky than those, requiring one to revert to "complex eigenvalues", to "repeated eigenvalues", to "undetermined coefficients", to techniques of "variation of parameters", to "Laplace Transforms", not to mention the many different tools involved in modelling and computer simulations when those are used to compute numerical solutions. The causal story that, in the case of traditional mechanistic explanations, provides explanatory force by rehearsing how the different entities and activities of the mechanism produce the explanandum phenomenon, appears to break down in the case of more complex and dynamic systems. Indeed, the mathematical techniques that are required to identify or even just approximate solutions in the case of sets of differential equations appear to block any attempt to causally relate the entities and activities of the mechanism to the explanandum phenomenon.[10]

The causal story condition (*CS*) therefore is not central in providing explanatory force to DME. Yet the question remains as to where this explanatory force comes from. Our proposal is that the explanatory force comes from the *mathematical derivation* – analytic or numerical – of the explanandum from the model of mechanism. In other words, not from the satisfaction of the condition (*CS*) but from the satisfaction of the condition (*MA*) that consists in providing a mathematical modelling of the model of mechanism such that some or all of its variables and terms correspond to properties of the components and activities of that mechanism, and such that, when analyzed or solved, it leads to a formal reproduction of the explanandum phenomenon. As can be seen in the case of glucose homeostasis, explaining why the concentration of glucose has the specific time-dependent values it has consists (1) in exhibiting a model of mechanism and (2) in asserting that the explanandum phenomenon fits with a mathematical solution of the set of ODE's that describe the mechanism.

The explanatory force of this second component of the explanation does not consist at all in a causal account of how the mechanism produces the phenomenon, but rather in showing that, given a certain set of equations, only specific solution-functions are possible that fit with the explanandum phenomenon. It is the mathematics that tell us how the system variables behave (typically as a function of time), and no causal story. In other words, we grant explanatory power to these solution-functions because they are the result of a mathematical treatment of the model: we are given *mathematical warrants* that the mathematical modelling necessitates the solution-functions. In other words, the facts doing the explaining explain in virtue of

---

[10]Even though the example here only involves ordinary differential equations, we argue that the same argument could be extended to mathematical and numerical methods used in process calculi, boolean networks, Bayesian networks, Petri nets, bipartite graphs, stochastic equations and other ways of formalizing biological models in general.

being mathematically necessary. Indeed, Lange characterizes the explanatory force of mathematical explanation by the type of modality that mathematical explanations exhibit (Lange 2012): mathematical explanations typically derive their explanatory force from being modally stronger than laws of nature might actually be, since mathematical explanations are true of any possible world, which is not the case of laws of nature. This is all the more the case for explanations that are solely mathematical in the sense of explaining in virtue of being mathematical in a different way from ordinary explanations that use mathematics.[11] Yet, even if DME are not solely mathematical since they involve not just mathematical models but also models of mechanisms, their explanatory force does come *in part* from being necessitated by mathematical facts.

6   Causally-Interpreted Model Explanations (CIME's)

Explanations of the alleged "mechanistic type" vary significantly in terms of the elements that make up their explanatory force, despite sharing the condition of displaying a mechanistic model (condition *MM*). Indeed, whereas ME's draw their explanatory force from rehearsing a causal story (condition *CS*), DME's do not lend themselves to a causal story telling that would be explanatorily relevant, but rather draw their explanatory force from a mathematical model and its resolution or simulation (condition *MA*). Therefore DME's are neither an extension nor a complement to ME's. Because they draw their explanatory force from a different element, they are rather a different type of explanation altogether.

Another way to see that these two types of explanations are indeed distinct is to notice that they are mobilized in different contexts. Given an explanandum, if the mechanistic model of the explanans is fairly simple and linear (or possibly with very few non-linear elements), then it will be enough for the explanans to complement the mechanistic model by a causal story, and this will do as an explanation. In this case, the explanation will be a mechanistic explanation of the ME type. Yet, if, given another explanandum, the mechanistic model of the explanans becomes complex, with numerous elements and with a high degree of non-linearity, then causal stories do not succeed in bridging the mechanistic model and the explanandum. Rather, they are to be dropped and replaced by a mathematical modelling of the mechanistic model assorted with its resolution/simulation. It is the analytic resolution or the numerical simulation of such mathematical model that enables one to successfully formulate an explanation of the explanandum phenomenon. In other words, the explanatory force of an explanans varies in nature depending on the complexity

---

[11]Examples of distinctively mathematical explanations include explanation of why no one ever succeeded in crossing all of the bridges of Königsberg exactly once as specified by Euler, or of why one cannot unknot a trefoil knot, or even of why certain cicada life-periods are prime. See for instance (Baker 2009, 2015, this volume; Lange 2012; Pincock 2007). See also Baetu (2015, this volume), and Breidenmoser and Wolkenhauer (2015, this volume).

of the model that it relies on. For simple models, the explanatory force may simply come from displaying a mechanistic model and rehearsing a causal story of how this mechanism works and produces the phenomenon of interest. But for complex models, the explanatory force will come from displaying a mechanistic model, and a corresponding mathematical treatment resulting in a reproduction of the phenomenon of interest. Because the elements that give DME's their explanatory force do not include the elements that make ME's explanatory, DME's are not an extension of ME's.

It is worth noticing that scientists themselves tend to use different terms to refer to these different types of explanations. As has been amply recognized (e.g. Machamer et al. 2000), the concept of "mechanistic explanation" is very much used in molecular biology. Yet often, mechanistic explanations of the ME type are offered in the context of fairly simple models of mechanisms. When the models become complex, typically with numerous elements and non-linearity effects, biologists no longer tend to refer to "mechanistic explanations": rather they couch their explanations in the vocabulary of "models". This distinction can be seen in the example of insulin research: the explanation of the activation of glucose transporters by insulin is often referred to as a mechanism (Cushman and Wardzala 1980; Hoffman and Elmendorf 2011; Krook et al. 2004); yet, when it comes to explaining the dynamic features of glucose homeostasis, "model" very often replaces "mechanism" (e.g. Nyman et al. 2011).[12]

At this point, one may still object that such model explanations are mechanistic in the weaker sense that they nevertheless rely on a model of mechanism as specified by the condition (*MM*). We could reply that the main argument of our paper was not to show that traditionally construed ME's and DME's did not share any explanatory element at all, but rather to show that they drew their explanatory force from *at least one different element*, a causal story (*CS*) in the case of ME, and a mathematical modelling (*MA*) in the case of DME.

We could also take a further step in the comparison of ME's and DME's. And indeed, we argue that they can be conceived as two extremes of a more encompassing type of explanation that we propose to call "Causally Interpreted Model Explanations" (or CIME's). To support this claim, a further analysis of the different elements of explanatory force that we have singled out above is required. The general idea is that these elements include heterogeneous ideas about formal models on the one hand, and about causal interpretation on the other, and that their reformulation in more homogeneous ways makes it possible to see how ME's and DME's draw their explanatory force differently.

Consider first the (*MM*) condition of displaying a model of mechanism:

(*MM*)   Displaying a model of mechanism that represents a real mechanism in S, with its entities, activities and spatio-temporal organization

---

[12]Because the explanatory force depends on the complexity of the model, and because complexity is relative to cognitive faculties, we acknowledge here the fact that the nature of the explanatory force will vary from cognitive context to cognitive context.

This condition actually includes two distinct elements. First, it includes a condition about displaying a *model* that has specific characteristics (namely a model with constituents that can be termed entities and activities, and which incorporates a specific spatio-temporal organization). And second, it includes a condition that requires the model (and its constituents and organization) to map onto a real mechanism (and its real constituents and organization). This second condition corresponds to an *interpretation* of the model; and because the entities and activities of mechanisms are individuated on the basis of their causal roles, it is a *causal* interpretation of the model.

Consider now the condition (*MA*) of displaying a mathematical model:

(*MA*)   Displaying a mathematical model such that some or all of its variables and terms correspond to properties of identified parts and operations of the mechanism, and providing mathematical warrants that its resolution/simulation fits P.

This condition of explanatory force also mixes requirements that concern a model on the one hand, and its interpretation on the other. The condition indeed requires that a *model* be presented, and this model be of a mathematical type. The condition also requires that some of the variables and terms of the mathematical model correspond to parts and operations – or entities and activities – of the underlying real mechanism that is supposed to produce the explanandum phenomenon. This second condition is again a condition of *interpretation* of the model, and this interpretation is typically required to be causal. Lastly, the condition requires that one should be given warrants that the mathematical solutions of the model follow, either analytically through deductions or numerically through simulations, from the model, and that these solutions fit the explanandum phenomenon. Let's call this last condition a condition of *derivation* of the explanandum phenomenon.

Lastly, consider the causal story condition (*CS*) that plays such a key role in ME's when conjoined with the model of mechanism condition (*MM*):

(*CS*)   Rehearsing a causal story that enumerates the cause-effect relationships taking place in the mechanism up to the production of the phenomenon P.

The requirements of (*CS*) can be articulated along two components. First (*CS*) requires the identification of causes and effects up to the explanandum phenomenon P. It therefore includes a *causal interpretation* of what is captured by the model of mechanism and that is supposed to happen in the real mechanism. Second, (*CS*) requires a "story" that links the different causes to their effects, and ultimately the mechanism to the explanandum phenomenon. This story typically includes the verbs that correspond to the activities of the entities of the mechanism. However, should a functional (mathematical) formulation be made available, such a story could easily be transcribed into simple mathematical (or even Boolean) expressions.[13] It is

---

[13]To see how this could be done, remember that a manipulationist account of causation is often assumed to be compatible with the mechanistic account of explanation. Such an account of causation is typically articulated around (causal) variables and makes it possible to transcribe the (cause-effect) relationships between these variables into mathematical functions. See for instance (Woodward 2003)

therefore entirely possible to rephrase the causal story as a succession of simple mathematical derivations, up to a mathematical expression of the explanandum phenomenon. For this reason, parts of the (*CS*) requirements indeed consist of some form of *derivation* requirement.

To sum up, the above unpacking of (MM), (MA) and (CS) shows that these requirements of explanatory force draw somehow differently on three main elements: a model, a causal interpretation and a derivation. We propose to define the notion of "Causally Interpreted Model Explanations" (CIME's) around these three elements as follows:

> A causally interpreted model explanation (CIME) explains a phenomenon (P) produced by a system (S) in virtue of the following three necessary and sufficient conditions:
>
> (*M*)   Displaying a model of S that includes variables and functions defined over these variables
>
> (*CI*)   Providing a causal interpretation of these variables and functions (for instance by means of a manipulationist account of causation)
>
> (*D*)   Deriving, through a mathematical treatment (be it analytical or numerical), the explanandum phenomenon P from the model.

So construed, CIME's explain a phenomenon neither in virtue of displaying a mechanism nor in virtue of providing a causal account, but in virtue of mathematically showing how the explanandum can be analytically or numerically derived from a model whose variables and functions can be causally interpreted.[14]

ME's and DME's can then be seen as two different limiting cases of CIME's, even if no sharp delineation may actually exist between the two. Indeed, a ME is a limit case of CIME in which (1) the model specified by (*M*) is relatively simple and the causal interpretation provided in (*CI*) makes it possible to construe the model as a mechanism, hence fulfilling the (*MM*) condition, and (2) the derivation of the explanandum from the model is also simple enough – typically involving linear relations – so as to be reformulated into a causal story – typically with verbs transcribing functional relations into activities, and other terms transcribing variables and their values into entities – thereby fulfilling (*CS*). Consider now a DME. We argue that a DME can be construed as a limit case of CIME in which (1) the model specified by (*M*) is relatively complex, involving in particular non-linear functions, (2) the causal interpretation provided by (*CI*) is likely to be less exhaustive compared to a ME, possibly involving non-interpreted parameters, and (3) the derivation of the explanandum from the model is done through relatively intricate analytical or numerical methods and, as a result, does not lend itself to a reformulation as a causal-story (see Fig. 12.9).

---

[14]This account is characterized in terms of models that rely on "variables" and "functions". We believe – though we do not show it here – that such characterization could be extended to cover other more specific ways of construing models, be they for instance in terms of set-theoretic predicates or phase-state. Note also that we do not mean that *all* variables in the model be causally interpreted. It is indeed possible that some of the variables do not receive any causal interpretation. This may also be the case for specific parameters and coefficients used in the model.

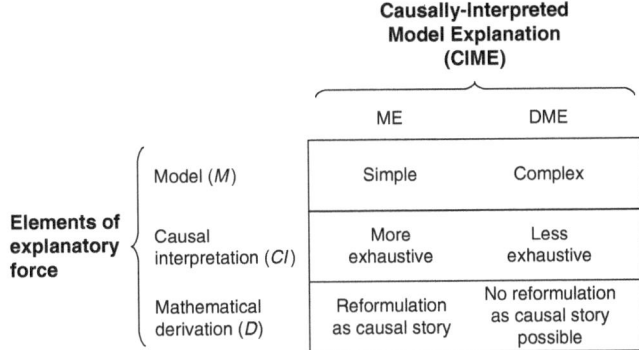

Fig. 12.9 Elements of explanatory force of Causally-Interpreted Model Explanations (CIME's), in relationship to Mechanistic Explanations (ME's) and Dynamic Mechanistic Explanations (DME's)

An important point is the fact that supplying a causal-story is no longer seen central in providing explanatory force: the key element of explanatory force is the mathematical derivation. While this derivation *can* be reformulated in causal language in some of the simpler cases, we have to acknowledge that in the more complex cases, such derivation involves many intricate mathematical treatments – including numerous theorems and proofs, approximating techniques, numerical simulation tools etc. – that make any causal retranscription hopeless. Yet, the derivation *does* convey the type of explanatory force that is expected from a proper explanation. And it does so in virtue of the mathematical warrants that it provides and that make the explanandum follow from the model with mathematical necessity.

We also argue that providing a mechanism per se is also not so central when it comes to explanatory force. What matters more generally is the presence of a model and of its causal interpretation. And while we recognize the fact that, in many relatively simple cases, such model and causal interpretation can be transcribed into a model of mechanism, we also want to underline the fact that, in numerous complex cases, the explanatory does not come from a model of mechanism but from a model – more generally construed – and the causal interpretation that accompanies it. The causal interpretation in particular is seen as providing the crucial link between the model and biological reality. But it is the analytical/numerical derivation that links the model to the explanandum, not a causal story.

7   Conclusion

In this contribution, we aimed at getting at the source of the explanatory force that is claimed to be that of mechanistic explanations. When looking at simple cases of mechanisms in biology, such as the insulin-induced relocalisation of glucose transport systems, it appears that this explanatory force can be understood as coming from a model of mechanism and a causal story that tells how the mechanism produces the explanandum phenomenon. However, when biological systems tend to become complex, as in the case of glucose homeostasis, mathematical modelling becomes a necessary explanatory component that turns mechanistic explanations into dynamic mechanistic explanations. It is important to note that the explanatory force of such explanations does not any longer come from the rehearsing of a causal story, but from mathematical warrants that the explanandum follows from the model. To account for this disparity of explanatory force, we have proposed a more general pattern of "Causally Interpreted Model Explanations" according to which explanations draw their explanatory force from a *model*, a *causal interpretation* that links the model to biological reality, and a *mathematical derivation* that links the model to the explanandum phenomenon. Mechanistic explanations and dynamic mechanism explanations can then be construed as different limit cases of causally interpreted model explanations. Nevertheless, providing causal stories is not as central to explanatory force as is mathematical derivation.

Acknowledgements The authors wish to thank the audience at the 2013 ISHPSSB meeting in Montpellier for very helpful discussion points. An earlier version of this paper also benefited from comments from William Bechtel and Pierre-Alain Braillard, as well from two anonymous reviewers. CM acknowledges financial support from UQAM research chair in Philosophy of science.

References

Baetu, T. (2015). From mechanisms to mathematical models and back to mechanisms: Quantitative mechanistic explanations. In P.-A. Braillard & C. Malaterre (Eds.), *Explanation in biology. An enquiry into the diversity of explanatory patterns in the life sciences* (pp. 345–363). Dordrecht: Springer.

Baker, A. (2009). Mathematical explanation in science. *The British Journal for the Philosophy of Science, 60*(3), 611–633. doi:10.1093/bjps/axp025.

Baker, A. (2015). Mathematical explanation in biology. In P.-A. Braillard & C. Malaterre (Eds.), *Explanation in biology. An enquiry into the diversity of explanatory patterns in the life sciences* (pp. 229–247). Dordrecht: Springer.

Baldan, P., Cocco, N., Marin, A., & Simeoni, M. (2010). Petri nets for modelling metabolic pathways: A survey. *Natural Computing, 9*, 955–989. doi:10.1007/s11047-010-9180-6.

Bechtel, W. (2011). Mechanism and biological explanation. *Philosophy of Science, 78*(4), 533–557. doi:10.1086/661513.

Bechtel, W., & Abrahamsen, A. (2010). Dynamic mechanistic explanation: Computational modeling of circadian rhythms as an exemplar for cognitive science. *Studies in History and Philosophy of Science Part A, 41*(3), 321–333. doi:10.1016/j.shpsa.2010.07.003.

Bechtel, W., & Abrahamsen, A. A. (2012). Thinking dynamically about biological mechanisms: Networks of coupled oscillators. *Foundations of Science.* doi:10.1007/s10699-012-9301-z.

Belsham, G. J., Brownsey, R. W., Hughes, W. A., & Denton, R. M. (1980). Anti-insulin receptor antibodies mimic the effects of insulin on the activities of pyruvate dehydrogenase and acetylCoA carboxylase and on specific protein phosphorylation in rat epididymal fat cells. *Diabetologia, 18*(4), 307–312. doi:10.1007/BF00251011.

Brännmark, C., Palmér, R., Glad, S. T., Cedersund, G., & Strålfors, P. (2010). Mass and information feedbacks through receptor endocytosis govern insulin signaling as revealed using a parameter-free modeling framework. *The Journal of Biological Chemistry, 285*(26), 20171–20179. doi:10.1074/jbc.M110.106849.

Breidenmoser, T., & Wolkenhauer, O. (2015). Explanation and organizing principles in systems biology. In P.-A. Braillard & C. Malaterre (Eds.), *An enquiry into the diversity of explanatory patterns in the life sciences* (pp. 249–264). Dordrecht: Springer.

Capotosti, F., Guernier, S., Lammers, F., Waridel, P., Cai, Y., Jin, J., Conaway, J. W., Conaway, R. C., & Herr, W. (2011). O-GlcNAc transferase catalyzes site-specific proteolysis of HCF-1. *Cell, 144*(3), 376–388.

Carracedo, A., & Pandolfi, P. (2008). The PTEN-PI3K pathway: Of feedbacks and cross-talks. *Oncogene, 27*(41), 5527–5541.

Combettes-Souverain, M., & Issad, T. (1998). Molecular basis of insulin action. *Diabetes & Metabolism, 24*(6), 477.

Craver, C. F. (2007). *Explaining the brain: Mechanisms and the mosaic unity of neuroscience.* Oxford: Oxford University Press.

Cushman, S. W., & Wardzala, L. J. (1980). Potential mechanism of insulin action on glucose transport in the isolated rat adipose cell. Apparent translocation of intracellular transport systems to the plasma membrane. *Journal of Biological Chemistry, 255*(10), 4758–4762.

Dalla Man, C., Rizza, R. A., Cobelli, C. (2006). Mixed meal simulation model of glucose-insulin system. In *Conference proceedings: 26th annual international conference of the IEEE Engineering in Medicine and Biology Society. IEEE Engineering in Medicine and Biology Society. Conference, 1*(10), 307–310. doi:10.1109/IEMBS.2006.260810.

De Jong, H. (2002). Modeling and simulation of genetic regulatory systems: A literature review. *Journal of Computational Biology: A Journal of Computational Molecular Cell Biology, 9*(1), 67–103. doi:10.1089/10665270252833208.

Deville, Y., Gilbert, D., van Helden, J., & Wodak, S. (2003). An overview of data models for the analysis of biochemical pathways. *Briefings in Bioinformatics, 4*(3), 246–259.

Gerhardt, C. C., Gros, J., Strosberg, A. D., & Issad, T. (1999). Stimulation of the extracellular signal-regulated kinase 1/2 pathway by human beta-3 adrenergic receptor: New pharmacological profile and mechanism of activation. *Molecular Pharmacology, 55*(2), 255–262.

Glennan, S. S. (2002). Rethinking mechanistic explanation. *Philosophy of Science, 69*(3), S342–S353.

Glennan, S. S. (2005). Modeling mechanisms. *Studies in History and Philosophy of Science Part C: Studies in History and Philosophy of Biological and Biomedical Sciences, 36*(2), 443–464. doi:10.1016/j.shpsc.2005.03.011.

Hoffman, N. J., & Elmendorf, J. S. (2011). Signaling, cytoskeletal and membrane mechanisms regulating GLUT4 exocytosis. *Trends in Endocrinology & Metabolism, 22*(3), 110–116.

Hubbard, S. R. (1997). Crystal structure of the activated insulin receptor tyrosine kinase in complex with peptide substrate and ATP analog. *The EMBO Journal, 16*(18), 5572–5581.

Hubbard, S. R., Wei, L., Ellis, L., & Hendrickson, W. A. (1994). Crystal structure of the tyrosine kinase domain of the human insulin receptor. *Nature, 372*(6508), 746.

Hunter, T. (1995). When is a lipid kinase not a lipid kinase? When it is a protein kinase. *Cell, 83*(1), 1–4.

Issad, T., & Kuo, M. (2008). O-GlcNAc modification of transcription factors, glucose sensing and glucotoxicity. *Trends in Endocrinology & Metabolism, 19*(10), 380–389.

Issad, T., Combettes, M., & Ferre, P. (1995). Isoproterenol inhibits insulin-stimulated tyrosine phosphorylation of the insulin receptor without increasing its serine/threonine phosphorylation. *European Journal of Biochemistry, 234*(1), 108–115.

Issad, T., Masson, E., & Pagesy, P. (2010). O-GlcNAc modification, insulin signaling and diabetic complications. *Diabetes & Metabolism, 36*(6), 423–435.

Kaplan, D. M., & Bechtel, W. (2011). Dynamical models: An alternative or complement to mechanistic explanations? *Topics in Cognitive Science, 3*(2), 438–444. doi:10.1111/j.1756-8765.2011.01147.x.

Kaplan, D. M., & Craver, C. F. (2011). The explanatory force of dynamical and mathematical models in neuroscience: A mechanistic perspective. *Philosophy of Science, 78*(4), 601–627.

Kasuga, M., Zick, Y., Blith, D. L., Karlsson, F. A., Häring, H. U., & Kahn, C. R. (1982a). Insulin stimulation of phosphorylation of the beta subunit of the insulin receptor. Formation of both phosphoserine and phosphotyrosine. *Journal of Biological Chemistry, 257*(17), 9891–9894.

Kasuga, M., Zick, Y., Blith, D., Crettaz, M., & Kahn, C. (1982b). Insulin stimulates tyrosine phosphorylation of the insulin receptor in a cell-free system. *Nature, 298*, 667–669.

Kiselyov, V. V., Versteyhe, S., Gauguin, L., & De Meyts, P. (2009). Harmonic oscillator model of the insulin and IGF1 receptors' allosteric binding and activation. *Molecular Systems Biology, 5*(243), 243. doi:10.1038/msb.2008.78.

Krook, A., Wallberg-Henriksson, H., & Zierath, J. R. (2004). Sending the signal: Molecular mechanisms regulating glucose uptake. *Medicine and Science in Sports and Exercise, 36*(7), 1212–1217.

Lacasa, D., Boute, N., & Issad, T. (2005). Interaction of the insulin receptor with the receptor-like protein tyrosine phosphatases PTPalpha and PTPepsilon in living cells. *Molecular Pharmacology, 67*(4), 1206–1213.

Lange, M. (2012). What makes a scientific explanation distinctively mathematical? *The British Journal for the Philosophy of Science, 0*, 1–27. doi:10.1093/bjps/axs012.

Machamer, P., Darden, L., & Craver, C. (2000). Thinking about mechanisms. *Philosophy of Science, 67*(1), 1–25.

Mandel, J., Palfreyman, N., Lopez, J., & Dubitzky, W. (2004). Representing bioinformatics causality. *Briefings in Bioinformatics, 5*(3), 270–283.

Moule, S. K., Welsh, G. I., Edgell, N. J., Foulstone, E. J., Proud, C. G., & Denton, R. M. (1997). Regulation of protein kinase B and glycogen synthase kinase-3 by insulin and β-adrenergic agonists in rat epididymal fat cells activation of protein kinase B by wortmannin-sensitive and-insensitive mechanisms. *Journal of Biological Chemistry, 272*(12), 7713–7719.

Nouaille, S., Blanquart, C., Zilberfarb, V., Boute, N., Perdereau, D., Roix, J., Burnol, A.-F., & Issad, T. (2006). Interaction with Grb14 results in site-specific regulation of tyrosine phosphorylation of the insulin receptor. *EMBO Reports, 7*, 512–518.

Nyman, E., Brännmark, C., Palmér, R., Brugård, J., Nyström, F. H., Strålfors, P., & Cedersund, G. (2011). A hierarchical whole-body modeling approach elucidates the link between in Vitro insulin signaling and in Vivo glucose homeostasis. *The Journal of Biological Chemistry, 286*(29), 26028–26041. doi:10.1074/jbc.M110.188987.

Pincock, C. (2007). A role for mathematics in the physical sciences. *Noûs, 41*(2), 253–275.

Smith, C., Rubin, C., & Rosen, O. (1980). Insulin-treated 3T3-L1 adipocytes and cell-free extracts derived from them incorporate 32P into ribosomal protein S6. *PNAS, 77*(5), 2641–2645.

Vazquez, F., & Sellers, W. R. (2000). The PTEN tumor suppressor protein: an antagonist of phosphoinositide 3-kinase signaling. *Biochimica et Biophysica Acta, 1470*(1), M21.

Virkamäki, A., Ueki, K., & Kahn, C. R. (1999). Protein-protein interaction in insulin signaling and the molecular mechanisms of insulin resistance. *Journal of Clinical Investigation, 103*(7), 931–943.

Wiekert, W. (2002). Modelling and simulation: Tools for metabolic engineering. *Journal of Biotechnology, 94*(1), 37–63.

Woodward, J. (2003). *Making things happen: A theory of causal explanation: A theory of causal explanation*. Oxford: Oxford University Press.

Part IV
The Role of Heuristics
in Biological Explanations

Chapter 13
Heuristics, Descriptions, and the Scope of Mechanistic Explanation

Carlos Zednik

Abstract The philosophical conception of mechanistic explanation is grounded on a limited number of canonical examples. These examples provide an overly narrow view of contemporary scientific practice, because they do not reflect the extent to which the heuristic strategies and descriptive practices that contribute to mechanistic explanation have evolved beyond the well-known methods of decomposition, localization, and pictorial representation. Recent examples from evolutionary robotics and network approaches to biology and neuroscience demonstrate the increasingly important role played by computer simulations and mathematical representations in the epistemic practices of mechanism discovery and mechanism description. These examples also indicate that the scope of mechanistic explanation must be re-examined: With new and increasingly powerful methods of discovery and description comes the possibility of describing mechanisms far more complex than traditionally assumed.

Keywords Mechanistic explanation • Scientific discovery • Evolutionary robotics • Mathematical representation • Dynamical systems theory • Systems neuroscience • Decomposability • Heuristics of explanation

1 Introduction

Many scientific explanations in biology and neuroscience are mechanistic explanations: they describe the mechanisms responsible for the phenomena being explained. Philosophers of science have sought to explicate mechanistic explanation by studying a handful of canonical examples. These include the mechanistic explanations of long-term potentiation (Machamer et al. 2000), the action potential (Craver 2006, 2007a), the citric acid cycle (Bechtel 2006), and edge-detection in vision (Bechtel 2008; Kaplan 2011). But although these examples have been enormously useful for developing a philosophical conception of what mechanisms

C. Zednik (✉)
Institute of Cognitive Science, University of Osnabrück, 49069 Osnabrück, Germany
e-mail: czednik@uos.de

© Springer Science+Business Media Dordrecht 2015
P.-A. Braillard, C. Malaterre, *Explanation in Biology*, History, Philosophy and Theory of the Life Sciences 11, DOI 10.1007/978-94-017-9822-8_13

are and how they can be discovered and described, it is questionable whether they actually reflect the epistemic practices that contribute to the discovery and description of mechanisms in contemporary scientific research. Thus, the canonical examples have been taken to suggest that mechanisms are discovered via the dual heuristics of decomposition and localization (Bechtel and Richardson 1993; Silberstein and Chemero 2013), and that they are generally described in iconic diagrams and simple animations (Bechtel and Abrahamsen 2005; Machamer et al. 2000). In contemporary biological practice, however, the practices of mechanism discovery and mechanism description are often grounded on computational and mathematical techniques that go beyond the well-understood principles of decomposition, localization, and diagrammatic representation.[1]

In order to provide a better reflection of contemporary research, this chapter introduces new examples of mechanistic explanation from evolutionary robotics and network approaches in biology and neuroscience. These new examples show how the canonical examples of mechanistic explanation fall short, and how mathematical and computational techniques effectively contribute to the discovery and description of mechanisms in hitherto unappreciated ways. Section 2 briefly reviews the core principles of mechanistic explanation as well as some of the canonical examples that illustrate these principles. Section 3 then shows how graph-theoretic measures and the evolution of simulated model organisms have been used as heuristic strategies to discover a possible mechanism of klinotaxis in *Caenorhabditis elegans* (Izquierdo and Beer 2013; Izquierdo and Lockery 2010). Subsequently, Sect. 4 shows how equations and analytic techniques from dynamical systems theory can be used to describe the organization, composition and activity of mechanisms (Beer 2003). These new examples show how contemporary epistemic practices of mechanism discovery and mechanism description go beyond decomposition, localization, and pictorial representation.

Finally, this chapter concludes with a brief exploration of the consequences of going beyond the traditional conception of mechanism discovery and mechanism description. Specifically, Sect. 5 considers the possibility that novel computational and mathematical techniques might increase the number and types of natural phenomena that can be explained in mechanistic terms. Indeed, the mechanisms discussed in Sects. 3 and 4 are of a kind that is often thought to be too large and complex to be effectively decomposed and described in pictures. Thus, the phenomena these mechanisms exhibit are often thought to lie beyond the scope of mechanistic explanation. As novel computational and mathematical techniques make the epistemic practices of mechanism discovery and mechanism description ever more powerful and sophisticated, however, these difficulties might eventually be overcome, thereby extending the scope of mechanistic explanation.

---

[1]For related discussions of the role of mathematical modeling in biology and its relation to mechanistic explanations see the contributions to this volume by Baetu (2015), Bechtel (2015), Braillard (2015), Mekios (2015), Brigandt (2015), and Issad and Malaterre (2015).

2 Mechanistic Explanation and Its Canonical Examples

Mechanistic explanations describe the mechanisms responsible for phenomena of explanatory interest. Therefore, a philosophical conception of mechanistic explanation is answerable to metaphysical as well as epistemological concerns: What are mechanisms and how do they relate to the phenomena being explained? How are mechanisms discovered and subsequently described in scientific practice?

Although the philosophical literature boasts several statements of what mechanisms are, a particularly influential one is due to Carl Craver, who states that a mechanism "is a set of entities and activities organized such that they exhibit the phenomenon to be explained" (Craver 2007, p. 5. See also: Bechtel and Abrahamsen 2005; Bechtel and Richardson 1993; Glennan 2002; Machamer et al. 2000). This definition captures three widespread ideas, each of which is exemplified by the molecular mechanism of the action potential in nerve cells (Craver 2006, 2007).

First, mechanisms consist of *entities* (or parts) on one hand and *activities* (or operations) on the other. Entities are structures or objects in the world with properties that change over time. Activities are what entities do: how entities change over time, and how such changes influence other entities. In the molecular mechanism for the action potential, component entities include sodium and potassium ions, among others, as well as dedicated ion channels on the cell membrane. The mechanism's component activities include the opening and closing of channels, and the passing of ions through corresponding channels.

Second, a mechanism's component entities and activities are *organized*—they are related to one another in a particular way. Whereas a mechanism's entities are often organized spatially, related to one another by physical distances and bearings, its activities are typically organized temporally, by occurring at a particular moment in, or for a particular length of, time. In the mechanism for the action potential, ion channels are situated on the cell membrane, and the opening of ion channels allows ions initially situated on one side of the membrane to pass to the other side.

Third and finally, Craver's statement of what mechanisms are specifies that the relationship between mechanisms and phenomena is one of *exhibiting*. Alternative locutions frequently used in this context include "producing" and "being responsible for". If a phenomenon can be understood as a particular pattern of changes to a particular set of properties over time (Bechtel and Abrahamsen 2010), a mechanism can be said to exhibit this phenomenon if its properties—i.e. the properties of its component entities and activities, as well as the properties of their organization—change in accordance with this pattern. Thus for example, the action potential—a period of rapid depolarization of the cell body followed by gradual repolarization—begins when the opening of sodium channels allows Na^+ ions to permeate the cell membrane, driving the voltage of the cell body toward the sodium equilibrium potential near $+30$ mV. In turn, repolarization occurs when potassium channels open to allow K^+ ions to leave the cell body, thus eventually allowing the voltage to return to its resting potential of approximately -70 mV.

Although metaphysical questions concerning the nature of mechanisms themselves often dominate philosophical discourse, an adequate philosophical conception of mechanistic explanation must also address epistemological questions that concern the representation of mechanisms in the scientific literature. To this end, it is useful to distinguish two distinct epistemic practices. In the first, *mechanism discovery*, a mechanism's component entities, activities, and organization are identified by studying both the system in which the mechanism is realized and the phenomenon for which the mechanism is deemed responsible. In the second, *mechanism description*, the mechanism's composition and organization are represented in a way that shows that (and ideally, shows how) the mechanism exhibits the target phenomenon.

Bechtel and Richardson's (1993) *Discovering Complexity* remains the single most comprehensive discussion of mechanism discovery. Bechtel and Richardson adopt a framework pioneered by Herbert Simon (1996), in which scientific discovery in general is likened to a search process through a space of possible solutions to a problem. Thus, the specific case of mechanism discovery is conceived as a search process through a space of possible descriptions of the mechanism for the target phenomenon—also known as "how possibly" models of the mechanism (Craver 2007). This search process ends when a true description has been identified—a "how actually" model of the mechanism. Although fallible, heuristic strategies greatly facilitate the search process by allowing scientists to specify and constrain the space of possible mechanisms, as well as to identify regions of the space that merit further exploration, to the exclusion of others.

Bechtel and Richardson place particular emphasis on two such heuristic strategies: *decomposition* and *localization*. Decomposition itself comes in two varieties. Structural decomposition involves breaking down a complex entity into a collection of simpler entities. Typically, the complex entity subject to structural decomposition is the system from which the target phenomenon arises, and which is therefore presumed to realize the mechanism responsible for that phenomenon. In contrast, functional decomposition involves breaking down a complex activity into a collection of simpler activities. The complex activity to which this strategy is usually applied is the phenomenon itself, and the aim is to show how that phenomenon results from the simpler activities whenever these are executed simultaneously or in a particular order (see also: Cummins 1983). Functional and structural decomposition are linked by the heuristic of localization. The aim of localization is to establish a mapping between entities and activities, which in ideal cases shows that the activities identified via functional decomposition are in fact performed by the entities identified via structural decomposition. Successful localization is frequently used as evidence that the component entities and activities of a particular mechanism have been successfully discovered.

The combined use of decomposition and localization in the service of mechanism discovery is clearly exemplified by the explanation of edge-detection in mammalian vision (see also: Bechtel 2008; Kaplan 2011). After the striate cortex had been identified as particularly relevant for visual processing in the early twentieth century, David Hubel and Torsten Wiesel (1959, 1968) applied the heuristic of structural

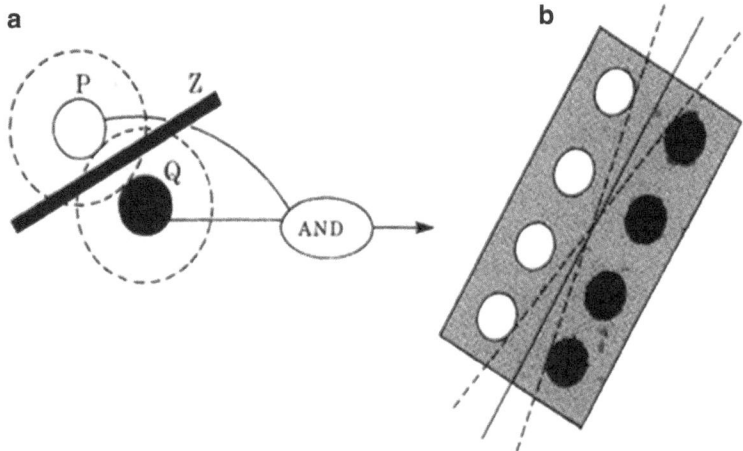

Fig. 13.1 The organization of simple cells that underlies edge-detection in vision (Reprinted from Marr and Hildreth 1980 by permission from the Royal Society). (**a**) A pair of cells with opposite response patterns, connected by a simple AND gate, can be used to detect zero-crossings in a blurred image. (**b**) An array of several cell-pairs can be used to detect edges

decomposition to characterize the different types of cells that composed it. Through single-cell recordings in cats and macaque monkeys, they distinguished three types of cells on the basis of their responses to visual stimuli: *simple* cells that respond to light at specific retinal locations; *complex* cells that respond to bars of light at a particular angle of orientation; and *hypercomplex* cells that respond to bars of light that span the full width of the receptive field. But although Hubel and Wiesel were thus able to identify some of the component entities of the mechanism for visual edge-detection, they described neither their corresponding activities, nor their spatiotemporal organization. Both of these shortcomings were remedied by David Marr and colleagues in the 1970s and 1980s. Specifically, Marr and Hildreth (1980) performed a functional decomposition of vision that showed that edge-detection could be achieved by a sequential process that blurs a visual image with a Gaussian filter, and then applies a Laplacian operator to detect those locations of the blurred image that have the highest changes in intensity—the so-called *zero-crossings*. Marr and Hildreth then showed that a pair of neighboring simple cells with complementary activation profiles could combine to detect a single zero-crossing (Fig. 13.1a), and that a larger arrangement of these cells could function as a single edge-detector (Fig. 13.1b). In this way, the heuristic of localization was used to link Hubel and Wiesel's structural decomposition of the striate cortex with Marr and Hildreth's functional decomposition of visual edge-detection to provide a description of (some of) the component entities and component activities of the mechanism for the target phenomenon.

The epistemic practice of mechanism discovery can be distinguished from the practice of mechanism description. Whereas the former typically involves heuristic strategies to identify the entities, activities, and organization of the mechanism to be described, the latter involves one or more descriptive media to represent these entities, activities and their organization. Notably, the distinction between mechanism discovery and mechanism description is conceptual rather than practical: they need not correspond to distinct periods of time, or be conducted by distinct individuals. Indeed, description and discovery may be mutually constraining, as when representing a mechanism in a certain way reveals errors or ambiguities to be remedied by identifying additional entities, activities or modes of organization.

As is implicit in the aforementioned examples, the descriptive media used to represent mechanisms often include verbal characterizations and iconic or schematic diagrams. In addition, mechanisms are often described as physical and simulated two or three-dimensional models (Bechtel and Abrahamsen 2005; Wright and Bechtel 2007). Depending on the nature of the mechanism being described, certain descriptive media tend to be more effective than others. For example, mechanisms in which the spatial relationships between components are crucial—such as the mechanism for visual edge-detection, in which the spatial organization of simple cells is paramount—are more easily represented in diagrams than in words. In contrast, mechanisms in which temporal relationships are critical may be more effectively described by animations that show the time course of relevant events. Finally, mechanisms in which physical details take a back seat to functional relationships between component activities are frequently described with schematic representations such as box-and-arrow diagrams.

No matter the medium, mechanism descriptions provide mechanistic explanations when they adequately represent the mechanism responsible for the phenomenon being explained. When exactly a description is adequate in this way remains controversial. Nevertheless, a widespread idea is that it should refer to those and only those component entities and activities that are actually relevant to the phenomenon being explained. Craver (2007) has elaborated on this idea by appealing to the notion of *mutual manipulability* (but for criticism see: Leuridan 2011). On Craver's account, a component is to be included in the description of a mechanism "when one can wiggle the behavior of the whole by wiggling the behavior of the component and one can wiggle the behavior of the component by wiggling the behavior as a whole" (Craver 2007, p. 153). Thus for example, the description of the mechanism for edge-detection in vision should refer to simple cells because Hubel and Wiesel used single-cell recordings to show that these cells are activated during episodes of visual edge-detection, and because they showed that interfering with these cells (e.g. through lesions) affects the organism's ability to detect edges (Hubel and Wiesel 1959, 1968).

The aim of this section has been to briefly review the metaphysical and epistemological principles of mechanistic explanation, as well as to introduce some of the canonical examples on the basis of which these principles are traditionally explored: the mechanisms for the action potential and for visual edge-detection. Although there are other such examples—the mechanisms of long-term potentiation (Machamer et al. 2000) and the citric acid cycle (Bechtel 2006) are notable omissions—their nature and means of discovery and description differ insubstantially from the examples reviewed here. As the following sections will demonstrate, however, these canonical examples provide an overly narrow glimpse on mechanistic explanation in contemporary scientific practice. Although the core principles of mechanistic explanation remain unchanged—mechanisms can still be viewed as organized collections of entities and activities, and mechanistic explanations are descriptions derived, in part, by the systematic application of heuristic strategies—the mechanisms that figure in contemporary scientific research are far more complex, and the heuristic strategies and descriptive media invoked by practicing scientists are far more numerous and sophisticated, than these canonical examples would suggest.

3 New Heuristics for Mechanism Discovery

The role of heuristic strategies in mechanism discovery is one of the most underexplored aspects of mechanistic explanation. Rather than question or elaborate on Bechtel and Richardson's (1993) discussion, most philosophical treatments assume that the heuristics of decomposition and localization are part-and-parcel of mechanistic explanation. Inspired by Bechtel and Richardson's extended discussion of situations in which the heuristics of decomposition and localization fail, it is often assumed that abandoning these heuristics is tantamount to abandoning the search for mechanisms (e.g. Chemero and Silberstein 2008). But this assumption is false: mechanism discovery is facilitated by any heuristic strategy that aids researchers to efficiently explore the space of "how possibly" models of a mechanism. This section introduces two such strategies: the *evolution of simulated model organisms* and *selective pruning*.

3.1 Evolving Simulated Model Organisms

In two separate but complementary studies, Eduardo Izquierdo and colleagues seek to discover the mechanism for klinotaxis in *Caenorhabditis elegans*. Klinotaxis is a form of goal-directed locomotion in which a chemical source is approached by repeatedly sweeping, and over the long run following, a chemical gradient whose concentration increases with proximity to the source (Fig. 13.2).

As is well-known, efforts to map the *C. elegans* nervous system have resulted in a detailed description of the organism's *connectome*: the 302 neurons and approximately 7,000 synaptic connections and gap-junctions that make up its nervous system (Varshney et al. 2011; White et al. 1986). Although incredibly detailed, this descriptive knowledge falls short of explanation: it is as of yet unknown which individual parts or properties of the connectome contribute to particular behavioral capacities, and exactly how they do so. By invoking the descriptive knowledge of the

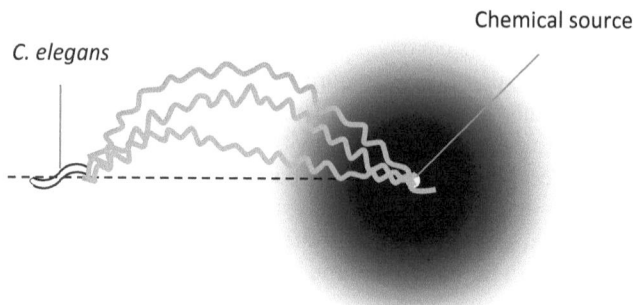

Fig. 13.2 *C. elegans* klinotaxis (Adapted from Izquierdo and Lockery 2010). The organism and its movement through a chemical gradient that increases in concentration with proximity to the source. The *dotted line* denotes the line of steepest ascent, from the organism's current position to the source; *wiggly lines* denote characteristic trajectories during individual klinotaxis episodes

C. elegans connectome to identify a possible mechanism for klinotaxis, Izquierdo and colleagues take a first step from description to explanation.

In the first study, Izquierdo and Lockery (2010) identify the possible component activities of the mechanism for klinotaxis. Previous ablation studies of the *C. elegans* connectome suggest the involvement of at least two kinds of chemosensory neuron (ASER and ASEL) that respond to increases and decreases of a chemical gradient in the environment, respectively, as well as two kinds of motor neuron (SMBD and SMBV) that control neck-muscle contractions on either side of the organism's body. What remains unknown is exactly how these chemosensory and motor neurons interact through mediating interneurons, and how motor action feeds back on chemosensation through the environment. Rather than address these questions through further ablation studies meant to isolate the contributions of specific interneurons, Izquierdo and Lockery adopt a simulation-based approach. Specifically, they invoke a simulated body model of *C. elegans* that is controlled by an artificial neural circuit (Fig. 13.3), and determine the range of circuit parameter values that enable the production of klinotaxis in a simulated environment.

This simulation relies on several simplifying assumptions. For example, the neural circuit does not include any interneurons, and approximates the nervous system's background activity as an oscillating signal that drives a snake-like motion characteristic of real-world *C. elegans*. In other respects, the neural circuit is true to biological detail. Specifically, it includes a pair of motor neurons corresponding to SMBD (dorsal) and SMBV (ventral) that determine muscle contractions on either side of the body model, and a pair of chemosensory neurons ("ON" and "OFF") that detect increases and decreases in the concentration of a chemical trace in accordance with the known response patterns of ASEL and ASER (Suzuki et al. 2008).

The aim of the study is to determine the functional relationships between chemosensation and motor action that contribute to klinotaxis in the simulated environment. To this end, Izquierdo and Lockery invoke an evolutionary algorithm

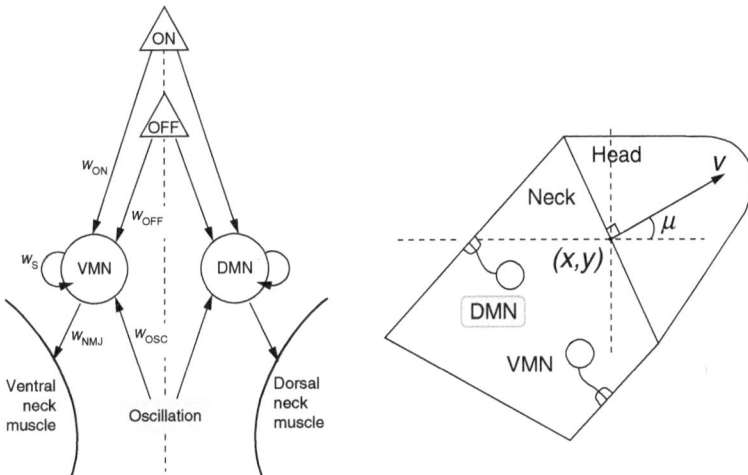

Fig. 13.3 *Left:* The *C. elegans* neural circuit. *Triangles* correspond to chemosensory neurons; *circles* to neck-motor neurons. *Arrows* indicate neural connections whose strength and direction is determined by the evolutionary algorithm (Reprinted from Izquierdo and Beer 2013). *Right:* The body model. Neck-motor neurons controlled by the neural circuit govern head angle, influencing the body's direction of motion in the simulated environment (Reprinted from Izquierdo and Lockery 2010 with permission from The Society for Neuroscience)

(Mitchell 1996) that, from a random "population" of neural circuits with distinct connectivity profiles, selects those circuits that lead to particularly efficient and reliable klinotaxis across varying environmental conditions. After evolving the population over several generations, the authors identify 77 successful neural circuits. Perhaps remarkably, although distinct in their neural connectivity profiles, all 77 circuits are observed to exhibit the same basic motor neuron response pattern to chemosensory stimulation: whereas stimulation of the ON cell (when the chemical gradient increases) reduces the differential activity of ventral and dorsal motor neurons and causes the body model to align itself with the direction of the source, stimulation of the OFF cell (when the chemical gradient decreases) has the opposite effect, causing the body model to turn away from the source. This motor neuron response pattern is one component activity of the mechanism for klinotaxis in the simulated environment.

A second component activity is environmental feedback from motor action back to chemosensory stimulation. Notably, Kaplan (2012) and Zednik (2011) have already argued that a mechanism may be physically distributed in this way, crossing the physical boundaries between brain, body and environment. In this particular simulation, the body's snake-like motion makes the head repeatedly oscillate about the line of steepest ascent to the source (see Fig. 13.2). This oscillation results in alternating stimulation of the ON and OFF cells, which in turn cause alternating to-and-fro movements with respect to the source. Thus, neural feed-forward processing and environmental feedback together produce an effective displacement in the direction of the source: klinotaxis.

There are reasons to believe that the two component activities of the mechanism for klinotaxis in simulation are also operative in the mechanism for klinotaxis in real-world *C. elegans*. For one, targeted lesions in the simulated neural circuit have behavioral effects very similar to those of corresponding lesions in the biological organism (Izquierdo and Lockery 2010, p. 12915). For another, the precise details of the sweeping motion observed in the simulated organism closely resemble those of real-world *C. elegans* in a range of environmental conditions (Izquierdo and Lockery 2010, p. 12912). Importantly, this behavioral correspondence emerges unexpectedly: the evolutionary algorithm only selects for the efficiency and reliability of the resultant behavior, not its similarity to real-world klinotaxis. At the end of their study, therefore, Izquierdo and Lockery hypothesize that the component activities of the simulated mechanism may in fact closely resemble those of the real-world mechanism. The "how possibly" model of the mechanism identified by evolving a simulated organism may turn out to be a "how actually" model of the mechanism for klinotaxis in biological *C. elegans*.

It may seem surprising that an empirical hypothesis about the mechanism for klinotaxis in a biological organism can be derived on the basis of a mere simulation. Barbara Webb (2009) has recently questioned the legitimacy of inferences from simulated to biological mechanisms. Indeed, evolutionary algorithms such as the one adopted by Izquierdo and Lockery are notoriously exploitative of specific details of the simulated environments in which they operate. In the present study, the fact that all 77 successful neural circuits realize the same pair of activities might just be an artifact of the specific details of the simulated neural circuit, body model, and environment; in the real world, klinotaxis might be performed by entirely different means, or be produced by a multitude of distinct but redundant mechanisms. But this line of reasoning does not imply that simulation-based strategies are useless, only that they are fallible. Fallibility is a signature feature of heuristic strategies, and is usually offset by the simplicity and speed with which such strategies can be deployed (Bechtel and Richardson 1993; Gigerenzer 1991). Indeed, exploring the space of possible mechanisms for klinotaxis via the evolution of simulated model organisms is likely to be a far more effective use of time and resources than exploring it via behavioral and lesion studies of the biological organism—especially insofar as evolutionary algorithms and simulations are particularly adept at identifying unintuitive and complex solutions that might otherwise be overlooked (see also: Wheeler 2005). Of course, conclusively determining whether or not any individual "how possibly" model of a mechanism is in fact a "how actually" model requires further testing, refinement, and eventual confirmation or falsification on the basis of empirical investigation. Still, viewed as a heuristic device for developing testable "how possibly" models of a mechanism, the heuristic role of evolving simulated model organisms is clearly significant.

3.2 Selective Pruning

Izquierdo and Lockery's simulation-based strategy identifies the (possible) component activities of the mechanism for klinotaxis in *C. elegans*, revealing these to be, on the one hand, particular motor responses to distinct types of chemosensory stimulation, and on the other hand, a specific kind of environmental feedback. What about the mechanism's component entities? As was mentioned above, the direct links between chemosensory and motor neurons in the simulated neural circuit are considerable simplifications of biological reality. In reality, several interneurons mediate between ASE chemosensory and SMB motor neurons. In a recent study, therefore, Izquierdo and Beer (2013) seek to determine which particular interneurons of the full *C. elegans* connectome contribute to the neural feed-forward processing described in Izquierdo and Lockery's earlier work.

To this end, Izquierdo and Beer invoke the heuristic strategy of *selective pruning*, in which the authors combine past experimental results with graph-theoretic measures to distinguish those elements of the connectome that are likely to be the component entities of the mechanism for klinotaxis from those that are not. The starting point of Izquierdo and Beer's study is a graph-theoretic representation that includes: all 12 chemosensory neurons known to detect concentrations of chemical gradients in the environment; all 28 head- and neck-motor neurons that determine the worm's movement; and all 234 interneurons, 6,246 chemical contacts and 890 gap junctions that make up the structural links between them (Fig. 13.4a).

Although the 274 elements of the graph in Fig. 13.4a fall short of the 302 neurons in the full *C. elegans* connectome, they include all neurons that are *potentially* relevant to the production of klinotaxis. The excluded neurons have no inbound connections to the relevant motor neurons, and can therefore be removed from

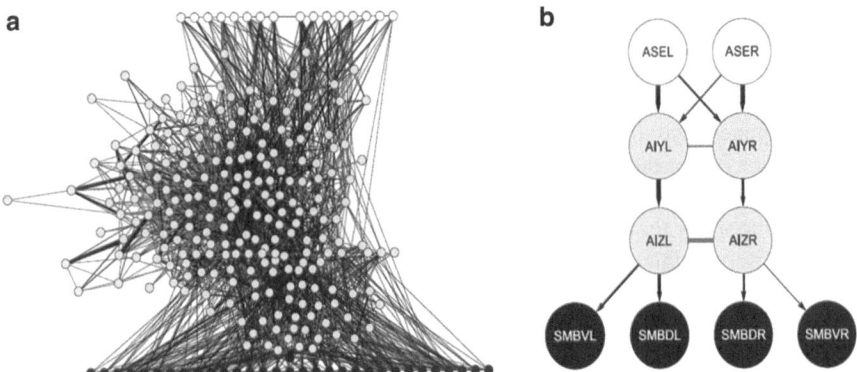

Fig. 13.4 (**a**) A subgraph of the *C. elegans* connectome depicting the network of neurons potentially relevant to klinotaxis. *Gray* units represent chemosensory neurons, *black* units represent motor neurons, *white* units represent interneurons. (**b**) The minimal network derived by selectively pruning the graph in figure (**a**) (Reprinted from Izquierdo and Beer 2013)

consideration. As the structural intermediaries between chemosensory and motor neurons, all interneurons represented in Fig. 13.4a are potentially relevant to the production of klinotaxis. However, not all of these interneurons need be functionally relevant: their contribution to klinotaxis may be negligible or redundant. In order to separate the relevant elements from the irrelevant ones, Izquierdo and Beer first appeal to previous experimental results to prune those chemosensory and motor neurons that have not previously been associated with klinotaxis (Bargmann and Horvitz 1991): only ASE chemosensory and SMB motor neurons remain under consideration. Subsequently, the number of interneurons that link ASE and SMB is reduced by applying several graph-theoretic measures: removing all weakly connected elements, such as those that have less than two outgoing connections, and removing all long-range pathways by excluding those interneurons that are not immediately adjacent to either a chemosensory or a motor neuron (Izquierdo and Beer 2013, p. 3). Through this kind of selective pruning, which is motivated by empirical as well as graph-theoretic considerations, the graph in Fig. 13.4a is reduced to the graph in Fig. 13.4b: the *minimal network* for klinotaxis by *C. elegans*.

The elements of this minimal network are probable candidates for the component entities of the mechanism for klinotaxis in biological *C. elegans*. First, the interneurons identified by the selective pruning strategy (AIZ and AIY neurons on the ventral and dorsal sides) are consistent with those identified in previous ablation studies of klinotaxis (Iino and Yoshida 2009; Kocabas et al. 2012). Importantly, this consistency was achieved despite the fact that the selective pruning strategy was not designed to reproduce this empirical result. Second, when used as an artificial neural circuit controller for a *C. elegans* body model similar to the one discussed above, the minimal network (fleshed out with appropriate connection weights) produces effective and realistic klinotaxis behavior closely analogous to the behavior observed by Izquierdo and Lockery (2010). Third, and perhaps most important, Izquierdo and Beer show that this klinotaxis behavior is produced by the same interdependence of neural feed-forward and environmental feedback described in the earlier study. Indeed, the interneuron response to chemosensory stimulation in this network is qualitatively identical to the pattern of neural activity described by Izquierdo and Lockery: the minimal network implements one of the two component activities of the mechanism described earlier.

Whether or not the minimal network in Fig. 13.4b actually describes (some of the) component entities of the mechanism for klinotaxis in biological *C. elegans*, what is important for current purposes is the strategy of selective pruning that was used to discover this network. This strategy can be clearly distinguished from the heuristic approaches to mechanism discovery discussed in Sect. 2. Recall that the heuristic strategy of decomposition involves breaking apart a complex system or entity into a collection of simpler entities. In contrast, selective pruning already presupposes that the relevant system has been decomposed and its component entities have been identified. In this particular example, from a previously available description of individual parts (the *C. elegans* connectome), Izquierdo and Beer identify a particular subset of these parts as the possible component entities of the mechanism responsible for the phenomenon being explained. Unlike many previous studies of the *C. elegans* connectome, Izquierdo and Beer thus take a first step from detailed description to genuine explanation. Although these studies generally agree that not all parts of the connectome are the component entities of particular behavioral mechanisms—hence the appeal of targeted ablation studies—it is hard to know how to isolate the relevant connectome elements from the irrelevant ones. Using graph theoretic measures to selectively prune the connectome in the way exemplified here is likely to be of great help.

The role of selective pruning in the epistemic practice of mechanism discovery is not limited to connectome research. Although technical advances in imaging, mapping, and computer modeling make it increasingly feasible to identify the individual components of different kinds of biological systems, it remains hard to know exactly which components actually contribute to the phenomena exhibited by such systems. For example, although researchers have successfully sequenced the genome of a variety of species, it remains unclear how best to systematically characterize the interactions between gene products, and thereby specify interdependencies in gene expression. Graph-theoretic methods similar to the ones invoked by Izquierdo and Beer have been used to separate strong protein interactions from weak ones (Schlitt and Brazma 2007), as well as to identify patterns of interaction between multiple proteins that are repeated throughout a genetic regulatory network: network motifs (Banks et al. 2008). Insofar as protein interaction networks can be viewed as mechanisms for gene expression, here again the heuristic strategy of selective pruning facilitates the discovery of biological mechanisms.

In summary, decomposition and localization are far from being the only useful heuristic strategies for mechanism discovery. The evolution of simulated model organisms and selective pruning can both be viewed as heuristic strategies that facilitate the identification of biological mechanisms and their components. But there are likely to be many others. Insofar as the true diversity of heuristic strategies remains unknown, the philosophical literature on mechanistic explanation is well-advised to consider more—and more recent—examples from biological research.

4 Beyond Pictures: Mathematical Mechanism-Descriptions

The extant philosophical conception of mechanistic explanation, bolstered by the canonical examples reviewed in Sect. 2, emphasizes the distinctly visual character of the epistemic practice of mechanism description (see e.g. Bechtel and Abrahamsen 2005; Bechtel and Richardson 1993; Wright and Bechtel 2007). In contrast to deductive-nomological explanations, which take the form of logical arguments that link linguistic propositions (Hempel 1965), the canonical examples of mechanistic explanation center on iconic or schematic diagrams. The prevalence of diagrams is due to the fact that most mechanistic explanations describe mechanisms that exist in space and time: their component entities have spatial properties, their component activities can be characterized in terms of changes to those spatial properties over time, and their overall organization is determined by the spatiotemporal arrangement of their components. Diagrams are ideally suited to represent this kind of spatiotemporal information because they are iconic in a way that other descriptive media are not: the spatial properties of diagrams can be used to "mirror" the spatial properties of the entities being represented, and diagram-sequences or animations can be used to visualize changes to the properties of a mechanism's components over time, such as the movement of an ion through a channel.

But mechanistic explanations are only contingently diagrammatic; mechanisms can also be described mathematically. The most straightforward way in which a mechanism might be described mathematically is by way of equations. Variables and parameters can be used to represent the properties of individual entities, such as their size, location, velocity, activation, or charge. Changes to these properties can be represented as changes in the values of the relevant variables over time, and relationships between individual entities or activities (e.g. their spatiotemporal or functional organization) can be captured in mathematical relationships between variables or coupled equations. Common examples once again include network models in cognitive neuroscience and in the study of protein interaction networks. Although such network models are often represented diagrammatically, as in Fig. 13.4, these diagrams are nearly always grounded on mathematical equations that precisely specify the interactions between elements of the network as well as the processing or transformation of information that occurs within individual elements.

Despite the existence of such examples, there is substantial disagreement concerning the suitability of mathematical equations for mechanism description. Consider:

> Equations do not offer the right kind of format, however, for constructing a mechanistic explanation—they specify neither the component parts and operations of a mechanism nor how these are organized so as to produce its the behavior. (Abrahamsen and Bechtel 2006, p. 171)

In later work, Bechtel and Abrahamsen (2010) explore the role of equations in mechanistic explanation, and conclude that these are generally used to complement, rather than provide, descriptions of mechanisms. For example, of the detailed mathematical models of the time course of circadian rhythms in *drosophila*, they say:

> the models are not proposals regarding the basic architecture of circadian mechanisms; rather, they are used to better understand the functioning of a mechanism whose parts, operations, and organization already have been independently determined. (Bechtel and Abrahamsen 2010, p. 322)

But although this analysis is surely true for certain examples, the claim that mathematical equations are generally unsuited for mechanism description is unfounded. As Craver (2006, 2007, 2008) and others have already argued, what matters is not how a mechanism is represented, but simply that it is represented (see also: Glennan 2002; Kaplan and Craver 2011; Machamer et al. 2000; Zednik 2011). Insofar as mathematical equations can be used to describe the very same properties as diagrams, there is no *a priori* reason to discount the former as vehicles of mechanism description.

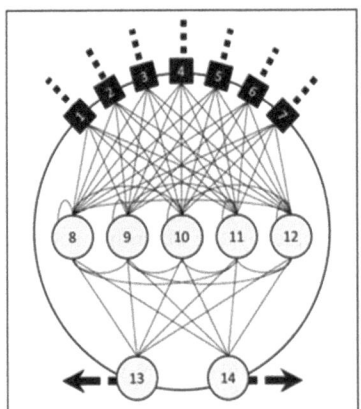 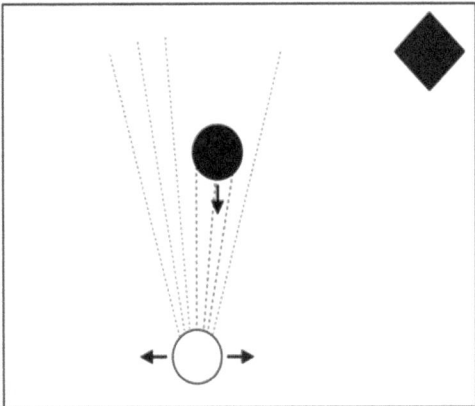

Fig. 13.5 (Adapted from Beer 2003). *Left:* The simulated agent and its continuous-time neural network brain. *Right:* The task environment

Given the possibility of mathematical mechanism-descriptions, when would such descriptions be beneficial? Whereas diagrams are particularly useful for representing mechanisms whose relevant properties are spatiotemporal, mathematical descriptions seem particularly useful for representing mechanisms whose relevant properties are distinctly mathematical. Indeed, some mechanisms are best understood by describing the abstract mathematical properties of their component activities, such as the limit of their activation, their probability of occurring, or the general tendency of their motion. As a concrete example, consider the role of different kinds of mathematical representation in Randall Beer's (2003) mechanistic explanation of perceptual categorization in a simulated brain-body-environment system (Fig. 13.5).

This simulated system consists of a single "minimally cognitive" model organism that is embedded in an environment that features a single circular or diamond-shaped object. The model organism is equipped with a continuous-time recurrent neural network brain that mediates between visual inputs and motor outputs. The system's behavior is determined by a set of 16 coupled differential equations:

$$\tau_i \dot{s}_i = -s_i + I_i\,(x, y; a)\, i = 1, \ldots, 7 \qquad (13.1\text{–}13.7)$$

$$\tau_i \dot{s}_i = -s_i + \sum_{j=1}^{7} w_{ji}\sigma\left(g\left(s_j + \theta\right)\right) + \sum_{j=8}^{12} w_{ji}\sigma\left(s_j + \theta_j\right) i = 8, \ldots, 12$$

$$(13.8\text{–}13.12)$$

$$\tau_i \dot{s}_i = -s_i + \sum_{j=8}^{12} w_{ji}\sigma\left(s_j + \theta\right) i = 13, 14 \qquad (13.13, 13.14)$$

$$\dot{x} = 5 \left(\sigma \left(s_{13} + \theta_{13} \right) - \sigma \left(s_{14} + \theta_{14} \right) \right) \tag{13.15}$$

$$\dot{y} = -3 \tag{13.16}$$

Equations (13.1–13.16) define the change over time in the brain's neural activity
($s_1 \dots s_{14}$), the organism's horizontal position (x), and the object's vertical position
(y). The brain's neural activity is continuously affected by the changing sensory
input vector I, a function of shape parameter α and of the relative positions of
organism and object. In contrast, neural parameters w, τ, σ, and θ are fixed by an
evolutionary algorithm that selects for successful categorization behavior in which
a falling object is classified according to its shape. Specifically, the organism is
evolved to "catch" circular objects by moving directly beneath them as they fall,
and to "avoid" diamond-shaped objects by moving horizontally to either side. As
an unexpected result of the artificial evolutionary process, successful organisms
perform perceptual categorization via an "active scanning" strategy: they repeatedly
move from side to side to "scan" the object before eventually settling on a position
either directly beneath it or away to one side.

Insofar as Eqs. (13.1–13.16) perfectly describe the parts of the brain-body-
environment system as well as their interdependencies, there is a sense in which
they explain the system's behavior by describing the mechanism responsible
for that behavior. Nevertheless, there is also a sense in which the description
provided by these equations is not particularly insightful: it remains quite unclear
exactly how the active scanning behavior arises from the interactions between
the individual parts of the system. Indeed, as Craver (2013) has already argued,
mechanistic explanation always involves choosing an appropriate level at which to
describe describing a particular mechanism's components. Thus, human circulation
is typically explained at the level of organs and tissue rather than at the level of
molecules, and long term potentiation is more easily illuminated by describing
mechanisms at the level of molecules than at the level of atoms. In much the same
way, Beer argues that it is far more insightful to decompose the system into two
interacting components at a level above the individual neurons: the brain, embodied
in the simulated agent (entity B), and the environment, defined by the relative
positions of agent and object (entity E). The activities that correspond to entities
B and E are, on the one hand, the influence of visual input on motor output, and on
the other hand, the sensorimotor feedback in which motor output at any given time
governs the accumulation of visual input at later times.

This way of verbally characterizing the two component entities and their
corresponding component activities provides a "sketch" (Machamer et al. 2000)
of the mechanism for perceptual categorization via active scanning. Beer turns this
sketch into a detailed mechanistic explanation by invoking analytic techniques from
dynamical systems theory. Specifically, *steady-state velocity fields* (shaded regions
in Fig. 13.6; see original color version in: Beer 2003) describe the activity of B, and
superimposed *motion trajectories* (lines in Fig. 13.6; see original color version in:
Beer 2003) describe the activity of E. Whereas the latter provide a straightforward

description of the relative positions of the model organism and the object over time, the former describe the organism's steady-state (or long-term) velocity for every possible pattern of perceptual input: the horizontal velocity the organism *would* achieve if its motion were stopped and its perceptual inputs were held constant for an extended period of time.

Gaining explanatory leverage from descriptions of a system's steady state-behavior, especially when that system's actual behavior is hard to describe, is a hallmark of dynamical explanation (Chemero 2009; Kelso 1995). Although dynamical explanations come in many different varieties (Zednik 2011), this is also true when a particular dynamical explanation describes the components of a mechanism. In the current example, the organism's steady-state velocity acts as a constraint that limits its instantaneous velocity, and can therefore be used to approximate the activity of entity *B*, the embodied brain. Consider the way in which the motion trajectories in Fig. 13.6 overshoot some shaded regions while reversing their direction over others. What determines whether a particular motion trajectory performs an overshoot or a reversal within any particular region is that region's shade (or rather, in the original version, its color) as well as the amount of time spent moving through it. Specifically, a motion trajectory of a particular shade performs a reversal whenever it is situated over a region of the opposite shade, and remains in that region long enough for the instantaneous velocity to

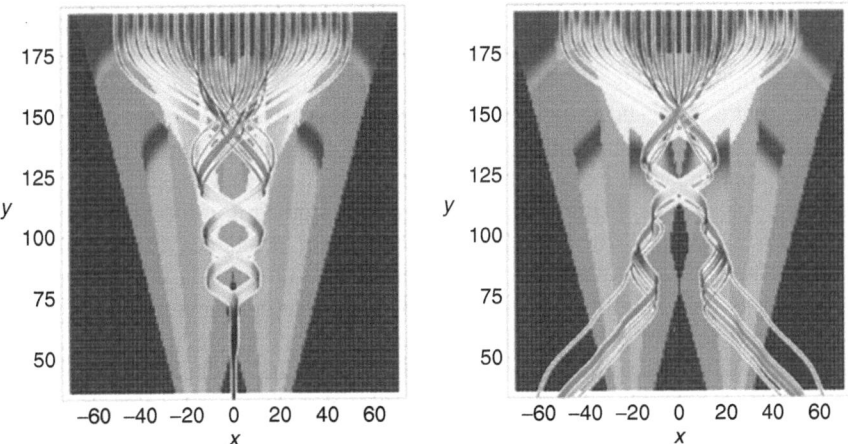

Fig. 13.6 Steady-state velocity fields with superimposed motion trajectories for "catching" circles (*left*) and "avoiding" diamonds (*right*) (Adapted from original color version in Beer 2003). Axes designate the relative positions of agent (*x*) and object (*y*). *Shaded regions* describe the activity of the embodied brain (*B*): the way perceptual input constrains the agent's horizontal motion. *Different shades* (colors in the original) indicates different steady-state velocities, directed either toward or away from the object. *Lines* describe the activity of the environment (*E*): the way the changing relative positions of agent and environment affect perceptual input. Their *shade* (color in the original) indicates the agent's instantaneous velocity; their *shape* indicates the way the relative positions of object and agent change over time

approach the steady-state velocity denoted by that region. Because active scanning is just a particular pattern of overshoots and reversals, it can be reconstructed from the specific details—shape and shade—of the motion trajectories and steady-state velocity fields in Fig. 13.6. Indeed, differences between the left and right side of Fig. 13.6 such as the different sizes of the central black region explain the differences in behavior between circle (catch) and diamond (avoid) trials. Notably, modifying the size, shape, or shade of particular regions (which can be done by e.g. changing certain network parameters) leads to novel and predictably correct or incorrect categorizations (Beer 2003, pp. 228–230). That is, this description of the activities of B and E renders the mechanism for active scanning amenable to mutual manipulation—Beer's mathematical description of the two-component mechanism for perceptual categorization is adequate for the purposes of mechanistic explanation (for further discussion see: Zednik 2011).

This example shows how equations, but also more sophisticated means of mathematical representation, can be used to describe the component entities and activities of mechanisms. Although Fig. 13.6 is of course also a diagram, it differs markedly from the diagrams invoked by the canonical examples described in Sect. 2 above. Whereas those diagrams typically describe a mechanism's spatiotemporal properties and "mirror" those properties in a relatively straightforward manner, the steady-state velocity fields in Fig. 13.6 describe a particular component entity's mathematical properties, and can be interpreted only on the background of the mathematical framework of dynamical systems theory. Irrespective of the perhaps unintuitive or hard-to-grasp nature of this description, what matters for current purposes is that it adequately describes (the component activities of) a mechanism.

In closing, there is of course no reason to believe that such mathematical descriptions can only be given for artificial examples like Beer's, and there is equally no reason to believe that only the framework of dynamical systems theory offers the right mathematical methods to describe mechanisms. Indeed, simulated model organisms analogous to Beer's have already been studied in information-theoretic terms (Williams and Beer 2010), and descriptive techniques from dynamical systems theory are regularly invoked to describe developmental mechanisms and mechanisms for spatial memory in cognitive and developmental psychology (e.g. Spencer and Schöner 2006). Moving even further beyond the traditional conception that mechanism descriptions are simple and diagrammatic in character will require paying closer attention to these and many other examples in which mathematical representations take center stage.

5   Extending the Scope of Mechanistic Explanation

Although the definition of mechanisms presented in Sect. 2 is intentionally broad—it includes all sets of "entities and activities organized such that they exhibit the phenomenon to be explained" (Craver 2007, p. 5)—not everything that satisfies this definition can actually figure in a mechanistic explanation of a natural phenomenon.

This is because mechanistic explanation is an epistemic activity that centers on the act of describing a mechanism, and because not all mechanisms can be feasibly discovered and subsequently described by practicing scientists. But what exactly distinguishes the mechanisms that can be feasibly discovered and described from those that cannot? Bechtel and Richardson (1993) invoke an influential analysis due to Herbert Simon (1996; see also: Wimsatt 1986), in which systems are classified according to the degree of interactivity between their components. Systems within which the degree of interactivity is "negligible" (Simon 1996, p. 207) are deemed *decomposable*: their behavior is an aggregation of the behavior of their components, and can typically be analyzed as such. Another class of systems—those in which the degree of interactivity between components is "weak, but not negligible" (ibid)—are deemed *nearly decomposable*. Although an analysis of a nearly decomposable system's behavior in terms of the behavior of its components will typically be approximate, it suffices to "understand, describe, and even 'see' such systems and their parts" (ibid). Thus, phenomena that arise from the activity of nearly decomposable systems are still amenable to analysis via the heuristics of decomposition and localization, and thus on Bechtel and Richardson's account, are subject to mechanistic explanation.

In Simon's classification, decomposable and nearly decomposable systems can be contrasted with *non-decomposable* systems, in which the degree of interactivity between components is on a par with the degree of activity within components. According to Bechtel and Richardson, non-decomposable systems resist the heuristic strategies of decomposition and localization, and thus, lie beyond the scope of mechanistic explanation. With respect to decomposition, this is because, absent independent criteria such as molecular composition or structure, interactivity is frequently used as a criterion to individuate system components. When the degree of interactivity is fairly uniform throughout a system, however, there may be no principled way to tell where one component ends and the next one begins. As concerns localization, although it is always possible to decompose systems arbitrarily, e.g. into equal-sized chunks of matter, it will be exceedingly difficult to identify each chunk's specific contribution to the activity of the system as a whole. In Craver's (2007) terminology, it will be difficult to show that such arbitrarily individuated parts of a system are in fact the *working* parts of a mechanism—those parts that perform particular component activities.

Despite the widespread appeal of Bechtel and Richardson's account of the scope of mechanistic explanation, its ties to Simon's classification of system interactivity seem ill-motivated. Whereas Simon's classification is metaphysical—it concerns the structure of systems in the world—Bechtel and Richardson's aims are distinctly epistemological: to distinguish those mechanisms that can be discovered and described from those that cannot. Notably, however, the success and failure of epistemic practices of discovery and description is dependent not only on the structure of the mechanisms being investigated, but also on the epistemic capacities of human investigators (see also: Glauer 2012). What the examples introduced in the preceding sections of this chapter show is that scientists' capacities to discover and describe mechanisms are continuously evolving, especially with the influx of increasingly powerful computer simulations and sophisticated methods of mathematical representation and analysis. Therefore, it seems fair to wonder whether such simulations and mathematical methods might be used to extend the scope of mechanistic explanation.

Consider again the mechanism for perceptual categorization via active scanning. This mechanism features dense reciprocal interactions not only within the agent's neural network brain (component entity B), but also between the brain and the agent's changing environment (component entity E). Thus, the system is a non-decomposable system in Simon's sense, whose behavior, on Bechtel and Richardson's account, lies beyond the scope of mechanistic explanation. Nevertheless, Sect. 4 above shows how the system can be decomposed into two interacting entities, and how analytic techniques from dynamical systems theory can be used to approximately but still adequately (for the purposes of mechanistic explanation) describe the corresponding activities. Therefore, Beer's mechanistic explanation of perceptual categorization via active scanning is a counterexample to the claim that Simon's notion of non-decomposability determines the scope of mechanistic explanation: some non-decomposable systems (in Simon's sense) might after all be decomposed (in the sense relevant to mechanistic explanation).

Might Bechtel and Richardson's account be rescued by divorcing it from Simon's classification of system interactivity, and making it entirely dependent on the success and failure of decomposition and localization? On such a modified account, although the mechanism for perceptual categorization via active scanning is non-decomposable in Simon's sense, it would still lie within the scope of mechanistic explanation just because it can be decomposed and its component activities localized in the way demonstrated by Beer. But this account leads to an overly narrow conception of scientific practice. Section 3 shows that many heuristic strategies other than decomposition and localization contribute to mechanism discovery. It is not difficult to imagine that some of these alternative strategies may succeed even when decomposition and localization fail. Of the novel heuristics introduced above, the evolution of simulated model organisms seems particularly promising. The practice of artificially evolving a mechanism to reproduce a phenomenon in a simulated environment has a rich history of yielding particularly unintuitive or complex examples (Harvey et al. 2005), many of which resist decomposition and localization but can nevertheless be described using sophisticated mathematical methods (Wheeler 2005). Although it remains to be seen to what extent such simulated mechanisms can be used to reason about mechanisms in the real world in the way exemplified by Izquierdo and Lockery's study of klinotaxis, this is an empirical question best resolved by scientific research rather than by philosophical reflection.

In summary, the fact that practicing researchers frequently invoke heuristic strategies other than decomposition and localization, together with the fact that they rely on descriptive techniques other than verbal characterization and simple diagramming or animation, suggests that the scope of mechanistic explanation extends beyond the boundaries specified by Bechtel and Richardson. Exactly how far beyond? It is unclear that this question can—or should—be answered *a priori.* Insofar as the scope of mechanistic explanation depends (at least partly) on practicing researchers' epistemic capacities, and insofar as these capacities are continuously evolving, answering this question will involve closely considering future development in the strategies, methods, tools and concepts of scientific research.

6 Conclusion

One reason for considering novel approaches to mechanism discovery and mechanism description is to develop an improved conception of contemporary scientific practice. To this end, Sect. 3 shows that mechanism discovery goes beyond the heuristics of decomposition and localization, and Sect. 4 shows that mechanism description goes beyond verbal characterizations and iconic or schematic diagrams. Of course, the number and heterogeneity of heuristic strategies and descriptive techniques that contribute to mechanistic explanation in the life sciences is likely to even go beyond the examples considered here. Insofar as the philosophical conception of mechanistic explanation seeks to capture this number and diversity, there is no way around considering more, and more recent, examples from actual scientific research.

A second reason for considering such novel approaches is to force a reconsideration of the scope of mechanistic explanation—to delineate the class of phenomena that can be explained by describing the mechanisms responsible for them from the class of phenomena that cannot. Might these novel approaches be used to discover and describe mechanisms deemed too complex or too large to be discovered and described by the heuristic strategies and descriptive techniques thus far considered in philosophical discourse? It appears so. At the same time, it is unclear that the scope of mechanistic explanation can be properly determined until after the discovery and description of particularly challenging mechanisms is actually attempted. Unsurprisingly, the question of which phenomena can or cannot be scientifically explained is probably best answered by scientists themselves.

References

Abrahamsen, A., & Bechtel, W. (2006). Phenomena and mechanisms: Putting the symbolic, connectionist, and dynamical systems debate in broader perspective. In R. J. Stainton (Ed.), *Contemporary debates in cognitive science* (pp. 159–185). Oxford: Blackwell.

Baetu, T. (2015). From mechanisms to mathematical models and back to mechanisms: Quantitative mechanistic explanations. In P.-A. Braillard & C. Malaterre (Eds.), *Explanation in biology. An enquiry into the diversity of explanatory patterns in the life sciences* (pp. 345–363). Dordrecht: Springer.

Banks, E., Nabieva, E., Chazelle, B., & Singh, M. (2008). Organization of physical interactomes as uncovered by network schemas. *PLoS Computational Biology, 4*(10), e1000203. doi:10.1371/journal.pcbi.1000203.

Bargmann, C. I., & Horvitz, H. R. (1991). Chemosensory neurons with overlapping functions direct chemotaxis to multiple chemicals in C. elegans. *Neuron, 7*(5), 729–742.

Bechtel, W. (2006). *Discovering cell mechanisms*. Cambridge: Cambridge University Press.

Bechtel, W. (2008). *Mental mechanisms: Philosophical perspectives on cognitive neuroscience*. London: Routledge.

Bechtel, W. (2015). Generalizing mechanistic explanations using graph-theoretic representations. In P.-A. Braillard & C. Malaterre (Eds.), *Explanation in biology. An enquiry into the diversity of explanatory patterns in the life sciences* (pp. 199–225). Dordrecht: Springer.

Bechtel, W., & Abrahamsen, A. (2005). Explanation: A mechanist alternative. *Studies in History and Philosophy of Science Part C: Studies in History and Philosophy of Biological and Biomedical Sciences, 36*(2), 421–441. doi:10.1016/j.shpsc.2005.03.010.

Bechtel, W., & Abrahamsen, A. (2010). Dynamic mechanistic explanation: Computational modeling of circadian rhythms as an exemplar for cognitive science. *Studies in History and Philosophy of Science Part A, 41*(3), 321–333. doi:10.1016/j.shpsa.2010.07.003.

Bechtel, W., & Richardson, R. C. (1993). *Discovering complexity: Decomposition and localization as strategies in scientific research*. Princeton: Princeton University Press.

Beer, R. D. (2003). The dynamics of active categorical perception in an evolved model agent. *Adaptive Behavior, 11*(4), 209–243. doi:10.1177/1059712303114001; discussion 244–305.

Braillard, P.-A. (2015). Prospect and limits of explaining biological systems in engineering terms. In P.-A. Braillard & C. Malaterre (Eds.), *Explanation in biology. An enquiry into the diversity of explanatory patterns in the life sciences* (pp. 319–344). Dordrecht: Springer.

Brigandt, I. (2015). Evolutionary developmental biology and the limits of philosophical accounts of mechanistic explanation. In P.-A. Braillard & C. Malaterre (Eds.), *Explanation in biology. An enquiry into the diversity of explanatory patterns in the life sciences* (pp. 135–173). Dordrecht: Springer.

Chemero, A. (2009). *Radical embodied cognitive science*. Cambridge, MA: MIT Press.

Chemero, A., & Silberstein, M. (2008). After the philosophy of mind: Replacing scholasticism with science. *Philosophy of Science, 75*(1), 1–27. doi:10.1086/587820.

Craver, C. F. (2006). When mechanistic models explain. *Synthese, 153*(3), 355–376. doi:10.1007/s11229-006-9097-x.

Craver, C. F. (2007). *Explaining the brain*. Oxford: Oxford University Press.

Craver, C. F. (2008). Physical law and mechanistic explanation in the Hodgkin and Huxley model of the action potential. *Philosophy of Science, 75*, 1022–1033.

Craver, C. F. (2013). Functions and mechanisms: A perspectivalist view. In P. Huneman (Ed.), *Functions: Selection and mechanisms*. Dordrecht: Springer.

Cummins, R. (1983). *The nature of psychological explanation*. Cambridge, MA: MIT Press.

Gigerenzer, G. (1991). From tools to theories: A heuristic of discovery in cognitive psychology. *Psychological Review, 98*(2), 254–267.

Glauer, R. (2012). *Emergent mechanism: Reductive explanation for limited beings*. Mentis: Münster.

Glennan, S. (2002). Rethinking mechanistic explanation. *Philosophy of Science, 69*(S3), S342–S353. doi:10.1086/341857.

Harvey, I., di Paolo, E. A., Tuci, E., Wood, R., & Quinn, M. (2005). Evolutionary robotics: A new scientific tool for studying cognition. *Artificial Life, 11*, 79–98.

Hempel, C. G. (1965). *Aspects of scientific explanation and other essays in the philosophy of science*. New York: Free Press.

Hubel, D., & Wiesel, T. (1959). Receptive fields of single neurones in the cat's striate cortex. *Journal of Physiology, 148*, 574–591.

Hubel, D., & Wiesel, T. (1968). Receptive fields and functional architecture of monkey striate cortex. *Journal of Physiology, 195*, 215–243.

Iino, Y., & Yoshida, K. (2009). Parallel use of two behavioral mechanisms for chemotaxis in Caenorhabditis elegans. *Journal of Neuroscience, 29*(17), 5370–5380.

Issad, T., & Malaterre, C. (2015). Are dynamic mechanistic explanations still mechanistic? In P.-A. Braillard & C. Malaterre (Eds.), *Explanation in biology. An enquiry into the diversity of explanatory patterns in the life sciences* (pp. 265–292). Dordrecht: Springer.

Izquierdo, E. J., & Beer, R. D. (2013). Connecting a connectome to behavior: An ensemble of neuroanatomical models of C. elegans Klinotaxis. *PLoS Computational Biology, 9*(2), e1002890. doi:10.1371/journal.pcbi.1002890.

Izquierdo, E. J., & Lockery, S. R. (2010). Evolution and analysis of minimal neural circuits for klinotaxis in Caenorhabditis elegans. *The Journal of Neuroscience, 30*(39), 12908–12917. doi:10.1523/JNEUROSCI.2606-10.2010.

Kaplan, D. M. (2011). Explanation and description in computational neuroscience. *Synthese, 183*, 339–373. doi:10.1007/s11229-011-9970-0.

Kaplan, D. M. (2012). How to demarcate the boundaries of cognition. *Biology & Philosophy.* doi:10.1007/s10539-012-9308-4.

Kaplan, D. M., & Craver, C. F. (2011). The explanatory force of dynamical and mathematical models in neuroscience: A mechanistic perspective. *Philosophy of Science, 78*, 601–627.

Kelso, J. A. S. (1995). *Dynamic patterns: The self-organization of brain and behavior.* Cambridge, MA: MIT Press.

Kocabas, A., Shen, C. H., Guo, Z. V., & Ramanathan, S. (2012). Controlling interneuron activity in Caenorhabditis elegans to evoke chemotactic behaviour. *Nature, 940*, 273–277.

Leuridan, B. (2011). Three problems for the mutual manipulability account of constitutive relevance in mechanisms. *The British Journal for the Philosophy of Science, 63*(2), 399–427. doi:10.1093/bjps/axr036.

Machamer, P., Darden, L., & Craver, C. F. (2000). Thinking about mechanisms. *Philosophy of Science, 67*(1), 1–25.

Marr, D., & Hildreth, E. (1980). Theory of edge detection. *Proceedings of the Royal Society of London. Series B, Biological Sciences, 207*(1167), 187–217.

Mekios, C. (2015). Explanation in systems biology: Is it all about mechanisms? In P.-A. Braillard & C. Malaterre (Eds.), *Explanation in biology. An enquiry into the diversity of explanatory patterns in the life sciences* (pp. 47–72). Dordrecht: Springer.

Mitchell, M. (1996). *An introduction to genetic algorithms.* Cambridge, MA: MIT Press.

Schlitt, T., & Brazma, A. (2007). Current approaches to gene regulatory network modelling. *BMC Bioinformatics, 8*(Supplement 6), S9. doi:10.1186/1471-2105-8-S6-S9.

Silberstein, M., & Chemero, A. (2013). Constraints on localization and decomposition as explanatory strategies in the biological sciences. *Philosophy of Science, 80*(5), 958–970.

Simon, H. A. (1996). *The sciences of the artificial* (3rd ed.). Cambridge, MA: MIT Press.

Spencer, J. P., & Schöner, G. (2006). An embodied approach to cognitive systems: A dynamic neural field theory of spatial working memory. In *Proceedings of the 28th annual conference of the Cognitive Science Society* (pp. 2180–2185), Vancouver.

Suzuki, H., Thiele, T. R., Faumont, S., Ezcurra, M., Lockery, S. R., & Schafer, W. R. (2008). Functional asymmetry in Caenorhabditis elegans taste neurons and its computational role in chemotaxis. *Nature, 454*(7200), 114–117. doi:10.1038/nature06927.

Varshney, L. R., Chen, B. L., Paniagua, E., Hall, D. H., & Chklovskii, D. B. (2011). Structural properties of the Caenorhabditis elegans neuronal network. *PLoS Computational Biology, 7*(2), e1001066.

Webb, B. (2009). Animals versus animats: Or why not model the real iguana? *Adaptive Behavior, 17*(4), 269–286. doi:10.1177/1059712309339867.

Wheeler, M. (2005). *Reconstructing the cognitive world.* Cambridge, MA: MIT Press.

White, J. G., Southgate, E., Thomson, J. N., & Brenner, S. (1986). The structure of the nervous system of the nematode Caenorhabditis elegans. *Philosophical Transactions of the Royal Society, B: Biological Sciences, 314*, 1–340. doi:10.1098/rstb.1986.0056.

Williams, P. L., & Beer, R. D. (2010). Information dynamics of evolved agents. In S. Doncieux, B. Girard, A. Guillot, J. Hallam, J.-A. Meyer, & J.-B. Mouret (Eds.), *From animals to animats 11: Proceedings of the 11th international conference on simulation of adaptive behavior* (pp. 38–49). Springer.

Wimsatt, W. C. (1986). Forms of aggregativity. In A. Donagan, A. N. Perovich, & M. V. Wedin (Eds.), *Human nature and natural knowledge: Festschrift for Marjorie Grene* (pp. 259–293). Dordrecht: Reidel.

Wright, C., & Bechtel, W. (2007). Mechanisms and psychological explanation. In P. Thagard (Ed.), *Philosophy of psychology and cognitive science* (pp. 31–79). New York: Elsevier.

Zednik, C. (2011). The nature of dynamical explanation. *Philosophy of Science, 78*(2), 238–263.

Chapter 14
Prospect and Limits of Explaining Biological Systems in Engineering Terms

Pierre-Alain Braillard

Abstract Systems biology represents an effort to develop new modelling approaches in molecular and cell biology, drawing inspiration from disciplines like physics, computer sciences and engineering. In particular, many scientists have called for a transfer of methods, models and concepts from engineering in order to analyze and explain biological systems in all their complexity. In this paper, I examine how such transfer can contribute to systems biology explanatory project. Model building in the context of post-genomic biology raises a number of difficult challenges, mainly due to the complexity of the processes studied and their intricate dynamical features. Engineering methods can be used to efficiently analyze quantitative data about systems behaviour and use them to build mathematical dynamical models, in a way that goes beyond classical mechanistic approaches. More generally, engineering has suggested adopting a modular framework, as a general approach of decomposition and explanation based on analogies between biological and engineered systems, which promises to identify intelligible principles in the complex organization of molecular networks. I discuss the nature of this explanatory framework and on what assumptions it rests.

Keywords Systems biology • Modularity • Reverse engineering • Mathematical modelling

1 Introduction

Functional biology has been deeply transformed during the last 20 years by the rapid development of genomics, functional genomics and systems biology. Many new experimental methods and modelling strategies have been explored in order to analyze and explain the daunting complexity of biological systems, from bacteria to animals. A striking tendency is the use of mathematical modelling, something that was still rare at the end of the 1990s and that can now be found in virtually every

P.-A. Braillard (✉)
Independent Scholar, Peyregrand, 12350 Drulhe, France
e-mail: brailla6@hotmail.com

© Springer Science+Business Media Dordrecht 2015
P.-A. Braillard, C. Malaterre, *Explanation in Biology*, History, Philosophy and Theory of the Life Sciences 11, DOI 10.1007/978-94-017-9822-8_14

biology journal issue. This has started to attract philosophical interest, because it raises several important questions concerning biological explanation, notably how these models relate to traditional mechanistic models that have been much discussed in the recent literature.[1]

Another notable feature of contemporary biology is its highly multidisciplinary nature. This is particularly true for systems biology, which emerged at the turn of the century from the convergence of methods and explanatory strategies from fields such as molecular and cell biology, biochemistry, physics, computer sciences, engineering, and applied mathematics, among others. It reminds us of the origins of molecular biology with the heavy contribution of physicists and the role played by idea coming from other fields. In the context of contemporary systems biology, an important question is to clarify the explanatory contribution of these various approaches more or less directly coming from other scientific domains.

In this chapter, my goal is to explore these two broad issues in order to contribute to a clarification of contemporary biology's (and particularly systems biology's) explanatory practices.[2] I will however restrict my attention to the contribution of engineering approaches and discuss the roles played by methods, modelling strategies and concepts that have been applied to biology. I will try to shed some light on how the mathematical tools that have been developed for the design and analysis of artificial systems are used to explain biological systems. It is important to understand their originality and how they can complement more traditional approaches.

This chapter is organized as follows. I begin by describing the challenge constituted by model building in the context of post-genomic biology. Biologists face a level of complexity that makes data analysis and model building very difficult to carry on. This is important in order to appreciate why help and inspiration have been sought from engineering approaches in order to tackle biological questions. In the second part of this paper, I discuss in some detail two examples illustrating how these approaches work in the analysis and modelling of biological systems and what their explanatory import is. In the third part, I extend the discussion beyond these examples and give an account of the general modular framework that underlies much of these engineering inspired approaches and show how it offers a strategy for decomposition and explanation. In particular I make explicit the different assumptions made about biological systems that justify the use of these explanatory strategies. In the last part, I point to some questions raised by these assumptions and to possible limits of a modular decomposition.

---

[1] These issues are discussed in this volume by Baetu (2015), Breidenmoser and Wolkenhauer (2015), Brigandt (2015), Issad and Malaterre (2015), and Zednik (2015).

[2] Note that I do not offer an explication of the notion of explanation. This might sound problematic in a volume about explanation, however instead of developing a general account of explanation in biology, I think that it is more interesting to analyze the various models that are considered as explanatory and how they relate to each other. Hence the analysis offered in this chapter should bring useful elements for an understanding of how engineering contributes to modify and enrich explanatory models in biology, even without a general account of what it means to explain.

2   From Data to Models: Reverse Engineering Biological Systems

As it has often been remarked, the technological revolution that started with the launching of huge genomics programs in the late 1980s (most notably the Human Genome Project) and that amplified with the development of functional genomics approaches – born from the necessity to functionally interpret DNA sequences, which are not very explanatory by themselves – has deeply transformed biology at different levels, both from experimental and theoretical points of view. These new experimental methods allow the study of biological processes at the system level, for example by enabling the monitoring of the expression of all the genes in parallel or the mapping of thousands of protein-protein interactions. This has resulted in a tsunami of data about biological systems, which constitutes an extraordinary opportunity to gain a system level understanding of cells and organisms and overcoming some of the limits of traditional reductionist explanatory models, which focus on small mechanisms in order to explain higher level function and biological phenomena, like development or cancer. However, they also bring serious challenges because biologists must be able to analyze these data and integrate them to build models of a scale and complexity totally unknown until recently in biology.

Another tendency in the past 20 years is to pay greater attention to dynamic behaviour of biological mechanism. It has been increasingly recognized that traditional mechanistic models are too coarse-grained and not quantitative enough representations of these processes for a good understanding of their functioning. In other words, if one wants to explain how biological functions emerge from the interplay of molecular components, it is necessary to analyze the precise dynamical properties of these mechanisms. Progress in the study of many molecular mechanisms in the last two decades has brought a lot of knowledge of these processes, including quantitative data, but how to determine how these details and data must be used in order to build mathematical models that are satisfactory from an explanatory point of view is far from straightforward.

Hence, model building in the post-genomic era involves facing these two broad challenges: scaling-up to a true systemic level and representing complex dynamical behaviours.

The general task of inferring biological systems' models from data about their behaviour and components interactions has come to be called reverse engineering (D'Haeseleer et al. 2000; Brazhnik et al. 2002), by analogy with what engineers sometimes do when they want to understand the design of a device made by a competitor. They can study and decompose the product but they don't have the blueprint and they ignore the design process. When one faces complex devices it is far from evident how one should understand how it is organized, how it functions and why it has some specific features. Biologists in the post-genomic era face an even harder task: how to identify and understand the way millions of interactions produce all the biological functions. Of course, molecular and cell biologists were also engaged in a somehow similar task, but their methods were quite different, because their ability to perturb and monitor cells and organisms were much more limited (I will come back to some of the differences below). The advent of genomics followed by functional genomics has stimulated the development of bioinformatics, which has given biologists powerful computational tools that facilitate the management and analysis of these very large sets of data. However, analyzing and integrating all these heterogeneous data (gene expression, protein interactions, etc.) is much more difficult than doing sequence alignment. Identifying correlations in gene expression data cannot easily lead to a good understanding of how these systems function. Moreover, despite impressive progress in computational power, the task of inferring the structure (or architecture) of molecular networks from these data cannot be done in a straightforward way. And even when a network has been relatively accurately characterized it is essential to clarify the relations between structural and functional properties, or in other words understand how a given structure produces a specific behaviour. Let us now explore some of these difficulties.[3]

Because systems biology has adopted a network framework (systems are modelled as networks of interacting genes, proteins and other molecular components) reverse engineering is usually equivalent to network identification (or network inference), which means inferring the underlying network of molecular interactions from the behaviour of the system. This identification can be done in several ways. The goal can be to identify the network structure without (or with only a little) dynamics, as represented by undirected graph, directed graph, signed directed graph or Bayesian networks. In other words, one tries to find out which components interact or are connected to which, without any detail of these interactions and how they determine the behaviour of the system. When a dynamical dimension is added (i.e. what are the temporal properties of the regulatory events), other types of models are necessary. These range from Boolean networks to systems of differential equations (in order of increasing detail). The variety of formal models of networks is reflected in a variety of tools used to build these models.

The amount and type of data necessary for such reconstruction, and the computational difficulties involved, naturally depend on the kind of identification (and model) one aims at. Analysis of gene expression of wild type and mutants can serve for structure identification. Regulatory interactions can be inferred from gene expression data, but these are only qualitative (which genes interact but not how the regulation works). Dynamics identification requires times-series data and higher accuracy of measurement. Fine-grained and dynamically precise models require still a larger amount of data.

What should however be clear is that despite the huge progress that has been made in experimental techniques, it is still an extremely difficult task. Generally speaking, there is a "mismatch between the available and required data to identify uniquely a model structure" (Gunawan et al. 2006, 224). If one wants precise and reliable data one needs to follow more bottom-up approaches, which take much

---

[3]For a discussion of the challenges and methods of reverse engineering, see for example Kell and Knowles 2006, Chapter 11.

more time (and are closer to classical molecular biology methods). There is thus some irony here, because while the flow of data produced by post-genomic biology is difficult to manage, it is often not sufficient to achieve its goals.

A related problem is that functional genomics is characterized by an inverse correlation between the accuracy and the amount of data. In other words, it remains impossible to perform highly accurate measurements at the whole system level. Systems biologists thus face a difficult trade-off, as Gunawan et al. acknowledge: "identification of models that embody both high complexity and details of a cellular network is an untenable problem, and thus, one major issue in reverse engineering as well as in data acquisition is to select the appropriate model structure that balances the network complexity and the detail of interactions" (Gunawan et al. 2006, 229).

The problem of parameters determination from experimental data is twofold. First, as many studies have recently shown, gene expression is inherently noisy and stochastic (Raser and O'Shea 2005). This introduces an important source of difficulty. But the problem is more fundamental, because even when assuming no noise in gene expression and prior knowledge of the network structure, complete identifiability is not straightforward and requires well-designed experimental procedures. When one thinks that real systems are noisy and that their structure is still largely unknown, one sees how challenging the task is. Moreover, the difficulty grows with the size and complexity of the network to be identified. Considering that molecular networks are composed of thousands of elements and hundreds of thousands of interactions, the goal seems impossible to achieve.

I have briefly described these difficulties, because they show us why an increasing number of scientists at the turn of the century started to look for solutions in other fields confronted with similar problems in modelling and explaining complex systems. Calls for building new bridges between biology and engineering were especially numerous, as I will describe below. For example, Csete and Doyle argued in a widely read paper that "the success of systems biology will certainly require modeling and simulation tools from engineering, where experience shows that brute-force computational approaches are hopeless for complex systems involving protocols and feedback." (Csete and Doyle 2002, 1668) Let us try to see exactly how engineering has contributed to systems biology project.

3   Using Engineering Tools and Concepts

In the same article, Csete and Doyle develop their diagnosis of post-genomics biology's challenges and their appeal to the development of engineering-inspired approaches.

> Realistic models of biological networks will not be simple and will require multiple feedback signals, nonlinear component dynamics, numerous uncertain parameters, stochastic noise models, parasitic dynamics, and other uncertainty models. Scaling to deal with large networks will be a major challenge. Fortunately, researchers in robust control theory, dynamical systems, and related areas have been vigorously pursuing mathematics and software tools to address exactly these issues and apply them to complex engineering systems. (Csete and Doyle 2002, 1668)

Indeed, since the 1990s, a growing number of engineers have been interested in bringing their expertise to help conducting this work, for which biologists were not prepared. Around the turn of the century, several articles and reviews advocating an important role for engineering methods in SB appeared in leading journals and certainly had a wide influence. Of course, contributions from other fields like physics were also expected, but engineering seemed naturally adapted to contribute to SB project.[4]

After the Second World War, part of engineering has been engaged in the design and management of complex functional systems, such as spacecraft, computer chip, robotics, and software integration. Engineering has certainly been one of the few domains where system thinking has found its most fertile and successful applications. The expression 'systems engineering' was first used in the 1940s (in Bell Telephone Laboratories). Systems engineering aims at the development of new methods and modelling techniques to understand, design and handle engineering systems as they grow more complex. A number of different approaches and methods have been developed during the last decades to manage this complexity: system dynamics (which study the behaviour of complex non-linear systems with feedbacks), optimization, system analysis, and control theory. Generally speaking their goal is a better understanding of the structure and function of complex systems (which can be of very different nature) in order to redesign them, improve them and better exploit them. Mathematical modelling and simulation play a central role as they allow testing hypotheses and theories about systems. Because the goal is often to be able to control these systems, mathematical models are used to predict input-output relations, based on the description of causal processes (these models are thus not only phenomenological). Here control theory is also playing a very important role, as we will see.

Coming back to post-genomic biology's challenges, engineering methods, tools and concepts can be useful in several ways. First, they are used in system identification (inferences from data, design of experiments, etc.). Second, the are helpful in the analysis of regulatory principles, which contributes to an understanding of the 'logic' of a mechanism, its robustness, its principles of control, etc. Third, they offer efficient ways of simplifying systems models, something that is essential for gaining understanding in the context of complex systems. These three aspects cannot be distinguished completely, but each has its importance if one wants to understand the multiple contributions of engineering in solving the complexity problem biology is facing. As I will discuss below, these three aspects are related to the emergence of a modular framework.

---

[4]Some of the modelling methods used in engineering are also used in other domains and are indeed very general, especially the whole framework of dynamic systems theory. But the point is that many of these methods, tough they are not purely belonging to engineering, have been developed and put to use by engineers in the study and design of complex systems.

My aim here is not to provide an overview of the variety of modelling methods inspired by engineering, but rather starting from two examples I will discuss some aspects of their contribution to explanatory strategies dealing with the complexity of biological systems.

3.1 System Identification

The field of system identification, which belongs to control engineering, consists in building mathematical models of dynamical systems from measured data with the aid of various mathematical methods. These experiments are usually precisely controlled perturbation experiments. Olaf Wolkenhauer sums up the contribution of system approach in the following terms:

> The first and probably most important lesson of systems theory is that we can only understand the behaviour of a system if we systematically perturb it and record its response. A systems approach is thus characterised by input/output descriptions and from this, the most important role of the modeller in systems biology is to support the design of stimulus/response experiments. The role of nonlinear systems and control theory is then to provide methodologies to encode interactions of genes/proteins in the structure of the mathematical equations that form a model. (Wolkenhauer 2006)

Of course, perturbation experiments have been central in molecular and cell biology for decades (they are at the hearth of genetics for example), but these are usually relatively crude interventions (e.g. gene deletion) and their effects are rarely measured quantitatively and dynamically. They are clearly not sufficient to untangle the complex and non-linear interactions that underlie the functioning of biological systems.

An important part of system identification is the optimal design of experiments so as to produce the most informative data. Not all data can bring the same amount of information about the system's structure and regulatory principles and hence one needs to determine which experiments are most useful. This cannot be done by intuition and mathematical methods are necessary. In engineering, what is called parameter identifiability "addresses the question whether a parameter of the model can be estimated given a set of experimental data (a priori identifiability), or to determine the accuracy which can be expected for each parameter (practical identifiability)" (Kremling and Saez-Rodriguez 2007). In other words, these methods clarify the conditions and limits of a reverse engineering approach in each case. Then experimental design's goal "is to define the most informative experiment, e.g. to distinguish between two or more model variants, to improve the validity of the model by reducing the parameter variances and to clarify the structure of the model" (Kremling and Saez-Rodriguez 2007, 336).

This looks quite different from fishing expeditions in functional genomics. Rather than collecting data without prior hypothesis and then hoping that some knowledge can be induced from them, here the goal is to determine which data will be useful in the construction of a specific kind of model based on several assumptions.

Let us see in some detail one example in order to appreciate how controlled perturbation experiments and dynamical analysis can lead to explanatory models and how these approaches are related to mechanistic methods.

The study of signal transduction cascade offers problems for which engineering tools seem particularly well suited. The effects of a cascade often depend on its precise dynamical behaviour, which often involves multiple feedback loops. For this reason, simple diagrams representing qualitative interactions between its molecular components are clearly not sufficient.

I will briefly describe the work of Mettetal et al. (2008) on the mechanism of osmo-adaptation in the yeast *Saccharomyces cerevisiae*. The mechanism of osmo-adaptation in yeast involves changes in the concentration of osmolyte glycerol in order to respond to osmotic pressure. The high-osmolarity glycerol (HOG) pathway is responsible for controlling this concentration.

The method followed by this group to gain a better understanding of the dynamical and regulatory properties of this cascade was not to study each component individually (for example by disrupting genes and looking at the result) or by building a detailed model in a bottom-up fashion. One of the reasons is that many interactions are poorly or not characterized in this system. This group instead developed an alternative strategy that has been described by commentators as a "top-down approach to the quantitative biology of a small size system" (Yildirim and Vidal 2008). The expression "top-down" is often used for modelling approaches that are based on global studies of system behaviour. However, in this case the target system is a classical mechanism, composed of a relatively small number of components. It can be considered as proceeding in a top-down direction, because its starting point is a quantitative study of the mechanism behaviour, without looking at any mechanistic details.

Indeed, Mettetal et al. started by considering the system as a black box. They then assumed it to be equivalent to a linear time-invariant (LTI) system. A single 'response function' characterizes such as system. If one can determine this response function, then it becomes possible to calculate the output for any arbitrary input. Because in this case the response function is unknown, the solution is to methodically apply different inputs and observe changes in output, in order to decode the response function (see Fig. 14.1). We don't need to discuss in detail here how this is done, but let us just note that it requires applying a series of oscillatory input signals each having a different frequency and look at how these oscillations propagate in the cascade. What is especially informative is the frequency dependences of the two following quantities: the amplitude of the response and the lag between the input stimulus and the output behaviour.

This system allows building relatively easily an experimental setup able to study input-output relations. They used a flow chamber with a computer-controlled valve to modulate the extracellular osmolyte (NaCl) concentration, thus creating square-wave osmolar shocks with variable frequencies. They measured the activity of mitogen-activated protein kinase Hog1 as their output signal. After osmotic shock, cytoplasmic Hog1 is activated and subsequently migrates to the nucleus.

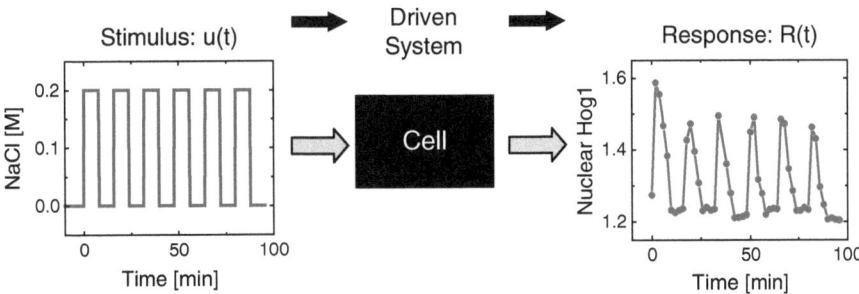

Fig. 14.1 Schema showing how square-wave oscillations in the input of extracellular NaCl (*blue line*) produce oscillations of Hog1-YFP translocation in a population of cells (*red circles*) (From Mettetal et al. 2008, Reprinted with permission from AAAS) (color figure online)

From these measurements, the authors could build a predictive model for the response to arbitrary osmotic input signals. This model was successfully tested, but at this point it does not yet throw much light on the functioning of the system. A pure black box model is not satisfactory from an explanatory point of view, because one wants to understand how the mechanism produces the behaviour (in this case the response function). As Lipan notes in his commentary of this work: "The ultimate goal, however, is a mathematical model for a white box, in which all the molecular components and their interactions are known. The road toward this goal is paved with intermediate gray-box models containing some biological inner structures. Toward this end, Mettetal et al. transform the black-box mathematical model into a gray one that successfully incorporates the first two of the three osmo-adaptation feedback loops described" (Lipan 2008, 417).

They thus gave an interpretation of their model in terms of the structure of the mechanism (using what was known about its functioning). The response function was interpreted as being the result of a combination of two negative feedback mechanisms, which control the levels of phosphorylated Hog1 and of intracellular osmolyte concentration. One feedback pathway depends on Hog1-induced glycerol accumulation, while the other pathway is Hog1-independent. These two loops have different response dynamics, which together control the response of the system (note that what they described does not reflect slower responses involving regulation of gene expression).

The main advantage of this method is that a good understanding of the dynamical properties of a system can be gained without the need for an extensive quantitative characterization of all individual parameters and molecular interactions of the system. The modelling method used reduces the underlying complex regulatory structure to its simplest form, and allows to identify and to model concisely the interactions that dominate system dynamics.

These methods greatly facilitate the study of dynamical properties of a sub-system composed of multiple feedbacks. Explaining a function in terms of feedback is certainly not new in biology (cf. the lac operon), but the important thing is that here the starting point in the analysis lies in black-boxing the mechanism, looking at the behavior and identifying an abstract dynamical function with mathematical tools.

What this example also shows is that there is not necessarily a tension with classical mechanistic models. In the end, the goal of such studies is to incorporate mechanistic details in order to explain how interactions between components produce the behaviour and thus the biological function. Thus, in a sense one can consider that it is essentially a matter of heuristics. However, it is important to realize that abstract dynamical models play a real explanatory role that is somehow independent from the mechanistic details, as I will discuss below.

3.2 Understanding Mechanisms' Regulatory Logic

I will now turn to a second benefit that can be gained from the application of engineering approaches, which is the ability to gain a better understanding of the "logic" of a complex mechanism – its principles of regulation[5] – and to capture it with a simplified model (the second and third points that I mention above). Here the starting point is a mechanism that has been relatively well characterized and the goal is to decompose it according to dynamical and engineering control principles, in particular with the aid of control theory. A good illustration of this strategy is found in the often-cited work of El-Samad et al. (2005; see also Tomlin and Axelrod 2005 and references therein).

The mechanism studied by this group is *E.coli* heat shock response system (hsr). The role of this mechanism is to respond adaptively to the stress created by an unusual increase in temperature, which can damage proteins by destabilizing their tertiary structure. The response involves the production of heat shock proteins, including molecular chaperones (which help the correct folding of proteins) and proteases (which degrade denatured proteins). Because denatured proteins can kill a cell, it is extremely important that the response is very rapid and also massive (i.e. the production of a lot of heat shock proteins). But at the same time such production in conditions when these proteins are not necessary would represent an important waste of energy for the cell. Hence, as for many mechanisms, a tight regulation is crucial.

The mechanism itself is quite well known since the 1990s and involves the RNA polymerase cofactor σ^{32}, which activates specifically the synthesis of the heat shock genes. After heat shock stress, a rapid and profound increase in σ^{32} activity produces a burst of protein expression. The regulation of σ^{32} activity depends on a feedforward loop that senses temperature and controls σ^{32} transcription and on feedback loops that sense the levels of denatured cellular protein and control the stability and activity of σ^{32}.

---

[5]What I mean by these expressions will become clear in the following discussion.

Given this mechanism, the question investigated through an engineering control perspective was to understand what allowed the system to respond rapidly, precisely (in terms of amount of proteins) and robustly.

The starting point of this study was to model in a relatively detailed way the interactions between the components, which led to a mathematical model composed of 31 equations and 27 parameters. Parameters were partly derived from the literature and partly inferred through computational techniques. However, even moderately large mathematical models become very rapidly intractable and cannot be exhaustively analysed, even with computational methods. Building fancy mathematical models, even when based on good empirical data, does not automatically lead to explanatory progress. To overcome this problem, the authors then applied a strategy of model reduction through a modular decomposition, as usually done by engineers: "Control and dynamical systems theory is a discipline that uses modular decompositions extensively to make modelling and model reduction of systems more tractable. Because biological networks are themselves complex regulation systems, it is reasonable to expect that seeking similarities with the functional modules traditionally identified in control engineering schemes can be particularly useful" (El-Samad et al. 2005, 2737).

A modular decomposition in engineering systems usually starts from the isolation of the process to be regulated. The other components of the system perform each a function that contributes to this regulation and they are described in terms of these functions. For example, there are sensing modules composed of sensing and detection mechanisms, and controller modules, which are mechanisms responsible for making decisions based on information provided by sensor modules. Then there are actuation modules, which transform the information-rich signal computed by the controller into a quantity of sufficient magnitude to drive the plant in the desired direction.

If we look at a heating system, we find sensors for measuring temperature, a thermostat acting as a controller, and heat fuel valve actuating the signal. The whole system is designed to control room temperature, despite fluctuations in external temperature. It forms a closed loop system. A crucial part is the feedback, because it allows the output to be fed into the input through the controller, so as to adjust the behaviour of the system. Engineers heavily rely on mathematical models to determine what kind of control can best produce the desired behaviour in a precise and robust way.

The authors explain how this control view can be applied to heat shock response system.

A similar modular decomposition can be carried in the hsr system if the protein-folding task is viewed as the process to be regulated. This plant is actuated by chaperones. The chaperone "signal" is produced by the high-gain transcription/translation machinery, which amplifies a modest σ^{32} input signal (few copies per cell) into a large chaperone output signal ($\approx 10'000$ per cell), much like an actuator. The σ^{32} control signal is the output of the computational or controller unit, which, based on the sensed temperature and folding state of the cell, modulates the number and activity of the σ molecules. The direct temperature measurement provided by σ^{32}-mRNA heat-induced melting, in addition to the indirect protein-folding

Fig. 14.2 Schema showing
the modular decomposition of
the hsr system proposed by
El-Samad et al. Each module
is composed of several
molecules and their
interactions. These details can
be ignored in the simplified
mathematical model, which
only represents interactions
between modules (From
El-Samad et al. 2005,
Copyright (2005) National
Academy of Sciences,
U.S.A.)

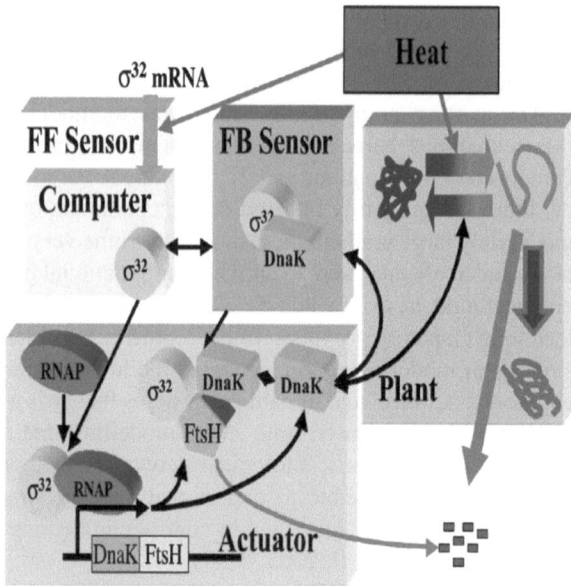

information, is assessed by the σ computational unit, producing an adequate control action
by adjusting the synthesis, degradation, and activity of σ^{32}. (El-Samad et al. 2005, 2737)

In this schema the controller is the level of σ^{32} activity, the feed forward is the
temperature dependent translational efficiency of σ^{32} synthesis, and there are also
two feedback signals, realized by sequestration and the degradation of σ^{32} from
measurement of the amount of denatured protein.

Decomposing the hsr mechanism in this way allowed the authors to build a
simplified mathematical model, composed of only 6 equations and 11 parameters.
Each equation describes the dynamics of one module (see Fig. 14.2). The advantage
of such a simplified model is that it can be more easily analysed mathematically
in order to understand how the system is regulated, or in other words what kind
of control is instantiated. Importantly, the simplified model does not completely
replace the detailed one, and it was used to guide numerical simulations with the
larger model.

In this study, mathematical modelling was used to explore the dynamical
properties of several variants of this regulatory structure, for example by removing
the feedbacks. It showed that the system can function without feedbacks, but then it
is less robust (i.e. very sensitive to parameter variations) and also less efficient (more
chaperones than required are produced). Adding the feedback loops dramatically
increases robustness, rapidity and efficiency. This analysis also discriminated the
role of each feedback, suggesting that the sequestration feedback is important for
robustness and efficiency while the degradation feedback increases rapidity of the
response.

We don't need to further discuss this study to see the explanatory benefits of using engineering approaches in the analysis of biological complexity. The two essential aspects that I would like to stress are the heuristics for building simplified models and the analysis of a mechanism's behaviour in terms of control architecture. The complexity of all the molecular interactions that constitute the mechanism can be advantageously replaced by a few equations able to capture the logic of the regulation, i.e. how the mechanism can perform the function precisely and robustly. It thus adds something important to purely mechanistic models, which would "only" describe how all the interactions produce the behaviour. Conceptualizing the system in terms of control highlights which interactions are crucial for each feature of the function (its precision, its rapidity, its robustness). From this point of view, it is similar to my previous example, but whereas the later shows how engineering methods can help build a dynamical model in the first place, the former illustrates how such modelling approaches offer powerful ways to simplify dynamical and functional analysis, which is necessary in the explanation of biological properties. Here we see the importance of intuition in the explanation of complex processes. For most biologists, even if we can reproduce the behaviour with a complex mathematical model based on molecular data, this will not be very satisfactory from an explanatory point of view, because they want to see which interactions contribute to specific regulatory properties. Here the simplified model plays an important explanatory role because it clarifies which are the most important interactions and which details can be ignored. It offers a complementary method of decomposition that helps avoiding being drawn in the details and complexity of these regulatory systems.

These two examples have offered a glimpse of how methods inspired by engineering have contributed to the explanation of biological systems, but we need to look at the more general explanatory framework that has emerged in the last 15 years in the context of systems biology. This framework is largely based on a modular view of biological systems. In the next section I will extend some of the remarks that I have made and try to give a more systematic account of the consequences for explanation in functional biology.

4   Modularity in Decomposition and Explanation

The modular framework that has been advocated in the context of systems biology has different aspects, but all of them are related to an effort to simplify the complexity of biological systems in order to explain how biological functions emerge from the multiple and often non-linear interactions between molecular components. It offers a general method of decomposition based on the idea that biological systems are composed of relatively autonomous units (sub-systems) with specific dynamical and functional properties. This kind of structure would thus be similar to the design of engineered systems and this would justify the use of modelling methods developed by engineers. To put this idea simply, if both kinds of systems share important resemblances they can legitimately be explained in similar terms. There is thus a very close link between the modular hypothesis and the use of various engineering methods in systems biology.

The idea of modularity is of course not completely new. There have been theoretical discussions about the role played by this kind of structures in natural systems long time ago, notably by Herbert Simon in the 1960s (Simon 1996 [1969]). Simon gave reason to think that complex evolved systems show some kind of modularity, and he also discussed implications for scientific analysis and explanation.

The concept of module has been much discussed in various fields during the last decade (Winther 2001; Schlosser and Wagner 2004; Callebaut and Rasskin-Gutman 2005; Mitchell 2005) but no simple and unifying definition has emerged. Some general features can nevertheless be recognized. From a structural point of view, a module is composed of elements that are more tightly interconnected than they are connected with components outside the module (i.e. intramodular connections are more frequent than intermodular ones). From a functional point of view, a module is conceived as a part of a larger system that does something (has a behavior or performs a function) relatively independently of the rest of the system.

Decomposing systems in structural parts that play relatively autonomous function is part of the general methodology in biology (Bechtel and Richardson 1993). Organ and proteins can be considered as modules. Yet, the modular approach that has been advocated since the turn of the century (Hartwell et al. 1999; Lauffenburger 2000; Segal et al. 2003; Tyson et al. 2003; Wolf and Arkin 2003) is different in several respects.

The main originality of this new modular approach is that modules are primarily analyzed and characterized dynamically, and that engineering provides the best tools and concepts both for explaining their properties and their contribution to biological processes and functions. The fundamental underlying idea is that fruitful comparisons can be established between biological modules and modules found in engineering.[6]

Engineered systems are largely modular, because this makes them easier to design, construct and modify. Artefacts are often composed of different parts with functional interfaces between them. A modular structure makes the modelling (design and understanding) much easier, because a module can be easily simplified as performing a relatively simple function. The inner working of the module (and its complexity) can be largely ignored, so long as its function (which is often an input-output relation) is described consistently with the data. Following this strategy, a complex system can be describes as a set of modules connected together. Also, a

---

[6]A word of caution: even if one restricts oneself to SB, the concept of module can be differently understood. The following analysis does not pretend to capture all of these uses and dimensions. Everyone in the field certainly does not hold some of the views and other modular approaches will not be discussed here.

module can be more easily modified without changing the overall structure of the system (one can redesign the module so that it better performs the desired function).

Several scientists have championed the view that biological systems (often conceived in terms of networks) are modular in a similar way. The following quote encapsulates this thesis:

> Complex molecular networks, like electrical circuits, seem to be constructed from simpler modules: sets of interacting genes and proteins that carry out specific tasks and can be hooked together by standard linkages. (Tyson et al. 2003, 221)

In one of the first articles to forcefully put forward this view, Hartwell et al. defended a similar idea:

> We argue here for the recognition of functional 'modules' as a *critical level of biological organization*. (Hartwell et al. 1999, C47; my emphasis)

> The properties of a module's components and molecular connections between them are analogous to the circuit diagram of an electrical device. (Hartwell et al. 1999, C50)

Analyzing biological systems in terms of modules is a powerful way to handle their complexity, as Rao and Arkin noted: "a modular decomposition provides a convenient abstraction for deducing the behavior of complex networks." (Rao and Arkin 2001, 392)

Modular approaches enable overcoming the limits in the modelling of complex networks described above, as recognized by many authors:

> Certainly, the computational limitations of dynamic gene network modeling are much easier to evade and an understanding of complex networks in terms of (higher-level) functional interactions is easier to achieve if a modular architecture underlies the network. (Szalli et al. 2006, 42)

> The reverse engineering of cellular networks represents a crucial aspect of a systemic approach for biological discovery in the post-genomic era. The major hurdle in this task is the high complexity of cellular networks, implying models with large numbers of nodes and interactions. Fortunately, the cellular networks appear to have structures that are shared with engineered systems. This is why the application of engineering methods, in particular systems identification, has shown to be fruitful in approaching these problems. (Gunawan et al. 2006, 241)

A quick look at the literature reveals that this modular framework has been widely adopted in the last decade (Lauffenburger 2000; Ihmels et al. 2002; Segal et al. 2003; Wolf and Arkin 2003). But how exactly does modularity facilitate the explanation of complex networks? Here it is important to see that in addition to modularity, it is the fact that biological modules can be modelled through engineering approaches that is explanatory relevant. The idea is that modellers can focus on this part of the system and capture through a relative simple mathematical model its essential dynamical properties. Here by essential, I mean the properties that contribute most to the biological function.

Of course, there are many different ways to model the dynamics of sub-system (bottom-up or top-down), but the goal is often the same: clarify the relation between a (regulatory) structure and its behaviour. A lot of work in the last 20 years has been devoted to rigorous analysis of behaviours produced by different kinds of regulatory structures, often feedback and feedforward loops.

At the most elementary level, we find models of network motifs, which can be considered the simplest modules.[7] Network motifs are constituted of few components (usually less than 5) connected in a specific way. Observations showing these structures were present in molecular networks with higher frequency than in random networks prompted interest in their study.

It has been hypothesized that motifs are characterized by specific dynamic functions, which underlie biological functions. For this reason they have been described as "biochemical building blocks" of biological networks.

At the beginning of last decade, motifs with noise filtering have received a lot of attention (Mangan and Alon 2003). Found in regulatory cascades, they would guarantee that only the "right" signals (e.g. with a particular duration) induce a response, while other signals and noise will not. Other motifs were described in terms of generic dynamical behaviour, like homeostasis, adaptation, oscillations, or switching.

In his book on biological circuits (Alon 2006), Uri Alon lists eight main network motifs, each defined by a specific information processing function. Feedforward loops can act as sign-sensitive filters, pulse generators, and response accelerators. Positive autoregulation can slow response time. Negative autoregulation can speed and stabilize responses. Single-input modules can generate temporal programs of expression.

The important point is that these models of network motifs describe biochemical mechanisms in terms of abstract dynamical functions. This is how simplified models are produced. In other words, and this is central from an explanatory point of view, a relatively complex biochemical model can be usefully replaced by a simple model describing the input-output function, like engineers designing electronic circuits can take advantage of a modular structure in the modelling of complex electronic circuits. In the same way that a large electronic circuit can be decomposed into many transistors, the hope underlying such approach is that complex molecular networks can be decomposed into many basic small circuits (motifs) each characterized by a simple dynamical function.

Alon describes the simplifying process in modular modeling as "the ability to treat regulatory circuits with simplified mathematical models that capture the essence of the behavior and have a certain degree of universality" (Alon 2006, 236).

Combined in different ways, these basic elements can explain higher-level functions. For example, the mechanism for eukaryotic cell cycle regulation can be decomposed in several such basic components – two toggle switches and an oscillator – each realizing one dynamical function:

> The network, involving proteins that regulate the activity of Cdk1–cyclin B heterodimers, consists of three modules that oversee the G1/S, G2/M and M/G1 transitions of the cell

---

[7]One might object that motifs are different from modules, but the explanatory goal is the same: model a small sub-system and capture its essential dynamical properties.

cycle. The G1/S module is a toggle switch, based on mutual inhibition between Cdk1–cyclin B and CKI, a stoichiometric cyclin-dependent kinase inhibitor. The G2/M module is a second toggle switch, based on mutual activation between Cdk1–cyclin B and Cdc25 (a phosphatase that activates the dimer), and mutual inhibition between CDK1–cyclin B and Wee1 (a kinase that inactivates the dimer). The M/G1 module is an oscillator, based on a negative-feedback loop: Cdk1–cyclin B activates the anaphase-promoting complex (APC), which activates Cdc20, which degrades cyclin B. (Tyson et al. 2003, 227)

Here we see how a typical molecular mechanism is described (and conceptualized) in a new way. Engineering thus offers a language that can describe biological systems precisely and adequately (Lazebnik 2002). The need for a new language in functional biology, more adequate than the diagrams and non-formal language usually found in molecular and cell biology, has often been stressed by the proponents of this modular framework, like in the review by Hartwell et al.:

To describe biological functions, we need a vocabulary that contains concepts such as amplification, adaptation, robustness, insulation, error correction and coincidence detection. (Hartwell et al. 1999, C47)

The most effective language to describe functional modules and their interactions will be derived from the synthetic sciences, such as computer science or engineering, in which function appears naturally. (Hartwell et al. 1999, C49)

This point is very important, because explaining is partly a matter of finding an adequate way of representing a phenomenon. Describing a sub-system as a filter or an amplifier (based on rigorous mathematical modelling) brings an explanatory gain because it captures economically some of its essential dynamical properties.

An important point to stress is that each module (or motif) is considered as relatively independent of the system in which it is embedded. This assumption justifies that the systemic context can be temporarily ignored in the modelling process. This is crucial to avoid the complexity problem.

Alon writes: "In many systems, network motifs appear to be connected to each other in ways that do not spoil the independent functionality of each motif, allowing us to understand the network, at least partially, based on the functions of individual motifs." (Alon 2006, 234) At a low level of complexity, Alon has analyzed how feedforward loops (FFL) can be combined to form multi-output FFL and he notes that "this pattern preserves the functionality of each three-node FFL (sign-sensitive filtering, etc.)." His conclusion is that "biological networks can be understood, to a first approximation, in terms of a rather limited set of recurring circuit patterns, each carrying out computations on a different timescale." (Alon 2006, 234) Hartwell et al. also stress this relative independence: "The notion of a module is useful only if it involves a small fraction of the cell components in accomplishing a relatively autonomous function." (Hartwell et al. 1999, C48)

In engineering, this is expressed through the principle that output nodes must have low impedance, which means that when an additional module is connected to the output this does not have much influence on it. Modules defined in this way "have the advantage that their signal transfer properties are independent of what they are connected to. Therefore, if a (simpler) substitute for a certain module can be found, it can replace it without altering the properties of the network as a whole, leading to a reduction of the complete system." (Kremling and Saez-Rodriguez 2007, 342)

Importantly, simplification in the description of modules' behaviour opens the way to a hierarchical modelling, which is essential in bridging the gap between the molecular and the systemic (cellular) levels. The basic idea is that these "simple" modules are linked through interfaces and are combined to perform more complex function. A system is thus composed of modules, which can themselves be decomposed in lower-levels modules, and so on until we reach the level of molecules. Going bottom-up, a combination of modules can be described in a simplified manner as one single module. This is potentially a very powerful modelling method for explaining the emergence of systemic properties. The authors we have already met have championed the "compositional" nature of these systems:

> From these components, nature has constructed regulatory networks of great complexity. With accurate mathematical representations of the individual components, we can assemble a computational model of any such network. By numerical simulation, we can compute the expected output of the network to any particular input. (Tyson et al. 2003, 229)

> Good engineering uses modularity and recurring circuit elements (network motifs) to build reliable and scalable devices from simpler subsystems. (Alon 2006, 238)

> The higher-level properties of cells, such as their ability to integrate information from multiple sources, will be described by the pattern of connections among their functional modules. (Hartwell et al. 1999, C48)

After these remarks, let us come back to the relation with more traditional methods of decomposition and explanatory strategies. One might think that there is nothing new compared to the hierarchical nature of mechanistic models (Craver 2006).

Although there is no serious incompatibility with approaches found in molecular and cell biology, an important difference is that modules and circuits are not primarily defined in terms of biological function or chemical properties, but rather either in terms of dynamical behaviour (oscillator, filter, signal amplification, etc.)[8] or in terms of control architecture (as described in El-Salam et al. study). Long time ago, Robert Rosen criticized molecular biology's decomposition methods, because they are based too heavily on structural properties.

> Thus when we apply a prespecified set of fractionation techniques to an unknown system, there is no reason why the fractions so obtained should be simply related to properties of the original system. Yet this is exactly what happens when a molecular biologist fractionates a cell and attempts to reconstruct its functional properties from the properties of his fractions. (Rosen 1972, 54)

---

[8]One might object that these are biological functions. However, the focus is for example on how a signal is transformed, regardless of the nature of the signal or its biological function. Of course, the goal is always to explain biological function, but I maintain that often in this kind of approach biological function becomes less important.

According to Rosen, what must be done is a decomposition based on models able to capture dynamical properties.

> The reason that the use of model systems is possible at all is that the same dynamical or functional properties can be exhibited by large classes of systems, of the utmost physical or chemical diversity. Two systems which are physically different but dynamically equivalent will be called *analogues* of one another [...]. If our interest is in the system dynamics, then this dynamics can be studied equally well (and often better) in any convenient system analogous to our original system. (Rosen 1972, 56)

> System analogy shows us that dynamical or functional properties can be studied essentially independently of specifics of physicochemical structure. (Rosen 1972, 56)

We can thus consider in a sense that SB's modular approaches try to realize the kind of dynamical decomposition Rosen was advocating, although contemporary systems biology rejects this disinterest for molecular properties.

Modules can thus be seen as an intermediate level between components (genes, proteins, etc.) and systemic properties. They can be considered as mechanisms, but they are identified and described with methods that differ from what is found in molecular and cell biology, and this results in different explanatory models. Their description in engineering terms provides a kind of *simplicity*, which helps avoiding sinking into complexity. The motto of these systems biologists might be "studying complexity but searching for simplicity". Engineers like simplicity and try to avoid complexity (or at least try to reduce it as much as possible) because it can rapidly become unmanageable. What they have brought to molecular biology is a new way to search for simplicity (molecular biologists were also searching for simplicity, but in a different way). This simplicity comes from the relative independence of the mechanism's behaviour from the details of its underlying interactions; i.e. from its robustness. The issue of robustness is also central to engineering approaches in biology, as I will discuss now.

5 Analyzing Robustness

The preceding examples have suggested that the concept of robustness constitutes an important part of engineering approaches to biology and modular approaches in particular. Literature on robustness in biology has grown very fast in recent years (see for example Kitano 2004; Wagner 2005; Carlson and Doyle 2002; Stelling et al. 2004) and this issue cannot be properly discussed here, but I will make some remarks that might be helpful to appreciate the contribution of engineering to biological explanations.

Although biologists have known for a long time that organisms are robust in the sense that they can endure many injuries and environmental changes, as well as internal perturbations like mutations, it has recently become the focus of much attention.[9] Biologists developed the feeling that this property had to be explained

---

[9]When talking about robustness, it is important to keep in mind that it is meaningless if not defined in terms of specific properties of the systems that are robust to specific changes (internal

rigorously. Of course, there are several classical explanations for robustness, like redundancy of parts (we have two kidneys and many genes exist in multiple copies) or homeostasis produced by simple feedback mechanisms (like Cannon's concept of homeostasis). Yet, this is not sufficient to account for all forms of robustness observed by biologists.

This recent interest in robustness is one of the clearest illustrations of new interactions between biology and engineering, as has been vigorously defended by several scientists, like Csete and Doyle: "A key starting point in developing a conceptual and theoretical bridge to biology [from engineering] is robustness, the preservation of particular characteristics despite uncertainty in components or the environment" (Csete and Doyle 2002, 1664).

Engineers can be naturally expected to have a strong interest in biological robustness. First, robustness has always be one of engineers' main preoccupations. They want to build the most possible robust systems (given various constraints, including economical considerations) and they have developed a rich theoretical framework to achieve this goal. Hence, it is reasonable to expect that some of this knowledge might be useful to biologists. Second, man-made systems are still far behind living systems in terms of complexity and robustness, and engineers might learn a lot from studying biology.

The first and most obvious relation between these concepts is the idea that modularity is part of the basic architectural requirements for designing a robust system. This is one of the arguments advanced in favour of the modularity of biological systems. In an influential review on this issue, Kitano (2004) considers modularity as one of the four principal mechanisms that ensure robustness of a system (along with system control, alternative mechanisms, and decoupling): "Modularity is an effective mechanism for containing perturbations and damage locally to minimize the effects on the whole system." (Kitano 2004, 828) Such an argument is seen by many as a strong support of the modularity hypothesis. Because biological systems must be robust, they are certainly modular and thus we can undertake a modular decomposition, or so the argument goes.

The second important idea is that modules themselves are robust. A module's robustness is its ability to perform the same function (input-output transformation) or have the same type of behaviour (for example oscillating) despite internal or external perturbations. Modelling techniques brought by engineering (and also physics) offer several ways to identify and explain a module's robustness. Basically what needs to be determined is the dependence of a dynamical or functional behaviour on the details of the components properties and interactions. The general idea is that the weakest the dependence the more robust the mechanism (see Gross 2015, this volume).

It is not necessary to describe in details these techniques to understand their relevance for biology. Let us briefly mention the most important ones.

---

or external). A feature of a given mechanism can be robust to one perturbation and very sensitive to a different one. Hence we must say that functional property P is robust to perturbation X.

- Sensitivity analysis allows a rigorous characterization of the dependency of a dynamical behaviour on a parameter's value. The goal is to determine how small changes in one parameter affect the behaviour of the system. In other words, it reveals the relations between the change in a biological model output and the perturbation on system parameters that cause this change. (SA can also be used to identify which parts of the system are important for its global behaviour and which ones do not have a strong influence. This can be very useful for medical applications, since it helps to define which components must be targeted).
- Bifurcation analysis allows determining how a qualitative behaviour can suddenly change with a parameter's value. By partitioning the parameter space into regions of different qualitative behaviour, it gives crucial information about the robustness of a system.
- Exploration of parameter space through simulations. Mathematical models that cannot be solved by analytical techniques are computed. Parameters are chosen and simulations are run. It is impossible to test all possible parameters, but a portion of the parameter space can be characterized. This gives an idea of the dependence of the behaviour on the parameters of the system. This method has been used in the often-cited work of Odell group on Drosophila segment polarity gene network (von Dassow et al. 2000; see also Keller 2003). They showed that a surprisingly high proportion of randomly chosen parameters led to the "correct" function, hence revealing the robustness of the mechanism.
- Control analysis, as already discussed in the El-Samad et al. study.

Engineering has also brought very important ideas concerning the existence of a deep link between complexity and robustness in functional systems. The hypothesis is that part of the observed complexity is present to provide the necessary robustness of biological functions (and their relative simplicity at the level of behaviour). Many biological functions could be performed with simpler mechanisms, but additional layers of regulation increase their reliability and guarantee that the function is produced despite noise at the level of the components. In the heat shock response system, the authors showed that a mechanism without the two feedback loops could also produce the desired behaviour, yet with lower rapidity, efficiency and robustness. Hence, this relatively complex control strategy is explained as a mechanism for increased robustness.

A consequence of this hypothesis is that such tendency has resulted in a spiralling complexity during biological evolution (Csete and Doyle 2002). Each new level of complexity introduces new fragilities and risks of failure, which call for additional control systems, etc. For instance, part of the electronic devices integrated in a car improve the efficiency of its function, despite difficult external conditions (rain, snow) and possible errors of the driver (urgent braking on icy road), but it creates new sources of possible failures.

Understanding the relation between robustness and the complexity of biological systems is a crucial part in the modelling effort. One cannot understand the structure of a mechanism or sub-system (or even whole systems) if one does not clarify the link between its complexity and the need for robust functioning. I think that this is a deep contribution of engineering approaches to explanation in biology.

From a methodological point of view, the concept of robustness plays an additional and primordial role in SB. The principle can be easily summarized: we know that biological mechanisms must have some degree of robustness, because they are working in conditions of internal and external noise, with different sources of environmental changes and internal perturbations; a mechanism that is too fragile will collapse and its bearer will die or at least its fitness will probably decrease. This simple idea provides a criterion for narrowing the set of possible mechanisms able to produce a given function. This is important because mechanistic models are always underdetermined in biology (i.e. many possible mechanisms can produce the phenomenon to be explained). We have seen that this is particularly the case in network inferences. Hence, the concept of robustness is central in SB's heuristics of RE. A mechanistic model that is not robust to parameters variations is not likely to be the good one. The ability to explore a model's robustness through various mathematical and computational tools can thus be very useful for model building. Model can be revised and improved according to this criterion. Stelling et al. recognize this when they write: "Since only a limited set of mechanisms establishes robustness in biological circuits, understanding robustness can provide a key for understanding cellular organization" (Stelling et al. 2004). Again, it is crucial to be able to specify which properties are robust, because obviously biological mechanisms and systems are not robust to all kinds of perturbations.

Morohashi et al. (2002) have used an example taken from Xenopus cell cycle oscillatory mechanism to show how robustness analysis can play this heuristic role.

Model building necessarily involves making choices between alternative explanations with apparently equivalent behaviors. We put forward an argument from first principles suggesting that robustness analysis can help distinguish between more and less plausible models, and pinpoint structural weaknesses in models. The proposal is predicated on the expectation that essential cellular processes that are conserved across multiple species must be functionally robust to mutational variations. (Morohashi et al. 2002, 29)

6   Limits and Questions

These recent developments are impressive and have stimulated a lot of research, but this enthusiasm should not obscure the fact that several of the assumptions underlying this modular framework are quite disputable. Since many different ways of defining modules have been proposed, can we consider that it is a precise scientific concept, or rather a vague analogy? What kind of entities are modules? Do they really exist or are they only useful scientific hypotheses (and thus instrumental entities), or worst, artifacts of decomposition methods? Sandra Mitchell asks: "Is nature modular, or do we impose modularity on it in order to understand it?" (Mitchell 2005, 101). Do they really have intrinsic properties or are these context-dependent? (Del Vecchio et al. 2008). Generally, two fundamental questions must be asked: (1) can biological systems really be decomposed in such relatively autonomous sub-systems? and (2) how good is the analogy with engineering modules?

The concept of modular design is borrowed from engineering, where systems are literally designed to be modular. But of course such assumption cannot be made in the case of natural systems, as Szalli et al. recognize: "advanced engineered systems are rather frequently modular in their overall design, but for evolved systems we do not even have the appropriate analytical tools to address the issue of modular decomposition" (Szalli et al. 2006, 44). These authors also stress a deep methodological problem: "a proper definition of biological modules would require dynamic models, for the development of which focusing on a small part of the cellular networks is necessary – a classical catch-22." (Szalli et al. 2006, 49)

These difficulties should not however be interpreted as a fatal flaw of the modular approach, but one must recognize its heuristic nature: "the operational definitions should be judged by their value in facilitating the development of dynamic models and by the extent they enhance our understanding of these systems." (Szalli et al. 2006, 44) If this heuristics proves to be fruitful, we can have some confidence in such modular analysis. Moreover, an iterative approach can give some robustness to module identification. Computational modelling helps refining and validating modular hypotheses, partly through suggesting new experiments.

We can then provisionally conclude that modules are partly relative and that biological systems are so complex (in the sense that many processes are entangled and intertwined) that it is probable that multiple decompositions are necessary to explain them (Rao and Arkin 2001; von Dassow and Meir 2004).[10] But modules are not completely arbitrary. The fact that some sub-networks seem to have intrinsic dynamical properties, which are essential for biological function, is certainly an objective matter. Eventually the modular framework (or hypothesis) should be judged by the progress it will favour, and it seems that so far its fruitfulness cannot be denied. If one does not worry too much about modules' strict definition or reality, they might prove to be very efficient conceptual and theoretical tools for systems biology.

7 Conclusion

After this quick overview of engineering-inspired approaches in biology, what can be concluded about their explanatory import? Should we consider that biological systems will soon be explained in engineering terms, like complex man-made systems? What is the difference with traditional mechanistic models?

---

[10]For one of the first philosophical discussions of the need for different types of decomposition in biology, see (Wimsatt 1974).

Looking at the literature of the last two decades, it seems that this is the project pursued by many scientists in systems biology. There is certainly not a strong incompatibility with molecular and cell biology's explanatory strategies, but I have suggested several reasons to consider them as deeply different. Although mechanistic decomposition methods and models have a genuine dynamic component (which is obvious in philosophical accounts such as Machamer, Darden and Craver), the decomposition strategy offered by engineering methods put abstract dynamical properties at the hearth of these explanatory models. Since many biological functions are produced by rich and complex dynamical behaviours, it seems that they need to be increasingly explained in terms championed by these systems biologists.

It should however be clear that this new framework has emerged as something very programmatic. It depends on many open hypotheses that are still difficult to confirm. As noted above, it is far from certain that biological and engineered systems are as similar as the proponents of these approaches claim. If it turns out that biological systems are not modular in the expected way, what will happen to this explanatory framework? It is likely that these modelling methods will remain useful, because they involve quite general mathematical tools that can be used to describe a lot of different systems. I think that part of this heuristics inspired by engineering does not completely depend on the validity of these assumptions. However, many explanatory models might appear as very misleading because they fail to acknowledge the influence of the systemic context. In that scenario, the project of explaining systemic properties in terms of semi-autonomous modules defined by specific dynamical properties and connected according to well-specified rules, might look to future biologists as a gross oversimplification of biological complexity, as fallacious as the mechanistic models from the seventeenth century, which depicted organisms as made of cogs and pulleys. Of course, it is never a good idea to make predictions about the future developments of science. What can be said so far is that the modular hypothesis has considerably enriched explanatory strategies in biology by building new bridges between disciplines and by encouraging scientists to develop new explanatory models.

Acknowledgments The ideas discussed in this paper have been presented at different occasions, mainly at the 2013 ISHPSSB meeting in Montpellier and at the IHPST Paris-CAPE Kyoto philosophy of biology workshop. I would like to thank the audiences for useful discussions. An earlier version of this paper also benefited from comments from Christophe Malaterre, as well from two anonymous reviewers.

References

Alon, U. (2006). *An introduction to systems biology: Design principles of biological circuits*. Boca Raton: Chapman & Hall.

Baetu, T. (2015). From mechanisms to mathematical models and back to mechanisms: Quantitative mechanistic explanations. In P.-A. Braillard & C. Malaterre (Eds.), *Explanation in biology. An enquiry into the diversity of explanatory patterns in the life sciences* (pp. 345–363). Dordrecht: Springer.

Bechtel, W., & Richardson, R. C. (1993). *Discovering complexity: Decomposition and localization as strategies in scientific research*. Princeton: Princeton University Press.

Brazhnik, P., Fuente, A., & Mendes, P. (2002). Gene networks: How to put the function in genomics. *Trends in Biotechnology, 20*(11), 467–472.

Breidenmoser, T., & Wolkenhauer, O. (2015). Explanation and organizing principles in systems biology. In P.-A. Braillard & C. Malaterre (Eds.), *Explanation in biology. An enquiry into the diversity of explanatory patterns in the life sciences* (pp. 249–264). Dordrecht: Springer.

Brigandt, I. (2015). Evolutionary developmental biology and the limits of philosophical accounts of mechanistic explanation. In P.-A. Braillard & C. Malaterre (Eds.), *Explanation in biology. An enquiry into the diversity of explanatory patterns in the life sciences* (pp. 135–173). Dordrecht: Springer.

Callebaut, W., & Rasskin-Gutman, D. (Eds.). (2005). *Modularity: Understanding the development and evolution of natural complex systems*. Cambridge, MA: MIT Press.

Carlson, J. M., & Doyle, J. C. (2002). Complexity and robustness. *Proceedings of the National Academy of Sciences, 99*(90001), 2538–2545.

Craver, C. (2006). When mechanistic models explain. *Synthese, 153*, 355–376.

Csete, M. E., & Doyle, J. C. (2002). Reverse engineering of biological complexity. *Science, 295*, 1664–1669.

D'Haeseleer, P., Liang, S., & Somogyi, R. (2000). Genetic network inference: From co-expression clustering to reverse engineering. *Bioinformatics, 16*, 707–726.

Del Vecchio, D., Ninfa, A. J., & Sontag, E. D. (2008). Modular cell biology: Retroactivity and insulation. *Molecular Systems Biology, 4*, 16110.

El-Samad, H., et al. (2005). Surviving heat shock: Control strategies for robustness and performance. *Proceedings of the National Academy of Sciences of the United States of America, 102*, 2736–2741.

Gross, F. (2015). The relevance of irrelevance: Explanation in systems biology. In P.-A. Braillard & C. Malaterre (Eds.), *Explanation in biology. An enquiry into the diversity of explanatory patterns in the life sciences* (pp. 175–198). Dordrecht: Springer.

Gunawan, R., Gadkar, K. G., & Doyle, F. J. (2006). Methods to identify cellular architecture and dynamics from experimental data. In Z. Szallasi, J. Stelling, & V. Periwal (Eds.), *Systems modeling in cell biology*. MIT Press.

Hartwell, L. H., et al. (1999). From molecular to modular cell biology. *Nature, 402*, C47–C52.

Ihmels, J., et al. (2002). Revealing modular organization in the yeast transcriptional network. *Nature Genetics, 31*, 370–377.

Issad, T., & Malaterre, C. (2015). Are dynamic mechanistic explanations still mechanistic? In P.-A. Braillard & C. Malaterre (Eds.), *Explanation in biology. An enquiry into the diversity of explanatory patterns in the life sciences* (pp. 265–292). Dordrecht: Springer.

Kell, D. B., & Knowles, J. D. (2006). The role of modeling in systems biology. In Z. Szallasi, J. Stelling, & V. Periwal (Eds.), *Systems modeling in cellular biology*. Cambridge, MA: MIT Press.

Keller, E. F. (2003). *Making sense of life*. Cambridge, MA: Harvard University Press.

Kitano, H. (2004). Biological robustness. *Nature Review Genetics, 5*, 826–837.

Kremling, A., & Saez-Rodriguez, J. (2007). Systems biology – An engineering perspective. *Journal of Biotechnology, 129*(2), 329–351.

Lauffenburger, D. A. (2000). Cell signaling pathways as control modules: Complexity for simplicity? *Proceedings of the National Academy of Sciences, 97*, 5031–5033.

Lazebnik, Y. (2002). Can a biologists fix a radio? – Or what I learned while studying apoptosis. *Cancer Cell, 2*, 179–182.

Lipan, O. (2008). Systems biology. Enlightening rhythms. *Science, 319*(5862), 417–418.

Mangan, S., & Alon, U. (2003). Structure and function of the feed-forward loop network motif. *Proceedings of the National Academy of Sciences of the United States of America, 100*, 11980–11985.

Mettetal, J. T., et al. (2008). The frequency dependence of osmo-adaptation in Saccharomyces cerevisiae. *Science, 319*, 482–484.

Mitchell, S. (2005). Modularity: More than a buzzword? *Biological Theory, 1*(1), 98–101.

Morohashi, et al. (2002). Robustness as a measure of plausibility in models of biochemical networks. *Journal of Theoretical Biology, 216*, 19–30.

Rao, C. V., & Arkin, A. P. (2001). Control motifs for intracellular regulatory networks. *Annual Review of Biomedical Engineering, 3*, 391–419.

Raser, J. M., & O'Shea, E. K. (2005). Noise in gene expression: Origins, consequences, and control. *Science, 309*, 2010–2013.

Rosen, R. (1972). Some systems theoretical problems in biology. In E. Lazlo (Ed.), *The relevance of general systems theory*. New York: George Braziller.

Schlosser, G., & Wagner, G. P. (Eds.). (2004). *Modularity in development and evolution*. Chicago: University of Chicago Press.

Segal, E., et al. (2003). Module networks: Identifying regulatory modules and their condition-specific regulators from gene expression data. *Nature Genetics, 34*(2), 166–176.

Simon, H. (1996 [1969]). *The sciences of the artificial* (3rd ed.). Cambridge, MA: MIT Press.

Stelling, J., et al. (2004). Robustness of cellular functions. *Cell, 118*, 675–685.

Szalli, Z., et al. (2006). On modules and modularity. In Z. Szallasi, J. Stelling, & V. Periwal (Eds.), *System modeling in cellular biology*. Cambridge, MA: MIT Press.

Tomlin, C. J., & Axelrod, J. D. (2005). Understanding biology by reverse engineering the control. *Proceedings of the National Academy of Sciences, 102*, 4219–4220.

Tyson, J. J., et al. (2003). Sniffers, buzzers, toggles and blinkers. *Current Opinion in Cell Biology, 15*(2), 221–231.

Von Dassow, G., & Meir, E. (2004). Exploring modularity with dynamical models of gene networks. In G. Schlosser & G. P. Wagner (Eds.), *Modularity in development and evolution*. Chicago: University of Chicago Press.

Von Dassow, G., et al. (2000). The segment polarity network is a robust developmental module. *Nature, 406*, 188–192.

Wagner, A. (2005). *Robustness and evolvability in living systems*. Princeton: Princeton University Press.

Wimsatt, W. C. (1974). Complexity and organization. In K. F. Schaffner & R. S. Cohen (Eds.), *Proceedings of the 1972 meeting of the Philosophy of Science Association* (pp. 67–86). Dordrecht: D. Reidel.

Winther, R. G. (2001). Varieties of modules: Kinds, levels, origins, and behaviors. *Journal of Experimental Zoology, 291*, 116–129.

Wolf, D., & Arkin, A. (2003). Motifs, modules, and games. *Current Opinion in Microbiology, 6*, 125–134.

Wolkenhauer, O. (2006). Engineering approaches: What can we learn from it in systems biology? In R. van Driel (Ed.), *Systems biology: A grand challenge for Europe European science foundation forward look report*. Strasbourg: European Science Foundation.

Yildirim, M. A., & Vidal, M. (2008). Systems engineering to systems biology. *Molecular Systems Biology, 4*, 185.

Zednik, C. (2015). Heuristics, descriptions, and the scope of mechanistic explanation. In P.-A. Braillard & C. Malaterre (Eds.), *Explanation in biology. An enquiry into the diversity of explanatory patterns in the life sciences* (pp. 295–317). Dordrecht: Springer.

Chapter 15
From Mechanisms to Mathematical Models and Back to Mechanisms: Quantitative Mechanistic Explanations

Tudor M. Baetu

Abstract Despite the philosophical clash between deductive-nomological and mechanistic accounts of explanation, in scientific practice, both approaches are required in order to achieve more complete explanations and guide the discovery process. I defend this thesis by discussing the case of mathematical models in systems biology. Not only such models complement the mechanistic explanations of molecular biology by accounting for poorly understood aspects of biological phenomena, they can also reveal unsuspected 'black boxes' in mechanistic explanations, thus prompting their revision while providing new insights about the causal-mechanistic structure of the world.

Keywords Scientific explanation • Quantitative-dynamic explanation • Mechanism • Mathematical model • Systems biology

1 Introduction

Inspired by the deductive-nomological tradition, some philosophers suggested that at least some mathematical models explain biological phenomena by applying the laws and theories of physics and chemistry to biological systems (Smart 1963; Weber 2005, 2008). Knowledge of the peculiarities of a given biological system supplies a list of parameters corresponding to the various parts of the system, their interactions and organization, as well as the initial and boundary conditions to which the system is subjected under experimental or physiological conditions. The laws and theories associated with the model provide a set of rules describing the relationships between parameters and how these relationships change with time. These rules, suitably expressed in mathematical language, play a key role in explanation by making possible the derivation of descriptions of phenomena by

T.M. Baetu (✉)
Programa de Filosofia, Universidade do Vale do Rio dos Sinos, Av. Unisinos, 950 Bairro Cristo Rei, 93022-000 São Leopoldo, RS, Brasil
e-mail: tudormb@unisinos.br

© Springer Science+Business Media Dordrecht 2015
P.-A. Braillard, C. Malaterre, *Explanation in Biology*, History, Philosophy and Theory of the Life Sciences 11, DOI 10.1007/978-94-017-9822-8_15

means of analytic and, more recently, computational methods. For example, using the Hodgkin and Huxley (1952) model as a case study, Marcel Weber (2005) argues that the mechanistic descriptions of molecular and cellular entities, activities, and organizational features (e.g., cell membranes, ions) specify how a physicochemical theory should be applied, while the explanatory burden falls on the regularities that describe the behavior of the system (in this case, a physicochemical law known as Nernst equation), and on how a description of the phenomenon of interest can be derived from these regularities.

Drawing on the highly influential view that biological phenomena are explained by showing how they are produced by mechanisms[1] (Bechtel 2006; Craver 2007; Darden 2006; Wimsatt 1972), Carl Craver (2006, 367) defends a very different view of the explanatory role of mathematical models: "[m]odels are explanatory when they describe mechanisms." In response to Weber, Craver points out that the Hodgkin and Huxley model also includes hypothetical assumptions and uninterpreted parameters whose purpose is to render the model empirically adequate. Empirical adequacy could have been obtained by appealing to a different set of assumptions and parameters (e.g., a different number of ion currents), while a different physical interpretation of certain parameters would have yielded a different mechanistic explanation of the phenomenon under investigation. Craver concludes that the Hodgkin and Huxley model is an incomplete 'how-possibly' account providing some preliminary insights about the possible mechanisms responsible for generating and propagating action potentials along axons, but should not be confused with the explanation of the phenomenon. In order to be explanatory, a model should provide a complete description of the mechanism actually responsible for a phenomenon. This description must "include all of the relevant features of the mechanism, its component entities and activities, their properties, and their organization" (Craver 2006, 367),[2] and "exhibit productive continuity without gaps from the set up to termination conditions" (Machamer et al. 2000, 3). In turn, a complete inventory of the explanatorily relevant mechanistic components and features, along with a specification of their causal-role and productive continuity provide an intuitive understanding of how phenomena are produced.[3]

---

[1]Machamer, Darden, and Craver define mechanisms as "entities and activities organized such that they are productive of regular changes from start or set-up to finish or termination conditions" (2000, 3). Alternatively, a mechanism is "a complex system that produces that behavior by the interaction of a number of parts, where the interactions among parts can be characterized by direct, invariant, change relating generalization" (Glennan 2002), or "a structure performing a function in virtue of its component parts, component operations, and their organization [. . .] responsible for one or more phenomena" (Bechtel and Abrahamsen 2005). McKay and Williamson (2011) propose a more generally applicable characterization, according to which a "mechanism for a phenomenon consists of entities and activities organized in such a way that they are responsible for the phenomenon."

[2]Explanatory relevance is equated to causal relevance and demonstrated by means of experimental interventions (Baetu 2012a; Craver 2007; Woodward 2003).

[3]"Intelligibility [. . .] is provided by descriptions of mechanisms, that is, through the elaboration of constituent entities and activities that, by an extension of sensory experience with ways of working, provide an understanding of how some phenomenon is produced" (Machamer et al. 2000, 22).

Both Weber and Craver agree that mathematical models can be explanatory, but they attribute explanatory value to very different features of these models. For Weber, the explanation lies specifically in the derivation of a description of phenomena from mathematically formulated law-like regularities, while for Craver a mathematical model is explanatory only to the extent it identifies the physical structures actually responsible for causing the phenomena.[4] Ultimately, propositions can be derived from other propositions whether or not these derivations reflect the causal structure of the world, while mechanistic structures can be identified experimentally without relying on any kind of conceptual derivation. This divergence about what counts as an explanation often translates into a direct contradiction: the same model may be deemed explanatory under a deductive-nomological approach, but not under a mechanistic one, as demonstrated in the case of the Hodgkin and Huxley model; conversely, qualitative descriptions of biological mechanisms count as explanations under a mechanistic approach, but have no explanatory value under a deductive-nomological approach.

I argue that despite this philosophical clash between deductive-nomological and mechanistic accounts, scientific practice can rely on an explanatory pluralism in which the two approaches are not only complementary, as recently argued by William Bechtel and Adele Abrahamsen (2010, 2011), but also dynamically integrated in a process of reciprocal validation. The view defended in this paper is that mathematical models play an explanatory role by attempting to provide an answer to the question "Can the proposed mechanism generate the phenomenon of interest in all its minute quantitative/dynamic details?" Using examples from the recent scientific literature, I show how the answer to this research question reveals both agreements and disagreements between mechanistic explanations and mathematical models of mechanisms involving derivations of descriptions of phenomena. Agreements are used to infer or reinforce the completeness of mechanistic explanations; conversely, disagreements between models and mechanisms prompt revisions of either or both models and mechanisms. I argue that, at least in some cases, attempts to reach an agreement between models and mechanisms generate progressive research programs, in the sense that these cycles of modifications and revisions reveal novel ways of thinking about the ontology of mechanisms, as well as surprising explanations to seemingly unrelated scientific puzzles. For additional discussions on the relationship between mechanistic and mathematical models see chapters in the present volume by William Bechtel (2015, Chap. 9), Pierre-Alain Braillard (2015, Chap. 14), Ingo Brigandt (2015, Chap. 7), Tobias Breidenmoser and Olaf Wolkenhauer (2015, Chap. 11), Fridolin Gross (2015, Chap. 8), Tarik Issad and Christophe Malaterre (2015, Chap. 12), and Frédérique Théry (2015, Chap. 6).

---

[4]Both views admit gradations in explanatory value. Models incorporating fundamental laws provide deeper explanations than models relying on more superficial regularities describing the behavior of a certain type of systems. Likewise, under a mechanistic approach, a model incorporating a more complete description of a mechanism is better than a model relying on a sketchier description. For an account of the completeness of mechanistic explanations, see (Baetu 2015)

The paper is organized as follows: I begin with a brief introduction of mathematical models of molecular networks in systems biology (Sect. 2), followed by a discussion of how mathematical models combine the application of laws, modeling and analysis strategies from chemistry, cybernetics, and systems theory (Sect. 3) with knowledge of mechanisms (Sect. 4). In Sect. 5, I show how mechanistic explanations and mathematical models complement each other. I then proceed in Sect. 6 to show that mechanisms and models don't always coexist in a state of static complementarily, but can also contradict each other. Based on an analysis of recent examples from science, I argue that mathematical models can provide criteria for assessing the completeness of mechanistic explanations and I show how the disagreements between models and mechanisms prompt important revisions of the current understanding of the molecular-mechanistic basis of biological activity. Finally, in Sect. 7, I summarize my findings and arguments.

2 A Brief Introduction to Mathematical Models of Molecular Networks

Molecular biology is one of the most important scientific achievements of the twentieth century. In conjunction with biochemistry and cell biology, it succeeded in explaining biological phenomena in mechanistic terms. Nevertheless, the approach pioneered by molecular biology has one important shortcoming: it yields primarily a qualitative understanding of mechanisms acting in isolation. The challenge for the twenty-first century biology is to integrate current knowledge of mechanisms into a conceptual framework that is "holistic, quantitative and predictive" (Kritikou et al. 2006, 801). This challenge was answered by an emerging field of investigation, dubbed 'systems biology.'

One strategy by means of which systems biologists hope to achieve their goals relies on mathematical modeling of molecular networks. Networks are abstract representations of physical systems consisting of parts connected by a web of relationships. Molecular networks represent molecular mechanisms broadly construed, including signal transduction pathways, metabolic pathways, as wells as more comprehensive systems composed of several mechanisms. The nodes of the network represent mechanistic entities (proteins, genes), while the connections between nodes represent mechanistic activities (chemical reactions rates, activation/inhibition). Other information about the modeled mechanisms (e.g., structural details of proteins, their tridimensional configuration, chemical mechanisms of reaction) is usually ignored, hence the abstract representational nature of molecular networks.

Mathematical modeling quantifies qualitative descriptions of networks, with most models falling in two distinct categories: discrete and continuous. Discrete quantification works on the assumption that the behavior of the network is determined by thresholds (e.g., a gene can be either expressed or not expressed). The same network can also be modeled in such a way that its nodes can take continuous values (e.g., the concentration of gene products can take any values within given intervals). Once the nodes are quantified, one can mathematically represent all the possible states of the network; that is, all the possible combinations of the values of the nodes. The next step is to determine how the state of the network changes over time because of its internal wiring and external inputs, something which is achieved by introducing transition rules for each node. In discrete networks, the rules by means of which nodes act on each other are represented by logical functions. In continuous networks, systems of differential equations represent the rates of change of the value of any given node in terms of the values of other nodes and external inputs.[5]

It is interesting to note that while Craver allows for mathematical models of mechanisms, we are never told what, if anything, quantification and mathematical formalism add to the explanatory value, the completeness, or the intelligibility of the model. We are now in a better position to answer this question. Since mathematical models incorporate very precise assumptions about how a system changes from one state to another, we can conclude that they allow for a more detailed description of productive continuity. The fact that this description deals with quantitative changes further suggests that mathematical models may be necessary in order to achieve precise quantitative manipulations of mechanistic components (as opposed to semi-quantitative ones, such as knockout or over-expression) and to design synthetic mechanisms that behave in quantitatively precise ways [e.g., the repressilator (Baetu 2015; Elowitz and Leibler 2000; Morange 2009)]. Finally, mathematical models reveal that networks can exhibit a number of unsuspected and perhaps physiologically significant properties such as attractors and steady states,[6] robustness and sensitivity,[7] adaptability,[8] and hysteresis[9] (for the evolutionary significance of these novel properties, see Brigandt, Chap. 7).

---

[5]For a more technical description, consult Shmulevich and Aitchison (2009). Additional assumptions are required in order to construct a model of a network. One has to choose between a synchronic and an asynchronous updating scheme, between a binary, multi-value, or stochastic logic, between different kinetic laws, between ordinary and partial differential equations, etc. Without these assumptions, it is impossible to model the dynamic behavior of the network.

[6]These states amount to long-term behaviors of networks and can be experimentally measured, thus allowing for precise quantitative predictions, as well as an assessment of the empirical adequacy of the model.

[7]Robustness is insensitivity to the precise values of biochemical parameters (changes in reaction rates, concentrations of substrates), thus allowing a system to function in a wide range of conditions and resist certain perturbations. Sensitivity denotes the contrary, namely a situation where a mechanism is operational only if the values of its parameters are fine-tuned to specific values. Robustness and sensitivity allow for optimization analysis, which is especially useful for identifying which mechanistic components should be targeted in order to achieve a desired result with maximal efficiency and minimal side effects.

[8]Ability to adapt to 'background noise': the smallest change in stimulus intensity that can be sensed (ΔS) increases with the background stimulus intensity (S), such that $\Delta S/S$ remains constant (Weber-Fechner law).

[9]A network may display more than one 'stable state', and it is possible that a change in the system's state caused by a transient stimulus (e.g., external input, temporary change in gene expression) is not followed by a return to the initial state when the stimulus is withdrawn. It has been hypothesized that such states may underlie developmentally differentiated cell types (Kauffman 2004) or physiological cell states [e.g., proliferating vs. apoptotic cells; (Huang 1999)].

3   Mathematical Models as Applications of Chemistry, Cybernetics, and Systems Theory

Early system theorists aimed to formulate a set of general principles governing the behavior of systems in all fields of scientific investigation. Most famously, von Bertalanffy (1976, 32) argued that "there exist models, principles, and laws that apply to generalized systems or their subclasses, irrespective of their particular kind, the nature of their component elements, and the relationships or 'forces' between them." This view is sometimes echoed by contemporary systems biologists. For example, in their discussion of a previously proposed continuous model of the bacterial chemotaxis signal transduction pathway revealing robustness, Baker et al. (2006, 190) claim that "[o]ne of the primary objectives of systems biology is to formulate biological laws that are akin to the laws of physics."[10] In turn, a generally applicable understanding of the relationship between a system's structure and its properties can yield 'design explanations' showing "why a given structure or design is necessary or highly preferable in order to perform a function or to have an important property like robustness" (Braillard 2010, 55); for a discussion of generalizable mechanistic explanations based on graph-theoretic considerations, see Bechtel, Chap. 9.

Under this approach, mathematical models in systems biology can be treated as applications of the modeling, analytic and computational methods of analysis pioneered by systems theory. This clearly amounts to a form of explanatory heteronomy of systems biology on systems theory,[11] although it is important to keep in mind that explanatory tools are indirectly borrowed from many fields of investigation, systems theory being more an interdisciplinary rendezvous point than a thoroughly unified science. For example, discrete modeling has its roots in cybernetics, computer science and engineering, while continuous models incorporate laws and mathematical models from chemical kinetics, some of which are themselves applications of statistical mechanics and thermodynamics to the field of chemistry (see also Braillard, Chap. 14).

It is also interesting to note that any given network may be mathematically modeled in more than one way, and, depending on the modeling strategy and associated assumptions, the same network may or may not be shown to possess

---

[10]The 'laws' to which they refer have a variable degree of generality, ranging from common features of networks displaying specific properties, such as robustness or adaptability, to more general properties of generic networks. Examples of the latter are found in Kauffman's (1993) seminal work on Random Boolean Networks (networks in which the connections between nodes are wired randomly). By investigating the behavior of such networks, some general principles emerged; for instance, networks become chaotic as the number of connections per node increases.

[11]Weber (2005) argues for an 'explanatory heteronomy' of biology on physics and chemistry, and spells out the sense in which the former reduces to the latter: biology relies on the laws of physics and chemistry in order to generate explanations, and, in this respect, can be viewed as applied physics and chemistry.

certain properties. In order to claim that the model explains certain aspects of biological phenomena, one needs to know to what extent a given set of modeling assumptions is true or approximately true of a given biological system; this determines the degree to with which the biological system is expected to behave as described by its corresponding model. As it turns out, mathematical models tend to incorporate several unverified, and potentially idealizing assumptions. Idealizations are problematic in many fields of investigation, yet the problem is more acute in systems biology. For example, in physics, it is usually clear what exactly is being idealized and to what extent; think of Newton's idealization of Earth as a homogenous sphere or as a material point. By contrast, in systems biology, very little is known about real time concentrations of substrates and real time kinetics of reaction, to the point that it is seldom clear to what extent a model idealizes these features, what are the potential drawbacks of these idealizations, and what can be done to correct the situation.[12]

This incertitude creates a paradoxical situation: mathematical models have the potential to explain, in Weber's deductive-nomological sense, yet it is often not clear to what extent they actually explain. Another way to frame this situation is to think of the explanatory value of a mathematical model as being contingent on additional evidence demonstrating that the model is a suitable surrogate of the physical system being modeled, such that knowledge generated by studying the model can be safely extrapolated to the target physical system and its ability to generate the phenomenon of interest (Baetu 2014).

Finally, mathematical modeling involves a certain degree of abstraction. Since a lot is known about the molecular basis of biological activity, the starting point of current modeling efforts is, more often than not, prior knowledge of molecular mechanisms. This prior knowledge determines the wiring of molecular networks, as well as the physical and causal interpretation of nodes and connections. However, not all that is known about mechanisms is incorporated in molecular networks. As a general rule, specific modes of action, the structure of mechanistic entities and other high-resolution biochemical details are not represented in molecular networks, which tend to amount to simplified schemas consisting of unstructured entities (proteins, genes) and generic activities (activation/inhibition, increase/decrease in concentration). Thus, simplified representations of mechanisms, known as 'mechanism schemas',[13] act as bridges mediating the transfer of knowledge from molecular

---

[12]For instance, current discrete modeling strategies assume that a network is either updated synchronously (the values of all its nodes are updated at the same time) or asynchronously (no two nodes are updated at the same time). Set aside the difficulty of finding out which of the two assumptions holds true of the particular system under investigation – a situation that makes it such that investigators simply test several models until they find one that simulates well characterized features of the system –, it is also possible that no real biological system will perfectly fall into one or the other of these two categories.

[13]"A mechanism schema is a truncated abstract description of a mechanism that can be easily instantiated by filling it with more specific descriptions of component entities and activities" (Darden 2006, 111–12).

biology to other fields of investigation, such as bioinformatics and systems biology (Baetu 2011b, 2012b).

On the one hand, abstraction generates networks simple enough to allow for computable solutions, while highlighting those features of biological systems most amenable to mathematical modeling (Baetu 2011a, 2012b). For example, a Gene Regulatory Network (GRN)[14] representation of signal transduction pathways does not tell us how a specific repressor protein regulates gene expression; the repressor may compete with a transcriptional activator for the same DNA binding site, it may bind the activator and cause a conformational change affecting the activity of the latter, or it may trap the activator in the cytoplasm. Removing such details makes it possible to reduce the total number of parameters in the model, thus increasing computability. At the same time, the network highlights specific features of the physical system it represents, in this case, the inputs and outputs of genome expression, showing how each gene module in the GRN behaves like a logic gate integrating several regulatory inputs in order to yield a single output, namely the presence or absence of its corresponding gene product.

On the other hand, however, models built from mechanism schemas cannot account for all the features of the phenomena produced by these mechanisms. Because specific modes of action and structural details of entities are abstracted, a mathematical model may tell us which mechanistic components should be targeted in order to achieve a desired result, but doesn't tell us how to operate changes in a physical system.[15] For this reason, it seems highly unlikely that mathematical models can replace current mechanistic explanations.

In summary, we can conclude that mathematical models in biology are not exclusively applications of what are traditionally deemed to be more fundamental sciences, such as physics and chemistry, but also of many other fields, such as systems theory and computer engineering. Second, as applications of other sciences and fields of investigation, mathematical models tend to yield putative explanations because it is usually not clear to what extent a given set of modeling assumptions apply to biological systems and to what extent they introduce idealizations. Finally, since details accounting for certain aspects of the phenomena under investigation are abstracted from the model, it seems unlikely that mathematical models can or are meant to replace mechanistic explanations indigenous to the biological sciences.

---

[14]GRNs are "hardwired genomic regulatory codes, the role of which is to specify the sets of genes that must be expressed in specific spatial and temporal patterns. [. . .] these control systems consist of many thousands of modular DNA sequences. Each such module receives and integrates multiple inputs, in the form of regulatory proteins (activators and repressors) that recognize specific sequences within them. The end result is the precise transcriptional control of the associated genes" (Davidson and Levine 2005, 4935).

[15]For example, in order to physically inhibit the activity of a repressor, detailed knowledge of its structure, such as a mapping of the amino acids responsible for DNA binding, is required; the repressor activity is tempered with by mutating specifically these DNA-binding amino acids.

4 The Mechanistic Content of Mathematical Models of Molecular Networks

Craver (2007) correctly points out that the mathematical model developed by Hodgkin and Huxley had only a partial causal interpretation, and that the model played primarily a role in discovery by guiding the subsequent elucidation of the mechanism responsible for generating and propagating action potentials along axons. Most notably, the model suggested the existence of transmembrane ion channels and provided some insights about their possible properties. In light of this kind of historical examples, it might be tempting to conclude that mathematical models are primarily instrumental constructs eventually superseded by mechanistic explanations. However, this is not always the case. The starting point of many modeling efforts, especially in today's practice, is prior knowledge of molecular mechanisms. This knowledge provides a straightforward physical interpretation of the nodes of the network (they correspond to genes and proteins) and the inter-nodal connections (they are activities of entities affecting other entities in the system/mechanism). Most modeling assumptions also refer to features of physical systems (e.g., reaction rates), and therefore have a clear physical interpretation; it is just that it cannot be easily determined to what extent the modeled systems actually possess these features, and how realistically these featured are modeled. Thus, many mathematical models incorporate substantial knowledge of the causal-mechanistic structure of the world and are not plagued by incomplete or problematic causal interpretations.

5 How Quantitative Models Complement Qualitative Descriptions of Molecular Mechanisms

The fact that many models are constructed in light of prior knowledge of molecular mechanisms raises an interesting question: Why would anyone bother designing and testing mathematical models when mechanistic explanations are already available? The simplest solution to the puzzle would be the claim that, despite their mechanistic content and unambiguous causal interpretation, mathematical models in systems biology are, after all, primarily meant to generate predictions. This view is certainly defendable: since it is often the case that realistic models outperform idealized models in terms of empirical adequacy, the fact that a mathematical model incorporates substantial knowledge about the causal-mechanistic structure of the world is not necessarily an indication that the model has explanatory ambitions. Nevertheless, a survey of the scientific literature suggests otherwise. In many cases, mathematical models are used not only to generate predictions, but also to account for certain anomalies and poorly understood aspects of phenomena known to be produced by mechanisms that have been already elucidated (Bechtel and Abrahamsen 2010, 2011). Building on the earlier suggestion that mathematical models complement current mechanistic explanations, I propose that the answer to the puzzle lies in a refutation of the assumption that a model is complete when it includes all the relevant features of the mechanism, their causal role, and productive continuity.

In order to understand how and why models of mechanisms fail to explain certain features of phenomena, let us consider the following example. Leukocytes exposed to antigens, inflammatory agents, and pathogens express a variety of genes required for mounting an immune response. After a brief period of activation, gene expression shuts down. The phenomenon to be explained amounts to a black-box correlation between input (cells are exposed to pathogens) and output (spike of gene expression) conditions (Fig. 15.1, top panel). A mechanistic explanation tells us what happens inside the black box: the spike of gene expression following stimulation is explained by a negative feedback regulatory mechanism whereby a transcriptional factor (nuclear factor κB, or NF-κB) is initially activated, then subsequently inactivated by an inhibitory protein (inhibitor of κB, or IκB) coded by a gene under its transcriptional control (Fig. 15.1, middle panel).[16]

The above explanation amounts to a qualitatively complete description of the regulatory mechanism responsible for the peak of gene expression. The causal contribution of the mechanism to the target phenomenon is well established, the mechanistic function of its various components is well understood, and there is a large body of evidence supporting the conclusion that there are no gaps in the causal chain linking input and output conditions. In sum, it is a typical description of a well understood mechanism one may very well find among the diagrams of a recent immunology textbook. Furthermore, the mechanistic explanation is also highly satisfactory relative to the pragmatic goals of molecular biology, such as treating illnesses and developing techniques for intervention and manipulation (Craver 2007, 38); for example, it shows how to achieve a loss of NF-κB activity (e.g., mutations in the DNA binding or nuclear localization domains).

Nevertheless, in some respects, qualitatively complete descriptions remain unsatisfactory (see also Brigandt, Chap. 7; Breidenmoser and Wolkenhauer, Chap. 17; Issad and Malaterre, Chap. 18). For instance, the above explanation supports the inference that, if the inhibitory protein IκB is synthesized too fast, there is no spike of activation, and conversely, if the inhibitor is synthesized too slowly or in insufficient quantity, gene expression is never turned off. However, this does not tell us whether the mechanism is sensitive or robust relative to the exact amounts of the inhibitor. If sensitive, the mechanism will malfunction in response to mutations that affect the amount and stability of the inhibitor; in contrast, if robust, the mechanism can adapt and continue to operate irrespective of changes in the concentration of the inhibitor. Likewise, the model doesn't tell us why there is an increase in the

---

[16]In resting cells, NF-κB is held in the cytoplasm by IκB (Huxford et al. 1998). When cells are stimulated (Fig. 15.1, middle panel, A), a chain of protein-protein interactions leads to the degradation of IκB (B); NF-κB is freed (C), translocates to the nucleus (D) where it binds κB sequences in the promoter regions of target genes drastically enhancing their transcription (Pahl 1999). NF-κB also binds the promoter of the IκB gene (E), and the newly synthesized IκB binds NF-κB, trapping it back in the cytoplasm (Sun et al. 1993).

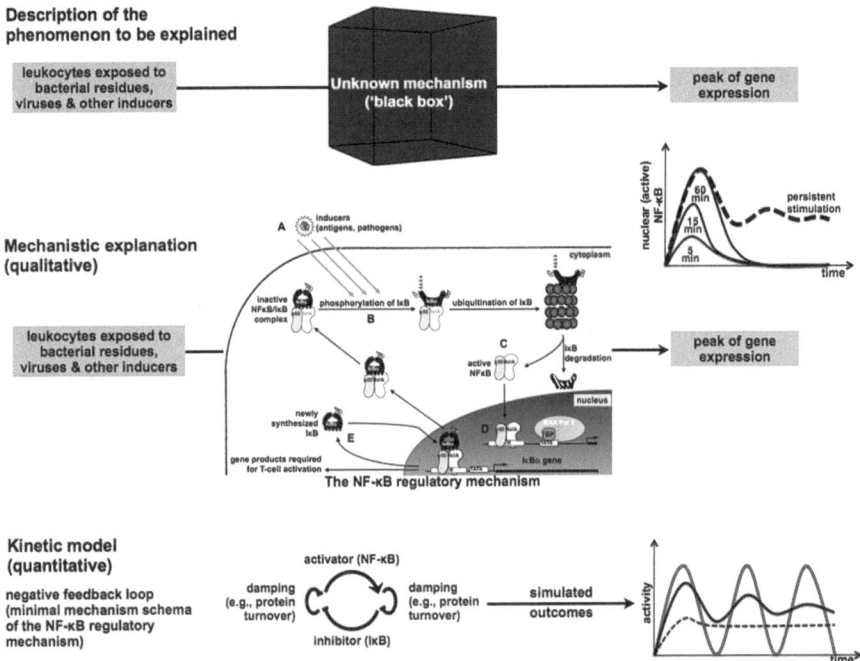

Fig. 15.1 The NF-κB regulatory mechanism (Adapted from Baetu and Hiscott 2002; Hoffmann et al. 2002)

intensity (but not the duration) of gene expression when cells are exposed to longer pulses of stimulation (Fig. 15.1, top right graph, solid curves), and only an increase in the duration (but not the intensity) of gene expression in the case of persistent stimulation (dashed curve). The failure to account for these minute features of the phenomenon under investigation stems primarily from a poor understanding of how the various parts of the mechanism change quantitatively over time.

In more general terms, the experimental data supporting the qualitative description of a mechanism demonstrates the necessary causal contribution of the proposed mechanism to the phenomenon of interest. This does not prove, however, that the mechanism can and does generate the phenomenon, exactly as it is measured in all its quantitative-dynamic details, thus leaving open the question whether the mechanism is also sufficient to generate the phenomenon. This constitutes a serious problem for a mechanistic account of explanation: if the mechanism doesn't generate the phenomenon, exactly as it is measured, then the mechanistic explanation fails. The immediate corollary is that quantitative parameters (e.g., concentrations of components), as well as their temporal dynamics (e.g., rates of reactions) must be included in an ideally complete mechanistic explanation, along with a complete list of mechanistic components, their organization and causal roles, and evidence for productive continuity. However, the inclusion of this additional information in a computationally useful way – that is, in a format that allows the derivation of the quantitative-dynamic features of the target phenomenon –, requires mathematical modeling, which has its drawbacks, namely the introduction of unverified or difficult to verify assumptions, potential idealizations, and the abstraction of some biochemical and molecular details. Thus, it is reasonable to conclude that qualitatively complete descriptions of mechanisms and their mathematical model counterparts stand in a relationship of explanatory complementarity rather than direct competition.

By the same token, this means that a quantitative model, such as the Hodgkin and Huxley model, cannot be replaced by a qualitatively more detailed description of a molecular mechanism. The Hodgkin and Huxley model, or rather a version of it benefiting from the retrospective hindsight of a more detailed causal interpretation, continues to play a very important explanatory role in contemporary neuroscience: it shows that the molecular mechanism qualitatively described by currently accepted mechanistic explanations can generate action potentials that closely fit quantitative experimental measurements. In other words, 'how-possibly' mathematical models are not only useful in guiding the subsequent elucidation of mechanisms or generating predictions, but also provide putative explanations for precise quantitative-dynamic aspects of biological phenomena.

Recent trends in biology support this interpretation. More and more studies published in leading journals complement qualitative descriptions of mechanisms supported by the experimental practice of molecular biology with quantitative models aiming to demonstrate that the proposed mechanisms can generate the quantitative-dynamic aspects of the phenomena of interest. For example, recent studies investigating the development of tubular organs rely on mathematical models in order to show that the proposed mechanisms can produce the phenomena under investigation in the right amount/intensity [e.g., precise allometric growth ratio of bronchioles in the lung (Tang et al. 2011)]. Furthermore, in some cases mathematical models are required in order to show that no additional mechanisms or mechanistic components are likely to be needed in order to generate the target phenomenon. For example, Taniguchi et al. (2011) propose that probabilistic biases in the distribution of adhesion proteins suffice to generate the right amount of twisting in the developing gut. Given the stochastic nature of the proposed mechanism, it is not possible to rely on commonsense mechanistic intuitions to establish that the mechanism can generate (even approximately) the target phenomenon (Horne-Badovinac and Munro 2011). The authors of the study reasoned that since the predictions of the mathematical model match the experimental data, a more complex model including additional parameters is not needed; since the model does not include physically uninterpreted parameters, it can be established that no corresponding mechanistic components are missing; hence a more complex mechanism, including additional entities and activities, or additional mechanisms are not likely to be needed to produce the target phenomenon (i.e., the proposed mechanism is necessary and sufficient to produce the phenomenon).

Due to uncertainties and potential idealizations associated with modeling assumptions, the above studies offer only putative explanations. The results they yield can be safely extrapolated to biological mechanisms and their ability to generate phenomena on condition that these models accurately describe their target biological mechanisms. Nonetheless, these putative explanations cover ground beyond the reach of qualitative descriptions of mechanisms. In order to determine whether a proposed mechanism can generate a phenomenon exactly as it is measured, down to minute quantitative-dynamic details, we usually cannot rely on qualitative descriptions of mechanisms and commonsense intuitions about how mechanisms work. Numerical computations are required, and this is precisely what mathematical models make possible.

6 The Integration of Mathematical Modeling and Mechanistic Explanations

In the previous section, I argued that mathematical models, viewed as applications of chemistry, cybernetics, and systems theory to biological phenomena, and mechanistic explanations, viewed as descriptions of causal-mechanistic structures, complement each other. As a rule of thumb, qualitative descriptions embody experimental evidence of the causal contribution of mechanisms (entities, activities, and their organization) to the target phenomenon, while mathematical models provide further 'proof or principle' demonstrations that the mechanisms in question can generate phenomena down to minute quantitative-dynamic details. The former rely on actual experimental control over the mechanism and its components, while the latter rely on the ability to mathematically/computationally derive close approximations of quantitative descriptions of phenomena.

I will now go a step further and argue that mathematical models and qualitative mechanistic descriptions don't always stand in a state of passive complementarity whereby each explains aspects of a phenomenon inaccessible to the other. Rather, there is a constant interaction between the two. Since molecular networks are abstract representations of molecular mechanisms, revisions of the latter may entail revisions of the former; in such situations, mathematical models of molecular networks are revised as well. Conversely, mathematical models can reveal anomalies and unsuspected explanatory holes in previously accepted mechanistic models, thus prompting their revision. This interaction is progressive,[17] in the sense that it addresses not only the immediate research aims – in this case, accounting for quantitative-dynamic aspects of phenomena produced by molecular mechanisms –, but provides unexpected explanations to seemingly unrelated and thus far unexplained phenomena.

Mathematical models can support current mechanistic explanations by showing that the proposed mechanisms can generate phenomena down to minute

---

[17]By analogy with Lakatos' (1978, 33) notion of 'progressive research programme' in which "each new theory [...] predicts some novel, hitherto unexpected fact."

quantitative-dynamic details. They can also reveal discrepancies and anomalies. Let us consider again the spike of gene expression in leukocytes. According to the mechanistic explanation depicted in the middle panel of Fig. 15.1, the spike is generated by a negative regulatory feedback loop mechanism (NF-κB activates IκB, and IκB inhibits NF-κB). However, a mathematical model of a generic negative feedback loop mechanism [e.g., (Goodwin 1963)] has three possible outputs, namely perpetual oscillation, damped oscillation, or a plateau of activation (Fig. 15.1, bottom right graph); none matches the observed spikes of gene expression (Fig. 15.1, top right graph). Hoffmann et al. (2002) interpreted this mismatch between the simulated results of the quantitative model and the experimental data as an indication that the currently accepted mechanistic explanation is inadequate.

It is not question of doubting the causal relevance of the NF-κB regulatory mechanism, or that this mechanism consists of a negative feedback loop. Experimental evidence clearly shows that the mechanism, as described in Fig. 15.1, is necessary for the generation of normal immune responses in vitro (in cell models) and in vivo (in animal models). In this particular case, it is also hard to doubt the results of the mathematical model or that the model employed is a suitable model of a negative feedback loop. Models of negative feedback loop systems have been extensively validated in physics, engineering and cybernetics. Thus, there is a direct explanatory conflict here. The explanandum is always the same, namely the characteristic spike of gene expression associated with an immune response, yet the mechanistic explanation shows that the NF-κB feedback mechanism is a necessary cause of this spike, while the mathematical model of the mechanism fails to accurately predict the quantitative-dynamic description of the spike, thus pointing to the opposite conclusion, that the phenomenon is not explained.

In order to solve the puzzle, researchers had to resort to a seemingly irrelevant piece of information. It was known for some time that there are three isoforms of IκB present in mammalian cells, and that only the gene coding for the α isoform is under the transcriptional control of NF-κB. The other two isoforms, β and ε, are expressed constitutively. There are, therefore, two overlapping versions of the NF-κB regulatory mechanism, one with a negative feedback loop, and one without. The β and ε isoforms were largely ignored because the results of knockout experiments suggested that they may play a different physiological role,[18] and, most importantly, the version of the mechanism without a feedback loop doesn't explain in any way how a spike of activation could be produced (one can only expect a plateau, like the one depicted at the bottom of Fig. 15.2).

Hoffmann et al. (2002) observed that cells that express only the α isoform generate a damped oscillation pattern of gene expression, closely matching the one predicted by the mathematical model. This suggested that when the β and ε isoforms

---

[18]In mice, IκBα$^{-/-}$ is associated with exacerbated inflammation and embryonic lethality, while IκBβ/ε$^{-/-}$ females have a shorter fertility span. Nevertheless, other experiments suggest that the three forms are partially redundant. For a review of the original scientific literature, consult Hoffmann (2002, 1241–42).

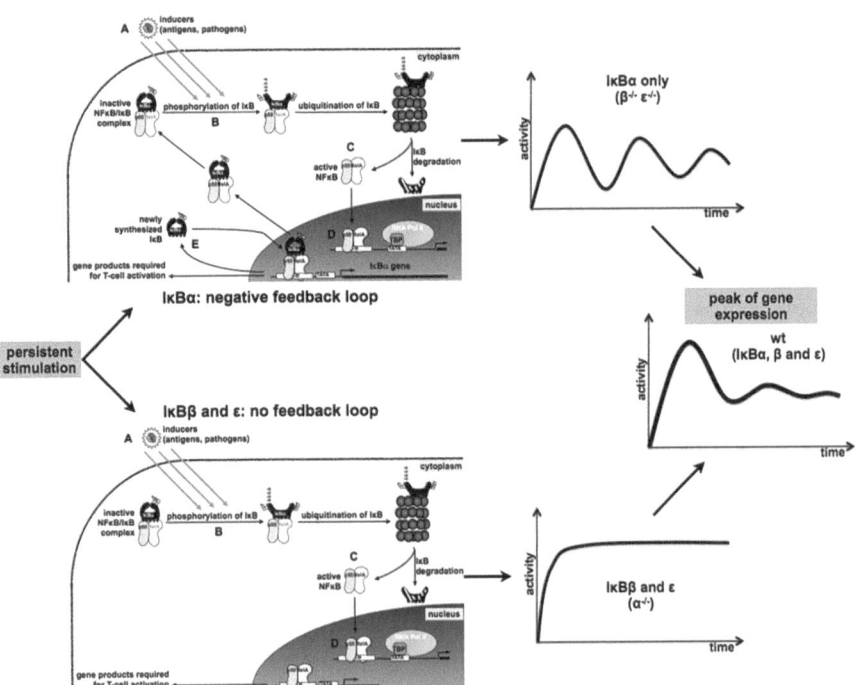

Fig. 15.2 The revised model of the NF-κB regulatory mechanism (Adapted from Hoffmann et al. 2002)

are also present, the damped oscillation pattern is somehow transformed into the characteristic wild-type spike of gene expression (Fig. 15.2). In order to test the combined contribution of two overlapping mechanisms, it is not possible to rely on a qualitative understanding of mechanisms. Precise numerical computation is necessary, and this requires mathematical modeling. Explorations with mathematical models showed that the observed spike, as well as a differential temporal dynamics of gene expression following pulse and persistent stimulation, can be derived from a mathematical model of a molecular network combining the two versions of the NF-κB regulatory mechanism, one involving a negative regulatory loop and the other a constitutive expression of the inhibitor IκB.

From a philosophical point of view, this is an extremely interesting result. It shows that mathematical models can prompt revisions of mechanistic explanations; in this case, the initial negative feedback loop mechanism was augmented to include a parallel pathway of activation not subjected to negative feedback. I take this two-way interaction whereby mathematical models are revised as mechanisms are elucidated in more detail, while mechanistic explanations are revised as a result of mathematical modeling to be a strong indication that a mixed, mechanistic and deductive-nomological approach is necessary in order to achieve more complete explanations, as well as to guide the discovery process in biology.

More important, the back and forth interaction between mechanistic and deductive-nomological approaches is progressive, in the sense that it generates new insights about the ontological status of mechanisms while offering unexpected explanations to seemingly unrelated and thus far unexplained phenomena. In the above example, the overlapping mechanisms explanation challenges our metaphysical intuitions about the nature of mechanisms. While we might be tempted to treat molecular mechanisms as neatly individuated objects, such as clocks and other man-made mechanical devices, they are in fact populations consisting of a large number of identical mechanisms collectively generating a biological phenomenon. As a general rule, each 'individual mechanism' in the population has a physiologically insignificant contribution to the target phenomenon; is ephemeral (Glennan 2010), not only because molecular components have a relatively short life-time, but also because, after having fulfilled its causal contribution, a component is randomly replaced with another copy of the same kind; and doesn't necessarily operate in perfect synchrony with other 'individual mechanisms.' The study by Hoffmann et al. further suggests that some mechanistic components come in several variants, with each variant acting along a partially distinct causal pathway, such that molecular mechanisms may in fact amount to heterogeneous populations of partially overlapping mechanisms.[19]

In turn, this renewed appreciation of the internal variability of molecular mechanisms suggests an unexpected explanation for a seemingly unrelated phenomenon. Many genes and gene products come in several copies displaying very small differences in terms of sequence, structure, and biochemical function. The physiological role of these seemingly redundant molecular components is poorly understood. The overlapping mechanisms explanation suggests that they are not mere spare parts or evolutionary accidents, but may in fact play a very important physiological role: generate partially overlapping versions of the same mechanism, collectively required in order to generate highly complex patterns of gene expression and associated phenotypes, adaptive responses to external stimuli, and other biological phenomena. Heterogeneous populations of partially overlapping mechanisms may

---

[19]Furthermore, there are cases when significantly distinct mechanisms responsible for distinct phenomena nevertheless share mechanistic components. Thus, in addition to a modular mode of organization whereby systems of mechanisms are organized serially or in parallel, the output of a mechanism serving as input for one or more other mechanisms (Bechtel 2006; Craver 2007; Darden 2006), significantly distinct mechanisms may also be firmly interlocked in the same manner as the partially overlapping mechanisms described above. The lessons learned form the NF-κB regulatory mechanism raise the possibility that non-modular sharing of mechanistic components plays a physiologically relevant role in adjusting quantitative-dynamic aspects of the phenomena produced by these mechanisms. If this turns out to be the case, then molecular mechanisms are unlike any man-made mechanisms, first because they are heterogeneous populations rather than individual objects, and second because they operate both in a modular and a non-modular fashion. In other words, there is a sense in which a cell or organism cannot be decomposed into a set of mechanism-modules, but is one integrated mechanism consisting of heterogeneous populations of mechanisms overlapping to various degrees.

also account for fine grained phenotypic variability, thus bridging the gap between variability required for evolution, and the seemingly monolithic explanations in molecular biology.

7 Conclusion

The emergence of systems biology is marked by a renewed interest in mathematical modeling. From a philosophical standpoint, this 'mathematical turn' in biology constitutes an excellent opportunity to investigate the relationship between deductive-nomological and mechanistic accounts of scientific explanation. I argue that mathematical models in systems biology combine substantial knowledge of molecular mechanisms with the application of laws, modeling and analysis strategies borrowed from chemistry, cybernetics and systems theory. Mechanism schemas obtained by abstracting high-resolution biochemical details act as bridges between molecular mechanistic explanations and mathematical models of networks. In turn, mathematical models account for poorly understood aspects of biological phenomena, most notably minute quantitative-dynamic features. Thus, in scientific practice, deductive-nomological and mechanistic approaches to explanation are not mutually exclusive, but complementary. Furthermore, mathematical models can reveal unsuspected 'black boxes' and motivate revisions of mechanistic explanations. This interplay between mechanistic explanations and their mathematical counterparts constitutes a progressive research approach that generates explanations of novel phenomena, and reveal strange properties of molecular mechanisms that have thus far escaped our attention.

Acknowledgments This work was supported by a generous fellowship from the KLI Institute. I would also like to thank the editors of the volume, Christophe Malaterre and Pierre-Alain Braillard, for their thoughtful comments on previous drafts of the paper.

References

Baetu, T. M. (2011a). A defense of syntax-based gene concepts in postgenomics: 'Genes as modular subroutines in the master genomic program'. *Philosophy of Science, 78*(5), 712–723.

Baetu, T. M. (2011b). Mechanism schemas and the relationship between biological theories. In P. McKay, J. Williamson, & F. Russo (Eds.), *Causality in the sciences*. Oxford: Oxford University Press.

Baetu, T. M. (2012a). Filling in the mechanistic details: Two-variable experiments as tests for constitutive relevance. *European Journal for Philosophy of Science, 2*(3), 337–353.

Baetu, T. M. (2012b). Genomic programs as mechanism schemas: A non-reductionist interpretation. *British Journal for the Philosophy of Science, 63*(3), 649–671.

Baetu, T. M. (2014). Models and the mosaic of scientific knowledge. The case of immunology. *Studies in History and Philosophy of Biological and Biomedical Sciences, 45*, 49–56.

Baetu, T. M. (2015). When is a mechanistic explanation satisfactory? Reductionism and antireductionism in the context of mechanistic explanations. In G. Sandu, I. Parvu, & I. Toader (Eds.), *Romanian studies in the history and philosophy of science*. Dordrecht: Springer.

Baetu, T. M., & Hiscott, J. (2002). On the TRAIL to apoptosis. *Cytokine & Growth Factors Reviews, 13*, 199–207.

Baker, M., Wolanin, P., & Stock, J. (2006). Systems biology of bacterial chemotaxis. *Current Opinion in Microbiology, 9*, 187–192.

Bechtel, W. (2006). *Discovering cell mechanisms: The creation of modern cell biology*. Cambridge: Cambridge University Press.

Bechtel, W. (2015). Generalizing mechanistic explanations using graph-theoretic representations. In P.-A. Braillard & C. Malaterre (Eds.), *Explanation in biology. An enquiry into the diversity of explanatory patterns in the life sciences* (pp. 199–225). Dordrecht: Springer.

Bechtel, W., & Abrahamsen, A. (2005). Explanation: A mechanist alternative. *Studies in History and Philosophy of Biological and Biomedical Sciences, 36*, 421–441.

Bechtel, W., & Abrahamsen, A. (2010). Dynamic mechanistic explanation: Computational modeling of circadian rhythms as an exemplar for cognitive science. *Studies in History and Philosophy of Science Part A, 41*, 321–333.

Bechtel, W., & Abrahamsen, A. (2011). Complex biological mechanisms: Cyclic, oscillatory, and autonomous. In C. A. Hooker (Ed.), *Philosophy of complex systems* (pp. 257–285). New York: Elsevier.

Braillard, P.-A. (2010). Systems biology and the mechanistic framework. *History and Philosophy of Life Sciences, 32*, 43–62.

Braillard, P.-A. (2015). Prospect and limits of explaining biological systems in engineering terms. In P.-A. Braillard & C. Malaterre (Eds.), *Explanation in biology. An enquiry into the diversity of explanatory patterns in the life sciences* (pp. 319–344). Dordrecht: Springer.

Breidenmoser, T., & Wolkenhauer, O. (2015). Explanation and organizing principles in systems biology. In P. A. Braillard & C. Malaterre (Eds.), *Explanation in biology. An enquiry into the diversity of explanatory patterns in the life sciences* (pp. 249–264). Dordrecht: Springer.

Brigandt, I. (2015). Evolutionary developmental biology and the limits of philosophical accounts of mechanistic explanation. In P.-A. Braillard & C. Malaterre (Eds.), *Explanation in biology. An enquiry into the diversity of explanatory patterns in the life sciences* (pp. 135–173). Dordrecht: Springer.

Craver, C. (2006). When mechanistic models explain. *Synthese, 153*, 355–376.

Craver, C. (2007). *Explaining the brain: Mechanisms and the mosaic unity of neuroscience*. Oxford: Clarendon Press.

Darden, L. (2006). *Reasoning in biological discoveries: Essays on mechanisms, interfield relations, and anomaly resolution*. Cambridge: Cambridge University Press.

Davidson, E., & Levine, M. (2005). Gene regulatory networks. *Proceedings of the National Academy of Science, 102*(14), 4935.

Elowitz, M., & Leibler, S. (2000). Synthetic gene oscillatory network of transcriptional regulators. *Nature, 403*, 335–338.

Glennan, S. (2002). Rethinking mechanistic explanation. *Philosophy of Science, 69*, S342–S353.

Glennan, S. (2010). Ephemeral mechanisms and historical explanation. *Erkenntnis, 72*, 251–266.

Goodwin, B. (1963). *Temporal organization in cells: A dynamic theory of cellular control processes*. London: Academic.

Gross, F. (2015). The relevance of irrelevance: Explanation in systems biology. In P.-A. Braillard & C. Malaterre (Eds.), *Explanation in biology. An enquiry into the diversity of explanatory patterns in the life sciences* (pp. 175–198). Dordrecht: Springer.

Hodgkin, A. L., & Huxley, A. F. (1952). A quantitative description of membrane current and its application to conduction and excitation in nerve. *Journal of Physiology, 117*, 500–544.

Hoffmann, A., Levchenko, A., Scott, M., & Baltimore, D. (2002). The I κB–NF-κB signaling module: Temporal control and selective gene activation. *Science, 298*, 1241–1245.

Horne-Badovinac, S., & Munro, E. (2011). Tubular transformations. *Science, 333*, 294–295.

Huang, S. (1999). Gene expression profiling, genetic networks, and cellular states: An integrating concept for tumorigenesis and drug discovery. *Journal of Molecular Medicine, 77*(6), 469–480.

Huxford, T., Huang, D.-B., Malek, S., & Ghosh, G. (1998). The crystal structure of the I κB/NF-κ B complex reveals mechanisms of NF- κB inactivation. *Cell, 95*, 759–770.

Issad, T., & Malaterre, C. (2015). Are dynamic mechanistic explanations still mechanistic? In P.-A. Braillard & C. Malaterre (Eds.), *Explanation in biology. An enquiry into the diversity of explanatory patterns in the life sciences* (pp. 265–292). Dordrecht: Springer.

Kauffman, S. (1993). *The origins of order: Self-organization and selection in evolution*. New York: Oxford University Press.

Kauffman, S. (2004). A proposal for using the ensemble approach to understand genetic regulatory networks. *Journal of Theoretical Biology, 230*(4), 581–590.

Kritikou, E., Pulverer, B., & Heinrichs, A. (2006). All systems go! *Nature Reviews Molecular Cell Biology, 7*, 801.

Lakatos, I. (1978). *Philosophical papers Vol I: The methodology of scientific research programmes*. Cambridge: Cambridge University Press.

Machamer, P., Darden, L., & Craver, C. (2000). Thinking about mechanisms. *Philosophy of Science, 67*, 1–25.

McKay, P., & Williamson, J. (2011). What is a mechanism? Thinking about mechanisms across the sciences. *European Journal for Philosophy of Science, 2*, 119–135.

Morange, M. (2009). Synthetic biology: A bridge between functional and evolutionary biology. *Biological Theory, 4*(4), 368–377.

Pahl, H. L. (1999). Activators and target genes of Rel/NF-κB transcription factors. *Oncogene, 18*, 6853–6866.

Shmulevich, I., & Aitchison, J. (2009). Deterministic and stochastic models of genetic regulatory networks. *Methods in Enzymology, 467*, 335–356.

Smart, J. J. C. (1963). *Philosophy and scientific realism*. New York: Humanities Press.

Sun, S.-C., Ganchi, P. A., Ballard, D. W., & Greene, W. C. (1993). NF-κB controls expression of inhibitor I κB α: Evidence for an inducible autoregulatory pathway. *Science, 259*, 1912–1915.

Tang, N., Marshall, W., McMahon, M., Metzger, R., & Martin, G. (2011). Control of mitotic spindle angle by the RAS-regulated ERK1/2 pathway determines lung tube shape. *Science, 333*, 342–345.

Taniguchi, K., Maeda, R., Ando, T., Okumura, T., Nakazawa, N., Hatori, R., Nakamura, M., Hozumi, S., Fujiwara, H., & Matsuno, K. (2011). Chirality in planar cell shape contributes to left-right asymmetric epithelial morphogenesis. *Science, 333*, 339–341.

Théry, F. (2015). Explaining in contemporary molecular biology: Beyond mechanisms. In P.-A. Braillard & C. Malaterre (Eds.), *Explanation in biology. An enquiry into the diversity of explanatory patterns in the life sciences* (pp. 113–133). Dordrecht: Springer.

von Bertalanffy, K. (1976). *General system theory: Foundations, development, applications*. New York: George Braziller.

Weber, M. (2005). *Philosophy of experimental biology*. Cambridge: Cambridge University Press.

Weber, M. (2008). Causes without mechanisms: Experimental regularities, physical laws, and neuroscientific explanation. *Philosophy of Science, 75*(5), 995–1007.

Wimsatt, W. C. (1972). Complexity and organization. In K. F. Schaffner & R. S. Cohen (Eds.), *PSA 1972, Proceedings of the philosophy of science association* (pp. 67–86). Dordrecht: Reidel.

Woodward, J. (2003). *Making things happen: A theory of causal explanation*. Oxford: Oxford University Press.

Part V
New Theories of Explanation
in Biology and Elsewhere

Chapter 16
Biological Explanations as Cursory Covering Law Explanations

Joel Press

Abstract There have been two main sorts of response to difficulties in applying the covering law model of explanation to biology. The first sort, which I call modified law accounts, more or less maintain the logical structure of covering law explanations, but weaken or alter the criteria of lawhood, so that inference from biological generalizations failing in one way or another to satisfy stricter criteria are still deemed explanatory. The second sort, which I call lawless accounts, involve a more wholesale rejection of the covering law model. According to these views, biological explanations are not inferences from natural laws at all. The new mechanist account is a promising example. I have been developing a third sort of account, which I call the cursory covering law model. According to this model, biological explanations can be accommodated within the covering law model without the weakening of the law constraint envisaged in the modified law accounts, provided it is permissible to employ approximating statements about laws as premises in the explanation. I argue that the cursory covering law model subsumes and explains the insights of both the modified law and lawless accounts. It can accommodate the apparent lack of strict biological laws while nevertheless explaining how less-strict biological generalizations can be the basis for cursory covering law explanations. It can also explain why biology significantly involves the discovery of biological mechanisms. Furthermore, since both the modified law and lawless accounts are consistent with the cursory covering law model, any difficulties in understanding biological explanation addressed by the newer approaches will be difficulties that can be addressed equally well within the covering law model.

Keywords Biological explanation • Covering law explanation • Deductive-nomological explanation • New mechanism • Mechanistic explanation • Hempel

J. Press (✉)
Department of Philosophy, Manderino Library 450G, California University of Pennsylvania, 250 University Ave., California, PA 15419, USA
e-mail: press@calu.edu

© Springer Science+Business Media Dordrecht 2015
P.-A. Braillard, C. Malaterre, *Explanation in Biology*, History, Philosophy and Theory of the Life Sciences 11, DOI 10.1007/978-94-017-9822-8_16

367

1   Introduction

Once upon a time, more or less, philosophers of science generally accepted that the covering law model of explanation, most fully developed by Carl Hempel, could accommodate all forms of scientific explanation. But, as will happen to grand unified philosophical theories, this consensus was soon challenged. Some, though not all, of the challenges arose when the model, which had been heavily influenced by reflections on physics, encountered difficulties when applied to biological explanations.

Despite considerable diversity, contemporary accounts of biological explanation can be broadly categorized into two sorts. The first sort more or less maintains the logical structure of covering law explanations, but weakens or alters the criteria of lawhood, so that inference from biological generalizations failing in one way or another to satisfy stricter criteria are still deemed explanatory. Robert Brandon and Sandra Mitchell both offer versions of this approach. I will refer to these as modified law accounts of biological explanation.

The second sort of response involves a more wholesale rejection of the covering law model. According to these views, biological explanations are not inferences from natural laws at all. A promising example of such an account is the new mechanist account of explanation developed by Peter Machamer, Lindley Darden, and Carl Craver, according to which biological explanations involve uncovering and exhibiting the biological mechanisms responsible for producing the phenomena to be explained. I will refer to these as lawless accounts of biological explanation.

I have been developing a third sort of account of biological explanation, which I call the cursory covering law model. According to this model, biological explanations can be accommodated within the covering law model without the weakening of the law constraint envisaged in the modified law accounts, provided it is permissible to employ approximating statements about laws as premises in the explanation. In other words, whereas modified law accounts involve inference from statements that fail to be strict laws, but nevertheless approximate strict laws, a cursory covering law explanation involves inference from premises that state approximately which strict laws are true by identifying the set of strict laws consistent with available evidence.

Before we proceed, a few words about the boundaries of this discussion are in order. First, it is worth pointing out that neither my own account, nor the others I discuss, are necessarily limited to biology. Other sciences, for example, social and behavioral sciences, raise many of the same issues. However, since this is a volume devoted specifically to biological explanations, I will simply refer to them as accounts of biological explanation.

Second, let me be clear that I have no intention of defending strict application of the covering law model, in every respect, to biology, or to any science. There are well-known objections to the covering law model that threaten its application to any science, but I am interested here only in the assumption that biology poses special problems for the covering law model not posed by the sciences it seems to fit best, such as physics. Whereas modified law accounts suggest that biology presents additional problems resolvable with relatively minor modification, and lawless accounts suggest that biology presents insurmountable additional problems, my claim is that biology does not present any additional problems at all.

In order to defend this thesis, I intend to show that explanations conforming to both the modified law approach and the lawless approach can be represented as what I am calling cursory covering law explanations. If so, then any difficulties in understanding biological explanation addressed by the newer approaches will be difficulties that can be addressed equally well within the covering law model.

The upshot will be that while both of these approaches offer important insights into certain distinctive features of biological explanation, there can yet be an underlying unity in all scientific explanations, as demonstrated by their adherence to the cursory covering law model.

Proponents of both the modified law account, and, perhaps especially, the lawless account have to some extent rejected the importance of such a unity. However, if we are to capture the sense in which biologists and other scientists are all engaged in the same activity, namely science, it seems to me important for philosophers of science to say what that similarity is. Of course, giving up on a unified account of scientific explanation would not necessarily undermine that project, since the similarity might lie elsewhere. But if unity can in fact be found in a single account of explanation, that seems to me a reason in its favor.

2 Reviewing the Covering Law Model

Since all of these models of explanation can be seen as reactions to (or reactions to reactions to . . .) the covering law model, it will be worthwhile to review a few of its more salient features. Hempel (1962, 1966) argued that scientists explain phenomena by presenting reasons for thinking that those phenomena could not have (or at least would not likely have) failed to occur, so long as the natural world works as we know it does. Presentations of reasons are arguments, so explanations are arguments.

If the explanans (the premises of the argument) is to support the explanandum (a conclusion stating that the explained phenomenon had to occur), the argument needs to include certain sorts of premises. The explanans may include other particular events. For example, in explaining the point of impact of a cannonball, we might present the explosion of the gunpowder and the mass of the cannonball as reasons. However, premises regarding such particular events cannot be the only premises, since, as Hume taught us, the premise that describes the explosion does not, on its own, or even along with other particular premises, entail that any other event will occur.

So, other premises are needed which logically link the particular premises with the explanandum. In order to serve as links, such premises need to assert something about the relations between the events in both the particular premises and the conclusion. Hempel calls the requirement for such links the requirement of explanatory relevance (1966, 48). Furthermore, if these linking premises are to be known, they will need to assert that events of these sorts are linked not only in this case, but in every case, or some high proportion of cases, so that the presence of the link can be supported by empirical evidence from other cases. This is the requirement of testability (1966, 49). Finally, if those other cases are to count as empirical evidence that the link applies in the context of this particular explanation, the link must be non-accidental.

Thus, Hempel concludes that scientific explanations are arguments that present reasons for expecting the explanandum to occur, with an explanans containing at least one premise asserting a law of nature, where laws of nature are non-accidental universal, or high-probability statistical, generalizations with empirical consequences.

3   Modified Law Accounts of Biological Explanation

There are several reasons why the role of laws in the covering law model seems particularly problematic in biology. In some cases, biologists seem not to formulate laws at all, yet seem no less successful in generating explanations. When biologists do use the term "law" to refer to certain generalizations, those generalizations tend to deviate in one way or another from the paradigm of a law by containing ceteris paribus clauses, or by being a priori mathematical truths (Sober 2000, 72–73). Furthermore, evolutionary theory seems hostile to universal biological generalizations. Any biological generalization that one might be tempted to call a biological law will be the result of a contingent process of natural selection. So the generalization has probably not held at times in the sufficiently distant past, before natural selection had generated the organisms exhibiting the generalization, or probably won't hold at times in the sufficiently distant future, at which point changing selective forces are likely to have changed the organisms so that they no longer exhibit the generalization. At any rate, even if some such generalizations were to turn out to be universal, they would seem to be accidentally so (Beatty 1995; Rosenberg 2001).

For this reason, the more conservative contemporary approach to biological explanation has been the modified law account. For example, Brandon (1997) notes that, in addition to being non-accidental empirical universal generalizations with explanatory force, laws must also be inferable from a small number of inferences. This is the property Goodman called projectability (1965). According to Brandon, it is an empirical fact that biological generalizations tend to be less projectable than physical ones. We can infer with high confidence that chemical elements measured to have certain properties in a few instances will have them always, that physical constants measured in a few instances will maintain their values always (i.e. they really are constants), and so on. By comparison, though some biological generalizations are more projectable than others, few if any approach the projectability of physical generalizations.

And yet, Brandon points out that biological generalizations are not entirely unprojectable. For example, he imagines a case in which a population of plants capable of both sexual and asexual reproduction is exposed to an environment which is observed for a few years to favor sexual reproduction. This is probably insufficient evidence to infer that the observed generalization, that sexual reproduction is usually favored, is true and non-accidental. However, if these conditions remain stable for a considerably longer span, and/or in a wider range of cases, we might then be prepared to, at least tentatively, do so. Even so, our confidence will be limited to situations very similar to those observed (1997, S455–S466).

Brandon then argues that this limited projectability implies limited explanatory power. Whereas we will generally be in a position to explain phenomena using physical laws when we are in possession of only a small amount of empirical evidence, biological generalizations will be sufficiently well-confirmed to be used in explanation only after considerably more empirical work has been done, and even then, only with limited confidence. Nevertheless, the logical form of biological explanation remains largely the same as it was in the covering law model. When, and to the extent that, the generalizations are expected to hold, we may infer that they have done so, and this explains the phenomenon. Brandon demurs from calling these generalizations with substantial but limited projectability laws, but since he admits that withholding the term "law" is terminological, I classify his view within the modified law approach.

Mitchell (1997) does call them laws, and offers pragmatic criteria for distinguishing which biological generalizations are laws. These pragmatic criteria are to be gleaned from an examination of the role such generalizations play in explanation and in other scientific activities, and may be relative to the scientific task at hand. A generalization of limited projectability might be explanatorily useful for making crude predictions about a system, but not for making detailed predictions. A simple but not very accurate generalization might be useful in education or engineering, but not in scientific discovery. When and to the extent that generalizations meet these pragmatic conditions, Mitchell endorses their lawhood.

Although this terminological difference may be of minimal importance in itself, it raises an important point. Whereas Mitchell endorses this pragmatic approach to determining lawhood, she contrasts this with views like Brandon's, or indeed Hempel's, which start out with a normative definition of "law" and then classify the generalizations of the various sciences accordingly. As she sees it, the pragmatic approach instead focuses on the various roles generalizations play, and accords lawhood to whichever generalizations satisfy the role. However, I think there is less disagreement between these approaches than may appear. Even Hempel starts with the idea that certain generalizations, and not others, play an explanatory role. He then attempts to discover what properties (relevance and testability) those generalizations would need in order to fill such a role. This might be called normative; he is attempting to discover the criteria that sort the generalizations with the explanatory goods from the ones that lack it. But then Brandon and Mitchell seem to be doing this too.

> The function of scientific generalizations is to provide reliable expectations...The tools we design for this are true generalizations [about the natural world]...The ideal situation would be, of course, if we could always detach the generalizations gleaned from specific investigations from their supporting evidence, carry these laws to all regions of spacetime, and be assured of their applicability...Nevertheless, we can and do develop appropriate expectations without the aid of general-purpose tools. (Mitchell 1997, S477)

Thus it seems that Mitchell, Brandon, and Hempel are all trying to determine what sorts of generalizations are necessary to biological explanation. But Mitchell and Brandon both conclude that the range of generalizations capable of doing the trick is wider than Hempel allows.

4 Lawless Accounts of Biological Explanation

Other philosophers have argued that accounting for biological explanation requires a more radical departure from the covering law model. The new mechanist approach of Machamer et al. (2000), for example, suggests that biological explanation, and especially explanation in molecular biology, differs from covering law explanations in focusing on how a phenomenon is produced, rather than why it occurs. Answering the former question requires, they say, specification of the mechanism whereby the phenomenon is produced. They define mechanisms as "entities and activities organized such that they are productive of regular changes from start...to finish...conditions," (2000, 3). So, for instance, explaining the phenomenon of DNA replication involves identifying the entities involved in the process (the initial DNA double helix, assorted enzymes, new base components, etc.), the sequence of activities in which they engage (unwinding, splitting, pairing, detection and correction of errors, etc.), and the properties of the entities in virtue of which they engage in those activities (molecular structures, spatial arrangement of weak charges, spatial orientation relative to each other, etc.).

Unlike Mitchell and Brandon, these authors specifically distance themselves from laws. They do require that the activities in which a mechanism's entities engage produce regular changes, and they suggest that, as with laws, this regularity of the activities supports counterfactual reasoning. However, they insist that regularity alone is not the primary source of explanatory power. Describing a mechanism does much more than establish the regularity of a process; it additionally illustrates the way in which the phenomenon was produced.[1] In biology, they say, explaining the phenomenon depends more on the latter than the former (2000, 7–8).

Machamer, Darden, and Craver further assert that full specification of a mechanism, and hence full satisfaction of their criteria for explanation, need not involve tracing the workings of a mechanism beyond entities and activities pragmatically

---

[1]For a complementary view on mechanisms, see Fagan (2015, this volume) who discusses the explanatory role of "jointness" in mechanistic explanations.

deemed to be relevant and noncontroversial by the scientists pursuing an explanation. They refer to these as "bottom out" entities and activities. So, for example, they assert that although explanations in molecular biology often involve specifying activities between entities involving chemical bonding, mechanical interaction (in the old sense of objects pushing and pulling each other), diffusion, etc., they rarely if ever require specifying the fundamental quantum-mechanical activities engaged in by sub-atomic particles. Of course, if at some point proposed mechanisms face anomalies that cannot be resolved without pushing beyond this bottoming out level, that level can change, but there is no absolute degree of detail necessary to the pragmatically complete specification of a mechanism (2000, 13–14).

The authors do grant that much important scientific activity involves merely partial specification of mechanisms. For example, scientists articulate mechanism schemata, which are abstract versions of particular mechanisms. Such schemata can be convenient ways to sum up important similarities between similar particular mechanisms, as well as a useful template for devising hypotheses about new ones (2000, 15–18). Scientists also devise mechanism sketches, which fill in some of the details of a mechanism, but fail to include all the details necessary to reach the aforementioned bottom out entities and activities. Instead, the sketch will leave some parts of the mechanism in a "black box," to be filled in after further inquiry (2000, 18). However, none of this activity is explanatory, they say, until a complete particular mechanism has been specified (Craver 2006).

5 The Cursory Covering Law Model of Biological Explanation

Both the modified law and lawless approaches are enlightening. However, I believe there is yet a better way of modeling biological explanation which incorporates many of the insights of both approaches without deviating from the covering law model nearly as much as either of them.[2]

Hempel was always quite clear that his covering law model is an idealization of scientific explanation, and took pains in many cases to discuss ways in which actual explanations could harmlessly deviate from it. The ideal covering law explanation would be a deductive-nomological explanation because this would maximally satisfy the joint requirements of relevance and testability. A deductive inference would be the strongest possible demonstration of relevance, and fully universal generalizations have testable consequences even in single cases. And yet, he allows that inductive-statistical explanations, with their weaker inferences and merely stochastic testability, satisfy these requirements well enough to be explanatory

---

[2] See also Issad and Malaterre (2015, this volume) who recognize the role that derivation plays in mechanistic explanations as well.

(1966, 67–69). He also allows that explanations may be elliptical, relying on known but unstated premises regarding laws or particular circumstances (1966, 52–53).

More importantly for what follows, Hempel allows what he calls partial explanations (1962, 16–18). These are explanations in which the conclusion deduced from the explanans is less specific than, but nevertheless includes, the explanandum. For example, suppose we know all the relevant laws of physics governing the firing of cannons, as well as various relevant initial conditions obtaining at the recent firing of some particular cannon, such as the angle at which it was fired, the mass of the cannonball, local atmospheric conditions, etc. But suppose that we do not know the exact mass of the cannonball. So long as we know the approximate mass, we will still be able to deduce an approximate landing point for the cannonball. If our actual measurement of the exact landing point falls within the expected range deduced from our approximation of the mass, Hempel says we will have partially explained why it landed where it did. The explanation is partially successful because we were able to show, not that the cannonball had to land where it did, but rather that it had to land either where it did, or somewhere pretty close.

Since even the most precise measurements have at least some margin of error, Hempel's acceptance of partial explanations would appear to be an acceptance of the fact that no explanation fully satisfies the covering law model in its ideal form, even if some come exceedingly close. In fact, given that most explanations will involve the specification of many initial conditions, as well as the explanandum, all of which will be known only approximately, the true logical form of the explanation will be multiply disjunctive. The explanans will, for example, state that the cannon was loaded with 8 ± 1 oz of gunpowder and a cannonball weighing 16 ± 1 oz, that the firing angle was $45 \pm 0.5°$, that wind velocity was 7 ± 5 miles per hour blowing north/northwest $\pm 6°$, and so on. The explanatory argument will be valid only if a conjunction of the laws of nature plus any possible combination of values within these ranges entails the explanandum measurement, within its own margin of error. If most but not all combinations of possible values entail explanandum values within the margin of error, this would seem to count as a new sort of inductive-statistical explanation, where the statistical character of the explanation lies in uncertainty about initial conditions rather than on inherently statistical laws.

It will be worth the trouble here to be explicit about the logical inference involved in partial covering law explanations. The logical form of ideal (i.e. non-partial) covering law explanations may be represented by the following schema.

$$L \vdash C \to E$$
$$\frac{C}{\therefore E}$$

Here L is a conjunction of all the laws of nature invoked by the explanation, C is a conjunction of all the particular initial conditions, and E is the explanandum. In this schema and the schemata to follow, italicized letters represent explicitly stated premises or conclusions of the explanatory argument. The point of specifying that L implies $C \to E$ in this schema is to call attention to two distinct, though thus far obvious, steps involved in the inference. First, $C \rightarrow E$, a conditional relating the particular initial conditions and explanandum involved in the explanation, is derived from L, which is composed of laws covering an entire range of conditions and explananda. Second, E is derived from C and $C \rightarrow E$. In the cannon example, the relevant laws apply to cannonballs of any mass, but it is the conditional applying to the actual mass of the fired cannonball that we need to complete the argument.

When we move to partial covering law explanations, explicitly recognizing these steps becomes important because the inference from laws is more complicated.

$$L \quad \vdash \forall_i \left(C_i \in \mathbf{C} \rightarrow \exists_j E_j \in \mathbf{E} \right)$$
$$\frac{C \quad \vdash \exists_i C_i \in \mathbf{C}}{\therefore E \vdash \exists_j E_j \in \mathbf{E}}$$

In partial explanation, the initial conditions and explanandum are stated only approximately, which we can represent by showing that the explicitly stated C and E are now approximating expressions implying only that the actual conditions and explanandum fall within defined sets of possible conditions and explananda ($\mathbf{C}$ and $\mathbf{E}$ respectively) consistent with those approximations. This then requires that the laws, L, imply an entire range of conditionals, so that no matter which of the possible initial conditions obtains, it will be possible to deduce that one of the possible explananda obtains as well. For example, if we only know the mass of the cannonball to within an ounce, the consequences of the laws for all the masses in that range are relevant.

Partial covering law explanations will not help with the problem of missing biological laws, since L is still an explicitly stated conjunction of laws. All that has changed so far is that a wider range of the consequences of those laws are relevant to the derivation. However what I have suggested in my previous work (Press 2009, 2011) is that if this sort of approximation of initial conditions is acceptable, the same should be true of approximations regarding the laws of nature. Suppose that instead of being uncertain about the mass of the cannonball, we are uncertain about the law governing gravity. Perhaps we think it is either Newtonian or Einsteinian, but are not sure which. From the premise that the law of gravity is either Newtonian or Einsteinian (along with the rest of the explanans) we cannot deduce the landing point of the cannonball, but we can deduce that it will fall within a very small range of locations.[3] If the measured landing point falls within that range, this explanatory argument will be just as valid as the partial explanation, and hence, if partial explanations are explanatory and count as covering law explanations, there seems to be no reason not to allow that this argument is also explanatory in virtue

---

[3]Clearly, in this example, unless our cannonball is being fired from the surface of a neutron star or in some similarly exotic environment, the difference between Newtonian and Einsteinian predictions will be immeasurably small. But this feature is not essential. The analogy with partial explanation would hold even if the two gravitational theories generated measurably different predictions.

of being a sort of covering law explanation. Indeed, such arguments are perhaps just another sort of partial explanation. However, since they are not explicitly discussed by Hempel, I have called explanations that include premises approximating the laws of nature (as well as, perhaps, initial conditions) cursory explanations.

Of course, there is nothing special about deduction from a disjunction between just two possible laws. Just as an approximation of initial conditions can specify a continuous range of values, a premise approximating the laws of nature could simply place outer bounds on what those laws might be. And in the case of laws, there might be many dimensions along which they might vary. For example, if we start with Newton's law of gravitation, one way to generate similar but competing laws would be to consider laws describing an attractive force inversely proportional to exponents of distance somewhat more or less than 2. Another would be to consider laws in which the gravitational constant is larger or smaller. Yet another would involve relativistic variations, and so on. A given premise in the explanans might put constraints on just one such dimension, or on several simultaneously.

The inference involved in this cursory covering law explanation will be even more complex than that involved in partial explanations, since in addition to dealing with approximation of initial conditions and explananda, we must now accommodate approximation of laws.

$$
\begin{array}{ll}
L & \vdash \exists_i L_i \in \mathbf{L} \ \& \ \forall_i \left[L_i \in \mathbf{L} \rightarrow \forall_j \left(C_j \in \mathbf{C} \ \rightarrow \exists_k E_k \in \mathbf{E} \right) \right] \\
\underline{C} & \underline{\vdash \exists_j C_j \in \mathbf{C}} \\
\therefore E & \vdash \exists_k E_k \in \mathbf{E}
\end{array}
$$

Just as C and E did above, our statement regarding laws, L, has now become an approximating expression implying only that the actual laws fall within a defined set of possible conjunctions of laws, $\mathbf{L}$. Furthermore, what must be inferred from this approximate claim about the laws of nature is that one of the possible explananda, $E_k \in \mathbf{E}$, should be inferable from any of the possible initial conditions, $C_i \in \mathbf{C}$, no matter which of the possible conjunctions of laws, $L_i \in \mathbf{L}$, is true. In the cannon example, any cannonball within an ounce of a pound should be expected to fall within the error bars of the observed distance whether gravity is Newtonian or Einsteinian.

One further complication is required to yield a fully general schema of cursory covering law explanations. In the previous example, all the conjunctions of laws, $L_i \in \mathbf{L}$, specified by L covered each of the possible initial conditions, $C_i \in \mathbf{C}$, specified by C. But this need not be the case. Suppose that we really are quite confused about the nature of gravity. Perhaps we think that some quasi-Aristotelian theory of gravity is a possibility, so that rather than its mass, it is the earthiness of the cannonball in virtue of which it endeavors to return to Earth (and the center of the universe) when fired. If so, then, for example, quasi-Aristotelian laws coupled with Newtonian initial conditions will fail to entail any of the possible explananda since the quasi-Aristotelian laws govern the behavior of earthy stuff, but Newtonian conditions would be specified in terms of mass.

Since we cannot in general count on just any old pairing of conjunctions of laws with conjunctions of conditions, (L_i, C_j), to entail an appropriate explanandum, $E_k \in \mathbf{E}$, the premises of our explanation need to jointly specify pairs that will work together. This is represented below by letting LC stand for an expression that specifies a set of conjunctions, $\mathbf{LC}$, the conjuncts of which include both all of the laws and all of the conditions necessary for a complete covering law explanans. Simultaneously, LC also asserts that each member of this set entails one of the possible explananda.

$$\frac{LC \quad \vdash \exists_i LC_i \in \mathbf{LC} \ \& \forall_i \left(LC_i \in \mathbf{LC} \to \exists_j E_j \in \mathbf{E} \right)}{\therefore E \quad \vdash \exists_j E_j \in \mathbf{E}}$$

In this schema, "LC", "LC", and "$\mathbf{LC}$" should be understood as simple, as opposed to compound, symbols. Essentially, each $LC_i \in \mathbf{LC}$ is a separate complete possible explanans, including both initial conditions and laws, and the explanation succeeds because LC asserts that one or another of these explanantia obtains, and that the obtaining of any one of them would provide a sufficient explanation of the explanandum.

So what sort of expression could perform the job assigned to expression LC? Of course, LC could just be an explicit disjunction of each of the possible explanantia. "Either *these* laws and initial conditions which imply the explanandum obtain, or *this other* set of laws and conditions which imply the explanandum obtain, or…" But it need not be anything so cumbersome or explicit. Suppose we have experimented with firing our cannon and perhaps throwing or catapulting the cannonball using a variety of devices on the surface of the earth, but have not investigated how any of these devices would work on other celestial bodies or with cannonballs filled with non-earthy elements such as air or water. LC could then simply be a recognition of whatever regularities we may have found in the way that the cannonball behaves in these limited circumstances, such that those regularities are consistent with Newtonian, Einsteinian, and quasi-Aristotelian explanations while ruling out at least some others. LC may not tell us much, but it may nevertheless tell us enough to ensure that the cannonball will land approximately where it in fact does, and hence to explain, at least to some extent, why it lands there. Furthermore, note that since these regularities are limited to a particular environment and particular launched object, LC does not itself explicitly state any law or laws.

In addition to being a valid argument, our schema for cursory covering law explanations still includes all the restrictions present in the original schema of an ideal covering law explanation. The cursory covering law argument will still involve demonstrating that the explanandum is implied by a set of true premises, some of which are laws of nature with all the features Hempel imparts to laws. The only difference is that with cursory covering law explanations, we do not know which set of such premises, which $LC_i \in \mathbf{LC}$, is actually true.

The introduction of cursory covering law explanations allows us to accommodate biological explanations that make no explicit reference to laws of nature because

LC can specify laws without itself stating any law of nature, and even without invoking the terminology of laws at all. In biological explanation, the premises placing these constraints will generally be any that describe known regularities, dispositions, capacities, etc. For example, an explanation of the evolution of zebra stripes might include the premise that lions tend to have more difficulty spotting striped prey than non-striped prey. This observation places constraints on, among other things, the laws of optics. To be sure, it comes nowhere close to telling us whether, for example, light travels as a ray, a particle, or a wave. But it does tell us that light travels in such a way that, over the sorts of distances at which lions typically interact with their prey, striped objects and non-striped objects at least differentially affect the eyes of lions. If this is all we need to know about the laws of optics in order to deduce that zebras' stripes enhance fitness, the simple statement that lions have trouble seeing striped objects, though not a law itself, can constrain the laws well enough to support a cursory covering law explanation.

Frequently, as the merging of *L* and *C* into *LC* suggests, these sorts of premises will simultaneously place constraints on several laws of nature, as well as on initial conditions. The observation that monkeys falling from trees tend to break bones will tell us at least something about laws governing gravity, about laws governing the forces accounting for the strength of bones, and about initial conditions describing bone composition. Furthermore, these constraints may be interdependent in complex ways. The fact that bones will break when subjected to a fall tells us that the forces (whatever they may be) that hold together the parts of monkey bones (whatever they may be) are insufficient to prevent breakage given the magnitude of gravitational forces (whatever they may be). But it does not tell us that the relevant parts are mostly calcium phosphate molecules, or that the relevant forces binding those molecules are electro-chemical bonds, or that gravity is relativistic. Had bones been composed of something else, perhaps even something other than molecules, different forces might have failed to hold them together upon impact. A sufficiently primitive biologist might know next to nothing about all this chemistry and physics, but if she knows that monkey bones break when monkeys fall from trees, she knows enough to rule out any possible chemistry or physics that predicts otherwise. And so, when the biologist goes on to explain why monkeys with prehensile tails, who use them to prevent falls, are fitter than those without, she can rely on her observation that such falls tend to lead to broken bones to constrain chemistry and physics enough for her explanatory purposes.

An important consequence of cursory covering law explanations for biology is that, even though knowledge about the laws of nature is necessary for such explanations, they can nevertheless be employed without knowing which such laws are true. At the root of these multiply disjunctive arguments we find the basic rule from propositional logic which usually goes by the name Constructive Dilemma or Separation of Cases.

$$\{(P \vee Q) \, \& \, (P \rightarrow R) \, \& \, (Q \rightarrow R)\} \vdash R$$

Logic students exposed to this rule for the first time sometimes find it to have an air of hocus pocus about it. How can we know the conclusion is true if we do not know whether it was P or Q that made it true? But of course we can. In fact, we can even know the conclusion without knowing all the possible ways of making it true, so long as we know that all the ways things could be are ways that would make it true. So, for example, we need not have envisaged all possible laws governing the transmission of light on the way to explaining why zebra stripes are adaptive, so long as our observation that lions have difficulty spotting striped animals would allow us to say, of any possible hypothetical law of optics, that the law is or is not consistent with the observation.

6 Modified Law Explanations as Cursory Covering Law Explanations

The relation between the modified law account and the cursory covering law model is fairly straightforward. The modified law account seems to be largely a response to the scarcity of biological laws, which would appear to undermine the covering law model. If one thinks that inference from laws, or something sufficiently like them, is necessary for biological explanation, but that there are no laws of the sort required by the covering law model, a natural response is to either broaden one's definition of law, as Mitchell does, or else at least allow, as Brandon does, that many biological generalizations that are not laws are at least law-like enough to get the job done. However, if the cursory covering law model is accepted, this maneuver is unnecessary, since it allows the possibility of biological covering law explanations even without biological laws. So long as various biological explanatory premises, including, but not necessarily limited to, the generalizations discussed by Mitchell and Brandon place constraints on strict laws of nature belonging to some other, perhaps more fundamental, science, there is no further requirement that those biological explanatory premises themselves be laws.

It does not necessarily follow that the generalizations Mitchell and Brandon discuss are not laws. However, the cursory covering law model relieves any pressure to accept them as laws simply in order to ensure that there are enough biological laws to serve in all biological explanations. Instead, we can be as persnickety as we like about which generalizations count as laws, basing our criteria on Hempel's original analysis of what is necessary to scientific explanation. If some biological generalizations pass Hempel's test, then some biological explanations will be inferences from biological laws. If in other cases biological generalizations are found wanting, they can still be the basis for cursory covering law explanations by placing explanatorily sufficient biological constraints on the laws of other sciences. But even if no biological generalizations are laws, biological explanation will be secure.

The cursory covering law model also naturally captures the limited projectability of biological generalizations discussed by Brandon. When we observe that sexual reproduction is favored in a particular population of plants in a particular environment, this generalization places limitations on various initial conditions and laws, but does not come close to providing detailed information about exactly what the initial conditions or laws are. For example, if sexual reproduction is favored by natural selection, it must be a heritable trait. As it happens, the correct explanation of heritability involves DNA, enzymes, and their behavior under the laws that govern molecular interactions. But even if we did not know this, as, for example, Darwin did not, the simple observation that sexual reproduction is a heritable trait does imply that there is some such explanation of heritability. This may be all we need to know about heritability in order to construct our cursory covering law explanation. Of course, in order to explain the explanandum that sexually reproducing plants are fitter than asexually reproducing plants we will also need a certain amount of information about how both sorts of plants are affected by their environments, how reproduction and development work in these plants, and so on. But this information can be similarly sketchy in the limitations it places on initial conditions and laws so long as it is sufficient to ensure that there is some complete covering law explanation of the phenomenon, even if our cursory explanation does not tell us exactly what it is.

In effect, the cursory covering law explanation offers us a huge set of possible covering law explanantia, and asserts that some undetermined one of them is the actual explanans. This may be enough to show that the explanandum was to be expected in this case, but not enough to give us much guidance in forming expectations about other cases. If we do not know which particular initial conditions held and were responsible in the first case, we will not know which conditions need to be present to produce the same result in other cases. If we do not know which laws were involved in linking the initial conditions to the explanans in the first case, we will not know which laws need to be operative to produce the same result in the other cases. However, as we make further observations, these may place firmer restrictions on both initial conditions and laws, so that our confidence in expecting similar results in other cases may grow.[4]

Returning to Brandon's example, if we did not know that heritability involved DNA in the originally examined plants, we could not immediately project our conclusion that sexual reproduction has higher fitness to cases involving other plant species, since we would not know that the presence of DNA in the new species, plus the expectation that the laws governing DNA reactions apply universally, would imply similar patterns of heritability. For all we would know, entirely different initial conditions and laws might be relevant. Thus, only after we had determined observationally that heritability operated within the same bounds in the new species

---

[4]This talk of observations by biologists placing constraints on laws of physics or chemistry may give the impression that biologists are themselves engaged in the formulation of laws at more fundamental levels. However this will not generally be the case because the biologist will almost certainly lose interest in more narrowly constraining such fundamental laws once she has constrained them enough to secure her cursory explanation. Only if her biological explanation required constraint of the physical laws beyond what had already been accomplished by physicists would she find herself actually needing to do physics. In most actual cases I would guess that the biologist does not even trouble herself to constrain the laws as much as existing knowledge of physics actually does. Few biologists find that they need to form professional opinions about relativity, much less string theory.

as in the old could we project such a conclusion. On the other hand, now that we have identified the actual mechanism of heritability, we can more quickly assure ourselves that heritability will work similarly in all terrestrial species simply by observing the relevant initial conditions, namely, the presence of DNA and its accoutrements. If we someday discover life elsewhere, how quickly we can project various generalizations about terrestrial biology will depend, in part, on whether the extraterrestrial organisms employ the same mechanisms. If they do, we will be able to quickly assure ourselves that the initial conditions relevant to heritability on earth will have the same consequences elsewhere. If they do not, we will have to rely on observation to discover how similar or different their patterns of heritability may be until we can discover the extraterrestrial mechanism of heredity.

Although I have presented cursory covering law explanations as a natural consequence of Hempel's own views, it is worth pointing out that he himself treated biological explanations in a way more in line with the modified law approach, though without admitting to any deviation from the covering law model in doing so. In one of his most famous examples, Semmelweis' explanation of childbed fever, Hempel asserts that inference from a biological generalization, albeit an elliptically stated one, is involved.

> The explanation makes no mention of general laws; but it presupposes that such contamination of the bloodstream generally leads to blood poisoning attended by the characteristic symptoms of childbed fever, for this is implied by the assertion that the contamination *causes* puerperal fever. (1966, 53)

Hempel's treatment of the case suggests that the explanation simply involves inferring the explanans from the generalization that bloodstream contamination leads to blood poisoning, and hence that this generalization is a biological law. By comparison, the cursory covering law model suggests that this generalization is probably best seen as placing limits on more fundamental laws of nature, quite possibly laws outside of biology, which nevertheless permit inference of the explanandum.

7 Mechanistic Explanations as Cursory Covering Law Explanations

Because the new mechanists make mechanistic explanation out to be something significantly different from inference from covering laws, it might seem harder to incorporate the insights of the new mechanists into the cursory covering law model. But there is in fact a very natural way of doing so. Recall the unwarranted, but entirely natural, complaint of the beginning logic student faced with Constructive Dilemma. How can I know R follows if I don't even know whether it follows from P or from Q? Of course, since both P and Q will get us to R, we don't need to know which so long as we know it must be one or the other, but still, it does seem like we would know more if we knew whether P or Q were responsible. And when it comes to explanation, surely we would grant that an explanation which not only shows that the explanandum was expected, but also lays bare the mechanism that produces it, is a better explanation than one that leaves out such details, even if it does show that the explanandum was to be expected without specifying the details.

According to the cursory covering law model, laying bare the mechanism corresponds to explicitly disambiguating all, or at least some, of the possible explanatia, $LC_i \in \mathbf{LC}$, specified by the explicit premises of a cursory covering law explanation, LC, and uncovering, or approximating, the actually obtaining explanans. Since Machamer, Darden, and Craven often use examples from neurobiology, consider this explanation of misplaced bodily sensation from Descartes' Sixth Meditation.

> When I feel pain in my foot . . . this happens by means of nerves distributed throughout the foot, and that these nerves are like cords which go from the foot right up to the brain. When the nerves are pulled in the foot, they in turn pull on inner parts of the brain . . . and produce a certain motion in them; and nature has laid it down that this motion should produce in the mind a sensation of pain, as occurring in the foot. But since these nerves . . . must pass through the calf, the thigh, the lumbar region, the back and the neck, it can happen that even if it is not the part in the foot but one of the intermediate parts which is being pulled, the same motion will occur in the brain as occurs when the foot is hurt, and so . . . the mind feels the same sensation of pain. (1642, 60)

Of course, in one sense, Descartes is simply wrong. Nerves are not pulled. However, he does seem to have explained, in a sense, why, for example, events in the loins or neck might sometimes be experienced as foot pain, whereas events in the arms are not.

If we set aside Descartes' hypothetical pulling mechanism, what remains is a cursory covering law explanation. The explanation's initial conditions include a description of the gross structure of the nervous system. However, this description places only rather broad limitations on what exactly nerves are composed of, and hence on which laws might be relevant. Whatever the ultimate constituents of nerves are, the description simply asserts that they are arranged along various lines connecting the brain to assorted parts of the body, and in such a way that sufficiently distant parts of the body tend to be connected to the brain via different lines. The explanation also relies on the observation that pain stimuli applied to the body at one end of these lines tends to regularly produce characteristic experiences at the other. This observation places weak constraints on both the laws of nature and the composition of nerves. Whatever nerves are composed of, the laws of nature are those that allow objects composed of such stuff to regularly transmit impulses from one end to the other. As the familiar literature on functionalism in the philosophy of mind repeatedly emphasizes, many possible compositions (i.e. electrical wires, hydraulic tubes, etc.), and many possible laws governing them (i.e. laws governing electrical currents, laws governing fluid dynamics, etc.), are consistent with nerves having such properties, at least grossly described. But the explanation is successful, as far as it goes, because whatever the details, within the constraints imposed by the observation that nerves transmit impulses, the explanandum will follow.

In order to replace this explanation with a mechanistic one, we need to identify the entities and activities actually involved in the production of the explanandum.

But doing this is simply a matter of narrowing the sets of possible initial conditions and laws. Suppose we discover that, rather than cords (entities) that contract (activity), nerves consist of chains of neurons (entities) that fire and then chemically stimulate subsequent neurons (activities). Identifying the entities as neurons rules out many alternative compositions that would have been disjunctively included in the original cursory covering law explanation. Similarly, identifying the activities of firing and chemical stimulation rules out as irrelevant many laws of nature that might have been involved in the transmission of nerve impulses had nerves been composed of other stuff.

Thus narrowed, have we now arrived at a mechanistic explanation? Probably not, as neurons, firing, and chemical stimulation are probably not bottom out entities and activities. When Machamer, Darden, and Craver discuss this example, neurons are further analyzed into their relevant parts (sodium gates, membranes, neurotransmitters, etc.), as are the activities (binding, diffusing, depolarizing, etc.) (2000, 8–11). But this further refinement is, once again, simply a matter of further narrowing the range of possible initial conditions and laws. Once the narrowing has reached entities and activities that scientists take to be unproblematic, we have reached the bottom out level, and we have what Machamer, Darden, and Craver call a mechanism.

However, even at this level, our cursory covering law explanation may still be disjunctive. Since Machamer, Darden, and Craver reject the idea that all explanations bottom out at the same fundamental level, or that this level necessarily contains laws, at least sometimes, and perhaps always, mechanistic explanations will remain cursory. For example, if the activities of the entities that comprise the mechanism are not themselves instances of the operation of laws of nature, the explanation will still be disjunctive about the laws governing, say, the behavior of the atoms that make up individual molecules, or the laws governing the forces that hold atoms together. So even the completed mechanistic explanation will be a merely cursory covering law explanation.

Consequently, it seems that any mechanistic explanation can be construed as a cursory covering law explanation. The protestations of the new mechanists notwithstanding, any description of a mechanism, in virtue of its description of regularly occurring processes, will place constraints on the laws of nature, whether or not the description of the mechanism mentions laws explicitly. And its explanatory force will derive from these constraints and their role in supporting a cursory covering law explanation.

However, proponents of mechanistic explanation might argue the real problem is that regardless of whether a proposed explanation satisfies the cursory covering law model, it is not truly explanatory unless it is also mechanistic. Craver (2006) makes this argument at length, arguing that mere mechanism sketches, which fail to fully describe bottom out entities and activities, are not true explanations. For example, even if Descartes' explanation above satisfies the cursory covering law model, since it fails to reach the bottom out level, it is a mere explanation sketch, and hence not explanatory.

However, this seems to me to place too much importance on the relatively arbitrary nature of the bottom out level. Even mechanistic explanations that reach this level will still generally be merely cursory covering law explanations. The only difference between a mechanism sketch and a mechanistic explanation will be the breadth of the approximations involved. The latter narrows that breadth to the bottom out level, while the former does not. But this binary difference is simply superimposed on differences in the breadth of approximation, which are a matter of degree. There seems to be no reason to think that explanatory force arrives suddenly when the bottom out level has been reached, rather than gradually, as the explanation becomes less and less cursory. So, while it certainly seems right to say that less cursory explanations (many of which will be mechanistic explanations) are better than, and even more explanatory than, more cursory explanations, we should still allow that all covering law explanations, no matter how cursory, are explanatory to at least some extent. We can still take on board the new mechanists' insight that uncovering mechanisms is central to explanation, since uncovering the entities and activities of mechanisms involves decreasing the breadth of approximation about both initial conditions and laws. As more and more detailed mechanisms are uncovered, cursory explanations can be made less and less cursory by eliminating additional explanantia consistent with less detailed mechanisms but inconsistent with more detailed mechanisms. With each refinement of the mechanism, we approach more and more closely to Hempel's covering law ideal where a single explanans is completely specified. How closely we need to approach that ideal may vary pragmatically from circumstance to circumstance, but Hempel's goal can still be the right one.

At the same time, we can also explain why pursuing an explanation towards this ideal beyond the bottom out entities and activities is often counterproductive, especially in biology. If most strict laws turn out to be relatively fundamental laws of physics or chemistry, fully explicit covering law explanations would generally be too complex to expect to uncover, and be uninformative to anyone but a Laplacean demon even if we did. A more cursory covering law explanation, perhaps in terms of mechanisms, that places just enough constraint on the details to provide a valid explanatory argument is likely to be more informative. Thus the new mechanists' pragmatic criteria for determining the bottom out level may turn out to be criteria that identify the sweet spot that balances approach toward the covering law ideal against an overabundance of unnecessary information.

8 Conclusion

The cursory covering law model subsumes and explains the insights of both the modified law and lawless accounts. It can accommodate the apparent lack of strict biological laws while nevertheless explaining how less-strict biological generalizations can be the basis for satisfactory cursory covering law explanations. It can also explain why biology significantly involves the discovery of biological mechanisms in a way that is not only consistent with the covering law model, but actually illuminates why the discovery and elaboration of mechanisms might improve explanatory quality. Since neither the modified law nor the lawless account is inconsistent with the universal application of the cursory covering law model, I conclude that the biological explanations they are meant to be accounts of can all be modeled as cursory covering law explanations. If so, then neither approach can succeed where the covering law model fails.

References

Beatty, J. (1995). The evolutionary contingency thesis. In L. D. Gottlieb & S. K. Jain (Eds.), *Concepts, theories, and rationality in the biological sciences* (pp. 45–81). Pittsburgh: University of Pittsburgh Press.

Brandon, R. (1997). Does biology have laws? The experimental evidence. *Philosophy of Science, 64*, S444–S457.

Craver, C. (2006). When mechanistic models explain. *Synthese, 153*(3), 355–376.

Descartes, R. (1642). Meditations on first philosophy. In J. Cottingham, R. Stoothoff, & D. Murdoch (Eds.), *The philosophical writings of Descartes* (Vol. II). Cambridge: Cambridge University Press.

Fagan, M. B. (2015). Explanatory interdependence: The case of stem cell reprogramming. In P.-A. Braillard & C. Malaterre (Eds.), *Explanation in biology. An enquiry into the diversity of explanatory patterns in the life sciences* (pp. 387–412). Dordrecht: Springer.

Goodman, N. (1965). *Fact, fiction, and forecast* (2nd ed.). Indianapolis: The Bobbs-Merrill Company.

Hempel, C. (1962). Explanation in science and history. In R. G. Colodny (Ed.), *Frontiers of science and philosophy*. Pittsburgh: University of Pittsburgh Press.

Hempel, C. (1966). *Philosophy of natural science*. Englewood Cliffs: Prentice Hall.

Issad, T., & Malaterre, C. (2015). Are dynamic mechanistic explanations still mechanistic? In P.-A. Braillard & C. Malaterre (Eds.), *Explanation in biology. An enquiry into the diversity of explanatory patterns in the life sciences* (pp. 265–292). Dordrecht: Springer.

Machamer, P., Darden, L., & Craver, C. (2000). Thinking about mechanisms. *Philosophy of Science, 67*(1), 1–25.

Mitchell, S. (1997). Pragmatic laws. *Philosophy of Science, 64*, S468–S479.

Press, J. (2009). Physical explanations and biological explanations, empirical laws and a priori laws. *Biology and Philosophy, 24*(3), 359–374.

Press, J. (2011). On the virtues of cursory scientific reduction. *Philosophy of Science, 78*(5), 1189–1199.

Rosenberg, A. (2001). How is biological explanation possible? *British Journal for the Philosophy of Science, 52*(4), 735–760.

Sober, E. (2000). *Philosophy of biology* (2nd ed.). Boulder: Westview Press.

Chapter 17
Explanatory Interdependence: The Case of Stem Cell Reprogramming

Melinda Bonnie Fagan

Abstract Stem cells are defined as undifferentiated cells that can produce both undifferentiated and differentiated (specialized) cells. The stem cell concept is thus intimately connected to core assumptions about the process of development. Making these assumptions explicit clarifies the general explanandum-phenomenon of stem cell biology: the branching pattern of cell development, from a single initiating 'stem,' through intermediate stages, to one or more termini. Importantly, the whole process, not only developmental termini (specialized cells of a mature multicellular organism), is the target of explanation. Explanations of cell developmental processes are revealed by experiments. Here I focus on one important kind of experiment: direct cell reprogramming, which manipulates the development of cells in artificial culture conditions. I then examine three accounts of biological explanation in light of this case: interventionist, gene-centric, and mechanistic. Though each offers some insight into explanations based on reprogramming, none is fully satisfactory. This motivates a modified account of mechanistic explanation, emphasizing interdependence among components.

Keywords Development • Experiment • Gene-centrism • Interventionism • Mechanistic explanation • Stem cells • Reprogramming

1 Introduction

This essay is concerned with explanations in a new and controversial field: stem cell biology. Because of its scientific and social interest, stem cell biology has been the subject of much commentary by bioethicists, cultural anthropologists, historians of science, political scientists, and sociologists.[1] Philosophers of science,

---

[1] E.g., Gottweis et al. (2009), Kraft (2009), Lau et al. (2008), and Maienschein et al. (2008).

M.B. Fagan (✉)
Department of Philosophy, University of Utah, Salt Lake City, UT 84105, USA
e-mail: mel.fagan@utah.edu

© Springer Science+Business Media Dordrecht 2015
P.-A. Braillard, C. Malaterre, *Explanation in Biology*, History, Philosophy and Theory of the Life Sciences 11, DOI 10.1007/978-94-017-9822-8_17

however, have so far contributed little to these discussions.[2] One reason for this is that stem cell research does not fit many philosophers' idea of what a science should be. Instead of general theories, laws, or mathematical models, its main accomplishments to date are experimental methods and their products: cell lines that can be induced to make many different kinds of human tissue, "knockout" mutant mice produced by germline manipulation, bone marrow transplantation to treat leukemia and other blood disorders, etc. But stem cell phenomena are not confined to laboratories and hospitals. The developmental processes that generate and maintain our bodies also involve stem cell phenomena.

Though our understanding of these phenomena, both natural and experimentally-induced, remains incomplete, some features of explanation in stem cell biology can be clearly discerned. First, the stem cell concept, carefully articulated, characterizes the field's primary explanandum-phenomenon: the branching pattern of cell development. Explanations of this general pattern and processes that conform to it are based on experiments that manipulate cell development. Of particular interest in this regard is "direct cell reprogramming." This method, recently the subject of a Nobel Prize (2012), transforms specialized cells of skin, muscle, nerves, etc. into cells that resemble those of an early embryo, with unrestricted developmental potential: "pluripotency." The reprogramming case offers an illuminating window onto emerging explanations in stem cell biology. I discuss the basic design of reprogramming experiments, strategies used to innovate the method, and the main experimental results to date.

I then consider three influential accounts of explanation in experimental biology: interventionist, gene-centric, and mechanistic. All three are based on Woodward's manipulability theory (2003), which offers an appealing analysis of causal relations revealed by experiment. The interventionist account, a direct application of Woodward's theory to explanation, states that explanations reveal patterns of causal dependency between values of variables. The gene-centric account, recently defended by Waters (2007), states that DNA sequences are "ontologically distinctive" causes of development, and therefore privileged in (some) biological explanations. According to the mechanistic account, explanations in experimental biology describe mechanisms: complex causal systems of multiple parts that perform some overall function or produce some end result (Bechtel and Abrahamsen 2005; Glennan 1996; Machamer et al. 2000). Though each offers some insight into the reprogramming case, none of the three fully accounts for emerging explanations based on cell reprogramming experiments, which I take to be representative of explanation in stem cell biology more generally. I then present an alternative, which makes *jointness*, or interdependence among components of a system, the centerpiece of explanation in stem cell biology.[3]

---

[2] For recent exceptions see Fagan (2013), Laplane (2011), and Leychkis et al. (2009).

[3] A complementary view is offered by Gross (2015, this volume) who defends the explanatory value of *non*-dependence relationships.

Before proceeding, a few clarifications are necessary. Though stem cell biology is a thoroughly experimental field, it does not lack for conceptual content, testable hypotheses, or abstract models. However, it does lack theories in the stricter sense of fundamental equations or general laws. Instead of abstract theories, the field is organized around exemplary methods and model systems. This organization allows for, and indeed requires, pluralism about experimental approaches, methods, standards – and explanations. Although the kind of explanation I discuss here is prevalent in stem cell biology, it does not exhaust explanation in this field, and it is possible that another kind of explanation will come to prominence in a few decades (or sooner). None of this obviates the need to understand the explanations most stem cell biologists seek today. It is upon these, after all, that the future of stem cell research, and its hoped-for clinical applications, will be built.

2 Explanandum-Phenomena

Though many biomedical fields investigate cell development, stem cell biology is distinctive in that its eponymous concept, 'the stem cell,' incorporates developmental ideas. Development is a complex causal process that involves growth, cellular change, and establishment of organismal form.[4] For stem cell research, cellular change is the main focus. The field's central explanatory task is to account for changes to cells in the process of development. This task and standards for its achievement are further clarified by examining the general definition of 'stem cell.'

Scientists define stem cells as undifferentiated cells that can give rise to (i) other undifferentiated cells and (ii) specialized cells of one or more kinds that make up the body of a mature multicellular organism, such as neurons, muscle cells, and white blood cells.[5] The phrase 'give rise to' refers to cell reproduction. Cells reproduce by binary division: a parent cell divides to produce two offspring cells.[6] Specialized cells cluster into kinds, each distinguished by a set of properties had by an individual cell and a functional role within a multicellular organism, played in virtue of having those properties. Undifferentiated cells are, roughly, cells that lack specialized cells'

---

[4]Much about the process of development remains controversial, including its start- and end-points (see Pradeu 2011, and related articles).

[5]Examples of such general definitions can be found in the most recent edition of *Essentials of Stem Cell Biology* (Melton and Cowan 2009, xxiv); the first issue of *Cell Stem Cell*, the official journal of the International Society for Stem Cell Research (Ramelho-Santos and Willenbring 2007, 35); and information pages on the websites of the US National Institutes of Health (http://stemcells.nih.gov/info/basics/basics1.asp) and the European Stem Cell Network (http://www.eurostemcell.org/stem-cell-glossary).

[6]There are two modes of cell division: mitosis and meiosis. In mitosis, the genome replicates once before the cell divides. In meiosis, the genome replicates once, but two rounds of cell division follow, yielding four offspring cells with half the complement of DNA. Stem cell phenomena involve mitosis, so only that mode of cell division is discussed here.

Fig. 17.1 Simple stem cell
model

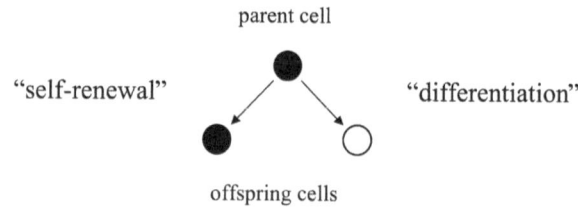

offspring cells

characteristic properties. 'Property-clusters' associated with each specialized cell type can be conceived as assignments of particular values for variable cell characters such as shape, size, biochemical activity and surface molecule expression; I shall use this vocabulary in what follows.

The general scientific definition of 'stem cell' has three features that deserve emphasis here. First, stem cells are not defined in the manner of specialized cells, but in terms of reproductive processes. Second, occurrence or realization of these processes is not required, but only the *potential* to realize them. A stem cell that can self-renew and differentiate need never actually do so. Third, these two reproductive processes involve roughly complementary comparisons across cell generations. Together, these three features shed light on the primary explanandum-phenomenon of stem cell biology. I discuss each in turn.

Because stem cells are defined in terms of cell reproductive processes, the stem cell concept is relational. More precisely, the stem cell concept presupposes a cell lineage *L*, comprised of two or more cell generations linked by one or more 'cycles' of cell division. The simplest example is a single cell that divides to produce two offspring cells, one resembling the parent and the other more specialized (Fig. 17.1). Such a cell exhibits both reproductive processes in a single cycle of cell division. However, a stem cell need not actually engage in either reproductive process. The potential to do so is enough.[7]

Any cell that *would* self-renew and differentiated, under appropriate environmental conditions, qualifies as a stem cell. The environmental conditions that matter vary depending on the cell lineage of interest. For example, rare cells in mammalian bone marrow, umbilical cord, and circulating blood are normally "quiescent," neither self-renewing nor giving rise to differentiated cells. But if transplanted to an immunodeficient host, these blood stem cells regenerate the entire blood and immune system in a few months. Wound-healing, in many cases, begins with mobilization of 'quiescent' stem cells in the damaged tissue. Stem cells in artificial culture, in contrast, are maintained by constant self-renewal as a continuously-growing 'line.' But such a line must have differentiation potential to qualify as a stem cell line. That is, when placed in an appropriate environment, cells of the line differentiate to produce one or more kinds of specialized cell. To summarize, stem

---

[7]Demonstrating this potential experimentally is another matter (see Fagan 2013).

cells are defined as having the potential to give rise, under appropriate conditions, to a lineage L of n cell generations, in which both self-renewal and differentiation occur.

These two reproductive processes are comparative and (roughly) complementary. Self-renewal is cell reproduction such that parent and offspring resemble one another. Since no two cells are similar or different with respect to *all* their properties, a scientifically-useful concept of self-renewal must be understood as relative to a set of variable cell characters C, values of which are compared across cell generations. Self-renewal is thus defined in terms of three variables, such that:

Self-renewal occurs in cell lineage L if and only if n generations in L have the same C-values.[8]

In practice, C includes characters that take the values that distinguish specialized cell types, such as size, shape, biochemical properties, and array of surface molecules. But because stem cells are defined as *undifferentiated*, self-renewal implies that the C-values that remain constant across cell generations differ from those of mature, specialized cells. Self-renewal and differentiation both involve comparison of character-values across cell generations; in this sense, they are comparative. For the former, what matters is sameness; for the latter, difference. So the two comparative reproductive processes are complementary.

Yet this is still too simple. Differentiation is not simply cell reproduction with change of character value across generations. Only changes in a particular 'direction' count. Differentiation requires that later cell generations be *more specialized* than earlier ones; that is, that offspring cells more resemble mature cells than their parents do.[9] For differentiation, the standard of comparison is not the initial parent, the first generation of L, but the end or terminus of cell development. Here there is a further complication: the end of development is not one, but many. Most multicellular organisms are composed of dozens, even hundreds, of distinct cell types, each with a different set of character-values C_m representing the property-cluster that defines that cell type. Differentiation can thus be defined as follows:

Differentiation occurs in L if and only if C-values in L become more similar to C-values of some members of M, where M is a set of sets of character values $\{C_1, \ldots C_m\}$, m is the number of mature cell types for L, and each member of M includes the C-values of one of those cell types.

---

[8]Many stem cell biologists use calendar time rather than number of cell generations as a time-parameter. The two are interchangeable if an estimate of the rate of cell division for L is available (which is usually the case).

[9]Another aspect of differentiation is *diversification*: a population of cells diversifies over n cell generations if and only if variation in C-values increases with successive generations. As the notion of specialization appears to predominate in stem cell research today, I focus on this aspect of differentiation.

Fig. 17.2 General stem cell model

Cell Lineage *L*

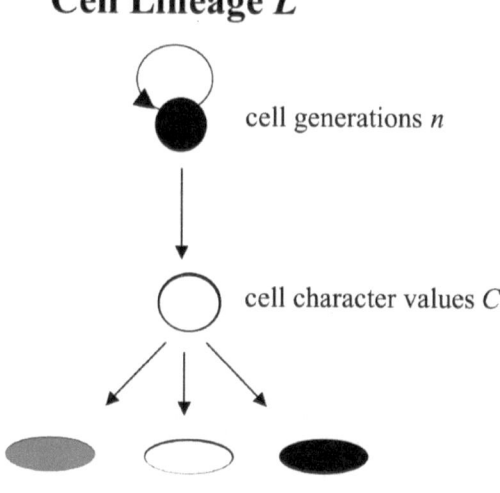

cell generations *n*

cell character values *C*

mature cell characters *M*

As with self-renewal, the characters included in *C* reflect the property-clusters that define mature cell types, which are just those included in *M*. Note that differentiation need not be complete, nor bear any specific relation to cell generations. Any change in the 'direction' of any member of *M* suffices for differentiation to occur. But the concept, in contrast to self-renewal, presupposes a set of developmental termini and their character-values as the standard for comparison within a cell lineage of interest.

Putting all these features together, the general definition of 'stem cell' is a formal representation of cell development: a single starting-point leading to one or more discrete end-points, connected by branching paths and structured into a genealogical hierarchy by reproductive relations and comparisons of character-values (Fig. 17.2).[10] It is akin to Waddington's landscape (1957), which represents exactly these formal features as a pattern of branching valleys down an inclined plane (Fagan 2012a, b). We can now offer a more rigorous general definition of 'stem cell:'

A stem cell is the initiating parent of cell lineage *L*, with maximal self-renewal and differentiation potential within *L* and relative to characters *C*, set of mature cell character-values *M*, and number of generations *n*.

Whether *L* is potential or actual, its structure is determined by values of these other variables.

---

[10]The limiting case is an un-branched line. Some scientists deny that cells that initiate lineages with this pattern count as stem cells. This view imposes an additional constraint on the minimal model; the basic components and relations are the same.

The above formulation clarifies the general explanandum-phenomenon of stem cell biology, as well as other fields that aim to understand cell development. This explanandum-phenomenon is the structure of L: a branching pattern of cell development, from a single initiating 'stem,' through intermediate stages, to one or more termini. For any particular case of interest, parameters C, L, M, and n must be specified. Absent some specification of these parameters, the stem cell concept is purely formal, bearing no clear relation to biological entities and processes. As there are many different ways to specify these parameters, there are many different kinds of stem cell; a cell may qualify on one set of specifications but not others. Any experiment involving stem cells specifies these parameters by choice of cellular material, timeframe of bioassays, and cell characters measured. Explanations of cell development are based on the collective results of many experiments (see Sect. 3). Before turning to these experiments, insights about the explanations sought can be gleaned from consideration of the explanandum-phenomenon itself.

For any cell or cell lineage of interest, the explanandum-phenomenon is a process that conforms to a pattern. The pattern is of branching tracks from a single initial stem to one or more termini, and the process is a developmental path from the stem to a terminus. An explanation in stem cell biology should account for both the pattern and the directed process. To account for the pattern, the explanation should include a description of the initial state (the stem) and conditions required to sustain it (the environment for 'stemness'), and similar descriptions of the end-state(s), intermediates (if any), and their respective environment(s). The description should account for relations between these various cell states; i.e., the arrangement of branching tracks that connect the stem and termini. To account for the process, the explanation should reveal how a developing entity (a cell or cell population)[11] proceeds via a particular path from stem to terminus. It should describe how cells transition from one state to another, and how the 'choice' of path is made at branch-points.

An explanation that accomplishes all this will necessarily be rather complicated and detailed. Not only are cells' intrinsic properties and relations featured, but also aspects of their environments. A comprehensive explanation of cell development would account for the 'topography' of the entire developmental 'landscape' that mediates between the zygote and all mature cell types that make up an organism's body. More modest explanations, of some piece or segment of overall organismal development, must still account for the structure of branching pathways, and cell trajectories along them. Such explanations are still works-in-progress, gradually being revealed by experiments. Given the character of the explanandum-phenomenon, experiments that manipulate cell developmental potential and pathways are particularly significant in this regard. The next section examines one such kind of experiment: "direct reprogramming," which manipulates cell development using a few key molecules.

---

[11]For simplicity, I focus on explanations of a single cell's development. In practice, however, cell populations are of equal importance (see Fagan 2013, Chap. 3).

3   Reprogramming Experiments

The term 'reprogramming' suggests a pre-existing plan for cell development. If not, then what are experimenters *re*-programming? But in fact the term means less than it seems (Brandt 2010). In stem cell biology, reprogramming refers to methods of changing a cell's differentiation potential – specifically, of expanding this potential to include every kind of cell in the body (*pluripotency*). Four such methods exist: nuclear transplantation, cell fusion, overgrowing cell cultures, and direct reprogramming with transcription factors (TFs; see below).[12] Only the last is discussed here. For brevity, I will sometimes refer to this simply as 'cell reprogramming' or 'reprogramming.' The method is simple, though slow and inefficient. Briefly, specific TFs are added to nuclei of cultured cells. A few cells treated this way exhibit altered developmental capacities, and become capable to differentiating to produce any of the body's specialized cells.

The concept of a 'transcription factor' needs some introduction. TFs are proteins that specifically bind to particular DNA sequences (Fig. 17.3). When these sequences are located on chromosomes within a cell's nucleus, TF binding affects the expression (mRNA or protein production) of nearby genes. The regulatory sequences that bind TF are distributed non-randomly on chromosomes, such that the same sequence often appears near genes with related functions. TF proteins can, accordingly, coordinate large-scale patterns of gene expression (Carroll 2005). One TF protein can change the expression of dozens, even hundreds, of other genes – including its own and other TF genes. In gene expression, then, TFs are involved as genes and as proteins. The term "factor" echoing Mendel's original description of cellular determinants of phenotype, is telling. But where Mendel's factors were a speculative hypothesis, the term "transcription factor" can refer to any of the molecular forms implicated in the Central Dogma: DNA, messenger RNA (mRNA) or protein. So in this context, "factor" is an inclusive catch-all term for these different molecules. With these preliminaries established, let us now examine reprogramming experiments in more detail.

3.1   Basic Procedure

Reprogramming experiments manipulate cell development by changing cells' environment and molecular constitution.[13] The first step is to extract specialized (differentiated) cells from an organism (usually human or mouse) and grow them in artificial cell culture. Next, a few specific TFs are inserted into cell nuclei. After

---

[12]The first two are 'cloning' techniques of the sort that produced Dolly the sheep. They are usually conceived as effects of cytoplasm on the nucleus, while the latter two are conceived as effects of external factors on whole cells.

[13]Details in Takahashi and Yamanaka (2006), Takahashi et al. (2007)

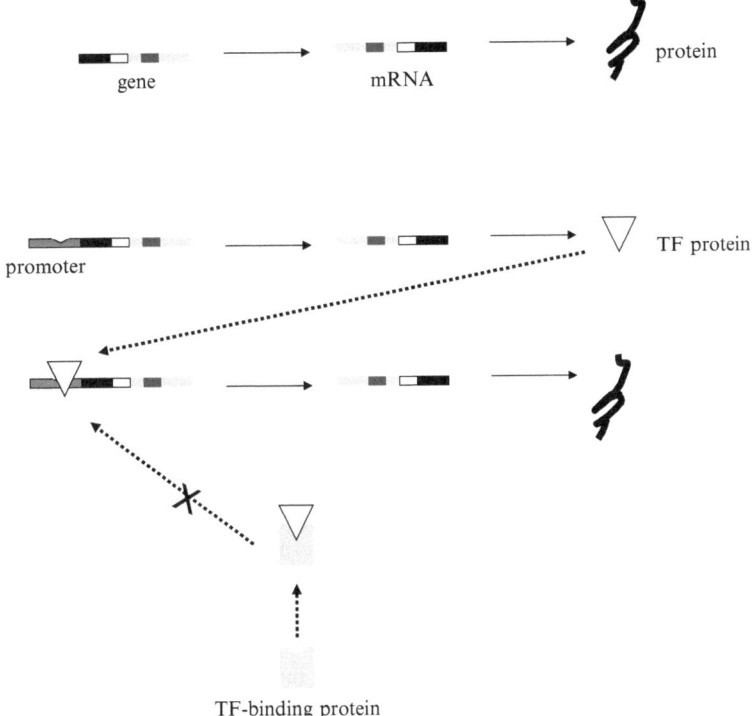

Fig. 17.3 Transcription factors and gene expression

several weeks, and many cycles of cell division, a few cells in these cultures (usually ≤0.05 %) transform in morphology, molecular traits, and developmental capacities to resemble embryonic stem cells (ESC). These developmental capacities, ability to self-renew and give rise to many different cell types, are the defining features of induced pluripotent stem cells (iPSC). Such cells can give rise to a cell lineage L with an unlimited number of generations and termini representing every cell type of a mature organism. iPSC thus satisfy the general definition of a stem cell.

3.2 Path to Discovery

The guiding hypothesis that led to the innovation of iPSC was that "factors that play important roles in the maintenance of ES cell identity also play pivotal roles in the induction of pluripotency of somatic cells [i.e., mature nongermline cells]" (Takahashi and Yamanaka 2006, 663). That is, features associated with stable ESC phenotypes in cell culture could also induce that phenotype in other cells. In 2000, Yamanaka's team in Kyoto made a list of 24 candidate factors. The list included proteins implicated in the developmental capacities of ESC and early embryos and genes known to be highly expressed as mRNA in ESC and tumors. DNA, RNA and proteins alike were all implicated in the original selection process. The 24 candidates were 'packaged' for delivery into cells using retroviruses: RNA viruses that infect hosts in DNA form, inserting their genome into that of the host. Engineered retroviruses delivered candidate factors to cells, singly and in combination. In the original experiments, then, genes (DNA sequences) were added to cultured cells.

These cultured cells were from mouse tail-skin. The first experiments used mice at early stages of development; later these experiments were repeated with skin cells from adult mice. But the original iPSC experiments did not systematically test all possible combinations of the 24 candidate factors. Instead, Yamanaka and colleagues used a more limited form of trial-and-error, ruling out unnecessary factors while seeking to maximize iPSC production. The experimental design was subtractive and step-wise. First, all 24 candidates were added at once to cultured cells. A few iPSC appeared, demonstrating that the entire slate of candidates together could induce pluripotency. Next, the Kyoto team added each factor to cells individually. This yielded no iPSC, showing that no single factor was sufficient to induce pluripotency (at the level of detection in these experiments). The next task was to whittle down the 24 candidates to a "core set." To do so, each candidate was individually subtracted from the sufficient set of 24. Collectively, these '23 factor' experiments identified 10 of the original 24 as individually necessary to produce iPSC. Yamanaka and colleagues then combined these 10 factors, and showed that all 10 together induced pluripotency – more efficiently than the original 24. This showed that factors' effects are non-additive: 24 together act differently than 10, with the difference reflected in the proportion of resultant iPSC.

The 'individual subtraction' strategy was then repeated with the set of ten, to identify the individually necessary factors for this combination. The results narrowed the list of candidates to four TF: Oct3/4, Sox2, Klf4 and c-Myc. These four TF genes induced pluripotency with the same efficiency (i.e., proportion of iPSC produced in a culture) as the ten 'semi-finalists.' No combination of two factors was sufficient, while some triples gave borderline results. These four TF were the first demonstrated "core set" of individually necessary and jointly sufficient factors for inducing pluripotency (or "stemness") in specialized cells. Once the four factors were identified, each was characterized at the molecular level, as DNA and protein (e.g., Yamanaka 2007, 43–45). Often, reprogrammers use the same designation for both, so which molecule is meant can be determined only in context, by the activities attributed to it.

3.3 Experimental Results

Since 2006, the original iPSC method has been reproduced and modified by hundreds of laboratories worldwide. This widely-dispersed experimental effort has

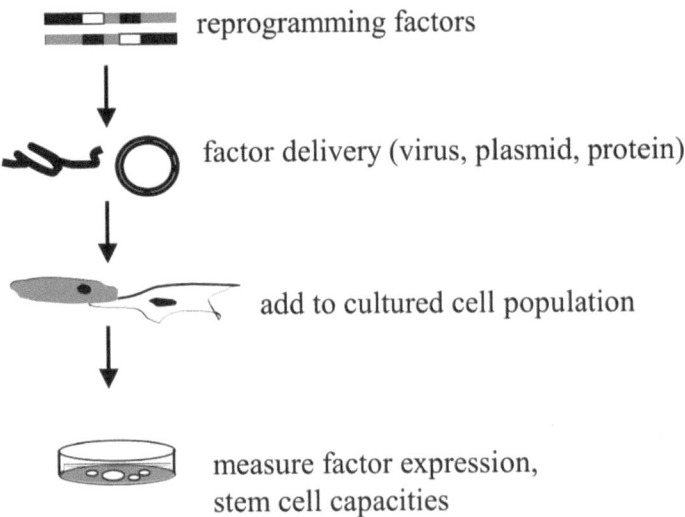

reprogramming factors

factor delivery (virus, plasmid, protein)

add to cultured cell population

measure factor expression,
stem cell capacities

Fig. 17.4 General scheme of cell reprogramming

produced many new iPSC lines, each with a detailed experimental history. It has also yielded some robust generalizations; for example, that reprogramming efficiency is inversely correlated with age of the source organism. That is, reprogramming embryonic or fetal cells tends to be more efficient than reprogramming cells from adults. But the spate of reprogramming experiments has *not* yielded a standardized procedure for producing iPSC. Many features of the original method have been revealed as unnecessary, including the core set of four TFs and use of DNA. Other jointly sufficient sets of TF can produce iPSC, notably the "core pluripotency network" consisting of Oct4, Sox2 and Nanog. Specific RNA and protein sequences, as well as DNA, can reprogram cells (reviewed in Maherali and Hochedlinger 2008). Different combinations of TF, 'delivery systems' to cell nuclei, organismal source, and culture conditions are more or less effective at producing iPSC, which themselves vary (often in subtle ways) in their molecular traits and developmental capacities.

The method of reprogramming today is not a single experimental procedure, but a sequence of decisions about the values of key variables: reprogramming factors, delivery method, starting cell population, and culture conditions (Fig. 17.4).[14] All four variables influence the outcome of a reprogramming experiment. Other key variables determine how the outcome is measured: criteria used to identify iPSC (characters C) and extent of differentiation potential of iPSC (character-values M). These are also the parameters that define iPSC as stem cells (Sect. 2). Any successful

---

[14]Reviewed in Maherali and Hochedlinger (2008), Hochedlinger and Plath (2009), Stadtfeld and Hochedlinger (2010), Okita and Yamanaka (2011), and Robinton and Daley (2012).

Fig. 17.5 Types of reprogramming, against the background of normal development (From Yamanaka 2009, Reprinted by permission from Macmillan Publishers Ltd: Nature, copyright (2009))

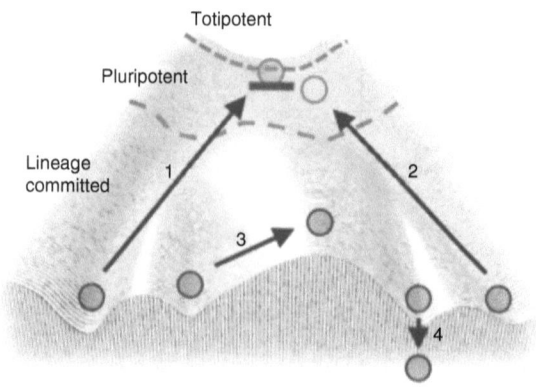

reprogramming experiment yields cells that qualify as stem cells in the context of that experiment, which can be expressed as a combination of the values of variables. For any particular experiment, values of variables are selected according to researchers' purposes, and mutually optimized insofar as current knowledge permits.

Results of reprogramming experiments are a starting point for explanations of cell development. Explanatory efforts proceed along at least three lines. First, outcomes of reprogramming experiments are grouped into major types, each of which bears a different relation to normal (unmanipulated) cell development: stable reprogramming, transient (partial) reprogramming, reprogramming without a pluripotent intermediate, or cell death. Waddington's landscape (see Sect. 2) is a useful device for representing these relations (Fig. 17.5).[15] In the first, normal development is reversed, and cells return to a stable pluripotent state. The second is the same, except that the pluripotent state is transient; normal development re-asserts itself. In the third, cells switch directly from one differentiated cell type to another, with no pluripotent intermediate. Finally, cell death is a frequent outcome of experimental manipulations, but also a 'programmed' response of cells in normal development. These general representations of reprogramming experiments abstract from key experimental details, such as organismal source of cells, ingredients of culture medium, specific genes or proteins added, methods of selecting reprogrammed cells. They set the stage for explanations of reprogramming, by relating these experiments to normal (unmanipulated) cell development.

Second, the details of reprogramming experiments and results characterize the different cell states in a lineage of interest L. In its most inclusive sense, "cell state" refers to values of six kinds of character: (1) developmental potential, (2) self-renewal capacity, (3) cell morphology, (4) regulation by "exogenous" cell signaling pathways, (5) gene and surface molecule expression, and (6) chemical properties

---

[15]The significance of Waddington's landscape for stem cell biology, and interfield relations with systems biology are discussed further in Fagan (2012a, b, 2013).

of DNA and chromosomes apart from sequence that affect gene expression (Hanna et al. 2010). These characters include both cellular and molecular properties, some intrinsic and others relational. Specifying values of these characters for a lineage L with a particular pattern of branching tracks is in keeping with the general explanatory aims of stem cell biology (see Sect. 2).

The basic structure of these explanations is revealed by a second, narrower use of "cell state," referring to a pattern of gene expression determined by a regulatory system of DNA, RNA, protein, and molecules serving as cofactors or signals (e.g., Hochedlinger and Plath 2009, 516). This narrower conception of cell state comprises only (4–6) of the inclusive conception above. Roughly speaking, these molecular aspects of cell state are explanans, cellular aspects (including stem cell capacities) the explananda. Focus on molecular characters is also evident in stem cell biologists' uptake of technologies that afford more precise and systematic measurement of specific RNA and protein molecules within and on the surface of cells; e.g., functional genomics and proteomics. The structure of 'incipient explanations' is further indicated by summaries of work in the field that explicitly correlate cell state and developmental potential on a developmental landscape (e.g., Hochedlinger and Plath 2009). All this suggests that stem cell biology aims at explanations of cell development that show how biochemical and genetic regulatory networks determine the stability (or lack thereof) of cell states; regulatory 'circuitry' accounts for the pattern of branching tracks that carves up the developmental landscape. Though we do not yet have such explanations, the distinction between molecular and cellular 'levels' of biological organization is a basic structural feature of models currently under construction.

Thirdly, reprogrammers are beginning to sketch molecular explanations of cell development. The relation to normal development, discussed above, offers a starting point for these explanatory sketches. One idea is that, the more differentiated a cell, the further 'uphill' it must travel to reach a pluripotent state, and the more likely it is that some unknown perturbing factor will block its path. This accounts for the lower efficiency of the iPSC method than that for producing ESC: more "force" is needed to "counteract gravity" (Hochedlinger and Plath 2009, 515–516). The difference between stable and unstable reprogramming is also readily explained in these terms: the former induces an "epigenetic bump" that keeps the cell from rolling back down the hill once the inducer is removed (Yamanaka 2009, 50). A molecular approach suggests the hypothesis that "epigenetic marks" accumulate on DNA during development (e.g., Zhou and Melton 2008, 386). The more these marks build up, the more difficult it is to achieve "reversal" via reprogramming. To test this hypothesis, stem cell researchers seek to identify specific biochemical modifications that influence normal and reprogrammed development.

Another question is whether the mechanism of reprogramming is "deterministic" or "stochastic" (e.g., Yamanaka 2009; Hanna et al. 2010). If the former, then pluripotency is induced in a distinct sub-population of cells within a culture, which are pre-disposed to respond to added TF. If this model is correct, the success rate of reprogramming is low because such "precursor" cells are rare in adult tissues, and the next task is to identify the character-values that distinguish them from other cells in a culture. In contrast, according to the stochastic model, the "epigenetic state" of any cell constantly fluctuates, as regulatory binding sites shift from 'open' to 'closed' positions and vice versa. Limits on these random shifts define the stable boundaries of cell types. If this model is correct, then cells induced to pluripotency are just those with the right regulatory binding sites 'open' at the right time. Any cell in the population could, in principle, be in such a state at a given time. Current evidence favors the stochastic model, and efforts to identify key regulatory binding sites and the molecules recruited to them are underway.

4 Philosophical Accounts of Explanation in Experimental Biology

The experimental successes of stem cell biology reveal much about the explanations sought in this field. I next consider three important philosophical accounts of explanation in experimental biology in relation to the stem cell case. All three presuppose Woodward's manipulability theory of causal relations (2003), so I begin by setting out the basics of this account.

4.1 Interventionism

According to Woodward's manipulability theory, X causes Y (where X and Y are variables that can take two or more values) if and only if there is a possible manipulation of some value of X under idealized experimental conditions, such that the value of Y changes. The meaning of causal claims is thus expressed by counterfactuals describing what *would* happen to the values of variables in an idealized experiment, or *intervention*. More precisely, an intervention I on variable X with respect to variable Y is a causal process that determines the value of X in such a way that, if the value of Y changes, then the change in Y occurs only in virtue of the change in X.[16] The concept of an intervention provides a regulative ideal for experiments aimed at discovering causal relations: to approximate the conditions of an idealized experimental intervention, using controls, standardization, shielding, etc. Insofar as they meet this standard, experiments reveal genuine patterns of counterfactual dependence among sets of entities and their properties. Woodward's theory thus offers an appealing analysis of causal relations revealed by experiment in many fields, including stem cell biology. The following sections presuppose this analysis. So, hereafter, the term 'cause' and its cognates should be understood to refer to the concept analyzed by Woodward's theory.

---

[16]Woodward's full analysis is more elaborate. But this simplified treatment will do for present purposes.

The interventionist account makes this theory the crux of explanation in general. According to this account, explanations of some phenomenon of interest Y show that the value of Y depends on the value(s) of some variable(s) X, thereby answering a range of questions about hypothetical values of X and Y. So the explanandum for an interventionist explanation is some phenomenon of interest Y, conceived as a variable that can take multiple values. But causal relations as such are only weakly explanatory. Because invariance is required only under some rather than all interventions, causal explanations need not include general laws. The range of interventions under which a dependency relation holds (i.e., the range of counterfactual circumstances for which answers can be given for questions about values of Y) may be broad or narrow. The broader this range, the greater the explanatory power (or "depth") of a causal relation. "Deeper" interventionist explanations reveal more of the variables that Y-values depend on (Hitchcock and Woodward 2003, 188). Such explanations describe a causal structure of multiple causal dependency relations, each invariant under some range of interventions.

The explanandum for stem cell biology, the branching pattern of cell development and processes conforming to this pattern, is not readily conceived as 'the Y' of an interventionist explanation. It is of course possible to take the general developmental pattern and associated processes as a single variable, and specific instances of this pattern as values of this variable. Y would then be 'the stem cell' and its developmental potential in the form of a lineage, and values of Y particular assignments of values to variables L, n, C and M (see Sect. 2). But this is not the kind of explanation sought in stem cell biology. The question of interest to reprogrammers and other stem cell researchers is not what makes a cell one kind of stem cell rather than another. Interventionism offers more insight into stem cell research if developmental *outcomes*, the stable termini of development to which a stem cell can give rise, are taken to be explananda. Reprogrammers do seek to understand causes of different developmental outcomes among cells.

On this explanatory approach, the phenomenon of interest Y is specialized cells resulting from processes of development. The different developmental termini, elements of M, are values of this variable. One can conceive reprogramming experiments as collectively revealing the causal factors on which developmental outcome depends. The resulting explanations would take the form of 'recipes' for producing a particular kind of specialized cell: neurons, heart muscle, blood cells, etc. Stem cell biologists do aim to articulate such methods; many laboratory protocols take exactly this form. But they are not the field's main explanatory goal. As the reprogramming case illustrates, explanatory efforts are concentrated on the relation between molecular and cellular character-values, and seek to reveal the molecular traits (character-values) that determine or change a cell's developmental potential. Interventionism does not readily capture this explanatory goal. All the causal variables that make a difference to developmental outcome are on the 'same level,' so to speak. In this way, interventionism subtly mischaracterizes explanation in stem cell biology.

Interventionism also offers insight if cell state (in the broad, inclusive sense) is the phenomenon of interest, with different values of this variable corresponding to combinations of character-values for features (1–6) above. Reprogrammers do aim to identify factors on which cell state depends, and given this explanandum, interventionism does offer a satisfying account. But reprogrammers also seek to understand how these factors interact, which molecules bind one another under what environmental conditions, and what results from these local interactions. By shifting explananda, interventionism can account for these explanatory efforts as well. But this does not clarify stem cell explanations as a whole, given the field's central explanandum-phenomenon. Interventionism can only illuminate fragments of stem cell explanations, depending on how the explanandum is conceived.

The problem is subtle, because the manipulability theory, on which interventionism is based, provides a satisfying account of individual causal relations among molecular components that are revealed by experiment. So an interventionist interpretation of any set of these relations is always available. Proponents of interventionism can accordingly claim to account for explanations based on cell reprogramming, as well as other experiments. But such a defense of interventionism's generality comes at the cost of understanding explanation in stem cell biology. To make progress toward the latter, we need to take seriously the general explanandum of the field, and current explanatory efforts exemplified by studies of reprogramming. This requires concepts beyond those of interventionism, which target the relation between cellular and molecular levels of organization.

4.2   Genes in Development

A second influential view of biological explanations gives genes (defined as DNA sequences encoding protein) a privileged explanatory role. To defend this idea for development, Waters (2007) extends Woodward's manipulability theory in two ways. First, he notes that "biologists are much more interested in the actual than the possible" (575). That is, explanations in biology aim to identify actual causal relations, not counterfactual connections between values of variables. Second, he argues that *specific* causal-effect relations are more important for biological explanation than a broad range of invariance under intervention. The paradigmatic example of such specificity is "DNA . . . as the cause of . . . linear sequences of nucleotides in RNA and amino acids in polypeptides" (553). Woodward (2010) further clarifies this concept of specificity in interventionist terms. On this sophisticated gene-centric account, biological explanations identify "actual, specific difference-makers" for variable characters of interest. If this is correct, then explanatory efforts in biology should target a subset of causal factors, notably DNA sequences that encode mRNA and protein sequences.[17]

---

[17]Waters does allow that some molecules other than DNA qualify as actual specific difference-makers, notably RNA splicing agents and micro-RNAs in eukaryotes (2007, pers. comm.). So his view is not *strictly* gene-centric. Nonetheless, explanatory privilege for these other molecules is

Waters defines actual specific difference-makers as follows. X is *the* actual difference-maker with respect to trait Y in population P if and only if (i) X causes Y in Woodward's sense, (ii) this relation is invariant with respect to other variables that actually vary in P, (iii) the value of Y actually varies among members of P, and (iv) actual variation in X fully accounts for the variation in Y in P.[18] X is *an* actual difference-maker for Y in P just in case X satisfies all of the above except (iv). Merely potential difference-makers, in contrast, satisfy (i) but not (ii) or (iii). For cell development, the effect to be explained is variation in the value of character Y in some population of cells P that is undergoing development. If differences in DNA sequence correlate with different values of Y in P, then DNA is an actual difference-making cause for Y.

DNA sequence is evidently not the only actual difference-maker for cell development. This motivates Waters' interest in *specific* difference-makers (273). The basic idea is that different specific changes in DNA sequence produce different specific changes in molecular products (RNA and protein), while other cellular components, such as RNA polymerase I, behave more like "on/off switches." Woodward (2010) defines this sense of specificity in terms of a mapping from states of a cause X to states of an effect Y, such that the more this mapping resembles a bijective function, the more specific the causal relation between X and Y (305).[19] I will refer to this as the "proportional influence" conception of specificity (PI-specificity). Relative to sets of alternative values of an effect variable and each of its causal variables, more PI-specific causes afford finer-grained control over the value of the effect. That is, if we can vary the state of the cause, then we can determine which state of the effect is realized, out of a range of alternatives. Non-specific causes are such that many different states of X map to the same state of Y, or the same state of X maps to many states of Y, or both.

Both the actual/potential distinction and PI-specificity provide a basis for discriminating among the various causes of an effect, which privileges the explanatory role of DNA. At first glance, cell reprogramming experiments seem to support this gene-centric view. After all, these experiments manipulate cell development by adding specific DNA sequences to nuclei of cultured cells. Adding a core set of TFs (such as Oct3/4, Sox2, Klf4, and c-Myc) to a population of cultured cells changes the value of many cell characters, including developmental potential, self-renewal capacity, morphology, gene expression, surface molecule expression, and biochemical properties of chromosomes. With appropriate controls, cell reprogramming approximates an intervention. So reprogramming experiments appear to support the

---

premised on their being causes of the same sort as DNA, and in this sense, Waters' view is gene-centric. Thanks to the editors for pushing me to clarify this point.

[18]Paraphrase of Waters (2007, 566–567). The final requirement is further explicated by several conditions pertaining to specificity, which are set aside here for simplicity.

[19]Woodward distinguishes two specificity concepts. As they are closely related, however, I discuss only the more rigorous here.

view that 'exogenous' DNA sequences are actual difference-making causes of cell development. However, there are several problems with this gene-centric account of reprogramming.

First, only a small percentage of DNA-treated cells are stably reprogrammed. This inefficiency spurs reprogrammers to characterize molecular entities other than DNA sequences: cell surface molecules, biochemical moieties on and proteins bound to chromosomes, etc. The widely-accepted stochastic model of reprogramming (see Sect. 3) entails that, though these additional, yet-to-be-characterized components do actually vary in cultured cell populations, their variations are transient, rapidly-fluctuating, and unsystematic. This rules out actual specific causal relations that are invariant under all but the slenderest range of interventions. Second, in the rare cells that are reprogrammed, DNA is not *the* actual difference-maker, but only one of many. Variants on the original method show that RNA or protein sequences can also reprogram cells (see Sect. 3). If reprogramming is conceived as a causal process that makes a difference to a population of cultured cells, then actual difference-makers include reprogramming TF, delivery method, and culture conditions. Different combinations of these variables yield different effects. In practice, they are treated as 'packages,' not distinct parameters. PI-specificity for reprogramming is a property of packages of these variables, not DNA sequence alone. Finally, even the original iPSC experiments did not target genes as actual difference-makers. Rather, Yamanaka and colleagues targeted sequences in multiple molecular forms: DNA, RNA and protein. Throughout the experiment, proteins and mRNA expression were attended to as difference-making factors, on par with DNA sequences.

The reprogramming case does support Waters' claim that actual difference-makers are privileged in biological explanations, but not the gene-centrism his account encourages. So this account does not advance us much beyond the manipulability theory. However, specificity in some sense is clearly important in explanations based on reprogramming experiments. These experiments show that TFs play a distinctive role as 'switches' for manipulating cell development. But this role is *not* PI-specific. The relation between TF sequence (whether conceived as DNA, RNA or protein) and cell state in the broad sense is not one-to-one, but many-to-many. Each TF has myriad 'downstream' effects, including some on other TFs. It is not uncommon for a single TF to produce conflicting outcomes, even in a constant environment. Moreover, no single TF suffices to reprogram cells. They operate in sets of two, three, or four, depending on other experimental variables. It is a "core network" of TFs that acts as a developmental switch for cultured cells. All this suggests that another specificity concept is more important for explanations based on cell reprogramming experiments than PI-specificity. TF are distinguished from other molecular components of cells, not because they *are* DNA sequences, but because (as proteins) they *bind to* DNA sequences. This specific DNA-protein binding makes a difference to gene expression (i.e., mRNA sequences produced), in turn influencing protein sequence and distribution within a cell.

Close examination of reprogramming experiments (Sect. 3) does not support a gene-centric view of explanations of cell development, nor a privileged role for

DNA as an actual PI-specific cause. Instead, emerging explanations based on reprogramming experiments suggest a crucial role for binding relations between different molecules. Rather than targeting PI-specific causal relations, cell reprogrammers focus on specific binding interactions among components of a molecular network. In such a network, DNA molecules are interactive partners with other molecules, not masters dominating minions.

4.3 Causal Mechanisms

A third account, supported by case studies from cell biology, molecular biology and neuroscience, begins with the idea that explanations describe causal mechanisms.[20] Mechanisms are defined as follows:

> A mechanism M consistsof multiple diverse parts ($x_1 \ldots x_n$) engaging in causal relations or activities ($\phi_1 \ldots \phi_n$) such that ϕ-ing x's are spatially and temporally organized so as to constitute M Ψ-ing (Fig. 17.6).[21]

Mechanistic explanations describe mechanisms: complex causal systems of multiple parts that perform some overall function. Causal relations among components are organized in a causal structure that is spatio-temporally localized and set within a wider context. The manipulability theory accounts for these causal relations and structures in a way that clearly reflects the role of experiment in mechanism-discovery (Woodward 2002; Craver 2007). It seems to follow that mechanistic explanations are elaborate causal explanations. But this does not account for their multilevel, constitutive aspect.[22]

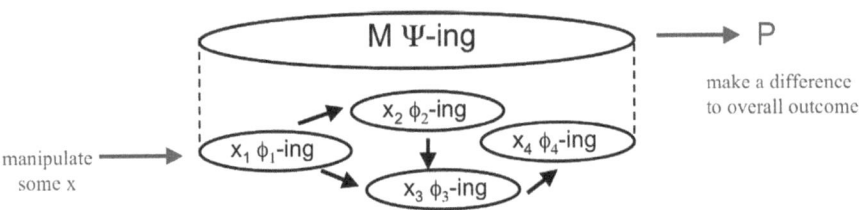

Fig. 17.6 Hierarchical structure of mechanisms (After Craver 2007)

---

[20]Influential case studies include: Bechtel (2006), Darden (2006), and Craver (2007).

[21]This terminology follows Machamer et al. (2000). There are several other important accounts of mechanistic explanation in biology, which overlap in many but not in all respects; these include Glennan (1996, 52) and Bechtel and Abrahamsen (2005, 423). Note that this definition is actually narrower than that of Machamer and colleagues, restricted to constitutive or multilevel mechanisms (see below).

[22]Experimental biologists often seek *underlying* mechanisms, to explain an overall process in terms of its working parts. This is part of the explanatory aim shared by many stem cell biologists.

Contrast between mechanistic and causal explanations appears in the questions they aim to answer. Causal explanations answer 'what-if-things-had-been-different?' questions about the values of variables within a particular range of invariance, by representing dependency relations among these values. Mechanistic explanations answer 'how-does-it-work?' questions about some complex causal system, by representing organized interactions among components. Of course, descriptions of causal relations are included in mechanistic explanations. But they are not the whole story. Craver's "causal-mechanical" account (2007) attempts to characterize the structure and norms of mechanistic explanation in causal terms. Though Craver's claims are limited to neuroscience, they generalize to other biological fields as well. Craver explicates the multilevel aspect of mechanistic explanation in terms of "constitutive relevance," a concept clarified in turn by "mutual manipulability" (2007, 152–160). The latter, as the name suggests, is an extension of Woodward's manipulability theory. Briefly, a working component (x ϕ-ing) is relevant to M Ψ-ing, if the latter can be manipulated by an intervention on x's ϕ-ing, *and* x ϕ-ing can be manipulated by an intervention on M Ψ-ing (ibid, 159–160). A component is irrelevant to a mechanism, if neither can be manipulated by an intervention on the other.

Mutual manipulability provides sufficient conditions for causal relevance across mechanistic levels: an overall system and its components. In this way, Craver's account explicates the multilevel structure of mechanistic explanation in causal terms. The mechanistic framework fits well with the structure of stem cell explanations: M corresponds to a cell lineage of interest, Ψ-ing to the process of cell development in the form of branching tracks from a common stem, x's to molecular components, and ϕ-ing to their causal activities. Stem cell biologists often state their explanatory goals in terms of underlying molecular mechanisms (e.g., Hanna et al. 2009, 596). In this field, explananda are complex cellular processes (development in a cell lineage of interest), to be explained by biochemical and genetic regulatory networks (which comprise cell state in the narrow sense; see Sect. 3). But the concepts of mutual manipulability and causal relevance do not account for the 'bottom-up' direction of these emerging explanations. Craver's conditions cut both ways; relevance relations are bottom-up *and* top-down. The causal-mechanical view leaves the direction of mechanistic explanation, in this case from molecules to cells, unaccounted for.

Another concept that engages the multilevel aspect of mechanisms is *modularity* (Woodward 2002). A component x of a mechanism M is modular just in case it can in principle be independently experimentally manipulated. That is, there is a conceivable intervention on x that changes x's ϕ-ing independently of other components of M.[23] If this condition is satisfied, then the generalization describing

---

[23]This formulation is not Woodward's own, though it captures what I take to be the main ideas of conception of modularity as it pertains to mechanistic explanation. Modularity in Woodward's own theory (2003) has a somewhat different significance.

a modular component's productive activity (x φ-ing) can change independently of generalizations describing the activities of other components. A modular explanation of M Ψ-ing would take the form of a list of modular causal generalizations, each describing one component's role in the overall mechanism. This modularity condition is extremely plausible. If we can conceptually individuate a working component of a mechanism (x φ-ing), then we can imagine an experiment that 'surgically' intervenes on x to change its φ-ing, without altering other components, and thus isolates the causal role of x within the mechanism of interest. If this condition is not met, there seems no reason to distinguish x as a component at all. So a modular explanation of cell development, conforming to the basic structure of mechanistic explanation, is a plausible explanatory approach. Explanations of cellular phenomena in systems biology often do take this form, with generalizations expressed as differential equations or Boolean rules. However, stem cell and systems biology are only beginning to connect with one another (see Fagan 2012a, 2013).

In stem cell biology, the prevailing style of explanation is not modular in the sense defined above. This is evidenced by the way core sets of TF were identified as switches for cell state and developmental potential (Sect. 3). From the first, these experiments targeted molecular factors that were characterized individually but work together. Experiments that did analyze the causal role of single candidate factors had negative significance; if reprogramming did not occur, then that factor was deemed essential for reprogramming. These results did not yield positive causal generalizations that could be concatenated into a modular explanation. Instead, reprogrammers focused on effects brought about by multiple interacting factors.

Subsequent reprogramming research maintains this focus. Since 2006, iPSC experiments have repeatedly shown that "drastic alterations of cell fate could be achieved with a combination of factors when no single factor would suffice" (Cohen and Melton 2011, 248). TFs influence cell state through "coordinated switching," which is brought about by a binding interaction between a TF protein and a regulatory DNA sequence. Emerging explanations of cell development based on reprogramming experiments describe binding interactions between DNA and protein, chromosomes and biochemical moieties, varieties of RNA, and more. Concepts of causal dependence, PI-specificity, mutual manipulability and modularity do not account for the focus on binding interactions, multilevel structure, and bottom-up direction of these molecular explanations. The next section sketches a view of mechanistic explanation in biology that accounts for all these features. This account takes interdependence rather than causal dependence as central. This is not to deny that causal relations are important for mechanistic explanation. Woodward's manipulability theory offers a satisfying account of causal relations described in mechanistic explanations – taken piecemeal. But to understand these explanations as a whole, and on their own terms, something more is needed.

5   The Joint Account[24]

As discussed above, descriptions of reprogramming mechanisms are rife with causal interactions involving two or more components working together; that is, *jointly*.[25] For example, TF and DNA molecules bind together via 'lock-and-key fit,' forming a complex held together by weak chemical bonds. Binding depends on these molecules' complementary shapes and biochemical properties, as well as certain spatio-temporal conditions. This lock-and-key fit, or "mesh," is the basis of jointness, which is distinct from causal dependence. The latter is a relation between a causal variable X and an effect variable Y, each of which can (at least in principle) take diverse values. Jointness, in contrast, holds between causal partners that together produce a particular effect. The relevant counterfactuals compare effects of multiple causal factors separately and together, not effects of different values of X on the value of Y. More precisely:

(J) Components x_1 and x_2 jointly ϕ if and only if (i) x_1 has properties that mesh with those of x_2, and vice versa, (ii) x_1 and x_2 form a complex x_1x_2 in virtue of their meshing properties, (iii) complex x_1x_2 ϕ's, and (iv) neither x_1 nor x_2 ϕ's individually.[26]

As noted above (Sect. 4), the manipulability theory can account for causal relations within a mechanism. Jointly-produced effects can be represented as a value (or range of values) taken by an effect-variable Y, and a meshing complex as a whole represented by cause-variable X. But such representations simply efface jointness, a concept that accounts for key aspects of mechanistic explanation. To see this, consider three features that follow from (J). First, jointness depends on properties of individual components, x_1 and x_2. The activity of a complex is characterized from the bottom-up. Second, meshing involves just those components; no exogenous initiating activity is required. The distinctive causal role of a complex x_1x_2 depends not on some incoming antecedent cause, but on complementarity of components and their ensuing interactions. In this sense, joint ϕ-ing 'arises from within,' out of interacting components. Third, joint ϕ-ing presupposes a permissive environment, a context in which the complex x_1x_2 can actually form. Most obviously, at least in biological mechanisms, components must be spatio-temporally near one another. But in any particular case, many other background conditions must also be satisfied. These three features are characteristic of complexes within biological mechanisms, as currently understood. Explanations of biological mechanisms show that (i–iv) are satisfied, by describing individual components and their activities, the properties

---

[24]This section summarizes the positive account from Fagan (2012c).

[25]This term is adapted from social action theory, where it describes actions of two or more agents (e.g., walking together, building a cathedral) and associated intentions.

[26]For simplicity, only the two-component case is presented. This makes condition (ii) redundant though it is necessary for complexes with three or more components.

that allow components to mesh and form complexes, spatio-temporal arrangements that determine what complexes form, causal activities of complexes, and (often) dissociation of complexes.

Jointness, as defined by (J), not only accounts for the content of mechanistic explanations, but their basic structure as well. An overall mechanism (M Ψ) is a complex of interacting parts (ϕ-ing x's). The concept of jointness provides a positive account of mechanistic organization, from the bottom-up. A mechanism's working, its overall behavior, can be defined as the joint activity of its interacting components. So the consensus view of mechanisms (Sect. 4.3) can be reformulated as a jointness condition:

(JM) Components $x_1, \ldots x_n$ jointly Ψ as mechanism M, if and only if (i) $x_1, \ldots$ x_n form causally-active complexes (ϕ_1-ing, $\ldots \phi_m$-ing) in virtue of meshing properties, (ii) these entities and activities are organized so as to constitute M Ψ-ing, and (iii) $x_1, \ldots x_n$ do not Ψ individually.

(JM) explicates the key features of mechanisms: multilevel structure, interactive organization, and an overall context fixed by boundaries of M. This condition also articulates a basic norm for mechanistic explanation: show that a mechanism of interest satisfies (JM) by describing how its components interact to jointly constitute the overall working mechanism and by describing the features in virtue of which they mesh, and so make up a complex whole, which plays a distinct causal role. Although description of causal relations is important, the crux of such an explanation is the constitution relation: constitutive *inter*dependence, rather than causal dependence. Because jointness depends on meshing properties of diverse components, mechanistic descriptions that conform to (JM) proceed from the bottom-up. This is exactly the kind of explanation that is emerging from experiments on cell reprogramming. Reprogramming mechanisms are (partly and provisionally) explained by description of molecular components that selectively associate into complexes, so as to jointly bring about a change in cell state.

6 Conclusion

I began this essay by explicating the general definition of 'stem cell,' revealing the significance of the stem cell concept for studies of development more generally. The concept is explicated as a model with four parameters, which define a structure: branching paths descending from a single 'stem' through a hierarchy of stages to multiple discrete end-points. The concept of a stem cell is thus intimately connected to basic assumptions about cell development. Making these assumptions explicit clarifies the central explanandum-phenomenon of stem cell biology: the branching pattern of cell development, from a single initiating 'stem' to multiple termini, the diverse cells that compose a mature multicellular organism. Cell reprogramming experiments manipulate this process to reveal the mechanisms underlying cell morphology, function, and developmental potential. They thus offer an illuminating case of explanations-in-progress, central for stem cell biology as well as being of general scientific interest.

The reprogramming case also highlights gaps in philosophical accounts of biological explanation. I examine three accounts here, all of which are relevant for explanations of cell development based on experiment. All presuppose Woodward's manipulability theory of causal relations. The interventionist account shares in the advantages of Woodward's theory, but fails to engage the general explanandum-phenomenon of stem cell biology. The gene-centric account, though offering insight into the corresponding explanans, does not capture its molecular diversity, nor the sense of specificity involved in transcription factor binding, the crux of the developmental switch in reprogramming. The mechanistic account best reflects the structure of stem cell explanations. However, a thoroughly causal view of this kind of explanation fails to account for key features, notably their bottom-up direction. These criticisms motivate a new account of explanation for stem cell biology: "the joint account," which preserves the insights of its predecessors but remedies their omissions.

With its newness, technical intricacy and lack of formal theory, stem cell biology presents a challenge for philosophers of science. But it is a challenge worth taking up. Stem cell research, perhaps more than any other field today, raises fundamental questions about the process of development, relations between cells and organisms, and the nature of selfhood and individuality. Understanding the structure, norms and experimental grounds of explanations in this field is vital to make sense of these questions, and to begin formulating answers. This essay, though necessarily brief, makes a start on this project.

Acknowledgements Many thanks to Pierre-Alain Braillard and Christophe Malaterre for the opportunity to contribute to this volume as well as helpful comments on the manuscript. Earlier versions of material in this paper were presented at workshops on Epistemology of Modeling and Simulation (Pittsburgh, PA, April 2011), Interdisciplinarity and Systems Biology (Aarhus University, Denmark, August 2011), and Individuals Across the Sciences (Sorbonne, Paris, May 2012), and at meetings of the Society for the Philosophy of Science in Practice (Exeter, UK, June 2011) and the American Philosophical Association, Pacific Division (Seattle, WA, April 2012). This essay has benefited from questions and comments of participants at all these events. Particular thanks are due to Hanne Andersen, Richard Grandy, Oleg Igoshin, Kirstin Matthews, Sean Morrison, Maureen O'Malley, Joe Ulatowski, Irv Weissman, and participants in the Fall 2010 graduate seminar in Philosophy of Science (Rice University). Support for this research was provided by the Humanities Research Center at Rice University's Collaborative Research Fellowship (2009–2010), and Faculty Innovation Fund (2010–2012).

References

Bechtel, W. (2006). *Discovering cell mechanisms*. Cambridge: Cambridge University Press.

Bechtel, W., & Abrahamson, A. (2005). Explanation: A mechanist alternative. *Studies in History and Philosophy of Biological and Biomedical Sciences, 36*, 421–441.

Brandt, C. (2010). The metaphor of 'nuclear reprogramming'. In A. Barahona, E. Suarez-Díaz, & H.-J. Rheinberger (Eds.), *The hereditary hourglass: Genetics and epigenetics, 1868–2000* (pp. 85–95). Berlin: Max Planck Institute for History of Science.

Carroll, S. (2005). *Endless forms most beautiful*. New York: W.W. Norton.

Cohen, D., & Melton, D. (2011). Turning straw into gold. *Nature Reviews Genetics, 12*, 243–252.

Craver, C. (2007). *Explaining the brain*. Oxford: Oxford University Press.

Darden, L. (2006). *Reasoning in biological discoveries*. Cambridge: Cambridge University Press.

Fagan, M. B. (2012a). Waddington redux: Models and explanation in stem cell and systems biology. *Biology and Philosophy, 27*, 179–213.

Fagan, M. B. (2012b). Materia mathematica: Models in stem cell biology. *Journal for Experimental and Theoretical Artificial Intelligence, 24*, 315–327.

Fagan, M. (2012c). The joint account of mechanistic explanation. *Philosophy of Science, 79*, 448–472.

Fagan, M. (2013). *Philosophy of stem cell biology*. London: Palgrave Macmillan.

Glennan, S. (1996). Mechanisms and the nature of causation. *Erkenntnis, 44*, 49–71.

Gottweis, H., Salter, B., & Waldby, C. (2009). *The global politics of human embryonic stem cell science*. London: Palgrave Macmillan.

Gross, F. (2015). The relevance of irrelevance: Explanation in systems biology. In P.-A. Braillard & C. Malaterre (Eds.), *Explanation in biology. An enquiry into the diversity of explanatory patterns in the life sciences* (pp. 175–198). Dordrecht: Springer.

Hanna, J., Saha, K., Pando, B., van Zon, J., Lengner, C., Creyghton, M., van Oudenaarden, A., & Jaenisch, R. (2009). Direct cell reprogramming is a stochastic process amenable to acceleration. *Nature, 462*, 595–601.

Hanna, J., Saha, K., & Jaenisch, R. (2010). Pluripotency and cellular reprogramming: Facts, hypotheses, unresolved issues. *Cell, 143*, 508–525.

Hitchcock, C., & Woodward, J. (2003). Explanatory generalizations, Part II. *Noûs, 37*, 181–199.

Hochedlinger, K., & Plath, K. (2009). Epigenetic reprogramming and induced pluripotency. *Development, 136*, 509–523.

Kraft, A. (2009). Manhattan transfer: Lethal radiation, bone marrow transplantation, and the birth of stem cell biology, 1942–1961. *Historical Studies in the Natural Sciences, 39*, 171–218.

Laplane, L. (2011). Stem cells and the temporal boundaries of development: Toward a species-dependent view. *Biological Theory, 6*, 48–58.

Lau, D., Ogbogu, U., Taylor, B., Stafinski, T., Menon, D., & Caulfield, T. (2008). Stem cell clinics online: The direct-to-consumer portrayal of stem cell medicine. *Cell Stem Cell, 3*, 591–594.

Leychkis, Y., Munzer, S., & Richardson, J. (2009). What is stemness? *Studies in History and Philosophy of Biological and Biomedical Sciences, 40*, 312–320.

Machamer, P., Darden, L., & Craver, C. (2000). Thinking about mechanisms. *Philosophy of Science, 67*, 1–25.

Maherali, N., & Hochedlinger, K. (2008). Guidelines and techniques for the generation of induced pluripotent stem cells. *Cell Stem Cell, 3*, 595–605.

Maienschein, J., Sunderland, M., Ankeny, R., & Robert, J. (2008). The ethos and ethics of translational research. *The American Journal of Bioethics, 8*, 43–51.

Melton, D., & Cowan, C. (2009). Stemness: Definitions, criteria, and standards. In R. Lanza et al. (Eds.), *Essentials of stem biology* (2nd ed., pp. xxii–xxix). San Diego: Academic.

Nobel Foundation (2012) The 2012 Nobel prize in physiology or medicine – press release. Nobelprize.org. 10 Oct 2012, http://www.nobelprize.org/nobel_prizes/medicine/laureates/2012/press.html

Okita, K., & Yamanaka, S. (2011). Induced pluripotent stem cells: Opportunities and challenges. *Philosophical Transactions of the Royal Society (London) Series B, 366*, 2198–2207.

Pradeu, T. (ed.) (2011). *Biological theory*. Special issue on 'Limits of development.' Springer.

Ramalho-Santos, M., & Willenbring, H. (2007). On the origin of the term 'stem cell'. *Cell Stem Cell, 1*, 35–38.

Robinton, D., & Daley, G. (2012). The promise of induced pluripotent stem cells in research and therapy. *Nature, 481*, 295–305.

Stadtfeld, M., & Hochedlinger, K. (2010). Induced pluripotency: History, mechanisms, and applications. *Genes & Development, 24*, 2239–2263.

Takahashi, K., & Yamanaka, S. (2006). Induction of pluripotent stem cells from mouse embryonic and adult fibroblast cultures by defined factors. *Cell, 126*, 663–676.

Takahashi, K., Tanabe, K., Ohnuki, M., Narita, M., Ichisaka, T., Tomoda, K., & Yamanaka, S. (2007). Induction of pluripotent stem cells from adult human fibroblasts by defined factors. *Cell, 131*, 861–872.

Waddington, C. H. (1957). *The strategy of the genes*. London: Taylor & Francis.

Waters, C. K. (2007). Causes that make a difference. *Journal of Philosophy, 104*, 551–579.

Woodward, J. (2002). What is a mechanism? A counterfactual account. *Philosophy of Science, 69*, S366–S377.

Woodward, J. (2003). *Making things happen*. Oxford: Oxford University Press.

Woodward, J. (2010). Causation in biology. *Biology and Philosophy, 25*, 287–318.

Yamanaka, S. (2007). Strategies and new developments in the generation of patient-specific pluripotent stem cells. *Cell Stem Cell, 1*, 39–49.

Yamanaka, S. (2009). Elite and stochastic models for induced pluripotent stem cell generation. *Nature, 460*, 49–52.

Zhou, Q., & Melton, D. (2008). Extreme makeover. *Cell Stem Cell, 3*, 382–388.

Chapter 18
Explaining Causal Selection with Explanatory Causal Economy: Biology and Beyond

Laura R. Franklin-Hall

Abstract Among the factors necessary for the occurrence of some event, which of these are selectively highlighted in its explanation and labeled as causes—and which are explanatorily omitted, or relegated to the status of background conditions? Following J. S. Mill, most have thought that only a pragmatic answer to this question was possible. In this paper I suggest we understand this 'causal selection problem' in causal-explanatory terms, and propose that explanatory trade-offs between abstraction and stability can provide a principled solution to it. After sketching that solution, it is applied to a few biological examples.

Keywords Causal selection • Causal explanation • Explanatory trade-offs • Robustness • Abstraction • Causal democracy

1 Introduction: Explanatory Sparseness and Systematicity

Our universe is dizzyingly complex, and everything that happens within it causally depends on innumerable other things. The living world in particular can appear almost horrifically complicated. Though some of this complexity remains beyond our grasp, scientists have unraveled ever-larger portions of it. The combination of this complex world and our increasingly sophisticated theories accounting for it should make two features of our causal-explanatory practice appear surprising: its *sparseness* and its *systematicity*.

Explanatory practice is *sparse* in that many apparently legitimate causal explanations are rather thin affairs, in which happenings are accounted for with only the tiniest sliver of information, and not by citing all, or even very many, of an event's

L.R. Franklin-Hall (✉)
Department of Philosophy, New York University, 5 Washington Place, New York, NY 10003, USA
e-mail: lrf217@nyu.edu

© Springer Science+Business Media Dordrecht 2015   413
P.-A. Braillard, C. Malaterre, *Explanation in Biology*, History, Philosophy and Theory of the Life Sciences 11, DOI 10.1007/978-94-017-9822-8_18

causal influences.[1] Explanatory practice is *systematic* in that those few morsels that sparse explanations feed to us do not seem to emerge higgledy-piggledy, as if they were the output of some 'explanatory lottery' in which all causal factors enjoyed equal odds. Instead, the features scientists judge explanatorily relevant follow regular contours, perhaps indicating that hidden principles govern their selection. More specifically, I will distinguish two dimensions of explanatory sparseness and systematicity: horizontal and vertical.[2]

Along the 'horizontal' dimension, we do not normally explain the occurrence of an event by citing all of the conditions *necessary* for its occurrence, that is, by describing what J. S. Mill called its 'total cause.' Instead, one or a few features are given special priority, with other factors relegated—for good or ill—to the status of 'background' or 'enabling' conditions. For instance, the life-threatening sickling of an individual's red blood cells during metabolic stress may be explained by citing a particular gene sequence—that coding for the hemoglobin protein—and not the equally necessary features of intra- and extra-cellular environments. More complicated, though still excruciatingly simple, explanations of biological development appeal to very spare gene regulatory architectures—like the double-repression network—in accounting for cell differentiation. These accounts also elide a profusion of essential cellular machinery. And moving from scientific to folk explanations, Hart and Honoré (1959: 10) provide a simple, and more famous, example of what is arguably the same phenomenon: it would be customary to explain the occurrence of a fire by citing the dropping of a lighted cigarette, but not by mentioning the environmental oxygen, even though its presence may have been equally necessary for the conflagration.

Along a 'vertical' dimension, those few features that do make the explanatory cut are often themselves quite 'high-level,' abstracting from sundry micro-details. For instance, a county's high fox population may be explained by the springtime boom in bunnies—a favorite prey item of the medium-sized canid—with the account remaining silent on the particular hoppings-about, matings, and eatings of bunnies across the rural landscape (activities still very consequential for the foxes' flourishing). The character of gene regulatory explanations in cell and developmental biology is identical; there, the production of all-important transcription factors is characterized in terms of coarse-grained concentrations, not by describing spatio-temporal details of molecular location. And, as above, Hart and Honoré's (1959) fire example can provide a folk window on the same phenomenon: even when circumstances conspire to make the presence of oxygen explanatorily

---

[1]The distinction between explanations as *communicative acts* and explanations construed in an 'ontic' mode as *sets of facts* will not loom large in this paper; throughout, I will presume that the content of communicative acts are the explanatorily relevant facts.

[2]Though many explanations are strikingly sparse, others—among them some so-called 'mechanistic' accounts—are less so. Though both sorts will be dealt with in this paper, as both are recommended by the explanatory theory that I will articulate, I begin by emphasizing explanatory sparseness because it is comparatively puzzling and in need of philosophical elucidation.

relevant to a fire's ignition—as when the inferno erupted in a manufacturing plant from which oxygen was normally evacuated—the explanation will invariably cite a rather high-level feature, such as the oxygen's non-trivial quantity, while omitting finer points, such as which particular oxygen molecules directly contributed to the blaze (even were such information miraculously available).

What accounts for these two varieties of selection—the ('horizontal') omission of background conditions and the ('vertical') omission of low-level physical detail? There is no consensus answer to this question in the philosophical literature on causation or causal explanation, it being widely believed that they represent independent dimensions and require different sorts of treatments.

On the one hand, vertical selection—that of the proper 'level of explanation'— is usually, though not invariably, understood as an objective matter. Thus, it is thought that an explanation might err in describing causes at the *incorrect* level, considering the explanatory target. For instance, defective explanations might be too low-level, including putatively irrelevant "gory details" (Kitcher 1984: 370); or, they might be too high-level, omitting objectively relevant details, as when a black-boxing explanation fails completely to "reveal underlying mechanisms" (Kaplan and Bechtel 2011: 442). Though there is agreement neither on just what the right level is, nor in virtue of what it would be correct, the hunch that there is a real phenomenon here, one amenable to a systematic treatment, is widespread.

Horizontal selection, on the other hand, is more often considered to be a purely pragmatic matter, and the omission of 'background factors' accounted for on quasi-Gricean principles. Mill, for instance, emphasized the "capricious" nature of causal selection, suggesting that factors are omitted "because some of them will in most cases be understood without being expressed" (1882: book 3, chapter 5) rather than on "any scientific ground."[3] Along the same lines, both Lewis (1986) and Hall (2004) put horizontal causal selection down to "invidious distinctions" between causes of equal ontological, and presumably also explanatory, mettle. Summarizing the general philosophical mood, Schaffer (2007) states that "selection is now generally dismissed as groundless."

2 Aim: To Explain Horizontal Sparseness and Selectivity

In the face of such skepticism, the central aim of this paper is to offer a causal-explanatory analysis of the *horizontal* dimension of selection (often called the 'causal selection problem,' a label I use interchangeably). My proposal will, I hope, cut a pleasing path between two unsightly extremes: first, the implausibly strong metaphysical claim—and one with appropriately few advocates—that background factors and those cited as causes in explanations are of entirely distinct

---

[3]I am somewhat simplifying Mill's discussion, in which he floats a number of more concrete proposals concerning causal selection, though in each case emphasizing their haphazard application.

ontological genera; second, the unsatisfyingly shallow proposal—one belied by a close look at our practice—that there is no interesting, objective structure to the cause/background conditions distinction, and that context-dependent pragmatics always reigns.

While the horizontal dimension of selection will be my special focus, to deal with it I must sketch an account of causal explanation—one general in aspiration, though designed with the biological sciences centrally in mind—from which my treatment of horizontal selection derives. According to this *Causal Economy Account*— described in more detail elsewhere (Franklin-Hall forthcoming)—complete explanations describe packages of causal factors that 'cost less' and 'deliver more,' making them, in metaphorical terms, maximally *economical*. Very briefly, 'cost' is equivalent to *total content of the explanation*—understood in terms of the number of ways the world might be that it excludes—while 'delivery' is equivalent to *stability that the explanans bestows on the explanatory target*—a factor tracking the ability of the factors cited in the explanans to make the explanandum event robust.

As will be detailed below, not only does this explanatory account offer a solution to the causal selection problem, but it also promises, in contrast with other proposals, to dispose of both sorts of causal selection—horizontal and vertical— at once, with background conditions and micro-details falling short in precisely the same way. Furthermore, the standard determinative of explanatory worth on the Causal Economy account—the *bang-for-your-buck principle*—processes these dimensions in concert, such that whether a factor is relegated to the status of background condition partly depends on its explanatory level. In consequence, my handling of horizontal selection cannot be completely insulated from the problem of vertical selection; I maintain that the common practice of separating them has been a mistake, as consideration of *background* and *level* must be kept in view simultaneously for either one to be seen clearly.

Beyond its solution to the causal selection problem, the Causal Economy account can address an issue more central to this volume: the well-advertised fact that scientists in different fields—ecology, evolutionary biology, and molecular biology, for instance—sometimes offer different explanations of the same event. While it has become customary to maintain that such differences result from the existence of substantially different explanatory norms governing the construction of explanations in different fields, the Causal Economy account can offer a more unified account of such differences, maintaining that a *single principle* guides explanation construction across the sciences, with context simply highlighting one or another equally acceptable accounts.[4] It is able to do this by embracing a variety of explanatory pluralism according to which there sometimes exist multiple, distinct yet correct explanations of particular individual events. As will be explained below, this pluralism is organic to the explanatory account itself, in particular a consequence of the fact that different candidate explanations of a single event can

---

[4]See Press (2015) (Chap. 16, this volume) for alternative views on explanatory unification.

be equally 'economical'—some costing and delivering much, others costing and delivering little, still others costing and delivering the same amount, but doing so in different ways.

The discussion unfolds as follows: in the next section I offer a brief overview of the causal selection problem and consider two responses to it. Following that, I sketch the Causal Economy explanatory account. Finally, I highlight its solution to the selection problem, while applying it to a few folk and biological examples, including to a dispute concerning the explanatory status of genetic and environmental causes of ontogeny.

3 Candidate Solutions to the Horizontal (or 'Causal') Selection Problem

The causal selection problem springs from two observations influentially highlighted by Mill (1882: book 3, chapter 5):

1. For any event (understood broadly to include states of affairs), there is an enormous set of factors—both positive and negative—necessary for its occurrence. We can think of these factors as those *on which the event depended,* or (perhaps equivalently) *which determined the event.* Call these an event's *determinants.*

2. For any event, a comparatively small set of factors—both positive and negative—is given special causal or explanatory priority, being labeled *causes* of the event (in the extreme, 'the cause'), and (perhaps equivalently) which are selectively cited in its explanation. Call these an event's *narrow causes.* Determinants that aren't narrow causes are *background factors.*

In its standard formulation, the problem of causal selection is the challenge of explaining the gap between the *determinants* of any event—which will all, technically speaking, be causes of it, on most any account of causation you may prefer—and its *narrow causes.* To do this in a substantive way requires offering a *selection principle* capable of telling the difference between narrow causes and background factors, as well as a rationale for the principle, in case the practice is thought justified. This principle might involve almost anything under the sun, possibly in complex combination: norms of communication, scientific convention, a causal metaphysics, or an account of explanation.[5]

---

[5]Should this principle be considered as part of the *semantics* or the *pragmatics* of causal-explanatory claims? That is, in cases in which a particular factor that is usually back-grounded (e.g., oxygen with the fire) is (apparently illegitimately) claimed to be *a* or *the cause* of the event, does this involve saying something *false* (the semantic view) or saying something strictly true, but falling short in some other respect (inappropriate, uninformative, irrelevant, etc.)? I prefer a semantic approach, but don't think the choice here makes any difference to the substance of my analysis; you should feel free to reconstruct the discussion on your preferred picture.

To set the stage for my own explanation-based proposal, I offer a whistle-stop tour of alternative takes on the causal selection problem, covering populational and frame-working accounts. These stops are chosen for their proximity to our final destination; like my account, they address the causal selection problem at a kind of 'intermediate level' of philosophical depth, neither via a deep causal metaphysics nor a shallow conversational pragmatics.

3.1   The Populational Account

On the *populational* approach, the narrow causes of an event—or phenomenon more generally—are distinguished from background conditions by their special status *relative to a population*. This view can be spelled out in a number of ways, depending on whether the population is actual, or merely hypothetical. Here I focus on a version of Waters' (2007) strategy, which appeals to an actual population, ignoring the many nuances irrelevant to my aims.

Waters' picture has two principal parts. The first is a counterfactual difference-making account of causation borrowed from Woodward (2003). This serves to define what I call the determinants. Broadly speaking, these are the factors that, had they been different, the phenomenon would have been different (or would have had a different probability).[6] Since the clash between accounts of causal selection does not concern the precise specification of this set—pre-emption puzzles aside, causal accounts agree enough on its composition—I will not explore this further.

The second part is a principle that picks the narrow causes—which Waters suggests we understand as *actual difference-makers*—from among the determinants. It is here that the populational element plays its part. Actual difference-makers are only definable relative to a population of entities, which might be "different actual entities" or "the same entity at different actual times" (2007: 566). They are the features that actually made a difference to the target event, across members of the population. Naturally, this population must be a mixed one with respect to the phenomenon in question, with some members of the population taking one state of it, and the other members some other. Relative to this population, some determinants (as they pertain to each member of the population) will be uniform, and others not. Actual difference-makers—if they exist, which isn't invariably the case—are those determinants whose actual variation in the population at least partly accounts for the actual difference in the target phenomenon in members of that population.

An example will illustrate. Consider a particular fruit fly with red eyes. What is the narrow cause of this redness? Prima facie, there are many factors—both intrinsic and extrinsic to the fly—that might be cited, since there are many factors

---

[6]I am glossing over numerous features of Woodward's account: how these differences are affected, that is, via interventions; what the causal relata themselves are, that is, variables, etc. Though important in other contexts, rehearsing these features would be only a confusing distraction here.

such that, had they been different, the eyes would not have been red. Yet Waters claims that biologists will often focus on a particular factor or factors in accounting for redness; it is the gap between the many determinants and the particular factors actually highlighted that makes this a problem of causal selection. This winnowing happens, on Waters' view, by conceptually embedding the red-eyed fly in a broader population, and considering what made a difference to eye color between *members of that population*.[7]

For instance, let us embed the red-eyed fly, which has genotype (+, pr), in a population of flies, some also with red eyes and that same genotype, and some lacking red eyes (possessing purple eyes, instead), with an alternative genotype (pr, pr). Virtually all other features of these flies are identical: they vary *just with respect to* whether they have a pr allele, or a wild-type (i.e., +) allele, at a particular locus. This is also among the many determinants of eye color. In this case, Waters says that the actual difference-maker for red eyes (in my language, its *narrow cause*) is the allele type (+ or pr) at that locus; all other determinants are mere background factors.

What if all the facts about flies and eyes had been just the same, but the red-eyed fly had been conceptualized as part of another actual population, one in which all flies shared the same (+, pr) genotype, but were cultivated at different temperatures? Importantly, this involves changing nothing about the causal order with respect to flies—it only changes *how the scientists think about the situation*. (Water says that he means 'population' in a statistical sense, so for the population to change, nothing about how flies interact, for instance, need be modified.) Assume that cultivation temperature is also a determinant of eye color. In this case, the narrow cause of red eyes—its actual difference maker—will change: now cultivation temperature, rather than allele-type. Thus, it is *our selection of a population*, on Waters' scheme, which ultimately explains causal selection.

3.2 The Frameworking Account

On a *frameworking* account, narrow causes differ from background factors relative to, not a population, but an 'explanatory framework' ('framework' henceforth). I will explore a version of this strategy from Strevens (2008: §§5.3 and 6.1). As above, this account has two parts. First comes a characterization of the determinants, the

---

[7]This embedding can be understood in one of two ways. On what appears to be Waters' preferred formulation and which I do not follow, when properly understood the real explanandum in this case is not the red eyes in a particular fly, but instead a difference in eye color in a particular population. Put in just this way, Waters would not actually be addressing the causal selection problem, since that is the problem of accounting for why particular determinants are cited in explanations of events, not in explaining differences in types of events across populations. (Were his solution to *that problem* it would be completely un-controversial.) So that Waters can be addressing the causal selection problem itself, I am articulating a version of his strategy that maintains the same explanandum, but makes the causal-explanatory claim relative to a population.

total set of an event's causes. These are difference-makers, though Strevens defines them via an optimization procedure rather than counterfactually. Subtleties aside, this procedure takes a fine-grained vertical model of the causal influences on an event and makes it as abstract as possible, while still requiring it be usable to derive, following the causal order, a statement of the occurrence of the explanatory target. Anything left in this model, post-optimization, is a difference-making cause. This model may look a sleek machine beside the more fine-grained one from which it was produced—but bulky it will still be, generously endowed with determinants.

The second element in this account of causal selection is the *framework*. This instrument, which Strevens modifies from Mackie's (1974: chapter 2) notion of a *causal field*, has many applications beyond causal selection, but its use there will be my exclusive focus. The framework specifies a state of affairs—one that must be veridical—that is assumed or held fixed in an explanation, and can be thought of as a distinct part of the explanatory request itself. When an explanation involves a framework, the explanation is not for some event—full stop—but rather for that event *given* whatever the framework specifies. The presence of the framework interacts with the optimization procedure described above, such that it only evaluates models in which this state of affairs holds. Furthermore, the state of affairs is not considered a difference-maker for (or explainer of) the target phenomenon, nor not one; it is simply a fact that the explanation takes for granted and evaluates difference-making relative to.

To illustrate, consider a frame-working approach to the fly example. On this view, if the particular allele is to be a narrow cause of red eyes, and temperature not, this will be because temperature, and not allele type, has been placed in the explanatory framework. In such a case, the explanatory target is not the *red eyes*, but, *red eyes, given that the fly was raised at such-and-such a temperature*. By the same token the reverse conclusion might be reached, that it is the *temperature* that is responsible for eye redness. For this result, simply place the fly's genotype in the explanatory framework (as well as, possibly, other difference-makers), but do not do this with respect to temperature. This picture generalizes to any example of causal selection (including those to which a populational account would not apply). So, while for Waters it was the explainer's selection of *an actual population* that did the work of causal selection, for Strevens it is the explainer's selection of a *framework*.

4   Evaluating Candidate Solutions to the Causal Selection Problem

Populational and frame-working accounts of causal selection are different on the surface, but share an essential commonality: *relativization*. On either view, an extra factor has been added to the explanatory request over and above the explanandum event. It is the selection of this factor that distinguishes narrow causes from background factors.

In relativizing in this way, the views just sketched are hardly unique. For instance, Schaffer (2012) offers a structurally similar account of causal selection (and causal semantics more generally) in which the relativizing utensils are *cause-and-effect-event-contrasts*. Similarly, Hesslow (1988) suggests that an *object of comparison* may be the proper relativizing device. Whatever the particular incarnation, interest in relativization is unsurprising in light of the fact that some explanations are explicitly relativized. It is in play, for instance, when sociologists ask outright what made a difference to educational achievement *in an actual population*, such as between students in a particular Los Angeles middle-school (Rumberger and Larson 1998). It is also in evidence when biologists ask why some individuals experience brain or liver damage, *given* (i.e., frameworking) that they were exposed to environmental lead (Onalaja and Claudio 2000). Accounting for relativization is thus a necessary part of any complete chronicle of our causal-explanatory practice, including in the account I will eventually recommend.

Nevertheless, if offered as the exclusive explanation for horizontal causal selection, relativizing strategies should not inspire you, for two reasons. First, there is the uncomfortable fact that many explanatory requests don't (apparently) make *any* reference to populations or frameworks at all, but despite this, selection seems to go off without a hitch. To illustrate with the simplest example: in the explanation of the breaking of a window, the throwing of the brick, not the absence of a wall between the brick and the glass (equally a determinant), will usually be prioritized. This selection is completely straightforward, even though in the normal circumstance neither population nor framework (nor any other relativizing implement) will be in view. This arguably indicates that such paraphernalia are not, at least ·invariably, responsible for selection.

Second, even in cases where the relativizer's apparatus is front-and-center, relativization-based solutions to the problem of causal selection provide no explanation for what is arguably the central issue: why certain kinds of causal factors are treated, by either the scientists or the folk, as narrow causes, and others as background factors. After all, *pace* Lewis and others who parade the capriciousness of selection, a good deal of actual selective practice is rather systematic. For instance, as noted in the introduction, transcriptional machinery is usually backgrounded in developmental explanations, and gas concentrations in explanations for forest fires. An illuminating account of causal selection will account for these patterns (as well as making sense of their exceptions). But relativizers don't do this. Instead, they out-source the explanatory task by suggesting that just what is selected is a consequence of which actual populations, frameworks, etc., are included in explanatory requests. And on the selection of *these* they remain silent.

Let me anticipate two responses. To the first point, the relativizer may counter that—while the relativizing apparatus isn't always noted out-right—in any given case a relativizing tool was *implicitly* included in the explanatory request, reaching out to do the selective work from just below the surface. To the second point, the relativizer may simply reject the explanatory demand. After all, on her view selection follows from *what explanatory requests are made*. And for the *origins of our questions*, perhaps no informative account—at least, none that it is the philosopher of science's job to provide—is possible.

These replies are difficult to decisively parry, but in a way revealing of their deeper deficiency: they combine to make the relativizer's theory, if not completely immune from potential counter-example, certainly very close to it. Let me explain. First, at the core of her account the relativizer offers a variable—the chosen relativizing instrument, of whatever nature—whose value, in any given case, establishes which determinants are narrow causes, and which not. Next, the relativizer states that the value of this variable will, often if not usually, be left implicit. Finally, she declines to provide a theory that might give us access to its setting, claiming that beyond her bailiwick.

In total, this makes the relativizer's solutions exceedingly prone to ad hoc maneuvers and just as suspect as a scientific theory of the same character. In any given case the relativizer can claim that the implicit variable's setting is—rather conveniently—fitting to the actual selective behaviors we observe. That such an account is incorrect, of course, doesn't follow. But cognizance of this characteristic can certainly motivate the search for a more systematic alternative, in particular one that is (1) more explicit, and (2) able to account for patterns in our causal-selective practice.

I hope my positive proposal can provide satisfaction on both of these counts. As noted already, it derives from an account of causal explanation whose attractions extend beyond a solution to the causal selection problem itself; for instance, as noted in the introduction, it is able to make sense of vertical selection as well as the pluralism of our causal-explanatory practice. In the next section I outline the heart of that account. Then the spotlight returns to the causal selection problem proper, eventually focusing on selective instances in biology in particular.

5   The Causal Economy Account

The philosopher of explanation—at least one without a revisionist bent—will take our actual explanatory practice as her datum and will devise a theory that both explains and rationalizes the principal features of that practice. Such an account will usually have two connected parts: (1) an articulation of just what gives an explanation its explanatory force (e.g., unifying disparate phenomena, answering what-if-things-had-been-different questions),[8] (2) a description of explanatory *form*, that is, of what a complete explanation consists in (e.g., a derivation of a statement of the occurrence of an event, a veridical causal model of a certain kind, etc.).

Over the last few decades, philosophical opinion has coalesced around a broadly causal explanatory picture, one according to which it is an event's *causes*, either some or all, that explain its occurrence. Yet, even with this constraint the explanatory

---

[8] See Issad and Malaterre (2015, Chap. 12, this volume) for a discussion of explanatory force.

tent must still be a large one, as causal enthusiasts harbor disagreements respecting both motivation and form. After all, while some see explanations as concerned with difference-making (Lewis 1986; Woodward 2003), with mechanistic models (Bechtel 2008; Glennan 2002; Machamer et al. 2000), or with a combination of the two (Craver 2007; Strevens 2008), others suggest that an ideal explanation will trace most or all of an event's causal influences—whether difference-making or no (Railton 1981; Salmon 1984). My own view, the *Causal Economy* account (Franklin-Hall forthcoming), is also a member of this causal-explanatory crowd. It has two principal features.

First, in contrast with many other theories, it is not tied to any particular causal metaphysics, but might be combined with alternative accounts of the causal relation, e.g., counterfactual (Lewisian or interventionist), regularity, conserved quantity, etc. This is possible because a metaphysics of the causal relation is not "where the action is" on the Causal Economy view, that is, the place from whence its proprietary explanatory constraints originate. Better then to leave metaphysical edification to the metaphysicians, and offer an explanatory account welcoming to all prospective partners.[9]

Second, Causal Economy offers a selection principle used to characterize the components of a complete explanation. According to this principle, good explanations—those deemed 'complete' and capable of providing understanding—are special for their *economy*: they 'cost less,' in virtue of being abstract, and 'deliver more,' in virtue of citing causal influences that make the event to be explained stable or robust. The motivation for this biggest bang-for-your-buck standard—and for the explanatory account more generally—is to isolate an event's *most important* causal factors, those in virtue of which it was, to a rather large degree, 'bound to happen.'

Before applying it to the causal selection problem, I elaborate on this account's two key aspects—the ecumenical causal metaphysics and the sectarian selection principle. To aid in doing so, bring to mind any candidate explanandum event. It may be a 'high-level' event, like the simple death of an animal, possessing coarse-grained identity conditions, or, less commonly, it may be perfectly concrete—a death individuated in terms of all of its intrinsic properties, down to the horrible particulars of a final gasping breath. Whatever the target event, many of features of our universe, both positive and negative, will be responsible for its happening *just as it did*. In fact, given the structure of at least two of the fundamental physical forces (gravitation and electromagnetism), which diminish rapidly with distance but never go to zero, virtually any other event in the backwards light cone of an explanatory target will be so implicated.

All such minimally responsible events—as well as the laws in virtue of which they are related—are, in my language, *causal influences*.[10] Granting the existence of this influential fabric, just what is its warp and woof? It is here that my metaphysical

---

[9]Here I am following Strevens (2008).

[10]This is a term of art, and should not be assimilated to Lewis' views in "Causation as Influence" (2000).

ecumenism rears its amiable head. I make no commitment respecting the real nature of causal influence, for instance, on whether it should be understood in terms of the *transfer of conserved quantities* via a relation of *in principle manipulability*, or something else entirely. I insist only on this: causal influence is fully physical.[11]

But why demand even this? Here are two reasons. The first reason involves our need to compare different candidate explanations—at different levels of abstraction, for instance—on the basis of their economy. To do this, they must be in some sense commensurate, something that is only straightforward if all explanations are constructed from the same physical starting materials. The second reason is tied to my commitment to physicalism itself. However significant high-level *explanatory* relations may be—and I strongly believe in their import and thus design my explanatory account to make room for them—there is ever more reason to think that there is at least *some sense* in which it is physics, if any science, that describes our universe's basic movers-and-shakers. So, though no *explanatory* fundamentalist, I will grant physics a precedence of some kind, one located in the character of causal influence itself.

It is the second element of the Causal Economy account, the selection principle, that stops this physicalist approach to causal influence from effecting an explanatory fundamentalism. Rather than an event's explanation offering up a total, fine-grained causal-influential chronicle—as would be provided by one of Railton's (1981) 'ideal explanatory texts'—a complete explanation should cite just a special part of that saga. The precise size and shape of this part—just which features it includes, and which it does not—will depend on the architecture of the run-up to the event in question. Yet, at least in many cases, the part judged explanatorily relevant will be *sparse* in both vertical and horizontal dimensions, omitting small influences, low-level details, and (what are often considered) background factors.

How does the selection principle do this? Space will only permit a sketch of its treatment of the simplest kind of explanation, what I will call a *direct explanation*. In a direct explanation, an event is explained by reference to other events or states of affairs, in concert with a causal law connecting these with the target event,[12] but without mention of intermediate events. This focus is apt because the causal selection problem is always posed in terms of direct explanations. Furthermore,

---

[11]Most transference accounts of the causal relation—such as Dowe's process theory—are already physically constrained, so my requirement is, on them, without effect. It has the most impact on counterfactual accounts, such as on a causal interventionism of the Woodward (2003) variety. If causal influence is cashed out in interventionist terms, I insist that the causal model to which interventionist causal claims are relativized be a fine-grained physical model, not a 'high-level' one (even though those would otherwise be kosher) (Here I follow Strevens (2008)).

[12]For lack of space, I am equally unable to say much about the content of these causal laws, and will focus on characterizing the *states of affairs* are to be explanatorily cited. But in brief, the content of such laws is determined—not by some independent 'high-level' account of causation, such as that provided by Woodward (2003)—but by the explanatory selection procedure itself. The causal law connecting state of affairs A and target event B asserts that A is a winner of the causal economy competition with respect to B.

though many of the most interesting scientific explanations are not of this kind, I see them as assembled from direct-explanation building blocks. So, in accounting for the nature of those blocks, I will still have the opportunity to describe a key component of the Causal Economy picture.

There are two conceptual steps in explanation assembly for direct explanations: *production* and *selection*. *Production* is a process in which an exceedingly large number of candidate explanations are manufactured from the complete causal-influential tale for a target event. That tale will be tellable in physical locution, given the nature of causal influence itself. Candidate explanations are produced by censoring the complete account via omission and abstraction. In the case of omission, particular causal influences are completely deleted. In the case of abstraction, causal influences are described in less detail, through *coarse-graining*, *amalgamation*, or *populational* transformation. In coarse-graining, a particular feature is described as falling in some range or exceeding some threshold (*over 30 miles per hour* replacing *35 miles per hour*). In amalgamation, multiple lower-level features applicable to a particular individual are combined in a more complex parameter, and the particular values of the components are thereby lost (*15 kgm/s momentum* replacing *5 kg mass at 3 m/s velocity*). In populational transformation, a population-level feature is cited rather than a set of parameters applicable to the individuals constituting the population (*temperature* of a gas, e.g., mean kinetic energy, replacing its constituent molecule's particular *kinetic energies* (themselves products of amalgamation, since kinetic energy $\approx \frac{1}{2}$ mass * velocity^2)).[13]

When all of these kinds of transformation are applied in different orders and degrees to the complete story for a particular event, there will result exceedingly many candidate explanations, each one a separate 'package of causal influence.' Some will include a good deal of the total chronicle, others small slices of it. Needless to say, most packages will in no way resemble the actual explanations scientists offer up. But within the rubbish nestle explanatory gems, special packages of causal influence that do appear explanatorily suitable.

These gems are identified in the selection step. As noted already, relative to a target event, an explanatory package maximizes the ratio of delivery to cost (or equivalently, maximizes the product of delivery and cheapness). The *cost* of a package is its *total content*, which I understand to be the number of ways that the world might be that it rules out. Other things equal, good explanations say very little about the catastrophically complicated run-up to a target event. The *delivery* of a package reflects the *stability boost* that it provides the explanatory target; this tracks the extent to which the package makes it the case that the target event would still have happened, even had circumstances been in various ways different.

---

[13]Both amalgamation and populational transformation are species of what is sometimes called *variable reduction*. Variable reduction isn't always considered a kind of abstraction, but I class it thus; it equally involves moving to a representation that leaves information out in contrast to the original.

I will amplify on both the cost and delivery aspects of the selection principle, in order. For *the number of ways that the world might have been that the explanation rules out*—my take on *cost*—to have any significance, it must be tied to scheme of world individuation. To maximize expressive potential, I assume a scheme that is fine-grained and physical.[14] Whether this scheme is itself objective—that is, whether it is rationally required for any explainer, rather than being in some way ultimately *dependent on us*—is something on which I remain agnostic. Whatever the metaphysical status of the individuation scheme at its heart, what pressure does cost put on complete explanations? In short: less is better. Other things equal, an explanation should be very abstract, trimming away as much of the causal-influential run-up as possible. This might involve deleting what we normally consider to be small causal influences—factors that aren't even among the *determinants* of an event. It might also involve removing (what we usually consider) *background factors* or *low-level details*, those extracted via omission or any of the kinds of abstraction noted above, thereby bringing about horizontal and vertical sparseness. Indeed, if we were considering cost alone, we would even omit the factors that *do* appear explanatorily relevant.

While cost considerations favor paring causal influences away as much as possible, delivery favors including factors in an explanatory account to the degree to which those factors make the target event stable or robust, providing what I will call a *stability boost*. Intuitively, a big booster of an event's stability is a factor that makes it the case that the event would have happened *even had many other things been different*. For instance, the impact of the Chicxulub asteroid provides a stability boost for the dinosaurs' demise, since, just so long as the impact itself occurred, many features—grazing patterns, immunological status—might have been different, yet the animals would still have perished.

To be somewhat more precise, the stability boost offered by a candidate explanans reflects the *additional stability* that the factors cited in the explanans contribute to the event, over and above the stability of the event simpliciter (the *baseline stability*). To measure this additional stability, we subtract the baseline stability from what I will call the *construct stability*. Thus,

$$\text{construct stability} - \text{baseline stability} = \text{stability boost}$$

Though it might be understood in a number of ways, I measure stability using a possible-worlds framework, which I can only sketch briefly here. On this approach, the baseline stability of an event is equivalent to the number of a privileged set of nearby possible worlds—worlds differing from our own via one or a number of simple, physical perturbations (effected via Lewisian small miracles)—

---

[14]Though not important to my task here, the cost measure should also be relativized to the size of the causal-influential fabric for the event at the time of the candidate influence (e.g., the span of the event's backwards light cone at the relevant time). After all, it is only as between properties of that material—one that gets rapidly larger earlier in time—that the selection principle must pick.

in which the target event nevertheless takes place.[15] More specifically, in each world experiencing a perturbation, let events unfold according to the actual laws until the time of the target event. Events are more stable simpliciter to the degree to which they occur across a greater range of these possible worlds. For instance, World War I might turn out surprisingly stable simpliciter: perhaps lots of things could have been different—among them the assassination that actually triggered the cataclysm—and yet a European war in that period still would have eventuated.

The *boost* provided by a candidate explanans is determined by how much stability, over and above the baseline stability, is due to factors cited in the explanans. To measure this, we must construct a new set of worlds, taking as our starting material the set of nearby possible worlds appealed to above. Consider those worlds at a time just after the perturbations that distinguish them from our own, and let them unfold, according to the laws, until the time at which the causal influences cited in the explanans should appear. Now in some of these worlds, the causal influences cited in the candidate explanans will probably be absent due to the particular perturbations found in those worlds. For every such world, let a second miracle occur, making it the case that the causal influences cited by the candidate explanans are in fact present.[16] (This second miracle will permit us to probe the relationship between the factors cited in the explanans, and the target event, in worlds somewhat different from our own.) Then, let all worlds continue to unfold according to the physical laws. The 'construct stability' is gauged by the number of worlds in which the target event (the explanandum) takes place. Generally, this value reflects both the degree to which target events are stable simpliciter, and the degree to which events become stable once we've fixed the influences cited in the explanans.

Having characterized baseline stability and construct stability, the stability boost is just the difference between the two. Broadly speaking, this factor reflects the stability that the explanans contributes to the target event.[17] One important consequence of the measure is this: packages of causal influence that offer substantial stability boosts—greater than that provided by competing packages—are those that are themselves unstable simpliciter, but such that, given their occurrence, the event to

---

[15]This set of worlds is one that I describe exclusively to make sense of our explanatory practice. I take no stand here on whether it is in any way an objective set, one that is special from a metaphysical point-of-view. See my (forthcoming) for more on the privileged set of nearby possibly worlds, produced by a basic set of perturbations, that is used to define the stability boost.

[16]Crucially, this will not usually just reverse the initial perturbation that had the consequence of disrupting the influences cited in the explanans. Many consequences of that original perturbation will persist, despite the re-enactment, since the disruption of the explanans influences will usually be but *one of many* down-stream effects of the initial perturbation.

[17]Because the baseline stability of any target event is a constant, subtracting it will not change the ordinal ranking of candidate explanans. The subtraction is nevertheless useful by allowing us to conceptually distinguish target events with high baseline stability from those only stable with the aid of factors cited in the explanans.

be explained is very stable.[18] Such influences are distinctive in two ways. First, they will be (what we would usually consider) difference-makers for the occurrence of the target event, setting them apart from the abundance of factors that make a difference to *how* it occurred, but not to *whether* it occurred.[19] Second, and more discriminatingly, they will, in most cases, constitute just a subset of the difference-makers—a feature that forms the heart of my solution to the causal selection problem. Excluded will be those factors that cannot contribute to an event's stability because they were themselves so stable. These factors cannot offer much of a stability boost because, in virtue of being so stable, they will remain present in many of the worlds in the privileged set (i.e., those that experience the original basic physical perturbations). And since they are so pervasively present, there is no role for a second miracle to institute them and in that way impact the occurrence of the target event. As I see it, such highly stable factors are mere channels through which the work of the real stabilizers has been transferred from elsewhere. Naturally, it is the originators of stability—not their envoys—that must be explanatorily relevant on any picture on which explanations tell *in virtue of what* an event was 'bound to happen.'[20]

Putting the pieces together: when the selection principle evaluates an event's candidate packages of causal influence—the output of the step producing the candidate explanans—it will extract those that jointly maximize abstractness and stability-boosting. There may, of course, be multiple packages that do this, some that are more concrete or even mechanistic in flavor, but which also make it the case that the target event is virtually bound to happen, and others that are more abstract while still offering a substantial stability boost. I submit that any of these optimal

---

[18]Predictably, it will be possible to contrive counterexamples to this gloss on my procedure, cases in which a baroque causal architecture thwarts my measure's ability to isolate factors that do have this property. When considering such cases, it is important to keep the spirit of the Causal Economy proposal in mind: its selection principle aims to informatively describe the broad principles influencing our explanatory practices, practices that I believe to be best defined for the causal systems that scientists normally encounter and on which their norms have been trained. The most important task for an account is to deal with those systems; after all, it is challenging enough to provide an informative description—that is, one that doesn't simply appeal to pervasive relativization—of principles guiding horizontal and vertical selection in those central cases.

[19]Fans of counterfactual difference-making accounts of causation may wonder about the nature of the base or contrast state relative to which difference-making is being implicitly evaluated. My algorithm, in effect, does not pick one 'default,' but instead surveys a large range of states, checks for difference-making relative to each of these states, and integrates over those results. In particular, big stability boosters are difference-makers relative to many or all of these alternative states, not just those present in the collection of worlds produced from our own via simple, physical perturbations at some prior time. This strikes me as better solution to the 'default problem' than privileging one such state, perhaps the one deemed (on subjective grounds) 'normal.'

[20]Though my procedure does, I hope, give special preference to the unstable difference-makers, those that seem particularly explanatorily relevant, I do not deny that even stable difference-makers might, in some contexts, have an explanatory role. In particular, if we expand the set of nearby possible worlds appealed to in the construction procedure—by allowing for more and more radical perturbations—even seemingly stable difference-makers will sometimes be absent in the privileged set, thus making them at least potentially explanatorily relevant.

packages, in concert with the causal laws connecting them with target events, can constitute *complete explanations*, and are scientifically acceptable accounts that yield real understanding.[21] But do such packages indeed resemble those direct explanations that we actually offer up? Yes, or so I will suggest in the next section in the process of addressing our central topic: an explanation-based solution to the problem of causal selection.

6 The Causal Economy Treatment of the Causal Selection Problem

Defending a substantive account of causal selection requires establishing the extensional adequacy, as well as rational defensibility, of its selection principle. Most importantly, it must be shown to pick *narrow causes* from *determinants* in a way at least broadly consistent with how those cuts are actually made. Three data streams might point us to the location of such cuts: (1) arm-chair philosophical intuition; (2) psychological studies, such as those probing the folk distinction between causes and 'background' or 'enabling' conditions (see Cheng and Novick 1991; McGill and Tenbrunsel 2000; N'gbala and Branscombe 1995); (3) the actual explanatory annals, that is, the sum total of extant explanatory texts. Absent space constraints, I would explore all three sources of evidence. Given constraints, establishing Causal Economy's consistency with the actual explanatory annals is my first priority. Yet, for the sake of getting out the basic move before bringing in scientific complexities, I start by treating a simple datum from stream (1): philosophical intuition.

6.1 *An Intuitive Example*

The most rehearsed case of causal selection concerns the explanation of a forest fire, an example mentioned already in the introduction. The determinants of such a fire extend to all the conditions, positive and negative, on which the fire depended, including: the fact that there was wood around; that the wood wasn't so wet as to be non-flammable; that there was oxygen present; that a lighted cigarette was dropped. Though all of these are in some sense *causes* of the fire, intuition suggests that only the last will invariably be mentioned in its explanation. It is the fire's *narrow cause*.

---

[21] Though not directly relevant to the causal selection problem—and thus not worth detailing here—Causal Economy requires a further constraint on abstraction to prevent a preference for disjunctive explanations that are contrived to be both abstract and stability boosting simultaneously. The constraint I prefer is a cohesion requirement—modeled on a standard from Strevens (2008)—on which a particular feature of the influential nexus cannot be made so abstract that it is impossible to move, in physical state space, from one possible realizer of it to another without moving through a realizer that is not an instance of it.

Agreeing with intuition, Causal Economy recommends an explanation in terms of the cigarette-dropping because that event is the feature of the fire's causal-influential run-up that maximizes causal economy: being a simple, local event, it is comparatively cheap to characterize; and it provides the target event, the fire, with a large stability boost. This is in virtue of the fact that the drop of the cigarette is a relatively unstable event, but one given which the fire is stable. In contrast, 'background factors'—the presence of oxygen, the fuel in the vicinity—are omitted because they are so stable that they can't much contribute to the stability of the fire.

Next, consider a variant case. Here the fire breaks out in a manufacturing plant in which facilities have been constructed to eliminate oxygen. Such a fire still required oxygen's presence. Thus, one feature of the run-up to the conflagration will have been a failure of the oxygen-evacuation mechanism. Many determinants of this fire will resemble those of the woodland blaze: the presence of a particular igniter (not a cigarette, perhaps, but some other spark source); that there was oxygen around; that flammable materials were not overly damp, etc. Yet intuition suggests that, in this case, the presence of oxygen *will be* (among) the fire's narrow cause(s).

How to make sense of oxygen's distinctive relevance? Crucially, on the Causal Economy view relevance is neither traced to some difference in explanatory presuppositions—as relativizers might have it—nor to the mere fact that oxygen is not 'normal' to the factory (though that is indeed true). Instead, its relevance is tied to an objective feature of the causal-influential architecture of the factory fire's run-up. Most notably, in the factory—and not in the forest—the presence of the oxygen was rather unstable. Thus, oxygen's presence provides the fire a stability boost that the oxygen in the forest could not provide to the forest fire.[22] In combination with the fact that oxygen's specification is itself cheap—given facts about fire propagation, oxygen gas must just exceed some concentration, but need not (for instance) have its constituent molecules arranged 'just so'—oxygen offers rather good economy.

I submit that the cigarette-dropping in the forest, oxygen's presence in the factory, and all other systematically selected features, share this in common: they are located at 'sweet spots' in the architecture of the run-up to the events that they explain. These are features at which two characteristics—cheapness and stability-boosting—somehow converge. A universe might have lacked sweet spots, given that these characteristics often trade-off: a sure way to maximize an event's stability is to give a full—and thus very costly—specification of its antecedents. And the reverse is also true: very cheap specifications usually deliver only minimal stability.

---

[22] A nuance: given that oxygen will infiltrate the factory following a failure of whatever mechanism was responsible for its evacuation, *that failure* itself may be an economical part of the causal-influential run-up. Thus the question arises as to whether it—and *not* the oxygen—might constitute the most economical—and thus complete—explanation. As far as I can tell, on the causal economy view both of these would make for good explanations, as both are unstable events that make the target event stable. They differ, of course, in their place in the temporal sequence, but causal economy will often find multiple co-equal (and equally complete) explanations that cite influences at different times (as well, as those at the same time). This, I hope, is such an instance.

Nevertheless, sometimes you get more than you pay for. And for this, cognitively limited creatures like ourselves should be thankful: it is on these grounds that complete explanations are, with any frequency, within our grasp.

6.2 Scientific Cases

With this core strategy in view, next I consider some selective examples in scientific explanatory practice. Saying anything punchy about something so vast is not easy, so I further focus attention on causal selection in biology. Being a science of exceedingly baroque systems, causal selection is pervasive in the life sciences, even among sophisticates. This makes it a gold mine for those wanting to identify the selective patterns to which a philosophical account is responsible. Even more narrowly, consider just two (somewhat overlapping) domains: *signaling systems* and *organismal development*.

6.2.1 Signaling Systems

Biologists use the term 'signaling system' to refer a variety of phenomena: visual and auditory animal communication mechanisms; immune system coordination; interactions between molecules within a single cell (Weng et al. 1999). Put aside the complex question of just what these systems have in common and consider particular explanations for events within them.

In explaining the hiding of a vervet monkey in a nearby bush its neighbor's short, grouped vocalization between 0.2 and 1 kHz—the "eagle alarm" (Seyfarth et al. 1980)—may be cited. In accounting for the growth of the shmoo (a membrane protrusion) in an alpha-type yeast cell, the membrane binding of **a**-factor pheromone, a molecule produced by a compatible **a**-type cell, will be named (Bardwell 2005). Or, to make sense of the breakdown of glycogen during your daily exercise, the release of epinephrine will be mentioned. It is this protein whose binding to 7TM receptors on liver cells begins a molecular cascade that eventuates in the glycogen's demise.

What do these accounts have in common? They omit any description of the inner workings of the signaling process and speak just of a system's production of a certain output in response to single, narrow input: the feature often labeled as the signal. This narrow focus is not a result of any target event depending exclusively on just such inputs, as in each case innumerable other conditions were necessary for the output to be explained; the vervet's hiding required his being awake, unrestrained, possessing sufficient oxygen to react, etc., and the shmoo's growth required an equally complex set of molecular requirements. But in these and other cases, the signal's presence will, at least judging by actual practice, be the explanatory relevant determinant.

This is just what we'd expect were the Causal Economy account correct: signaling events possess the properties that make for good explanations. First, they are cheap. Though presumably not a conceptual requirement on something's being a signal, each actual signal's presence is low in content (e.g., a relatively high-level phonological pattern, or a single molecule type's coarse-grained concentration). Second, they deliver, providing stability-boosts, in virtue of the following: (1) their presence is a difference-maker for the target event (e.g., vervet hiding, shmoo growing, glycogen destruction); (2) their presence is itself unstable, making both 'on' and 'off' states among the nearby possibilities; (3) other in-principle difference-makers for the target event *are* (comparatively) stable.[23] Given this combination, signaling states usually win the explanatory competition.

6.2.2 Organismal Development

Next consider organismal development. For most multicellular creatures that repro-duce by way of a single-celled bottleneck, development is a process by which that single cell—the zygote—transforms into a multi-cellular and functionally integrated organism via coordinated growth, morphological change, and cellular differentiation. Though complicated, development is an important phenomenon for a theory of causal selection to treat in the light of an ongoing debate in the philos-ophy of biology concerning the developmental importance of 'genetic causes,' the nucleotide sequences that constitute an organism's DNA. Though controversy over the status of genetic causes is not always understood against the backdrop of the problem of causal selection, it often is, and I will take it in that light here.

Trivially, every aspect of an organism's developmental trajectory will depend on (or be determined by) a variety of factors, positive and negative, near and far. Some of these features are genetic; others are not. Non-genetic developmen-tal determinants include non-nucleic organismal properties, as well as properly 'environmental' features, those that originate in processes fully extrinsic to the organism. Though there is broad agreement on the simple fact that genetic and non-genetic features, both necessary for development, interact in complex ways to produce organismal traits, talk of 'a genetic blue-print' and 'genes-for' particular traits is common enough in scientific and (even more) in popular discussions that some have wondered whether there might be reason to judge genetic factors *more important* or *relevant* to developmental outcomes than other factors. Ignoring all sorts of subtleties, there are two families of reactions to such a suggestion. Some, like Weber (forthcoming) and Rosenberg (2006), believe that genetic factors are,

---

[23]To spell out in more detail how the stability boost measure applies to a particular case, consider a specific signaling event, such as a particular monkey's flight to the bush. To consider the stability boost of a signal from a neighboring animal, rewind the tape to some time before the signal was sent and perturb the actual world in a variety of ways, thus producing the set of privileged nearby possible worlds. Some number of these perturbations would prevent the signaling animal from emitting the call. And since the call itself is required for the hiding to occur (and because other factors equally required are themselves very stable), an explanans that includes the call's production will substantially augment the stability of the hiding event.

at least to a first approximation, privileged developmental explainers, features that scientists very properly prioritize. The project of these 'gene-centric' philosophers is to say in just what respect genes are so causally special. Others, like Gannett (1999), follow a broadly Millian line and deny that there is a principled and across-the-board difference between the roles of different kinds of developmental determinants, whether genetic, environmental, or otherwise. Their burden is to provide an 'error theory' of the genetic preference one sometimes finds in scientific explanatory practice; Gannett, for instance, points here to the scientists' financial motives.

The degree to which Causal Economy is in tension with these alternatives so briefly sketched will depend on precisely how they are spelled out (Stegmann 2012). Yet, at least when viewed in their broadest outlines, Causal Economy steers between the parity and gene-centric views, rejecting one aspect of each. First, it straightforwardly discards the Millian idea: because of the existence of sweet spots in the causal architectures of developing organisms, for many developmental explananda there will be principled explanatory—and not merely pragmatic—differences between developmental determinants, elevating some as narrow causes, while relegating others to the background. Second, Causal Economy rejects an across-the-board gene-centrism.[24] Though in *some cases* genetic features might be uniquely explanatory of developmental events, in other cases environmental features offer the best pay-off, and in still other circumstances features falling neatly in neither category (e.g., non-nucleic, molecular features, such as methylation patterns) should be (and are) prioritized on explanatory grounds.[25] In this respect the Causal Economy approach is consistent with critiques of gene-centrism, like that offered by Stotz (2006), which claim that pervasive prioritization of genetic features as developmental explainers is unmotivated by the causal complexities of living systems.

Let me illustrate this middle position via an example: the explanation of phenotypic sex, that is, for why particular organisms are either phenotypically male or phenotypically female. While in mammals and birds sex is usually given a genetic explanation—mammal sex is explained by an aspect of the father's chromosomal contribution and avian sex by the mother's—in turtles it is explained environmentally. During a particular period (the aptly named *thermosensitive period*), the temperature of an egg-bound turtle embryo is said to determine its sex, with higher temperatures making females, and lower temperatures, males. Of course, as per a biological application of the Millian insight, the fact that any particular organism—tortoise, cat, or cockatoo—develops the phenotype (typical) of one sex or the other will strictly depend on *both* environmental and genetic

---

[24]To be fair, it is unclear whether anyone really holds an across-the-board gene-centrism; even broadly 'pro-gene' philosophers, such as Weber and Rosenberg, articulate positions that are considerably more nuanced. If they too want to reject gene-centrism, our disagreement would concern the subtler issue of the *standard* by which a subset of an event's necessary conditions—sometimes genetic ones and sometimes not—might be explanatorily privileged.

[25]There might also be cases in which multiple factors—some genetic and some not—are equally causally economical, and would each constitute complete explanations.

factors, among others. To see this, note that there are changes to a developing mammal's intra-uterine environment that would lead even a genetically male fetus to develop a characteristically female phenotype. And similarly, certain genetic features of a developing turtle are necessary for low temperatures to eventuate in male offspring; after all, in other amniotes (such as crocodiles), the relationship between temperature and sex is different, with moderate temperatures yielding males, and extremes, females.

In light of this pervasive multi-causal determination of sex, why is it said to be 'genetic' in mammals and birds, and 'environmental' in turtles? This difference in selection follows neatly from the different economy of the genetic vs. environmental explanations across these cases, something well defined even absent relativizing maneuvers. The economy of such patterns vary in light of the fact that—just as the causal run-up to the forest fire was objectively different from the run-up to the factory's blaze—turtles and mammals (among others), as embedded in their environments, are complex systems with different causal architectures. In the case of the developing turtle, low temperature is the cheap and stability-boosting determinant of male phenotype, while in mammals, presence of the Y chromosome is. After all, a low incubation temperature is itself an unstable event—had the egg been just somewhat differently positioned in the subterranean nest, that temperature would have been different—but one in virtue of which the turtle's sex is stable. Yet this particular kind of temperature sensitivity *does not hold* of a mammalian juvenile, though the presence of the sex-determining chromosome has an equivalent character. This difference in architecture exists in spite of the fact that, as emphasized already, every animal's (phenotypic) sex strictly speaking has genetic, environmental, and other determinants besides.

7 Conclusion: Relativization Revisited

Though I hope that the inherent plausibility of Causal Economy's selection principle speaks somewhat in its favor, the theory's ability to account for selective patterns in scientific explanatory practice is the ground on which I recommend it here. Beyond the few examples that space permitted me to explore, the ambition is that Causal Economy could negotiate *all clear instances* of selection. It is thus vulnerable, in principle, to counterexample. And just what sorts of cases would be most threatening? Naturally, those in which systematically explanatory cited factors were not causally economical. This would be because, in contrast to alternatives, they were comparatively (1) costly, or (2) not stability-boosting of the events they were offered to explain. For instance, what a ticklish situation it might have been for Causal Economy had herpetologists accounted for the sex of Lonesome George—the last Galapagos tortoise from the Isle of Pinta—on genetic grounds!

I know of no such clear counterexamples to Causal Economy's recommendations. Yet a different kind of case may appear to make trouble more obliquely. By way of conclusion I will respond to an instance of this kind: the car crash, as presented by (among others) Hanson (1958: 54) and Carnap (1995, 1966: 191–92). In the run-up to this crash, an angry driver, depressing the gas pedal in a fit of furry, speeds down a country lane in a rainstorm. Hitting a bump, the car spins out of control, colliding with a barn wall. What accounts for the crash? Naturally, this event has many determinants, including: any property of the mechanism connecting pedal depression with vehicular acceleration, the car's speed, the lack of steer stabilizing equipment, the road's wetness, and the driver's anger. Carnap claims that—rather than expecting any consensus on which of these was *the cause* of the crash— different factors will be highlighted by different individuals, each one "looking at the total picture from his point of view" (192). The policeman, for instance, may put the crash down to the speed; the psychologist to the anger of the driver; the engineer to the road's wetness.

Of course, the mere suggestion that this event may have multiple explanations, which may be cited by different parties, is not in itself a problem for the Causal Economy view. The selection principle allows for multiplicity in virtue of simply requiring an explanation to maximize the ratio of delivery to cost, and, as noted already, there may often be multiple ways of achieving this maximum. In particular, whenever there are multiple sweet spots we should expect there to be multiple maxima, and thus multiple co-equal explanations. And though the focus of this paper has been on competition between 'horizontal' alternatives, all of which have been relatively high-level, legitimate alternatives can also vary vertically: low-level explanations, such as those often called 'mechanistic,' may cost and deliver a lot, but offer the same economy as high-level ones that cost little and deliver proportionality. Alternatives may even vary along both horizontal and vertical dimensions simultaneously—as Potochnik (2010) has noted is characteristic of alternative scientific explanations for the same phenomenon.

However, in the case at hand it seems that merely noting the possibility of multiple explanations will not provide a satisfactory response, as none of the multiple explanations of the car crash that Carnap mentions will win the Causal Economy competition. After all, in order to make the crash stable, some confluence of these factors, each independently unstable, must occur. Thus, Causal Economy will require a complete explanation to cite that collection of unstable difference-makers as a package. Why then does it seem that the policeman and the psychologist, for instance, will just state part of the full explanation for the crash, and a part that seems guided by their unique circumstances? Granting this datum for the purposes of argument, my response is simple: there are some contexts—including extremely practical ones—in which partial explanations may be offered, and pragmatics (by which I mean either conversational pragmatics, or implicit relativization of some kind) then do determine which part is cited, and thus determine which cause is selected.

In exploiting the relativizer's line here, have I sapped the Causal Economy account of its advertised explanatory pay-off? I don't believe so. First, even if there are instances, such as the case of the crash, in which relativization is called for, the Causal Economist uses this strategy far less frequently than does the full-blooded relativizer. After all, there are many circumstances—such as respecting the biological events examined above—in which the architecture of the systems involved yields winners in the explanatory competition that are singular factors, rather than a combination of factors. Whenever this is the case, the Causal Economist can move from causal architecture to selection on objective, explanatory grounds. Second, even when the Causal Economy does bring relativization into play, it does less heavy lifting than it does for alternative views. This is because the selection principle has still wildly narrowed the space of possible determinants that might be explanatorily cited, and thus the role of pragmatics in accounting for the selected feature is proportionality diminished.

Even granting that relativization does not eradicate Causal Economy's explanatory offerings, it should still be acknowledged to limit them. There will now be a variety of selective questions on which it has nothing to say except the following: A was selected, and not B, because B was (for any of a variety of unspecified reasons) frameworked. Yet perhaps this is fitting, given the data before us. Let me explain. The discussion opened by emphasizing two features of explanatory practice: its sparseness and its systematicity. Over the course of the paper, I suggested that these were the offspring of two parents: first, an account of causal explanation according to which uniform principles determine selection, along both horizontal and vertical dimensions—principles that prefer sparse explanations; second, the fact that the kinds of causal structures responsible for the events that scientists target—particularly in the life sciences—have features that make for clear explanatory winners across a range of cases, accounting for the systematicity that we observe.

Yet this systematicity is not pervasive, and those stressing the haphazard and interest-relative nature of selection are on to something; at times causal selection is influenced by practical considerations. Best for an account of selection to acknowledge this, and to neither shoe-horn the data to suite an overly systematic theory, nor stipulate that penumbral cases should be awarded as "spoils to the victor." Instead, the victor should be she who both accounts for those patterns in selective practice that actually exist—as I hope the Causal Economy account does—and makes room for an absence of systematicity elsewhere.

Acknowledgements For helpful comments on this paper, thanks to David Frank, Maria Kronfeld-ner, Michael Strevens, David Velleman, two referees, and to (this volume's editors) Pierre-Alain Braillard and Christophe Malaterre. Though I was not able to directly address all of the useful suggestions I received, they uniformly aided me in the development and presentation of my argument.

References

Bardwell, L. (2005). A walk-through of the yeast mating pheromone response pathway. *Peptides,* *26*(2), 339–350.

Bechtel, W. (2008). *Mental mechanisms: Philosophical perspectives on cognitive neuroscience.* New York/London: Routledge.

Carnap, R. (1995, 1966). *An introduction to the philosophy of science.* Mineola, NY: Dover Publications.

Cheng, P. W., & Novick, L. R. (1991). Causes versus enabling conditions. *Cognition, 40*(1–2), 83–120.

Craver, C. F. (2007). *Explaining the brain: Mechanisms and the mosaic unity of neuroscience.* Oxford: Clarendon.

Franklin-Hall, L. R. (forthcoming). The causal economy approach to scientific explanation. In C. K. Waters (Ed.), *Minnesota studies in the philosophy of science.* Minneapolis: University of Minnesota Press.

Gannett, L. (1999). What's in a cause? The pragmatic dimensions of genetic explanations. *Biology and Philosophy, 14*(3), 349–373.

Glennan, S. (2002). Rethinking mechanistic explanation. *Philosophy of Science, 69*, S342–S353.

Hall, N. (2004). Two concepts of causation. In J. Collins, N. Hall, & L. A. Paul (Eds.), *Causation and counterfactuals* (pp. 225–276). Cambridge, MA: MIT Press.

Hanson, N. R. (1958). *Patterns of discovery: An inquiry into the conceptual foundations of science.* Cambridge: CUP Archive.

Hart, H. L. A., & Honoré, T. (1959). *Causation in the law.* Oxford: Clarendon Press.

Hesslow, G. (1988). The problem of causal selection. In D. J. Hilton (Ed.), *Contemporary science and natural explanation: Commonsense conceptions of causality* (pp. 11–32). New York: New York University Press.

Issad, T., & Malaterre, C. (2015). Are dynamic mechanistic explanations still mechanistic? In P.-A. Braillard & C. Malaterre (Eds.), *Explanation in biology. An enquiry into the diversity of explanatory patterns in the life sciences* (pp. 265–292). Dordrecht: Springer.

Kaplan, D. M., & Bechtel, W. (2011). Dynamical models: An alternative or complement to mechanistic explanations? *Topics in Cognitive Science, 3*(2), 438–444.

Kitcher, P. (1984). 1953 and all that. A tale of two sciences. *The Philosophical Review, 93*(3), 335–373.

Lewis, D. (1986). Causal explanation. *Philosophical Papers, 2*, 214–240.

Lewis, D. (2000). Causation as influence. *The Journal of Philosophy, 97*(4), 182–197.

Machamer, P., Darden, L., & Craver, C. F. (2000). Thinking about mechanisms. *Philosophy of Science, 67*, 1–25.

Mackie, J. L. (1974). *The cement of the universe.* Oxford: Clarendon Press.

McGill, A. L., & Tenbrunsel, A. E. (2000). Mutability and propensity in causal selection. *Journal of Personality and Social Psychology, 79*(5), 677.

Mill, J. S. (1882). *A system of logic, ratiocinative and inductive: Being a connected view of the principles of evidence and the methods of scientific investigation.* London: Longmans, Green.

N'gbala, A., & Branscombe, N. R. (1995). Mental simulation and causal attribution: When simulating an event does not affect fault assignment. *Journal of Experimental Social Psychology, 31*(2), 139–162.

Onalaja, A. O., & Claudio, L. (2000). Genetic susceptibility to lead poisoning. *Environmental Health Perspectives, 108*(Suppl 1), 23.

Potochnik, A. (2010). Levels of explanation reconceived. *Philosophy of Science, 77*(1), 59–72.

Press, J. (2015). Biological explanations as cursory covering law explanations. In P.-A. Braillard & C. Malaterre (Eds.), *Explanation in biology. An enquiry into the diversity of explanatory patterns in the life sciences* (pp. 367–385). Dordrecht: Springer.

Railton, P. (1981). Probability, explanation, and information. *Synthese, 48*, 233–256.

Rosenberg, A. (2006). *Darwinian reductionism, or, how to stop worrying and love molecular biology.* Chicago: University of Chicago Press.

Rumberger, R. W., & Larson, K. A. (1998). Toward explaining differences in educational achievement among Mexican American language-minority students. *Sociology of Education, 91*, 68–92.

Salmon, W. (1984). *Scientific explanation and the causal structure of the world.* Princeton: Princeton University Press.

Schaffer, J. (2007). The metaphysics of causation. In E. N. Zalta (Ed.), *Stanford encyclopedia of philosophy.* Stanford, CA: The Metaphysics Research Lab.

Schaffer, J. (2012). Causal contextualisms: Contrast, default, and model. In M. Blaauw (Ed.), *Contrastivism in philosophy*. New York, NY: Routledge.

Seyfarth, R. M., Cheney, D. L., & Marler, P. (1980). Vervet monkey alarm calls: Semantic communication in a free-ranging primate. *Animal Behaviour, 28*(4), 1070–1094.

Stegmann, U. (2012). Varieties of parity. *Biology and Philosophy, 27*(6), 903–918.

Stotz, K. (2006). Molecular epigenesis: Distributed specificity as a break in the central dogma. *History and Philosophy of the Life Sciences, 28*(4), 533.

Strevens, M. (2008). *Depth*. Cambridge, MA: Harvard University Press.

Waters, C. K. (2007). Causes that make a difference. *Journal of Philosophy, 104*, 551–579.

Weng, G., Bhalla, U. S., & Iyengar, R. (1999). Complexity in biological signaling systems. *Science, 284*, 92–96.

Woodward, J. (2003). *Making things happen: A theory of causal explanation*. Oxford: Oxford University Press.